AF774846

Springer Series in

ADVANCED MICROELECTRONICS 27

Springer Series in
ADVANCED MICROELECTRONICS

Series Editors: K. Itoh T. Lee T. Sakurai W.M.C. Sansen D. Schmitt-Landsiedel

The Springer Series in Advanced Microelectronics provides systematic information on all the topics relevant for the design, processing, and manufacturing of microelectronic devices. The books, each prepared by leading researchers or engineers in their fields, cover the basic and advanced aspects of topics such as wafer processing, materials, device design, device technologies, circuit design, VLSI implementation, and subsystem technology. The series forms a bridge between physics and engineering and the volumes will appeal to practicing engineers as well as research scientists.

18 **Microcontrollers in Practice**
By I. Susnea and M. Mitescu

19 **Gettering Defects in Semiconductors**
By V.A. Perevoschikov and V.D. Skoupov

20 **Low Power VCO Design in CMOS**
By M. Tiebout

21 **Continuous-Time Sigma-Delta A/D Conversion**
Fundamentals, Performance Limits and Robust Implementations
By M. Ortmanns and F. Gerfers

22 **Detection and Signal Processing**
Technical Realization
By W.J. Witteman

23 **Highly Sensitive Optical Receivers**
By K. Schneider and H.K. Zimmermann

24 **Bonding in Microsystem Technology**
By J.A. Dziuban

25 **Power Management of Digital Circuits in Deep Sub-Micron CMOS Technologies**
By S. Henzler

26 **High-Dynamic-Range (HDR) Vision**
Microelectronics, Image Processing, Computer Graphics
Editor: B. Hoefflinger

27 **Advanced Gate Stacks for High-Mobility Semiconductors**
Editors: A. Dimoulas, E. Gusev, P.C. McIntyre, and M. Heyns

Volumes 1–17 are listed at the end of the book.

A. Dimoulas E. Gusev
P.C. McIntyre M. Heyns
(Eds.)

Advanced Gate Stacks for High-Mobility Semiconductors

With 292 Figures

Springer

Dr. Athanasios Dimoulas
National Center for Scientific Research
DEMOKRITOS, Patriarchou Grigoriou &
Neapoleos, 15310, Aghia Paraskevi,
Athens, Greece
E-mail: dimoulas@ims.demokritos.gr

Evgeni Gusev
QUALCOMM Inc.
5775 Morehouse Dr., San Diego, CA 92121, USA
QUALCOMM MEMS Technologies
2581 Junction Ave. San Jose, CA 95134, USA
E-mail: egusev@qualcomm.com

Professor Paul C. McIntyre
Stanford University
Department for Materials Science
McCullogh Building, Stanford, CA 94305, USA
E-mail: pcm1@stanford.edu

Professor Marc Heyns
IMEC
Kapeldreef 75
3001 Leuven, Belgium
also at Katholieke Universiteit Leuven,
MTM Department
E-mail: marc.heyns@imec.be

Series Editors:

Dr. Kiyoo Itoh
Hitachi Ltd., Central Research Laboratory, 1-280 Higashi-Koigakubo
Kokubunji-shi, Tokyo 185-8601, Japan

Professor Thomas Lee
Stanford University, Department of Electrical Engineering, 420 Via Palou Mall, CIS-205
Stanford, CA 94305-4070, USA

Professor Takayasu Sakurai
Center for Collaborative Research, University of Tokyo, 7-22-1 Roppongi
Minato-ku, Tokyo 106-8558, Japan

Professor Willy M. C. Sansen
Katholieke Universiteit Leuven, ESAT-MICAS, Kasteelpark Arenberg 10
3001 Leuven, Belgium

Professor Doris Schmitt-Landsiedel
Technische Universität München, Lehrstuhl für Technische Elektronik
Theresienstrasse 90, Gebäude N3, 80290 München, Germany

ISSN 1437-0387

ISBN-10 3-540-71490-1 Springer Berlin Heidelberg New York
ISBN-13 978-3-540-71490-3 Springer Berlin Heidelberg New York

Library of Congress Control Number: 2007931596

This work is subject to copyright. All rights are reserved, whether the whole or part of the material is concerned, specifically the rights of translation, reprinting, reuse of illustrations, recitation, broadcasting, reproduction on microfilm or in any other way, and storage in data banks. Duplication of this publication or parts thereof is permitted only under the provisions of the German Copyright Law of September 9, 1965, in its current version, and permission for use must always be obtained from Springer-Verlag. Violations are liable to prosecution under the German Copyright Law.

Springer is a part of Springer Science+Business Media.

springer.com

© Springer Berlin Heidelberg 2007

The use of general descriptive names, registered names, trademarks, etc. in this publication does not imply, even in the absence of a specific statement, that such names are exempt from the relevant protective laws and regulations and therefore free for general use.

Camera-ready by the Author and SPi, Pondicherry
Cover: eStudio Calmar Steinen

Printed on acid-free paper SPIN: 12029037 57/3180/SPi - 5 4 3 2 1 0

Preface

The continuous miniaturization of information processing and storage units has always been at the heart of advances in modern electronics. A large part of these advances is based on the evolution of bulk CMOS technology. Further progress is inhibited mainly by poor scaling of the transistor gate which causes short channel effects and results in overall performance loss. Part of the problem could be fixed by introducing SOI and/or multiple-gated devices (e.g., FinFETs, planar double gated, or tri-gated) which results in better electrostatic control of the channel. Further improvements could be made by using high mobility materials. In part, this has already been implemented since mobility enhancing strained Si is considered to be an irreplaceable part of next generation devices. By introducing high mobility semiconductors such as germanium (Ge) or III–V compounds it may be possible to enhance significantly the device performance for future generation nanoelectronics.

To develop a viable Ge MOS technology is a very challenging task. First, it is necessary to engineer compliant germanium-on-insulator (GeOI) substrates to ensure volume production at low cost. Second, it is important to develop appropriate surface passivation methodologies and high-k dielectrics in order to combine good electrical behavior with potential for gate scaling to equivalent oxide thickness less than 1 nm. Finally, it is necessary to master Ge processing to fabricate MOSFET devices with high $I_{\mathrm{ON}}/I_{\mathrm{OFF}}$ ratio and enhanced channel mobilities. Since the first demonstration of functional Ge pMOSFETs with high-k dielectrics five years ago, there has been a lot of progress in bulk Ge transistors mainly using Si passivating layers and compressive strain which enhance p-channel mobility several times above Si/SiO_2 universal. On the other hand, there are also concerns that due to small energy gap, leakage current at source and drain as well as band-to-band tunneling will generate high OFF-state current especially in aggressively scaled Ge devices. Despite of this, with the right choice of device architecture (e.g., double-gated thin Ge films) and with the help of circuit design, power management and control should be possible so that junction leakage is not expected to be a serious obstacle. At the present time the biggest concern is that only Ge pMOSFETs perform satisfactorily, while nMOSFETs either underperform or do not function at all for reasons which are not fully understood. The puzzling point is that all Ge surface passivating methods which benefit pMOSFETs have only minor influence on nMOSFETs so that channel mobility and ON-state currents in these devices remain low. This implies that there may be a fundamental materials problem which goes beyond surface passivation. Although this sets an interesting research scene in materials science and physics of devices it has also serious technological consequences. It implies that in future implementations of CMOS technology based on high mobility materials, the nMOS part should be made of materials other than Ge (e.g., strained Si or III–V compound semiconductors) co-integrated with Ge pMOS devices on the same complex engineered substrate.

The use of III–V channel materials for nMOS (instead of strained Si) is an attractive option due to their very high electron mobility. This means that III–V MOS technology must be developed and indeed not in competition to Ge but in compliance with it in a dual-channel CMOS approach. It is well-known that III–V transistor technology in the form of MESFETs and HEMTs exists since many years. However, this technology has been developed at the micron level and is appropriate for low density, low-noise analog circuits for niche market LSI applications. To transform this into a new III–V MOS technology which will follow the aggressive scaling rules of extremely dense mainstream circuits for ULSI digital applications is an extremely challenging task. Unlike the case of Ge, processing of III–V compounds in a standard or slightly modified semiconductor line using toolset and know-how similar to those applying for Si is very difficult. Issues related to self-aligned gate definition, implantation and high temperature activation annealing, etching and contact resistance must be addressed before we come any close to a viable MOS technology. In parallel, more fundamental materials and device architectural issues must be addressed. Channel materials (e.g., GaAs vs. InGaAs) and device structures (e.g., surface channels vs. buried channels) must be carefully selected for optimum performance. The device layer structure will also determine to some extent the operation mode (inversion, depletion or enhancement mode). Surface passivation of III–V compounds is a long-standing problem with no satisfactory solution yet. The main reason is the strong Fermi-level pinning at the oxide/semiconductor interface which is not fully characterized and quantified and, for that reason, not very well understood at the present time.

This book is a collection of review articles written by some of the key players in Ge and III–V research and development. The articles describe what could be considered as established knowledge after the renewal of interest in Ge and III–V MOS technology during the last five years of research. It is divided in four parts covering all areas from high mobility substrates, up to surface passivation and high-k gate preparation and characterization as well as field effect transistor fabrication and testing.

In chapters 1 through 3 the reader will find a review of mobility enhancing channels including strained Si and alternative orientations substrates.

Emphasis is given on (110)-oriented Si substrates with enhanced hole mobility making it particularly useful for pMOS devices. In addition, a review on the progress of GeOI substrates is given with a special emphasis on wafer bonding and layer splitting technique.

Chapters 4 through 7 describe Ge surface preparation, passivation, and gate dielectric emphasizing the characterization of interfaces between high-k dielectrics and semiconductors in an attempt to elucidate their role in the electrical behavior of the whole gate stack. Ab initio theoretical studies of oxide growth on semiconductors complement our knowledge about atomic configuration and binding principles at interfaces which determine band offsets.

In chapters 8 through 12 we present a number of analytical methodologies, including structural, chemical, physical, and electrical characterization of high-k oxides on Ge, GaAs, and Si. This is complemented by first principles calculations of dielectric properties (κ-values) and their correlation to the crystal symmetry and electronic structure. Point defects, band offsets interface reactions and interdiffusions are all related to the electrical behavior of gate stacks.

In chapters 13 through 17 we focus on the fabrication and characterization of field effect transistors made of Ge, III–V compounds and Si. Although basic transistor characteristics including channel mobility are studied using long-channel transistors, the scalability and manufacturability of Ge FETs is tested on the basis of short channel deep submicron transistors processed in a pilot Si line. In addition, the issues of Ge nanodevices are thoroughly discussed in connection with alternative architectures which will allow performance gain in future aggressively scaled devices based on Ge.

Athens, Greece,
July 2007

Athanasios Dimoulas

Contents

Contributors

V.V. Afanas'ev
Department of Physics and Astronomy, University of Leuven
Celestijnenlaan 200D
3001 Leuven, Belgium

T. Akatsu
SOITEC, Parc Technologique des Fontaines
F-38190 Bernin, France

S.R. Amy
Laboratory for Surface Modification, Department of Chemistry and Chemical Biology,
Rutgers University
Piscataway, NJ

C.A. Ashman
Naval Research Laboratory
Washington DC 20375, USA

S.K. Banerjee
Microelectronics Research Center, University of Texas at Austin
Austin, TX 78758, USA

S. Bedell
IBM Semiconductor Research and Development Center (SRDC)
Research Division, T.J. Watson Research Center
Yorktown Heights, NY 10598, USA

T. Billon
CEA/LETI, 17, Rue des Martyrs
F-38054 Grenoble, France

P.E. Blöchl
Institute for Theoretical Physics, Clausthal University of Technology
Leibnitzstrasse 10,
38678 Clausthal-Zellerfeld, Germany

R. Bonzom
IMEC, Kapeldreef 75
B-3001 Leuven, Belgium

D. Brunco
Intel affiliate at IMEC
Belgium

M. Caymax
IMEC, Kapeldreef 75
B-3001 Leuven, Belgium

I. Cayrefourcq
SOITEC, Parc Technologique des Fontaines
F-38190 Bernin, France

G. Celler
SOITEC, Parc Technologique des Fontaines
F-38190 Bernin, France

Y.J. Chabal
Laboratory for Surface Modification,
Department of Chemistry and
Chemical Biology,
Rutgers University
Piscataway, NJ

J.J.-H. Chen
Microelectronics Research Center,
University of Texas at Austin
Austin, TX 78758, USA

J.O. Chu
IBM Semiconductor Research and
Development Center (SRDC)
Research Division, T.J. Watson
Research Center
Yorktown Heights, NY 10598, USA

C.O. Chui
Department of Electrical
Engineering, Stanford University
Stanford, CA, USA

T. Clarysse
IMEC, Kapeldreef 75
B-3001 Leuven, Belgium

T. Conard
IMEC, Kapeldreef 75
B-3001 Leuven, Belgium

S. De Gendt
IMEC, Kapeldreef 75
B-3001 Leuven, Belgium

B. De Jaeger
IMEC, Kapeldreef 75
B-3001 Leuven, Belgium

A. Delabie
IMEC, Kapeldreef 75
B-3001 Leuven, Belgium

P. Delugas
Sardinian Laboratory for
Computational materials Science
and Department of Physics, Cagliari
University, Cittadella Universitaria
09042 Monserrato (CA), Italy

Philips Research Laboratories,
Leuven, Belgium

W. Deweerd
IMEC, Kapeldreef 75
B-3001 Leuven, Belgium

M. Fanciulli
CNR-INFM MDM National
Laboratory, via C. Olivetti 2
I-20041 Agrate Brianza (MI), Italy

A. Filippetti
Sardinian Laboratory for
Computational materials Science
and Department of Physics, Cagliari
University, Cittadella Universitaria
09042 Monserrato (CA), Italy

V. Fiorentini
Sardinian Laboratory for
Computational materials Science
and Department of Physics,
Cagliari University,
Cittadella Universitaria
09042 Monserrato (CA), Italy
vincenzo.fiorentini@dsf.unica.it

J. Fompeyrine
IBM Research GmbH, Zurich
Research Laboratory
8803 Rűschlikon, Switzerland

C.J. Först
Departments of Nuclear Science and
Engineering and Materials Science
and Engineering, Massachusetts
Institute of Technology

77 Massachusetts Avenue
Cambridge, MA 02139, USA

M.M. Frank
IBM Semiconductor Research and Development Center (SRDC)
T.J. Watson Research Center
Yorktown Heights, NY 10598, USA

B. Ghyselen
SOITEC, Parc Technologique des Fontaines
F-38190 Bernin, France

M. Gribelyuk
IBM Semiconductor Research and Development Center (SRDC)
Microelectronics Division
Hopewell Junction, NY 12533, USA

K.W. Guarini
IBM Semiconductor Research and Development Center (SRDC)
Research Division, T.J. Watson Research Center
Yorktown Heights, NY 10598, USA

S. Guha
IBM Thomas J. Watson Research Center
Yorktown Heights, NY 10598, USA

E.P. Gusev
IBM Semiconductor Research and Development Center (SRDC)
Research Division, T.J. Watson Research Center
Yorktown Heights, NY 10598, USA

D. Hellin
IMEC, Kapeldreef 75
B-3001 Leuven, Belgium

M. Heyns
IMEC, Kapeldreef 75
B-3001 Leuven, Belgium

M. Hong
Departments of Physics and Materials Science and Engineering,
National Tsing Hua University
Hsin Chu, Taiwan, Republic of China

M. Houssa
IMEC, Kapeldreef 75
B-3001 Leuven, Belgium

M. Ieong
IBM Semiconductor Research and Development Center (SRDC)
Research Division, T.J. Watson Research Center
Yorktown Heights, NY 10598, USA

T. Janssens
IMEC, Kapeldreef 75
B-3001 Leuven, Belgium

B. Kaczer
IMEC, Kapeldreef 75
B-3001 Leuven, Belgium

D.Q. Kelly
Microelectronics Research Center,
University of Texas at Austin
Austin, TX 78758, USA

M. Kennard
SOITEC, Parc Technologique des Fontaines
F-38190 Bernin, France

K. Kita
Department of Materials Engineering,
The University of Tokyo
Japan

J. Kwo
Departments of Physics and Materials Science and Engineering,
National Tsing Hua University
Hsin Chu, Taiwan, Republic of China

D.L. Kwong
University of Texas at Austin
Austin, TX, USA

Yu. Lebedinskii
Moscow Engineering Physics
Institute 31, Kashirskoe chaussee
115409 Moscow, Russian Federation

S.J. Lee
National University of Singapore
Singapore

F. Letertre
SOITEC, Parc Technologique des
Fontaines
F-38190 Bernin, France

F. Leys
IMEC, Kapeldreef 75
B-3001 Leuven, Belgium

J.-P. Locquet
IBM Research GmbH, Zurich
Research Laboratory
8803 Rűschlikon, Switzerland

R. Loo
IMEC, Kapeldreef 75
B-3001 Leuven, Belgium

C. Marchiori
IBM Research GmbH, Zurich
Research Laboratory
8803 Rűschlikon, Switzerland

K. Martens
IMEC, Kapeldreef 75
B-3001 Leuven, Belgium

C. Mazure
SOITEC, Parc Technologique des
Fontaines
F-38190 Bernin, France

P. W. Mertens
IMEC, Kapeldreef 75
B-3001 Leuven, Belgium

M. Meuris
IMEC, Kapeldreef 75
B-3001 Leuven, Belgium

H. Nomura
Department of Materials Engineering,
The University of Tokyo
Japan

T. Ohmi
New Industry Creation Hatchery
Center, Tohoku University
Japan

B. Onsia
IMEC, Kapeldreef 75
B-3001 Leuven, Belgium

K. Opsomer
IMEC, Kapeldreef 75
B-3001 Leuven, Belgium

J.A. Ott
IBM Semiconductor Research and
Development Center (SRDC)
Research Division, T.J. Watson
Research Center
Yorktown Heights, NY 10598, USA

L. Pantisano
IMEC, Kapeldreef 75
B-3001 Leuven, Belgium
pantisan@imec.be

A.M. Pourghaderi
IMEC, Kapeldreef 75
B-3001 Leuven, Belgium

G. Raskin
UMICORE, Watertorenstraat 33
B-2250 Olen, Belgium

Z.M. Rittersma
Philips Research Leuven
Leuven, Belgium

K.C. Saraswat
Department of Electrical Engineering,
Stanford University
Stanford, CA, USA

A. Satta
IMEC, Kapeldreef 75
B-3001 Leuven, Belgium

G. Scarel
CNR-INFM MDM National
Laboratory, via C. Olivetti 2
I-20041 Agrate Brianza (MI), Italy

T. Schram
IMEC, Kapeldreef 75
B-3001 Leuven, Belgium

K. Schwarz
Institute of Materials Chemistry,
Vienna University of Technology
Getreidemarkt 9/165-TC
1060 Vienna, Austria

G. Seguini
CNR-INFM MDM National
Laboratory, via C. Olivetti 2
I-20041 Agrate Brianza (MI), Italy

H. Shang
IBM Semiconductor Research and
Development Center (SRDC)
Research Division, T.J. Watson
Research Center
Yorktown Heights, NY 10598, USA
shuiling@us.ibm.com

E. Simoen
IMEC, Kapeldreef 75
B-3001 Leuven, Belgium

S. Sioncke
IMEC, Kapeldreef 75
B-3001 Leuven, Belgium

J. Snow
IMEC, Kapeldreef 75
B-3001 Leuven, Belgium

M. Sousa
IBM Research GmbH, Zurich
Research Laboratory
8803 Rűschlikon, Switzerland

S. Spiga
CNR-INFM MDM National
Laboratory, via C. Olivetti 2
I-20041 Agrate Brianza (MI), Italy

A. Stesmans
Department of Physics and
Astronomy, University of Leuven
Celestijnenlaan 200D
3001 Leuven, Belgium

S. Takagi
Graduate School of Frontier
Science, The University of Tokyo
MIRAI Project, Advanced
Semiconductor Research Center
(ASRC), National Institute of
Advanced Industrial Science and
Technology (AIST)

A. Teramoto
New Industry Creation Hatchery
Center, Tohoku University
Japan

V. Terzieva
IMEC, Kapeldreef 75
B-3001 Leuven, Belgium

A. Toriumi
Department of Materials Engineering,
The University of Tokyo
Japan

M. Toyama
Department of Materials Engineering,
The University of Tokyo
Japan

W. Vandervorst
IMEC, Kapeldreef 75
B-3001 Leuven, Belgium

S. Van Elshocht
IMEC, Kapeldreef 75
B-3001 Leuven, Belgium

E. Van Moorhem
IMEC, Kapeldreef 75
B-3001 Leuven, Belgium

J. Van Steenbergen
IMEC, Kapeldreef 75
B-3001 Leuven, Belgium

X. Wang
IBM Semiconductor Research and Development Center (SRDC)
Microelectronics Division
Hopewell Junction, NY 12533, USA

C. Wiemer
CNR-INFM MDM National Laboratory, via C. Olivetti 2
I-20041 Agrate Brianza (MI), Italy

G.D. Wilk
ASM America, 3440 East University Drive
Phoenix, AZ 85034, USA

G. Winderickx
IMEC, Kapeldreef 75
B-3001 Leuven, Belgium

P.D. Ye
School of Electrical and Computer Engineering and Birck Nanotechnology Center, Purdue University
West Lafayette, IN 47906, USA

A. Zenkevich
Moscow Engineering Physics Institute 31, Kashirskoe chaussee
115409 Moscow, Russian Federation

C. Zhu
National University of Singapore
Singapore

P. Zimmerman
Intel affiliate at IMEC
Belgium

1

Strained-Si CMOS Technology

S. Takagi

Summary. Improvement in performance of Si MOSFETs through conventional device scaling has become more difficult, because of several physical limitations associated with the device miniaturization. Thus, much attention has recently been paid to the mobility enhancement technology through applying strain into CMOS channels. This chapter reviews this strained-Si CMOS technology with an emphasis on the mechanism of mobility enhancement due to strain. The device physics for improving drive current of MOSFETs is summarized from the viewpoint of electronic states of carriers in inversion layers and, in particular, the subband structures. In addition, recent experimental results on implementing strain into CMOS channels are described.

1.1 Introduction

A guiding principle of performance enhancement in Si CMOS has been the scaling law over 30 years. Under 90 nm technology node and beyond, however, the performance enhancement of CMOS through the device scaling such as shrinking the gate length and thinning the gate oxide has become more and more difficult, because of several physical limitations in miniaturization of MOSFETs. For example, thinning the gate oxide, needed to reduce the supply voltage, leads to the rapid increase in gate tunnelling current. Also, the increase in substrate impurity concentration, needed to suppress short channel effects, leads to the reduction in carrier mobility and resulting decrease in the on-current [1]. As a result, simple device scaling encounters a trade-off relationship among the current drive, the power consumption and the short-channel effects, all of which are quite important factors for LSI applications.

Thus, the device technologies using new channel structures and new channel materials, which mitigate the stringent constrains regarding the device design, have recently stirred a strong interest, in addition to high-k gate insulator technologies. These device technologies, called "Technology Boosters" in ITRS 2004 edition [2], include strained Si channels, ultrathin SOI, metal gate electrodes, multigate structures, ballistic transport channels, metal source/drain junctions, and so on. Among them, strained-Si channels [3–6] have been recognized as a technology applicable to near term technology nodes, thanks to the recent progress in so-called "local strain techniques", and have actually been included in 90 nm logic CMOS technologies [7]. The mobility enhancement obtained by applying appropriate strain, can provide higher carrier velocity in MOS channels, resulting in higher current drive under fixed supply voltage and gate oxide thickness. This means that thicker gate oxides and/or lower supply voltage can be used under a fixed current drive, leading to the mitigation of the trade-off relationship among current drive, power consumption and short-channel effects. As a result, the strain engineering and resulting increase in channel mobility has been regarded as a device technology mandatory in future technology nodes, as well.

This chapter reviews the principle and the device application of this strained-Si CMOS technology with an emphasis on the physical mechanism of mobility enhancement due to strain.

1.2 Impact of Mobility Enhancement on Current Drive of Short-Channel MOSFETs

In short channel MOSFETs, the modelling of the current drive is not straightforward, because of the co-existence of the velocity saturation effect due to high lateral electric field and the non-stationary transport effect, caused by the fact that carriers in ultra-short channels travel from source to drain without encountering sufficient scattering events. Furthermore, it is expected that ballistic transport [8], where carriers have no scattering in channels, can be realized in extremely-short channels less than 10 nm. Thus, quasi-ballistic transport models [9,10] to describe the current drive by considering a small number of scattering events have been proposed on a basis of full ballistic motion.

Figure 1.1(a, b) shows the schematic diagrams of factors that dominate current drive under a classical drift model and a quasi-ballistic model, respectively. In both models, the drive current is described by the product of surface carrier concentration and velocity near the source region. Since the surface carrier concentration is constant under fixed values of gate insulator thickness, threshold voltage and gate voltage, the increase in the carrier velocity near source region is needed for the enhancement of the drive current.

In the drift model, the velocity near source region is strongly affected under non-stationary transport by low-field mobility near source region, while the velocity near source region in the quasi-ballistic model is determined by the injection velocity from source and the back-scattering rate into source [9, 10], which are also given by low-field mobility near source region. As a consequence, the increase in low-field mobility near source region can lead to the enhancement of the drive current in short channel devices, in both models.

Actually, it has been reported from the simulation results of carrier velocity and drive current in strained-Si n-MOSFETs with gate length of 100 nm that

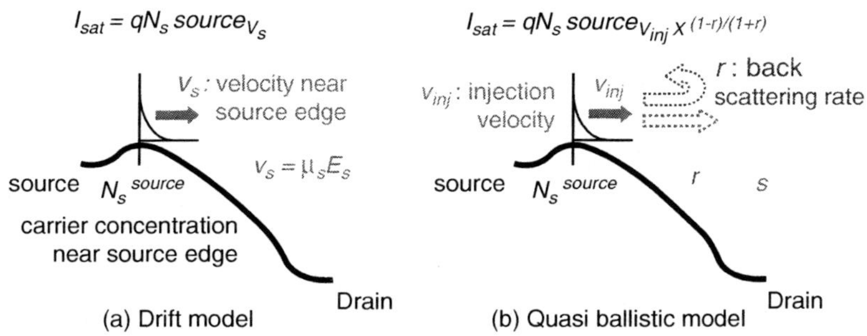

Fig. 1.1. Schematic diagrams to show factors to dominate current drive, I_{sat}. Fig. 1.1(**a**) and (**b**) show the diagrams based on a classical drift model and a quasi-ballistic model, respectively. E_s, q and μ_s mean the lateral electric field, the elementary charge and mobility near source region, respectively

the increase in mobility can provide the increased velocity and resulting higher drive current under a constant saturation velocity model [11]. Also, recent experimental and theoretical results [9,12,13] have shown that the drive current of MOSFETs with gate lengths of 100–50 nm is roughly proportional to the square root of low-field mobility. These results strongly suggest that low-field mobility is still important for the current drive in short-channel MOSFETs.

On the other hand, carrier velocity is also affected by the scattering probability of high energy carriers, typically reflecting in the energy relaxation time. As described below, strain induces band splitting, which can lead to longer energy relaxation time and resulting higher velocity [11]. Thus, device simulations accurately taking non-stationary transport effects and detailed band structures into account are mandatory for quantitative understanding of the current drive of short-channel MOSFETs.

1.3 Physical Mechanism of Mobility Enhancement in Strained-Si n- and p-Channel MOSFETs

1.3.1 Physical Origin of Mobility Enhancement in n-Channel MOSFETs

Before explaining the physical origin of mobility enhancement due to strain, it is necessary to describe the electronic properties of Si MOS inversion layers. Figure 1.2 schematically shows the equi-energy surfaces of inversion-layer electrons in the two-dimensional subband structure on a (100) surface and the characteristics of the valley structures. The conduction band in bulk Si is composed of six equivalent valleys. In inversion layers, on the other hand, these six valleys are split into the twofold valleys locating at a central position in two-dimensional k-space and the fourfold valleys locating on k_x and k_y axes, because of two-dimensional quantization. Three-dimensional

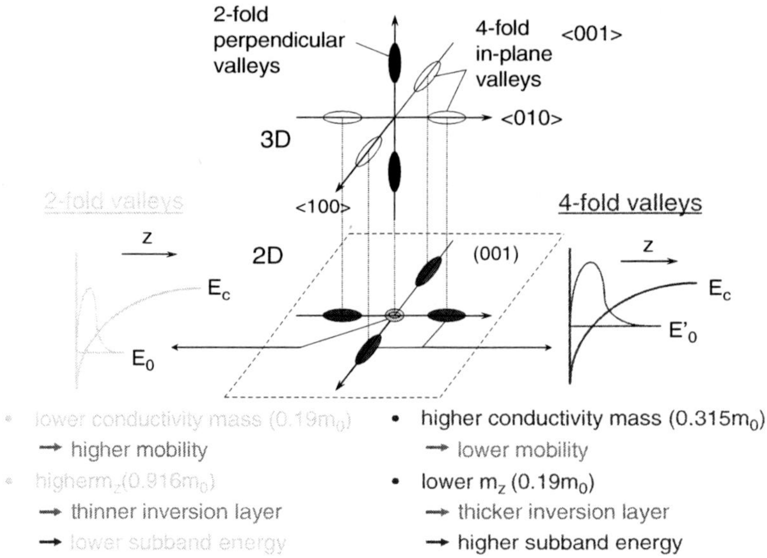

Fig. 1.2. Schematic diagram of characteristics of the two- and four-fold valleys in two-dimensional electrons on a (100) surface

electrons have an anisotropy in the effective mass, composed of light transversal effective mass, $m_t (= 0.19\, m_0)$, where m_0 is the electron mass in free space, and the heavy longitudinal effective mass, $m_l (= 0.916\, m_0)$. As a result, the twofold degenerate valleys have the effective masses of m_t in parallel and m_l in perpendicular to MOS interfaces, while the fourfold degenerate valleys have the effective masses of m_t and m_l in parallel and m_t in perpendicular to MOS interfaces.

This difference in the effective mass leads to a variety of differences in physical properties between the twofold and the fourfold valleys. For instance, the conductivity mass parallel to MOS interfaces is lower in the twofold valleys than in the fourfold valleys and, thus, the mobility of electrons in the twofold valleys becomes higher than that in the fourfold valleys. Also, since the thickness of inversion layers and the subband energy are determined by the effective mass perpendicular to MOS interfaces, the thickness of the inversion layers is thinner and the subband energy is lower in the twofold valleys having higher effective mass perpendicular to the MOS interfaces than in the fourfold valleys.

The impact of strain on the electron mobility in n-channel Si MOSFETS can also be understood in terms of this subband structure or valley structure [14]. Figure 1.3 schematically shows the effect of tensile strain on the subband structures. The electron occupancy of the twofold and the fourfold valleys at room temperature is almost the same without any strain. This is because the lower subband energy of the twofold valleys is compensated by the higher density-of-states of the fourfold valleys having the higher valley

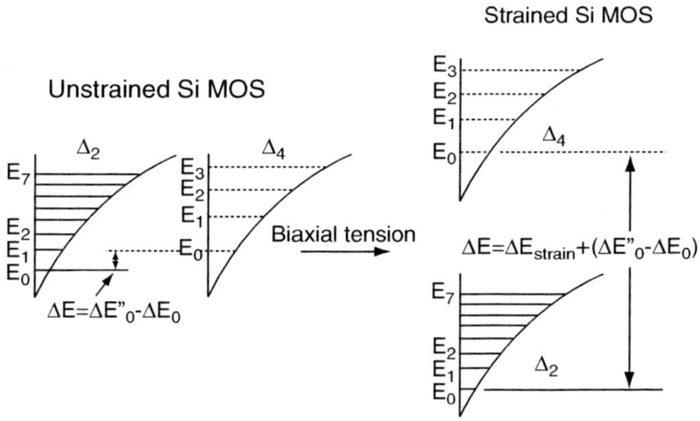

Fig. 1.3. Energy lineups of the Si conduction band in the inversion layer with and without tensile strain

degeneracy and the higher density-of-state mass. When tensile strain parallel to MOS interfaces or compressive strain perpendicular to MOS interfaces is applied to MOSFETs, the conduction band edge in the fourfold valleys becomes higher than that in the twofold valleys and this splitting energy is added to the subband energy difference caused by the surface quantization. As a result, the subband energy between the two valleys significantly increases.

This increase in the subband energy splitting yields an increase in the inversion-layer mobility through the following two mechanisms. One is the increase in the averaged mobility due to the increase in the occupancy of electrons in the twofold valleys having higher mobility. The other is the suppression of inter-valley scattering between the twofold and the fourfold valleys. This is because, when the splitting energy between the twofold and the fourfold valleys is higher than the phonon energies associated with inter-valley scattering, the transition of electrons in the twofold valleys through a phonon absorption process cannot occur, resulting in the reduction in the scattering probability. Since the inter-valley scattering has a large contribution to the total scattering rate of Si MOSFETs at room temperature and the influence becomes larger with an increase in temperature, this increase in the mobility due to tensile strain is more effective in enhancing LSI performance during real operation at temperatures higher than room temperature.

On the other hand, when compressive strain parallel to MOS interfaces or tensile strain perpendicular to MOS interfaces is applied to MOSFETs, the electron mobility tends to decrease. This is attributable to the increase in the occupancy of the electrons in the fourfold valleys having lower mobility.

The change in the conductivity in Si by applying mechanical strain is well known as the piezo-resistance effect and the experimental data for bulk Si has been reported 50 years ago [15]. The experimental results of the piezo-resistance effect on Si MOSFETs have also been reported extensively [16–18]. Here, the characteristics of mechanical strain are, in general, that the amount

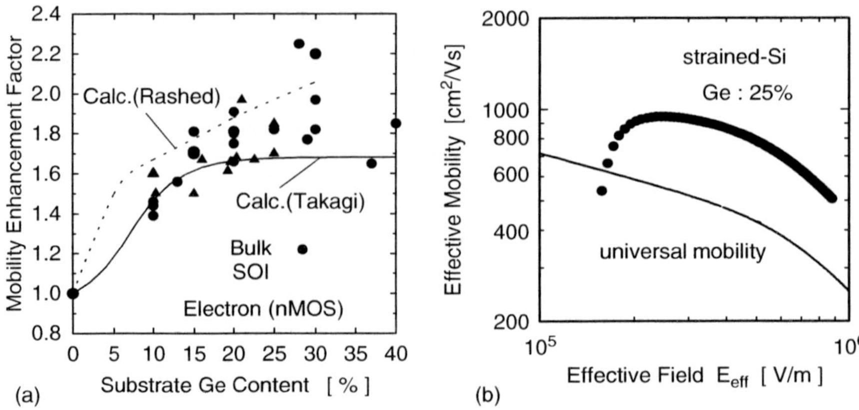

Fig. 1.4. Mobility characteristics of bi-axial strained-Si n-channel MOSFETs (**a**) Mobility enhancement factor as a function of Ge content in SiGe substrates, which is in proportion to strain. Strained Si on relaxed SiGe with Ge content of 24 at % has strain of 1%. *Closed circles* and *triangles* show the experimental values in bulk and SOI MOSFETs, respectively. *Solid* [14] and *dash* [32] *lines* mean the results of theoretical calculations. (**b**) Effective field ($E_{\rm eff}$) dependence of electron mobility in bi-axial strained-Si nMOSFETs

of the strain is small and the strain configuration is uni-axial. It has been shown for n-channel MOSFETs that mechanical tensile strain leads to a mobility increase [17, 18], also attributed to the subband energy splitting. As a consequence, since a primary parameter for the mobility enhancement in n-channel MOSFETs is the subband energy splitting between the twofold and the fourfold valleys, there exists no essential difference in physical mechanism for mobility modulation due to bi-axial and uni-axial strain, though a quantitative difference in the amount of the enhancement is seen.

The relationship between electron mobility in n-channel MOSFETs and bi-axis tensile strain parallel to MOS interfaces has been systematically investigated by using strained-Si MOSFETs fabricated on relaxed SiGe substrates. Figure 1.4(a) shows the experimental results for the mobility enhancement factor [19–31], defined by the ratio of the mobility in strained-Si MOSFETs to that in conventional (unstrained) Si MOSFETs, as a function of Ge content in SiGe substrates. Here, strain in Si on relaxed SiGe with Ge content of 24 at % amounts to strain of 1%. Results of the enhancement factor theoretically calculated on the basis of phonon scattering are also shown [14, 32]. Agreement between the experimental and theoretical results is fairly good. It is found that maximum an enhancement factor of roughly two is obtained.

Figure 1.4(b) shows the effective field ($E_{\rm eff}$) dependence of electron mobility in n-channel MOSFETs at room temperature with and without tensile strain [29, 33]. It is found that the mobility enhancement factor is almost constant, irrespective of $E_{\rm eff}$. Since the mobility in moderate $E_{\rm eff}$ region is known to be dominated by phonon scattering, the mobility enhancement in this region can be explained by the mechanisms described above. In high E_{eff} region, on the other hand, almost all the electrons can occupy the twofold valleys even without any strain, because of the increased confinement caused by strong surface electric field. Since this fact suggests that the band splitting might have much less influence on the mobility in high E_{eff} region, the high enhancement factor experimentally observed in high E_{eff} region has been attributed to the reduction in the probability of surface roughness scattering [34], which dominates the mobility in high E_{eff} region. However, the physical origin is still unclear because of the lack of direct evidence for the decreased surface roughness scattering.

1.3.2 Physical Origin of Mobility Enhancement in p-Channel MOSFETs

Compared with n-channel MOSFETs, the impact of strain on hole mobility in p-channel MOSFETs is complicated and the physical mechanism has not been fully and quantitatively understood yet. Also, it has recently been recognized in p-channel MOSFETs that the effects of uni-axial and bi-axial strain on the hole mobility are significantly different [7, 35–37], which is in contrast to n-channel MOSFETs. It has been pointed out that uni-axial compressive strain perpendicular to channel direction and bi-axial tensile strain are effective in enhancing the hole mobility in p-channel MOSFETs [38].

Figure 1.5 shows the results of theoretical calculations of the three-dimensional Si valence band structure near the Γ point with uni-axial compressive strain and bi-axial tensile strain [38]. Here, assuming a MOSFET channel direction as parallel to ⟨110⟩, the strain directions are taken to be parallel to (001) surface for bi-axial strain and in the ⟨110⟩ direction for uni-axial strain. The right and the left directions of are horizontal axes in the figures in the wave vectors perpendicular to the MOS interface (along ⟨001⟩ direction) and parallel to the channel (along ⟨110⟩ direction), respectively. While, without any strain, the heavy hole band and the light hole band degenerate at the top of the valence band, the application of strain leads to the band splitting and shifting the light hole band upward. As a result, the top of the valence band is composed of the light hole band. In addition, the modulation of the curvature of the bands due to strain provides a change in the effective mass and the anisotropy in the effective mass parallel and perpendicular to the MOS interface. As a consequence, hole mobility enhancement due to strain is attributable to the following three mechanisms: (1) reduction in the effective mass of occupied bands; (2) suppression of inter-subband scattering due to the subband energy splitting; (3) increase in the occupancy of subbands having higher mobility.

Figure 1.6(a) shows experimental results of the hole mobility enhancement factor in bi-axial tensile-strained Si p-MOSFETs fabricated on relaxed SiGe substrates as a function of Ge content in the SiGe substrates [20, 22, 26, 28–30, 39–45]. The theoretically calculated results of the enhancement factor are

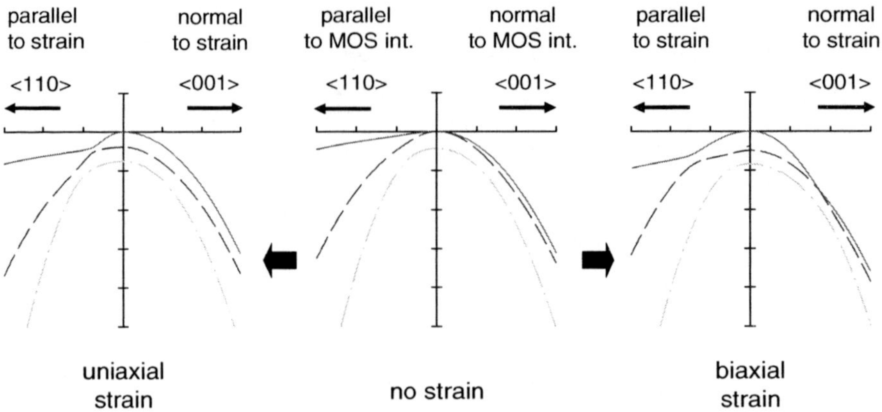

Fig. 1.5. Results of theoretical calculations of the change in the Si valence band structure near the Γ point with uni-axial compressive strain and bi-axial tensile strain [38]. The strain directions of the uni-axial and the bi-axial strain are parallel to ⟨110⟩ direction and (001) surface, respectively. The channel direction is assumed parallel to ⟨110⟩ direction. The right and the left directions of the horizontal axes correspond to wave vectors perpendicular to the MOS interface (⟨001⟩ direction) and parallel to the channel (⟨110⟩ direction), respectively. Without any strain (the center figure), *solid, dash* and *dotted dash lines* correspond to heavy hole band, light hole band and split-off band, respectively

also shown [46, 47]. It is found that, with a Ge content of 30% or higher, an enhancement factor of roughly two can be obtained, while the enhancement factor is small with low Ge content. Figure 1.6(b) shows the experimental E_{eff} dependence of hole mobility in bi-axial tensile strain Si p-MOSFETs at room temperature. It is found [29, 44] that the hole mobility enhancement factor decreases with an increase in E_{eff}. Note here that the values of the enhancement factor plotted in Fig. 1.6(a) are the maximum ones in the lower E_{eff} region. Since the mobility at high E_{eff} is important for practical applications it is necessary to use tensile-strained Si films with high Ge content and a resulting high strain for bi-axial tensile strained Si p-MOSFETs.

It has recently been recognized [7, 35–37] that, when uni-axial compressive strain is applied along ⟨110⟩ direction to p-MOS channels parallel to ⟨110⟩, the hole mobility enhancement is higher for rather small strain magnitude and is not significantly reduced by increasing E_{eff}. Figure 1.7 shows the experimental E_{eff} dependence of hole mobility in Si p-channel MOSFETs with uni-axial compressive and bi-axial tensile strain [36]. It is confirmed that a higher enhancement factor in the high E_{eff} region can be maintained for uni-axial strain.

It has been, on the other hand, reported [16, 18] in measurements of the piezo-resistance of (100) surface Si p-channel MOSFETs by using uni-axial mechanical strain that compressive strain parallel to the channel direction and tensile strain parallel to channel width increase the hole mobility. These

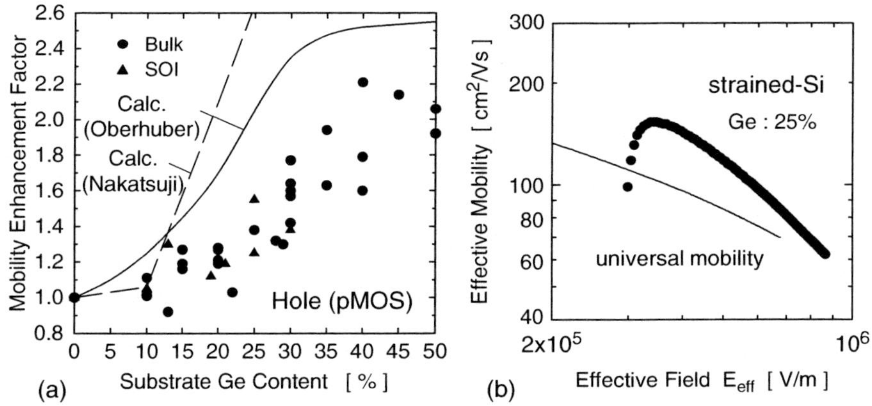

Fig. 1.6. Mobility characteristics of bi-axial tensile strained Si p-channel MOSFETs (**a**) Mobility enhancement factor as a function of Ge content in SiGe substrates. *Symbols* show experimental results. *Closed circles* and *triangles* show the values in bulk and SOI MOSFETs, respectively. *Solid* [46] and *dash* [47] *lines* mean the results of theoretical calculations. (**b**) E_{eff} dependence of hole mobility in bi-axial tensile-strain Si p-channel MOSFETs

complicated dependencies of strain on hole mobility in p-channel MOSFETs can be roughly summarized by Fig. 1.8, which have been obtained by the recent theoretical calculations [38]. For small strain magnitude, typically seen in piezo-resistance measurements by applying mechanical strain, compressive strain parallel to ⟨110⟩ channel direction (tensile strain parallel to channel width) increases the hole mobility. On the other hand, when the amount of

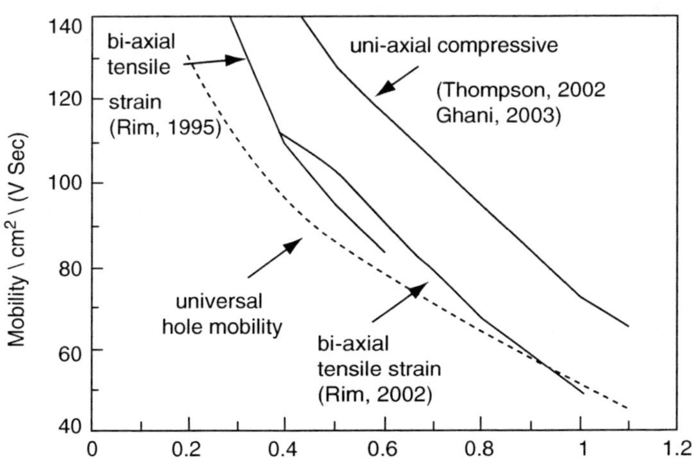

Fig. 1.7. Experimental results of the E_{eff} dependence of hole mobility in Si p-channel MOSFETs with uni-axial compressive and bi-axial tensile strain

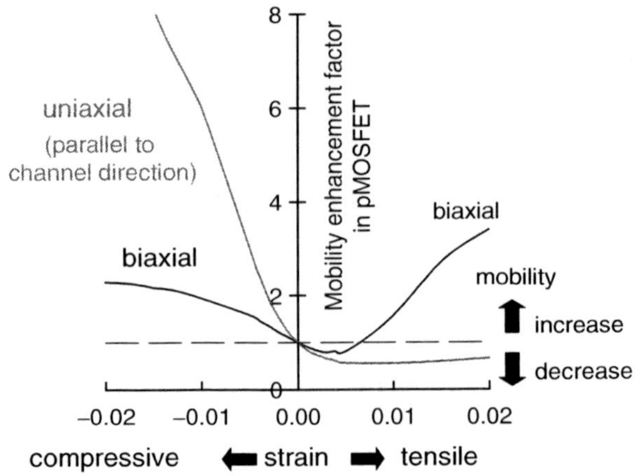

Fig. 1.8. Calculated results of the mobility enhancement factor for inversion-layer holes in Si p-channel MOSFETs with uni-axial compressive and bi-axial tensile strain [38]. The value of $E_{\rm eff}$ is taken to be $1\,{\rm MV\,cm^{-1}}$

strain increases to some extent, both compressive and tensile bi-axial strain also increase the hole mobility and, in particular, the mobility enhancement by tensile bi-axial strain becomes higher. It can be understood from these results that uni-axial compressive strain parallel to $\langle 110\rangle$ channels is most effective for the hole mobility enhancement in p-channel MOSFETs and, if the amount of strain is large, bi-axial tensile strain are effective.

The difference in the $E_{\rm eff}$ dependence of the hole mobility in Si p-channel MOSFETs between uni-axial compressive and bi-axial tensile strain has been explained by the difference in the physical mechanism for hole mobility enhancement for the two strain configurations. The hole mobility enhancement for bi-axial tensile strain has been attributed mainly to the suppression of inter-band scattering due to the band splitting between heavy hole and light hole bands and less to the contribution of the change in the effective mass due to strain [36, 37]. In addition, the effective mass perpendicular to MOS interfaces in the subband originating in the light hole band, which is lower in energy, is lighter than that in the subband originating in the heavy hole band, as seen in the right figure of Fig. 1.5. As a result, the increase in the subband energy due to carrier confinement at MOS interfaces is higher in the light hole band than in the heavy hole band. Since this increase in the subband energy due to confinement reduces the strain-induced energy difference between the light hole and heavy hole bands, the total amount of band splitting reduces with an increase in $E_{\rm eff}$ and, finally, the heavy hole band becomes higher in energy than the light hole one. This change in the band splitting has been regarded as a main cause for the decrease in the mobility enhancement in bi-axial tensile strain p-MOSFETs with increasing $E_{\rm eff}$. [36–38, 48–50]

In contrast, the hole mobility enhancement by uni-axial compressive strain has been attributed both to a decrease in the effective mass associated with the strain and to the suppression of inter-band scattering [36–38, 51]. In addition, contrary to bi-axial tensile strain, the effective mass perpendicular to MOS interfaces in the subband originating in the light hole band is heavier than that in the heavy hole band, as seen in the left figure of Fig. 1.5. Therefore, the increase in the subband energy difference between the light hole and the heavy hole subband due to carrier confinement is added to the strain-induced energy difference, leading to a further increase in the subband energy difference and a resulting increase in the occupancy of the lowest subband having the lower effective mass. Since this effect becomes more evident with increasing E_{eff}, higher hole mobility enhancement is maintained for uni-axial compressive strain. As a result, the difference in the E_{eff} dependence of hole mobility between uni-axial compressive and bi-axial tensile strain has been ascribed to the difference in the influence of the uni-axial and the bi-axial strain on the band structure, particularly, to the difference in the effective mass perpendicular to MOS interfaces.

While these interpretations are based on the recent band calculations, they have not been fully established yet, because of the complicated valence band structure in Si and the differences in the interpretations [47,51] existing among the various calculation models. Further investigations of the transport properties of inversion-layer holes in strained-Si MOSFETs are clearly needed, from both the theoretical and experimental viewpoints.

1.4 Implementation of Strain into MOSFETs

1.4.1 Global Strain Technology

As a device structure with a bi-axial tensile strained channel, MOSFETs on strained-Si layers epitaxially grown on relaxed SiGe substrates, which have a larger lattice constant than Si, have been extensively studied [3–6,52]. Furthermore, a variety of new substrates and device structures such as strained-Si-On-Insulator (Strained-SOI) MOSFETs [6, 53, 54], where strained-Si/relaxed SiGe layers are formed on buried oxides, and Strained-Si-directly-On-Insulator (SSDOI) MOSFETs [55, 56], where straind-Si layers are directly bonded to buried oxides, have been proposed and demonstrated as modified versions of bulk strained-Si MOSFETs. Typical substrates and device structures using these bi-axial tensile strained films are schematically shown in Fig. 1.9. The technologies to fabricate MOSFETs on wafers over which strained-Si layers are formed have recently been called "*Global strain technology*".

As for MOSFETs using these global strain Si substrates, the research and development on device optimization have currently been conducted for applications to 45 nm technology and beyond. Figure 1.10 shows a TEM photograph of one example of strained-SOI MOSFETs with gate length of 32 nm [31]. Many research groups have already reported improvement in on-current of around 10–25% with global strain Si MOSFETs, with short gate lengths less

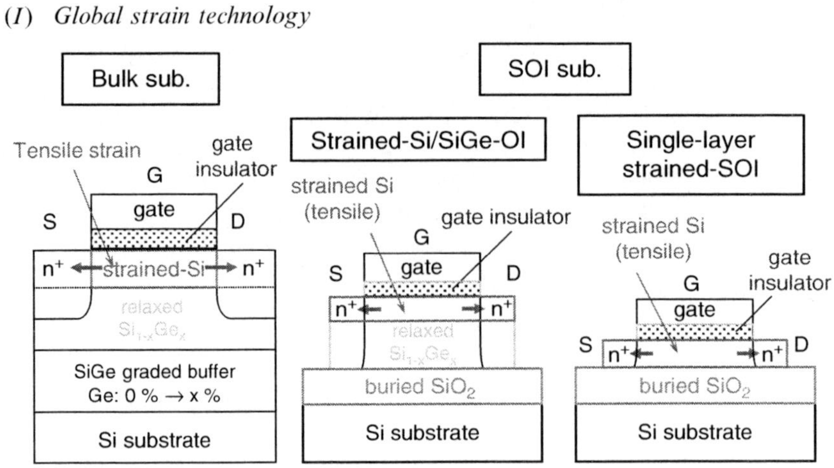

Fig. 1.9. Schematic cross-sections of typical MOS structures using global bi-axial tensile strain

than sub-100 nm [31, 56–63]. The successful operation of CMOS with gate length of 25 nm [58] and the integration of strained-Si MOSFETs with high-k gate insulators [64] and metal gates [59] have also been demonstrated.

Advantages of global strain Si MOSFETs are listed as follows: (1) large and uniform strain can be realized; (2) conventional CMOS fabrication processes can be applied with minimal modification for global strain substrates supplied from wafer vendors. On the other hand, there are still many issues for practical

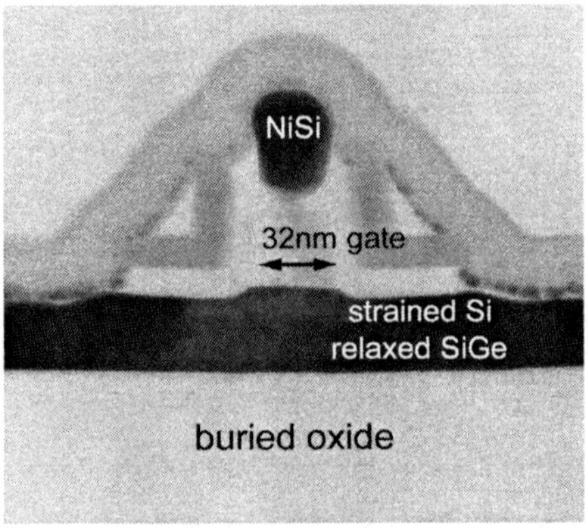

Fig. 1.10. TEM photograph of cross-sectional view of a strained Si-On-Insulator MOSFET with gate length of 32 nm [31]. A trained-Si thin film, where a channel of the MOSFET is formed, and a relaxed-SiGe thin film are fabricated on a buried oxide layer. The gate electrode is made of poly-Si and NiSi formed in source/drain region and on the top of the poly-Si gate

use of this technology. For example, (1) limited performance improvement in p-channel MOSFETs with small or moderate strain (2) wafer quality including defects and dislocations (3) wafer cost (4) increase in junction leakage current. Also, the importance of reducing the parasitic resistance in source/drain regions has been pointed out for higher performance enhancement in ultra-short gate lengths [31, 63].

1.4.2 Local Strain Technology

As a technology to solve the above issues associated with global strain techniques, "*local strain technology*", which introduces structures and materials to induce strain into channels locally inside MOSFETs, has recently stirred keen interest. Figure 1.11 schematically shows a variety of local strain technologies. In particular, the following two methods are quite typical.

(1) *SiGe source/drain.* SiGe is epitaxially grown selectively in source/drain regions, where are etched off toward substrates. These buried SiGe regions induce uni-axial compressive strain into channels, applied to p-channel MOSFETs [7, 35]. Recently, n-channel MOSFETs with tensile strain induced by selectively growing SiC in source/drain regions instead of SiGe have also been reported [65].
(2) *SiN capping layer.* SiN capping layers with intrinsic stress, deposited on MOSFETs as interlayer films, induce strain into MOS channels [66–68]. In many cases, SiN films with tensile strain have been applied to n-channel MOSFETs, while SiN films with compressive strain have recently been developed and used for p-channel MOSFETs [69].

Besides these approaches local strain from isolation regions like shallow trench isolation [70], poly-gate electrodes [71], silicide regions [72], etc. can also be used. Judging from the fact that /SiGe source/drain and strained SiN capping layers has been applied to logic LSI processes under 90 nm technology

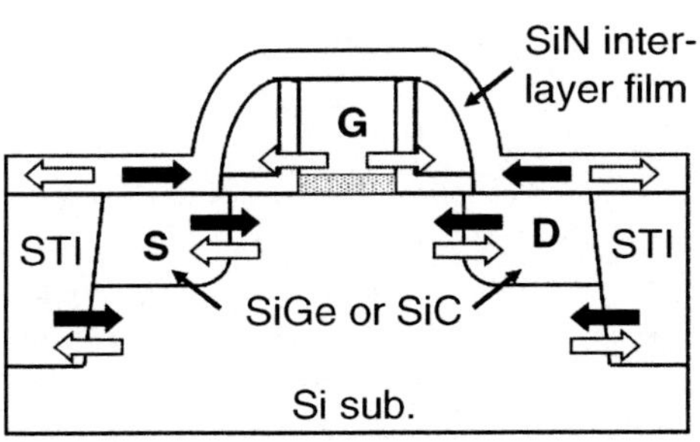

Fig. 1.11. Schematic illustration of a device structure using a variety of local strain techniques. STI in the figure means regions of Shallow Trench Isolation. *Black and white* arrows indicate compressive and tensile strain, respectively

node for mass production [7, 35], these local strain technologies have already become quite practical.

It should be noted that most of these techniques tend to produce high strain along a specific direction (uni-axial strain). Actually, the experimental results of uni-axial strain in Fig. 1.7 have been obtained from MOSFETs with SiGe source/drain, above described [7, 35–37]. Advantages of the local strain technologies are listed as follows: (1) since the standard CMOS processes can be used with minor changes and novel wafers are not needed, the implementation is easy and the production cost is low; (2) By using uni-axial strain, high performance enhancement of p-channel MOSFETs can be obtained even with comparatively small amount of strain.

Owing to these advantages, introduction and the optimization of local strain techniques are, at present, being conducted extensively for near term technology nodes. Furthermore, a variety of new structures and new techniques that enables one combine local strain technology with future CMOS structures remain under investigation.

As for local strain technology, on the other hand, issues to be solved are: (1) the amount of strain and the resulting performance boost could be limited; (2) since the strain profile in the channels is non-uniform and the amount of strain is strongly dependent on device geometries and dimensions, variation in electrical characteristics could be large; (3) the circuit design is not easy, because of the geometry dependence of strain.

In addition, an important concern in any strained Si CMOS technologies including the global and the local ones is strain relaxation, which may come from subsequent processing or from the device geometry. It is, in the simplest case, well known that the strain in strained films patterned into isolated areas is relaxed from their edges. Actually, uni-axial strain has been created on bi-axial global strain substrates by utilizing this phenomenon [73, 74]. However, any unintentional strain relaxations have to be avoided by carefully designing the process conditions and the device geometry, though the robustness against strain relaxation can be strongly dependent on the strain-application techniques. It should be noted here that local strain evaluation techniques that probe the strain distribution inside small devices with high spatial resolution are of paramount importance for this purpose.

1.5 Conclusions

The strained-Si CMOS technology has been regarded as mandatory for future technology nodes, because of the necessity to maintain high current drive. On the other hand, a future important goal for strained-Si technology is to establish methods of applying strain that are compatible with a variety of other technology boosters such as high k/metal gate technology and multi-gate structures. Thus, a future direction for research and the development includes the optimal design of strain profiles in future CMOS structures and their realization through global strain techniques, local strain techniques or their combination. Device/process designs for the robustness against performance variation and avoidance of reliability problems can are also other important priorities. In order to accomplish these tasks, further comprehensive and quantitative understanding of the effects of strain on electrical characteristics and fabrication processes as well as the metrology of strain with high resolution are strongly needed.

Acknowledgements

The author would like to thank N. Sugiyama, T. Numata, T. Mizuno, T. Tezuka, N. Hirashita, T. Maeda, T. Irisawa, K. Usuda, S. Nakaharai, Y. Moriyama, A. Tanabe, J. Koga, Y. Miyamura and E. Toyoda for their cooperation and M. Hirose, T. Kanayama, S. Kawamura, T. Masuhara and N. Fukushima for their continuous support. This work was partly supported by the New Energy and Industrial Technology Development Organization (NEDO).

References

1. D.A. Antoniadis, Proc. VLSI Symposium, Honolulu, p. 2 (2002)
2. 2004 Edition of the ITRS, http://public.itrs.net/
3. F. Schaffler, Semicond. Sci. Technol. 12, p. 1515 (1997)
4. C.K. Maiti, L.K. Bera and S. Chattopadhyay, Semicond. Sci. Technol. 13, p. 1225 (1998)
5. C.K. Maiti, N.B. Chakrabarti and S.K. Ray, "Strained silicon heterostructures: materials and devices" – (IEE circuits, devices and systems series; No. 12), The Institute of Electrical Engineers, London, UK, 2001
6. M.L. Lee, E.A. Fitzgerald, M.T. Bulsara, M.T. Currie and A. Lochtefeld, J. Appl. Phys. 97, 011101 (2005)
7. S. Thompson, N. Anand, M. Armstrong, C. Auth, B. Arcot, M. Alavi, P. Bai, J. Bielefeld, R. Bigwood, J. Brandenburg, M. Buehler, S. Cea, V. Chikarmane, C. Choi, R. Frankovic, T. Ghani, G. Glass, W. Han, T. Hoffmann, M. Hussein, P. Jacob, A. Jain, C. Jan, S. Joshi, C. Kenyon, J. Klaus, S. Klopcic, J. Luce, Z. Ma, B. McIntyre, K. Mistry, A. Murthy, P. Nguyen, H. Pearson, T. Sandford, R. Schweinfurth, R. Shaheed, S. Sivakumar, M. Taylor, B. Tufts, C. Wallace, P. Wang, C. Weber, and M. Bohr: Int. Electron Devices Meet. Tech. Dig., San Francisco, p. 61 (2002)
8. K. Natori, J. Appl. Phys., 76, p. 4879 (1994)
9. M.S. Lundstrom, IEEE Electron Device Lett. 18, p. 361 (1997)
10. M.S. Lundstrom, IEEE Electron Device Lett. 22, p.293 (2001)
11. K. Rim, J.L. Hoyt, and J.F. Gibbons, IEEE Trans. Electron Devices, 47, p. 1406 (2000)
12. R. Ohba and T. Mizuno, IEEE Trans. Electron Devices 48, p. 338 (2001)
13. A. Lochefeld and D.A. Antoniadis, IEEE Electron Device Lett., p. 591 (2001)

14. S. Takagi, J.L. Hoyt, J.J. Welser, J.F. Gibbons, J. Appl. Phys. 80, p. 1567 (1996)
15. C.S. Smith, Phys. Rev. 94, p. 42(1954)
16. D. Colman, R.T. Bate, and J.P. Mize, J. Appl. Phys. 39, p. 1923 (1968)
17. G. Dorda, J. Appl. Phys. 42, p. 2053 (1971)
18. A. Hamada, T. Furusawa, N. Saito and E. Takeda, IEEE Trans. Electron Devices 38, p. 895 (1991)
19. J.J. Welser, J.L. Hoyt, S. Takagi and J.F. Gibbons, Int. Electron Devices Meet. Tech. Dig., San Francisco, p. 373 (1994)
20. T. Mizuno, N. Sugiyama, H. Satake and S. Takagi, Proc. VLSI Symposium, Honolulu, p. 210 (2000)
21. T. Mizuno, S. Takagi, N. Sugiyama, H. Satake, A. Kurobe, and A. Toriumi, IEEE Electron Device Lett. 21, 230 (2000)
22. K. Rim, S. Koester, M. Hargrove, J. Chu, P.M. Mooney, J. Ott, T. Kanarsky, P. Ronsheim, M. Ieong, A. Grill and H.-S.P. Wong, Proc. VLSI Symposium, Kyoto, p. 59 (2001)
23. L.-J. Huang, J.O. Chu, S. Goma, C.P. D'Emic, S.I. Koester, D. F. Canaperi, P.M. Mooney, S.A. Cordes, J.L. Speidell, R.M. Anderson and H.-S.P. Wong, Proc. VLSI Symposium, Kyoto, p. 57 (2001)
24. Z.-Y. Cheng , M.T. Currie, C.W. Leitz, G. Taraschi, E.A. Fitzgerald, J.L. Hoyt and D.A. Antoniadis, IEEE Electron Device Lett. 22, 321 (2001)
25. M.T. Currie, C.W. Leitz, T.A. Langdo, G. Taraschi, E.A. Fitzgerald and D.A. Antoniadis, J. Vac. Sci. Technol, B19, 2268 (2001)
26. N. Sugii, D. Hisamoto, K. Washio, N. Yokoyama and S. Kimura, Int. Electron Devices Meet. Tech. Dig., Washington, D.C., p. 737 (2001)
27. T. Tezuka, N. Sugiyama, T. Mizuno and S. Takagi, Proc. VLSI Symposium, Honolulu, p. 96 (2002)
28. K. Rim, J. Chu, H. Chen, K.A. Jenkins, T. Kanarsky, K. Lee, A. Mocuta, H. Zhu, R. Roy, J. Newbury, J. Ott, K. Petrarca, P. Mooney, D. Lacey, S. Koester, K. Chan, D. Boyd, M. Ieong and H.-S. Wong, Proc. VLSI Symposium, Honolulu, p. 98 (2002)
29. T. Mizuno, N. Sugiyama, T. Tezuka, T. Numata, and S. Takagi, Proc. VLSI Symposium, Honolulu, p. 106 (2002)
30. S. Takagi, T. Mizuno, T. Tezuka, N. Sugiyama, T. Numata, K. Usuda, S. Nakaharai, Y. Moriyama, J. Koga, A. Tanabe, N. Hirashita and T. Maeda, Int. Electron Devices Meet. Dig., p. 57 (2003)
31. T. Numata, T. Irisawa, T. Tezuka, J. Koga, N. Hirashita, K. Usuda, E. Toyoda, Y. Miyamura, A. Tanabe, N. Sugiyama and S. Takagi, Int. Electron Devices Meet. Tech. Dig., San Francisco, p. 177 (2004)
32. M. Rashed, W.K. Shih, S. Jallepalli, T.J.T. Kwan and C.M. Maziar, Int. Electron Devices Meet. Tech. Dig., Washington, D.C., p. 765 (1995)
33. T. Mizuno, N. Sugiyama, T. Tezuka, and S. Takagi, IEEE Trans. Electron Devices 49, p. 7 (2002)
34. M.V. Fischetti, F. Gamiz, and W. Hansch., J. Appl.Phys. 92, p. 7320 (2002)
35. T. Ghani, M. Armstrong, C. Auth, M. Bost, P. Charvat, G. Glass, T. Hoffmann, K. Johnson, C. Kenyon, J. Klaus, B. McIntyre, K. Mistry, A. Murthy, J. Sandford, M. Silberstein, S. Sivakumar, P. Smith, K. Zawadzki, S. Thompson and M. Bohr, Int. Electron Devices Meet. Tech. Dig., Washington, D.C., p. 978 (2003)

36. S.E. Thompson, M. Armstrong, C. Auth, S. Cea, R. Chau, G. Glass, T. Hoffman, J. Klaus, Z. Ma, B. McIntyre, A. Murthy, B. Obradovic, L. Shifren, S. Sivakumar, S. Tyagi, T. Ghani, K. Mistry, M. Bohr and Y. El-Mansy, IEEE Electron Device Lett.25, p. 191 (2004)
37. S.E. Thompson, G. Sun, K. Wu, J. Lim and T. Nishida, Int. Electron Devices Meet. Tech. Dig., San Francisco, p. 221 (2004)
38. M. Uchida, Y. Kamakura, and K. Taniguchi, Proc. Int. Conf. Simulation of Semiconductor Processes and Devices, p. 315 (2005)
39. K. Rim, J.J. Welser, J.L. Hoyt and J.F. Gibbons: Int. Electron Devices Meet. Tech. Dig., Washington, D.C., p. 517 (1995)
40. D.K. Nayak, K. Goto, A. Yutani, J. Murota and Y. Shiraki, IEEE Trans. Electron Devices 43, p. 1709 (1996)
41. C.K. Maiti, L.K. Bera, S.S. Dey, D.K. Nayak and N.B. Chakrabarti, Solid State Electron.41, p. 1863 (1997)
42. T. Tezuka, N. Sugiyama, S. Takagi and A. Kurobe, Abs. 25th International Conference on the Physics of Semiconductors, Osaka, H-270, p. 629 (2000)
43. T. Mizuno, N. Sugiyama, A. Kurobe, and S. Takagi, IEEE Trans. Electron Devices 48, p. 1612 (2001)
44. T. Mizuno, N. Sugiyama, T. Tezuka, and S. Takagi, IEEE Trans. Electron Devices 50, p. 988 (2003)
45. C.W. Leitz, M.T. Currie, M.L. Lee, Z.-Y. Cheng, D.A. Antoniadis and E.A. Fitzgerald, J. Appl. Phys. 92, p. 3745 (2002)
46. R. Oberhuber, G. Zandler and P. Vogl, Phys. Rev. B 58, 9941 (1998)
47. H. Nakatsuji, Y. Kamakura and K. Taniguchi, Int. Electron Devices Meet. Tech. Dig., San Francisco, p. 727 (2002)
48. T. Mizuno, N. Sugiyama, T. Tezuka, T. Numata, T. Maeda and S. Takagi, Int. Electron Devices Meet. Tech. Dig., San Francisco, p. 31 (2002)
49. T. Mizuno, T. Mizuno, N. Sugiyama, T. Tezuka, T. Numata and S. Takagi, IEEE Trans. Electron Devices 51, p. 1114 (2004)
50. M.V. Fischetti, Z. Ren, P.M. Solomon, M. Yang and K. Rim, J. Appl. Phys. 94, p. 1079 (2003)
51. L. Shifren, X. Wang, P. Matagne, B. Obradovic, C. Auth, S. Cea, T. Ghani, J. He, T. Hoffman, R. Kotlyar, Z. Ma, K. Mistry, R. Nagisetty, R. Shaheed, M. Stettler, C. Weber, and M.D. Giles: Appl. Phys. Lett.85, p. 6188 (2004)
52. J.J. Welser, J.L. Hoyt and J.F. Gibbons, IEEE Electron Device Lett. 15, p. 100 (1994)
53. T. Mizuno, S. Takagi, N. Sugiyama, H. Satake, A. Kurobe, and A. Toriumi, IEEE Electron Device Lett. 21, p. 230 (2000)
54. S. Takagi, T. Mizuno, N. Sugiyama, T. Tezuka and A. Kurobe, IEICE Trans. Electron. E84-C, p. 1043 (2001)
55. T.A. Langdo, M.T. Currie, A. Lochtefeld, R. Hammond, J.A. Carlin, M. Erdtmann, G. Braithwaite, V.K. Yang, C.J. Vineis, H. Badawi, and M.T. Bulsara, Appl. Phys. Lett. 82, p. 4256 (2003)
56. K. Rim, K. Chan, L. Shi, D. Boyd, J. Ott, N. Klymko, F. Cardone, L. Tai, S. Koester, M. Cobb, D. Canaperi, B. To, E. Duch, I. Babich, R. Carruthers, P. Saunders, G. Walker, Y. Zhang, M. Steen, and M. Ieong, Int. Electron Devices Meet. Tech. Dig., Washington, D.C., p. 49 (2003)
57. B.H. Lee, A. Mocuta, S. Bedell, H. Chen, D. Sadana, K. Rim, P. O'Neil, R. Mo, K. Chan, C. Cabral, C. Lavoie, D. Mocuta, A. Chakravarti, R.M.

Mitchell, J. Mezzapelle, F. Jamin, M. Sendelbach, H. Kermel, M. Gribelyuk, A. Domenicucci, K.A. Jenkins, S. Narasimha, S.H. Ku, M. Ieong, I.Y. Yang, E. Leobandung, P. Agnello, W. Haensch, and J. Welser, Int. Electron Devices Meet. Tech. Dig., San Francisco, p. 946 (2002)
58. J.-S. Goo, Q. Xiang, Y. Takamura, H. Wang, J. Pan, F. Arasnia, E. N. Paton, P. Besser, M.V. Sidorov, E. Adem, A. Lochtefeld, G. Braithwaite, M.T. Currie, R. Hammond, M.T. Bulsara and M.-R. Lin: IEEE Electron Device Lett. 24, p.351 (2003)
59. Q. Xiang, J.-S. Goo, J. Pan, B. Yu, S. Ahmed, J. Zhang, and M. -R. Lin, Proc. VLSI Symposium, Kyoto, p. 101 (2003)
60. J.R. Hwang, J.H. Ho, S.M. Ting, T.P. Chen, Y.S. Hsieh, C.C. Huang, Y.Y. Chiang, H.K. Lee, A. Liu, T.M. Shen, G. Braithwaite, M.T. Currie, N. Gerrish, R. Hammond, A. Lochtefeld, F. Singaporewala, M.T. Bulsara, Q. Xiang, M.R. Lin, W.T. Shiau, Y.T. Loh, J.K. Chen, S.C. Chien, S.W. Sun and Frank Wen, Proc. VLSI Symposium, Kyoto, p. 103 (2003)
61. H.C.-H. Wang, Y.-P. Wang, S.-J. Chen, C.-H. Ge, S.M. Ting, J.-Y. Kung, R.-L. Hwang, H.-K. Chiu, L.C. Sheu, P.-Y. Tsai, L.-G. Yao, S.-C. Chen, H.-J. Tao, Y.-C. Yeo, W.-C. Lee, and C. Hu, Int. Electron Devices Meet. Tech. Dig., Washington, D.C., p.61 (2003)
62. T. Sanuki, A. Oishi, Y. Morimasa, S. Aota, T. Kinoshita, R. Hasumi, Y. Takegawa, K. Isobe, H. Yoshimura, M. Iwai, K. Sunouchi and T. Noguchi, Int. Electron Devices Meet. Tech. Dig., Washington, D.C., p. 65 (2003)
63. H. Kawasaki, K. Ohuchi, A. Oishi, O. Fujii, H. Tsujii, T. Ishida, K. Kasai, Y. Okayama, K. Kojima, K. Adachi, N. Aoki, T. Kanemura, D. Hagishima, M. Fujiwara, S. Inaba, K. Ishimaru, N. Nagashima and H. Ishiuchi, Int. Electron Devices Meet. Tech. Dig., San Francisco, p. 169 (2004)
64. K. Rim, E.P. Gusev, C.D. Emic, T. Kanarsky, H. Chen, J. Chu, J. Ott, K. Chan, D. Boyd, V. Mazzeo, B.H. Lee, A. Mocuta, J. Welser, S.L. Cohen, M. Ieong and H.-S. Wong, Proc. VLSI Symposium, Honolulu, p. 12 (2002)
65. K.W. Ang, K.J. Chui, V. Bliznetsov, A. Du, N. Balasubramanian, M.F. Li, G. Samudra and Y.-C. Yeo, Int. Electron Devices Meet. Tech. Dig., San Francisco, p. 1069 (2004)
66. F. Ootsuka, S. Wakahara, K. Ichinose, A. Honzawa, S. Wada, H. Sato, T. Ando, H. Ohta, K. Watanabe, and T. Onai, Int. Electron Devices Meet. Tech. Dig., San Francisco, p. 575 (2000)
67. S. Ito, H. Namba, K. Yamaguchi, T. Hirata, K. Ando, S. Koyama, S. Kuroki, N. Ikezawa, T. Suzuki, T. Saitoh, and T. Horiuchi, Int. Electron Devices Meet. Tech. Dig., San Francisco, p. 247 (2000)
68. A. Shimizu, K. Hachimine, N. Ohki, H. Ohta, M. Koguchi, Y. Nonaka, H. Sato, and F. Ootsuka, Int. Electron Devices Meet. Tech. Dig., Washington, D.C., p. 433 (2001)
69. H.S. Yang, R. Malik, S. Narasimha, Y. Li, R. Divakaruni, P. Agnello, S. Allen, A. Antreasyan, J.C. Arnold, K. Bandy, M. Belyansky, A. Bonnoit, G. Bronner, V. Chan, X. Chen, Z. Chen, D. Chidambarrao, A. Chou, W. Clark, S.W. Crowder, B. Engel, H. Harifuchi, S.F. Huang, R. Jagannathan, F.F. Jamin, Y. Kohyama, H. Kuroda, C.W. Lai, H.K. Lee, W. Lee, E.H. Lim, W. Lai, A. Mallikarjunan, K. Matsumoto, A. McKnight, J. Nayak, H.Y. Ng, S. Panda, R. Rengarajan, M. Steigerwalt, S. Subbanna, K. Subramanian, J. Sudijono, G. Sudo, S. Sun, B. Tessier, Y. Toyoshima, P. Tran, R. Wise, R. Wong, I.Y. Yang, C.H. Wann, L. T.

Su, M. Horstmann, T. Feudel, A. Wei, K. Frohberg, G. Burbach, M. Gerhardt, M. Lenski, R. Stephan, K. Wieczorek, M. Schaller, H. Salz, J. Hohage, H. Ruelke, J. Klais, P. Huebler, S. Luning, R. van Bentum, G. Grasshoff, C. Schwan, E. Ehrichs, S. Goad, J. Buller, S. Krishnan, D. Greenlaw, M. Raab, and N. Kepler, Int. Electron Devices Meet. Tech. Dig., San Francisco, p. 1075 (2004)

70. S. Tiwari, P.M. Mooney, M.V. Fischetti and J.J. Welser, Int. Electron Devices Meet. Tech. Dig., Washington, D.C., p. 939 (1997)
71. K. Ota, K. Sugihara, H. Sayama, T. Uchida, H. Oda, T. Eimori, H. Morimoto and Y. Inoue, Int. Electron Devices Meet. Tech. Dig., San Francisco, p. 27 (2002)
72. Z. Krivokapih, V. Moroz, W. Maszara, M.-R. Lin, Int. Electron Devices Meet. Tech. Dig., Washington, D.C., p. 445 (2003)
73. H. Yin, K.D. Hobart, R.L. Peterson, S.R. Shieh, T.S. Duffy, and J.C. Sturm, Appl. Phys. Lett. 87, 061922 (2005)
74. T. Irisawa, T. Numata, T. Tezuka, K. Usuda, N. Hirashita, N. Sugiyama, E. Toyoda and S. Takagi, Proc. VLSI Symposium, Kyoto, p. 178 (2005)

2

High Current Drivability MOSFET Fabricated on Si(110) Surface

A. Teramoto and T. Ohmi

Summary. This chapter demonstrates CMOS device characteristics on the Si(110) surface by using a surface flattening process and radical oxidation. By forming a MOS device on the Si(110) surface, high-speed and low flicker noise p-MOSFETs can be realized. Furthermore, the current drivability of p-MOS and n-MOS which are balanced in the CMOS (balanced CMOS) on Si(110) surface can also be realized. These devices are very useful for application to analog/digital mixed signal circuits.

2.1 Introduction

Miniaturization of metal oxide silicon field effect transistor (MOSFET) has been able to increase the integration and performance of LSI devices. However, miniaturization of the critical dimension of integrated circuits is accompanied by a decrease in thickness of gate insulator films for MOS transistors. Leakage currents are mainly composed of leakage current through the gate insulator films and the drain leakage current. Suppression of the leakage current in ULSI devices is one of the crucial requirements for an improvement of ULSI devices. Improvement of current drivability of MOSFETs without shrinkage of the device scale is also very important for increasing LSI performance. Efforts to increase device performance using strained silicon [1–4], Fin-FETs [5–7], etc. have recently been reported. In addition, Si(110) surface channel devices have also been investigated [8–14]. It has been reported that the hole mobility in the channel on Si(110) is largest compared with other low-index surfaces [15]. Figure 2.1 shows the dependence of the effective mobility of electrons and holes in the channel on Si surface orientation and channel direction [15]. Altering the crystalline orientation of the Si surface channel has the advantage that it is possible to increase in the current drivability without changing the material and the device structure. However, present-day gate formation technology cannot form high quality insulator films on the Si(110) surface. We reported that very high quality gate insulators can be formed on every silicon

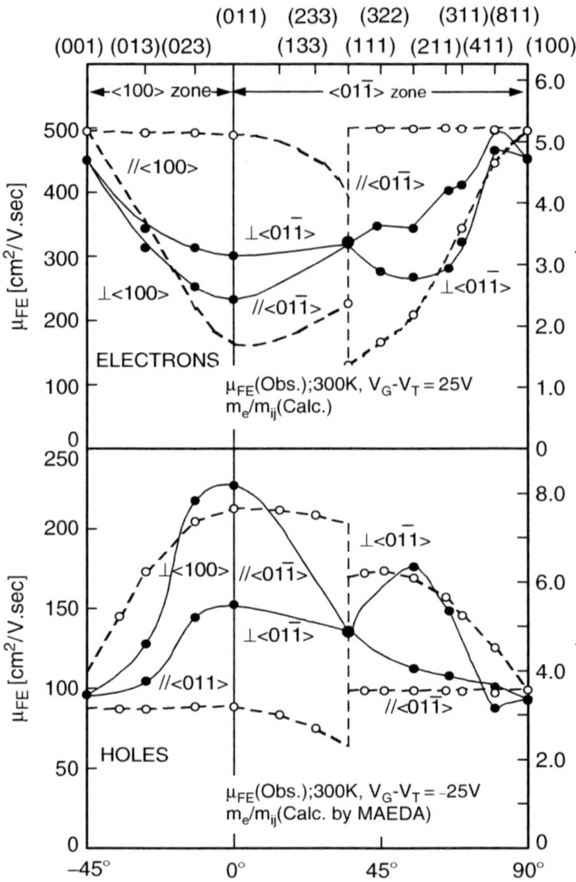

Fig. 2.1. Dependence of the effective mobility of electrons and holes in the channel on the Si surface orientation and channel direction [15]

surface by microwave-exited high-density plasma oxidation/nitridation, and very low $1/f$ noise is realizing by using this oxidation/nitridation technology [8, 16].

In this chapter, we demonstrate that low noise balanced CMOS that is very useful for analog/digital mixed signal circuits can be fabricated on a very smooth Si(110) surface by using a newly developed five-step room temperature clean and the microwave-excited high-density plasma oxidation process.

2.2 Experimental

Dual gate-MOSFETs on Cz–Si(100) and Cz–Si(110) surfaces are employed in these experiment. Five nanometer gate oxides are formed by the

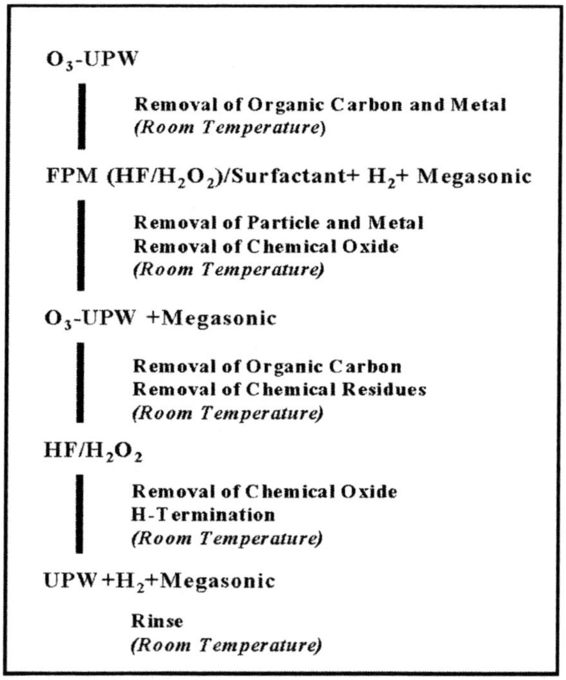

Fig. 2.2. Five-steps room temperature wet cleaning [20]

microwave-excited high-density plasma oxidation (radical oxidation) at 400°C after modified RCA cleaning [17–19] and five-steps room temperature wet cleaning [20] (shown in Fig. 2.2). As^+ and BF_2^+($4 \times 10^{15}\,cm^{-2}$) ions are implanted to the gate poly-Si (300 nm) and source/drain regions after gate formation for n-MOSFET and p-MOSFET, respectively. After the formation of an aluminium interconnect, hydrogen sintering is performed at 400°C in $N_2/H_2 = 9/1$.

Surface micro-roughness and dissolution of Si atoms into water also evaluated. The silicon surfaces are treated by five-steps room temperature cleaning and the RCA cleaning following diluted HF dipping/UPW rinse. As a result, native oxide was not formed on both of the silicon surfaces studied. Surface Micro-roughness of the silicon surfaces is evaluated by vacuum scanning tunnelling microscopy (STM) and the atomic force microscopy (AFM) after the RCA and five-steps room temperature cleaning. The surface micro-roughness and Si dissolution into water measured by the inductive coupled plasma Auger electron spectroscope (ICP-AES) are also evaluated after immersion in water containing various dissolved oxygen concentrations of 0 (N_2 ambient), 8 ppm ($O_2/N_2 = 1/4$), and 32 ppm (O_2 ambient).

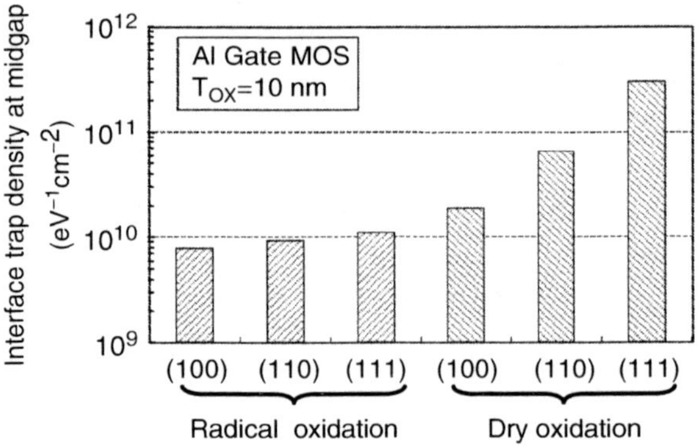

Fig. 2.3. SiO_2/Si interface trap density formed by radical oxidation at 400 °C and thermal oxidation at 900°C [14]

2.3 Results and Discussions

2.3.1 MOSFET Characteristics on the Si(110) Surface

Static Characteristics of MOSFET Fabricated on Si(110)

Figure 2.3 shows the interface trap density at the Si/SiO_2 interface formed by the radical oxidation (400°C) and by conventional dry oxidation (900°C) on Si(100), Si(110) and Si(111) oriented surfaces [14]. It is well-known that high quality SiO_2 films and Si/SiO_2 interface having a low interface trap density can be realized by the thermal oxidation only on Si(100) surface as shown in Fig. 2.3. However, as shown in Fig. 2.3, radical oxidation using the microwave excited high density plasma can form high quality SiO_2 films and Si/SiO_2 interface having a low interface trap density on every silicon surface orientation. This means that every silicon surface can be applied to LSI formation

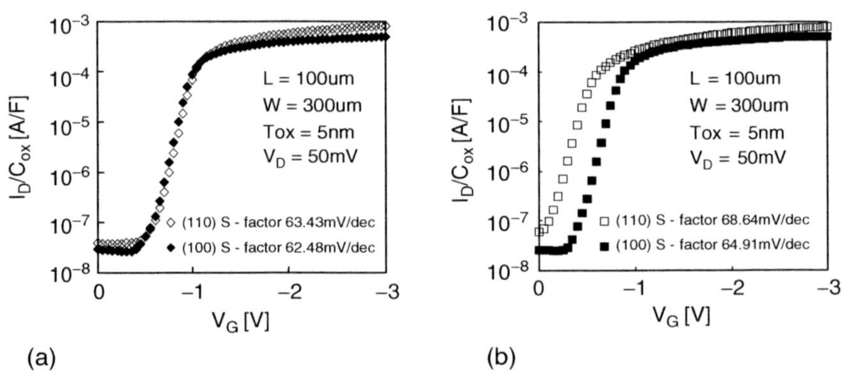

Fig. 2.4. $I_D - V_g$ characteristics of p-channel MOSFETs. (**a**) The gate oxide was formed by radical oxidation. (**b**) The gate oxide was formed by the dry oxidation

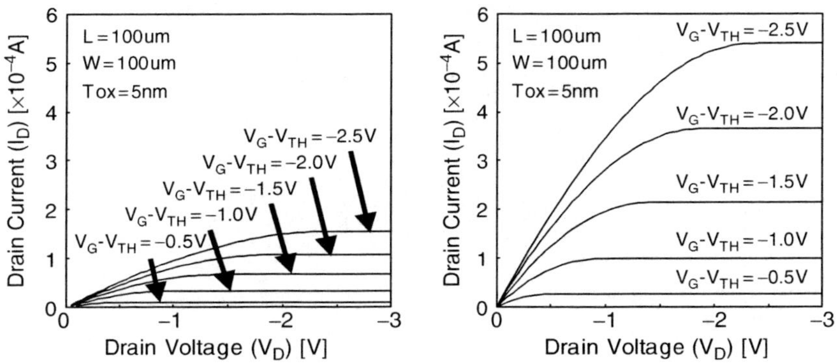

Fig. 2.5. $I_D - V_D$ characteristics of p-MOSFETs on Si(100) and Si(110) surfaces. The current drivability of p-MOSFET on Si(110) is three times larger than that on Si(100)

by using radical oxidation. Figure 2.4 shows the drain current (I_D)-gate voltage (V_G) characteristics (drain voltage (V_D) = 50 mV) of p-MOSFETs having gate oxides formed by (a) radical oxidation and (b) dry oxidation. In the case of dry oxidation, a difference in threshold voltage appears on Si(100) and Si(110), as a result Si/SiO_2 interface traps and fixed charge in the gate oxide. No threshold voltage shift is observed between the MOSFETs on Si(110) and Si(100) having gate oxide formed by the radical oxidation. This indicates that radical oxidation can be used for gate oxide formation on Si(110) surfaces. Figure 2.5 shows the $I_D - V_D$ characteristics of p-MOSFETs on (a) Si(100) and (b) Si(110). The current drivability of p-MOSFET on Si(110) is three times larger than that on Si(100). Figure 2.6 shows the effective channel mobility (μ_{eff}) as a function of effective electric field (E_{eff}) in the p-channel MOSFET. E_{eff} is defined by $(Q_B + Q_i/\eta)/\epsilon_{si}$, where η of n- and p-MOSFET

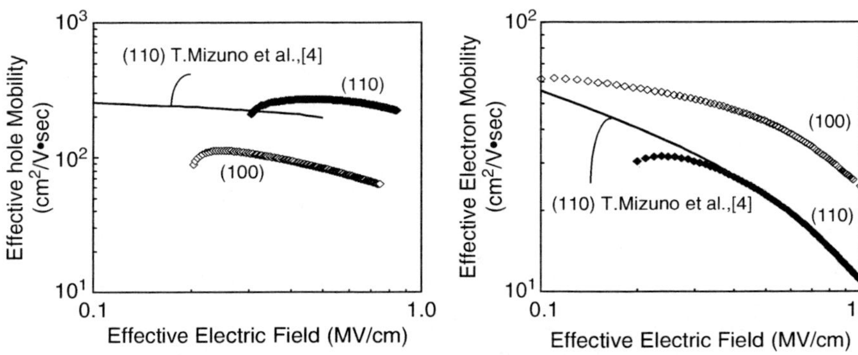

Fig. 2.6. Effective channel mobility (μ_{eff}) as a function of an effective electric field (E_{eff}). μ_{eff} of p-MOSFET on Si(110) is much larger than that on Si(100) and is also larger than that reported μ_{eff} on Si(110) [4]

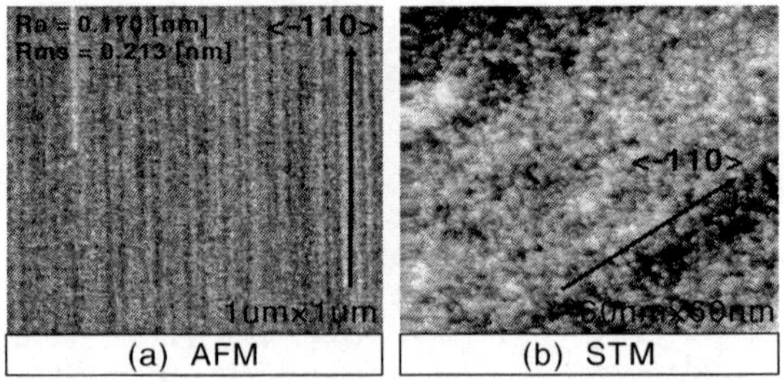

Fig. 2.7. (**a**) AFM image and (**b**) STM image of (110) silicon surface after UPW final rinse at RCA cleaning. The lines like trough to the $\langle -110 \rangle$ direction are observed

are taken to be 2 and 3, respectively [4] and ϵ_{si} is dielectric constant of a silicon. μ_{eff} of the p-MOSFET on Si(110) is much larger than that on Si(100) and is also larger than that reported μ_{eff} on Si(110) [4,21,22]. This may result from the formation of higher-quality oxides on the Si/SiO_2 interface using radical oxidation. However, μ_{eff} of the n-MOSFET on Si(110) is the same as that reported previously for Si(110) [4] and less than that for Si(100).

Figure 2.7 shows (a) AFM and (b) STM images of the Si(110) surface after UPW final rinse at RCA cleaning and diluted HF treatment. Lines parallel to the $\langle -110 \rangle$ direction are observed. This suggests that surface micro-roughness on Si(110) surface is caused by RCA cleaning resulting in degradation of the channel mobility for n-MOSFETs. Figure 2.8 shows the average surface micro-roughness (Ra) change for the Si(110) surface after wet oxidation at 1,000°C or radical oxidation at 400°C. Both high temperature wet oxidation and radical oxidation are isotropic oxidation processes. Therefore, both oxidations have a surface flattening effect. Figure 2.9 shows the (a) AFM and (b) STM images of the Si(110) surface after H_2-UPW+ megasonic rinse at the end of the five-step cleaning [20] (shown in Fig. 2.2) for either wet oxidation or radical

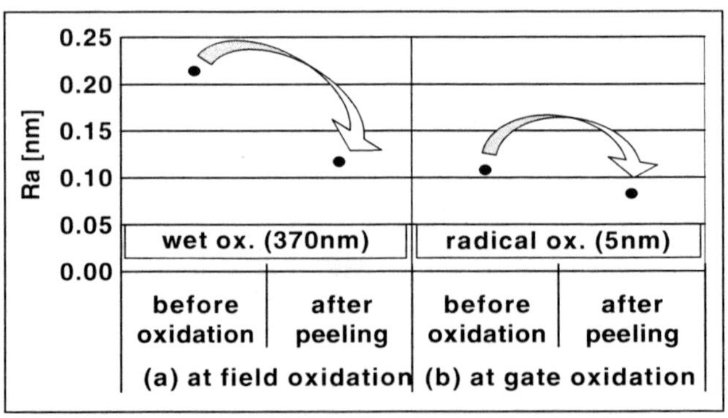

Fig. 2.8. Average surface micro-roughness (Ra) change for Si(110) surface by the wet oxidation at 1,000°C and the radical oxidation at 400°C

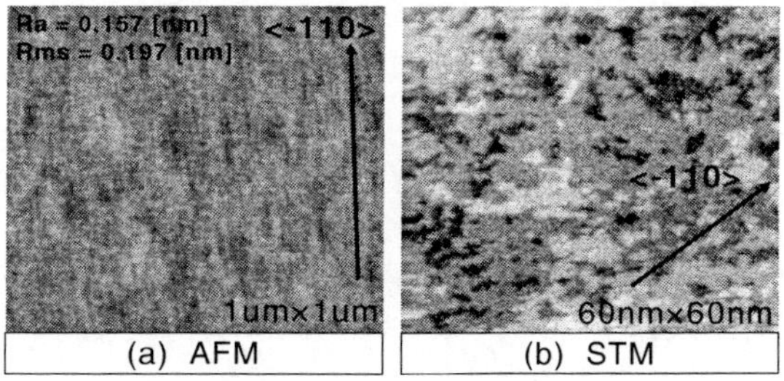

Fig. 2.9. (**a**) AFM and (**b**) STM images of Si(110) surface after H_2-UPW + megasonic rinse at five-steps cleaning

oxidation. Wide terraces are seen in the STM images. This indicates that the micro-roughness of the Si(110) surface can be suppressed by a combination of a flattening process (e.g. wet and radical oxidations) and a cleaning technology that does not generate surface micro-roughness (e.g. the five-steps room temperature wet cleaning). Figure 2.10 shows μ_{eff}-E_{eff} characteristics of conventional and Si/SiO_2 flattened n-MOSFETs. The μ_{eff} value can be improved by Si/SiO_2 flattening. Figure 2.11 shows the channel direction dependency of the normalized I_D on Si(110). Normalized I_D is defined as $I_D/(\text{conventional } I_D\langle 110\rangle)$. I_D of flattened n-MOSFETs is larger than that of conventional devices for every channel direction. On the other hand, the drain currents have a strong dependence on the channel direction, and channel directions giving a maximum current for n-channel MOSFETs differs from that of p-channel MOSFETs by 90°. Figure 2.12 shows the noise power as a function of frequency (f). The noise power is proportional to $1/f$. The $1/f$ noise of p-MOSFET on Si(110) is 1 order of magnitude smaller, although current drivability is the as same as for n-MOSFETs on Si(100). This is

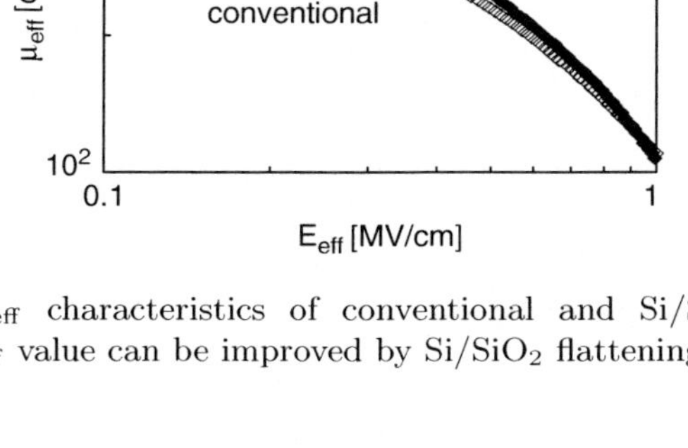

Fig. 2.10. μ_{eff}-E_{eff} characteristics of conventional and Si/SiO_2 flattened n-MOSFETs. The μ_{eff} value can be improved by Si/SiO_2 flattening

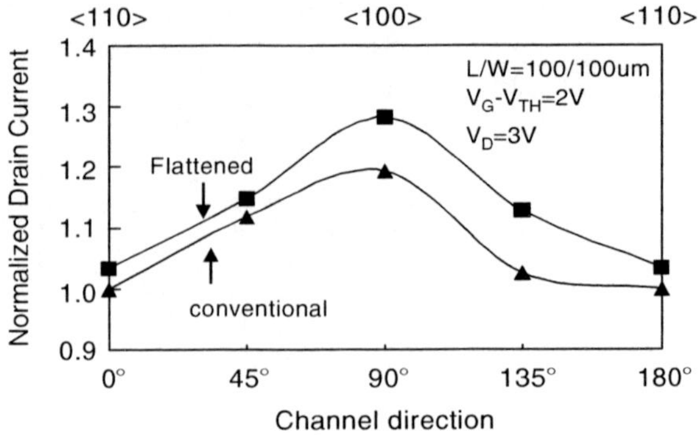

Fig. 2.11. Channel direction dependency of normalized I_D on Si(110). Normalized I_D is defined as $I_D/(\text{conventional } I_D\langle 110\rangle)$

consistent with the observation that the Si/SiO_2 interface is very smooth after processing [16].

Dependence of the Noise Characteristics of MOSFETs on Micro-Roughness [23]

Typical results of electronic noise measurements performed on p-MOSFETs fabricated on Si(100) are shown in Fig. 2.13. Several measurements were made for all combinations of cleaning and oxidation procedures. The combination of radical oxidation [8, 13] and the five-step room temperature clean [20] is very effective in reducing the $1/f$ noise in MOSFETs compared with conventional thermal oxidation and RCA clean [17–19]. This suggests that the surface roughness of Si(100) before oxidation and after the conventional RCA cleaning is larger than that for the alkali-free five-step room-temperature clean described above. The improvement factor of surface micro-roughness after the five-step clean is around 10% compared with a conventional RCA clean.

The same procedures described above has been carried out on Si(110) p-MOSFETs. Looking at the results of $1/f$ noise measurements shown in Fig. 2.15 and at the AFM images presented on Fig. 2.16, and then comparing them to those obtained for the conventional orientation, we can achieve a larger reduction of low frequency noise and a better improvement of the micro-roughness by once more simply changing the pre-gate formation cleaning to the five-steps room temperature clean. The cleaning process appears to have an impact that is much more pronounced on Si(110) than on Si(100). The degradation of the surface resulting from RCA cleaning shown in Fig. 2.16a, which is far worse compared to that shown in Fig. 2.14a, can be explained by the fact that the alkali solution etch rate of Si(110) is faster than that of Si(100) [24]. This significant change in the noise level has its origin in the cleaning process and the resulting surface quality of the Si/SiO_2 interface and in particular the change in the surface micro-roughness.

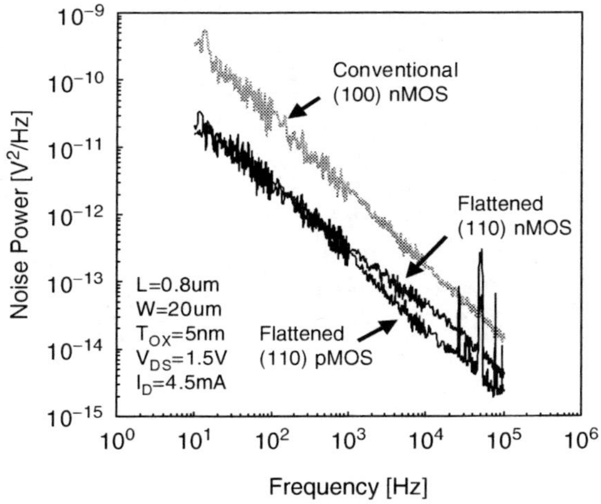

Fig. 2.12. Noise power as a function of frequency (f). The noise power is proportional to $1/f$. The $1/f$ noise of p-MOSFET on Si(110) is 1 order of magnitude smaller, although current drivability is the as same for the n-MOSFET on Si(100)

2.3.2 Suppression of Surface Micro-roughness

In the previous section, reducing the surface micro-roughness was demonstrated to be very important for improvement of current drivability and suppress of $1/f$ noise. In this section, we describe how the etching and roughening characteristics of Si(110) surface in solutions used in pre-gate

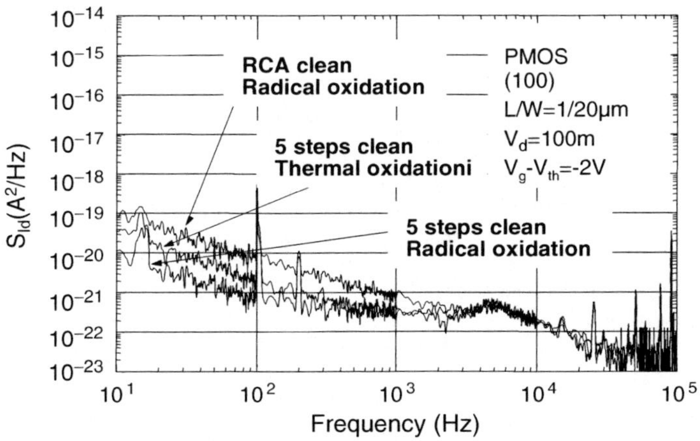

Fig. 2.13. Evolution of the spectral density of drain current in a p-MOSFET fabricated on Si(100) for two different cleaning processes and two different oxidation techniques

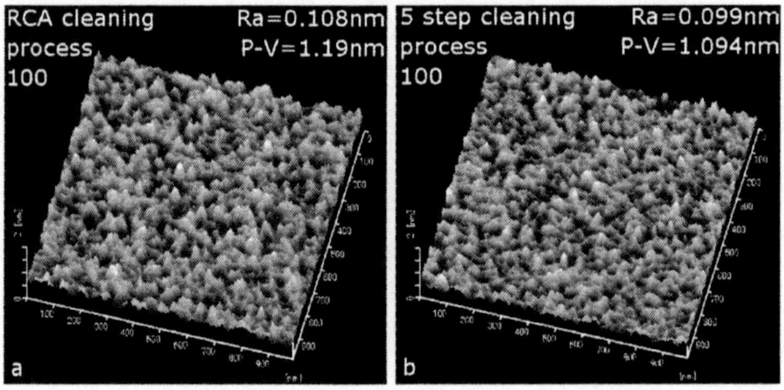

Fig. 2.14. Micro-roughness Ra and peak-valley maximum amplitude P–V after the RCA clean (**a**) and the five-step clean (**b**) for Si(100)

oxide water cleaning compare with the characteristics of the Si(100) surface [25]. The Si(110) surface is etched and roughened easily by solutions containing OH^- ions compared to the observed etch rate of Si(100).

Hydrogen termination of the Si surface is often used to passivate it against contamination, native oxidation, and so on. The silicon surface terminated by hydrogen is chemically stable [26]. In this section, hydrogen termination of Si(110) surface is also described. The hydrogen terminated Si(110) surface is very weakly passivated against native oxidation compared with the hydrogen terminated Si(100) surface, and as a result, precise control of the process atmosphere is essentially required for high performance devices to be fabricated on the Si(110) surface.

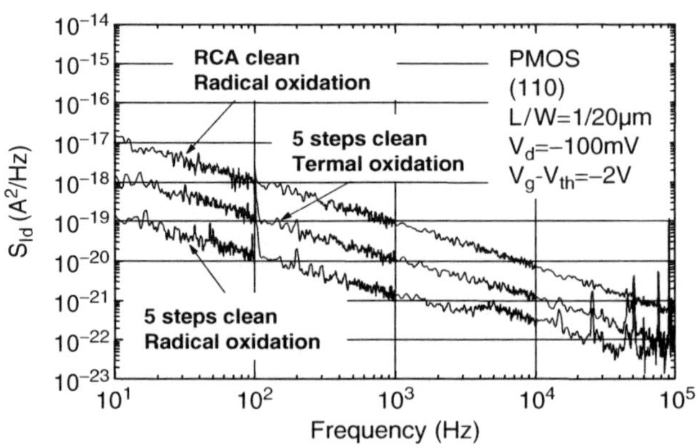

Fig. 2.15. Evolution of the spectral density of drain current in p-MOSFET fabricated on Si(110) for two different cleaning processes and two different oxidation techniques

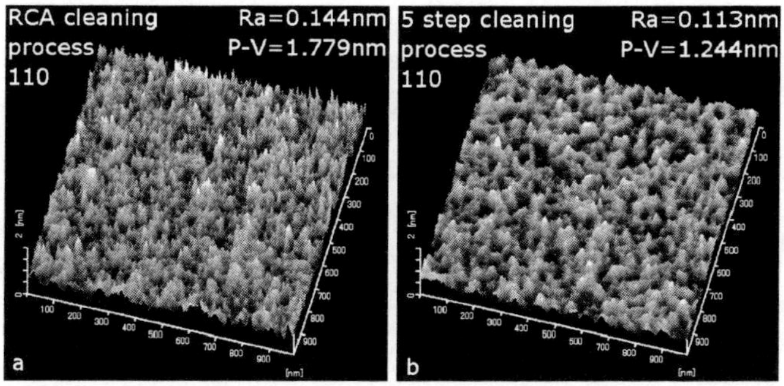

Fig. 2.16. Micro-roughness Ra and peak-valley maximum amplitude P–V after RCA clean (**a**) and the five-step clean (**b**) for Si(110)

Surface Roughening in Ultrapure Water

Ultrapure is usually used for wafer cleaning in LSI fabrication. However, the hydrogen terminated silicon surface is etched even in ultrapure water. Consequently, a silicon surface which is exposed to ultrapure water is roughened. The surface micro-roughness of Si(100) and Si(110) after treatment in ultrapure water for 1, 3 and 24 h in nitrogen ambient are shown in Fig. 2.17 and the AFM images of Si(110) surface after treatment in ultrapure water for 1, 3 and 24 h in nitrogen ambient are shown in Fig. 2.18. The amount of dissolved silicon atoms present in the water of exposure is shown in Fig. 2.19. The surface micro-roughness and silicon dissolution increase as the treatment time becomes longer. The surface microroughness (Ra) increases from 0.12 to 1.16 nm for 24 h exposure of on Si(100) surface. On the Si(110) surface, it increases from 0.12 to 4.58 nm. Dissolved silicon amounts for 24 h

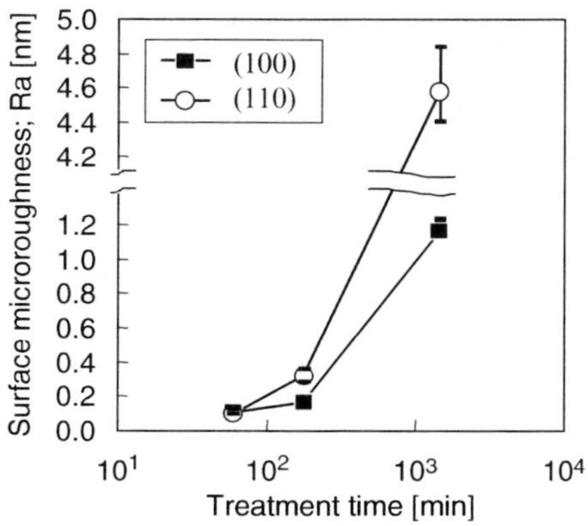

Fig. 2.17. Ultra-pure water treatment time dependence of surface micro-roughness

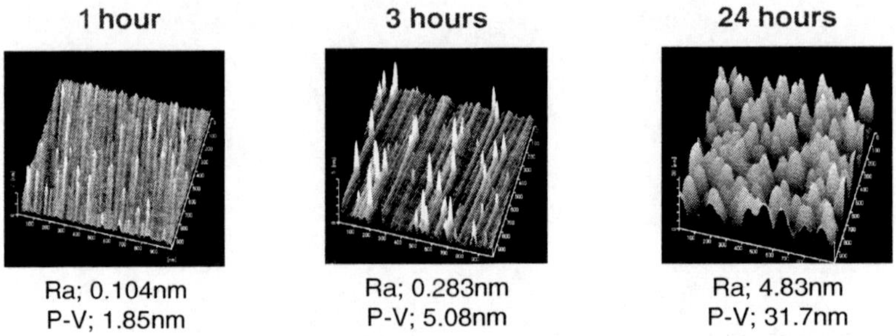

Fig. 2.18. AFM images of Si(110) surface after ultra-pure water treatment of 1, 3, 24 h

from Si(100) and Si(110) correspond to 91 and 200 atomic layers removed, as determined. The results from the dissolved silicon data are consistent with the surface micro-roughness data. Therefore, the surface micro-roughness degradation is attributed to silicon dissolving into ultrapure water during longterm exposure. More silicon atoms are dissolved into ultra-pure water from the Si(110) orientation wafer compared with the Si(100) orientation. The silicon dissolution reaction is assumed to result from oxidation of the surface by hydroxide ions (OH^-). OH^- ions in ultrapure water can etch silicon atoms as alkali solutions [25]. Silicon atoms are dissolved as $HSiO_3{}^-$ and $SiO_3{}^{2-}$ ions in aqueous solution. Although the treatment time examined in these experiments is longer than industrial rinsing processes, which are typically only a few minutes, dissolving of silicon atoms during the rinsing process for even short times can increase the surface micro-roughness. However, suppression of surface micro-roughness is one of the most important issues for improvement

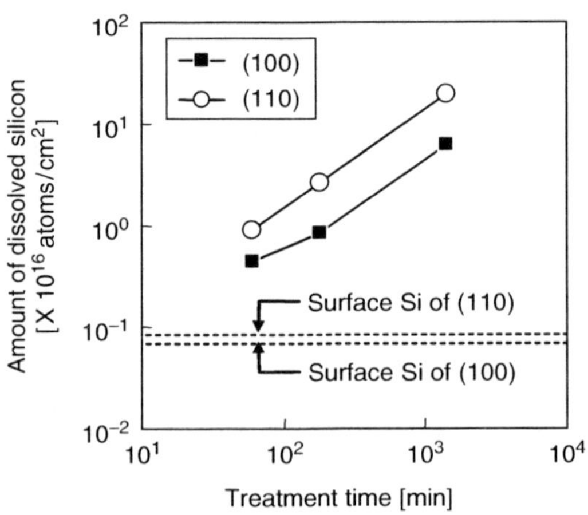

Fig. 2.19. Ultrapure water treatment time dependence of amount of dissolved silicon.

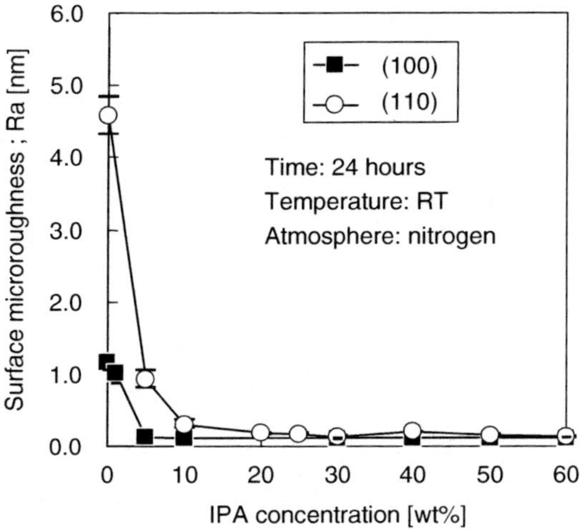

Fig. 2.20. Dependence on IPA concentration in UPW of the surface micro-roughness

of the channel mobility in the MOSFET. Therefore rinsing solution which do not etch and roughen the Si surface are required.

Suppression of Surface Micro-roughness Using IPA/UPW Solutions

The surface micro-roughness after 24 h treatment in Isopropyl alcohol (IPA) containing solutions (< 60 wt%) are shown in Fig. 2.20. The typical AFM images are shown in Fig. 2.21 and the resulting amounts of dissolved silicon atoms in each solution are shown in Fig. 2.22. The surface roughening is suppressed with an increase of the IPA concentration for both of Si(100) and Si(110) surfaces. For the Si(100) surface, the suppression saturates at 5 wt% IPA concentration while it saturates at 30 wt% IPA concentration for Si(110). The amount of dissolved silicon atoms is also decreased with an increase of the IPA concentration. These results indicate that 5 and 30 wt% IPA addition are sufficient to suppress the surface roughening for (100) and (110) orientation silicon, respectively. At those compositions, the dissolved silicon atoms decreased to an amount corresponding to a few atomic layers for both (100) and (110) orientation. These values are very small compared with the situation for ultrapure water without IPA addition. In 5 and 30 wt% IPA added to ultrapure water, the molecular ratios (water:IPA) of water and IPA are about 63:1 and 7.8:1. Although water molecules are still the major component in each solution, the reactions to dissolve silicon atoms are prevented at the silicon surface by the hydrophobic property of IPA. As a result, the surface roughening is suppressed with IPA added to ultrapure water.

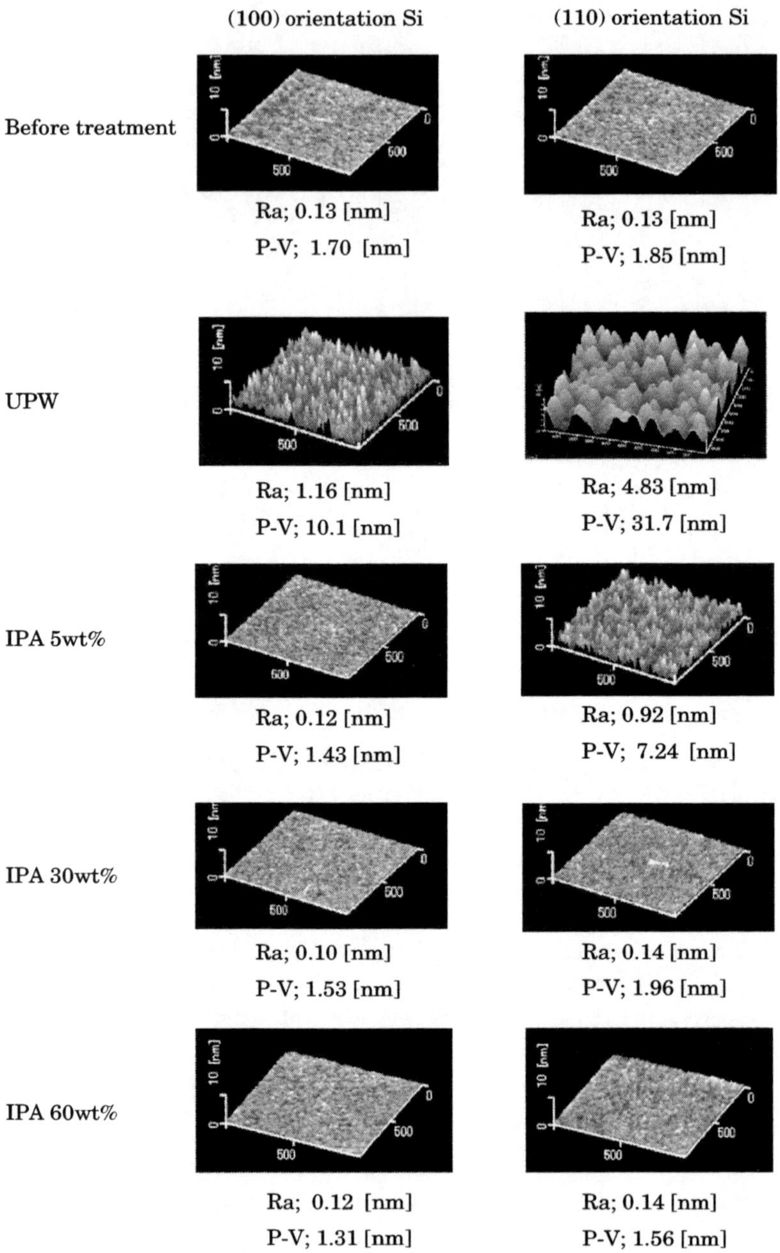

Fig. 2.21. AFM images of the initial Si(100) and Si(110) surface and those after IPA/UPW (0–60 wt%) solution 24 h treatment

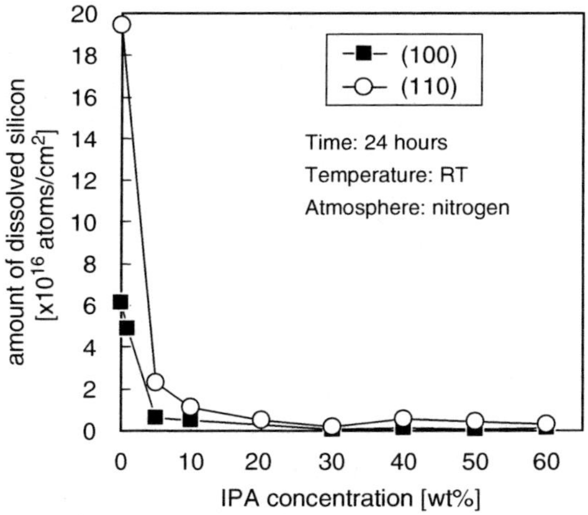

Fig. 2.22. Amount of dissolved silicon into IPA/UPW solution as a function of IPA concentration

Hydrogen Termination of Silicon Surfaces

On the Si(100) surface, hydrogen termination is chemically stable [26]. Figure 2.23 shows the characteristic thermal desorption spectroscopy (TDS) results for hydrogen from a Si(100) surface treated by dilute HF and UPW rinsing. Two peaks at 380 and 520°C are observed in this graph [28]. A silicon atom in the Si(100) surface is terminated by two hydrogen atoms [29] (shown in Fig. 2.24). A dangling bond from a silicon atom, which is generated by desorption of one of two hydrogen atoms bonds gives rise to the feature at 380°C. The remaining hydrogen is desorbed from the Si(100) surface at

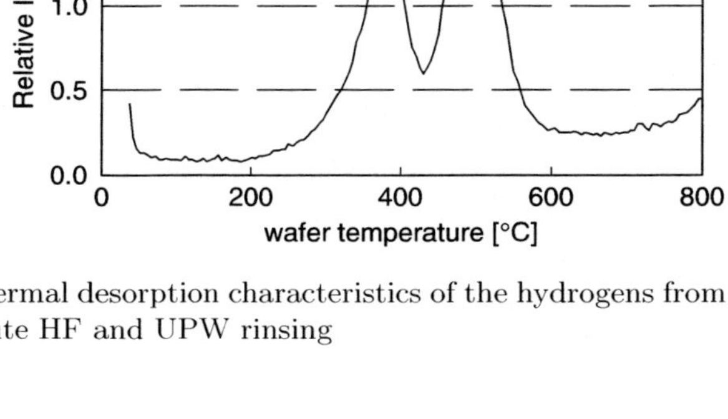

Fig. 2.23. Thermal desorption characteristics of the hydrogens from Si(100) surface treated by dilute HF and UPW rinsing

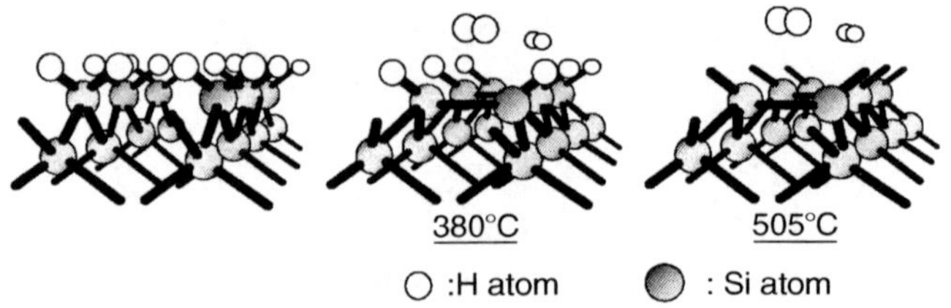

Fig. 2.24. Schematic molecular model of the hydrogen-terminated Si(100) surface

520°C [28]. Hydrogen is desorbed from the Si(110) surface only at 530°C (shown in Fig. 2.25).

Figure 2.26 shows the characteristics of the TDS characteristics of (A) Si(100) and (B) Si(110) surfaces just after dilute HF treatment and UPW rinsing and after exposure to air for 2 and 12 h after dilute HF treatment and UPW rinsing. Only the peak from hydrogen desorbed at 530°C decreases for the Si(100) surface after exposure to the air for 2 h. Both desorption features for hydrogen at 530 and 380°C decrease after exposing the passivated Si(100) surface to the air for 12 h. Figure 2.27 shows the quantity of Si–O bonds as evaluated by an attenuated total reflectance Fourier transform infrared (ATR/FT-IR) spectrometry. While the extent of hydrogen termination decreased, the Si–O bonds on the Si(100) surface increased. This means that the Si(100) surface is oxidized as desorption of the surface hydrogen termination occurs. Figure 2.28 shows a schematic structure of (A) the Si(100) surface terminated by two hydrogens a silicon atom, (B) the Si(100) surface partially terminated by hydrogen after desorption at 380°C, and (C) the Si(110) surface terminated by hydrogen. In order to oxidize the surface (A), the oxygen atom must replace the terminating hydrogen atoms on the surface, or oxygen must enter into the back bond of the surface silicon atoms. Since the Si–H

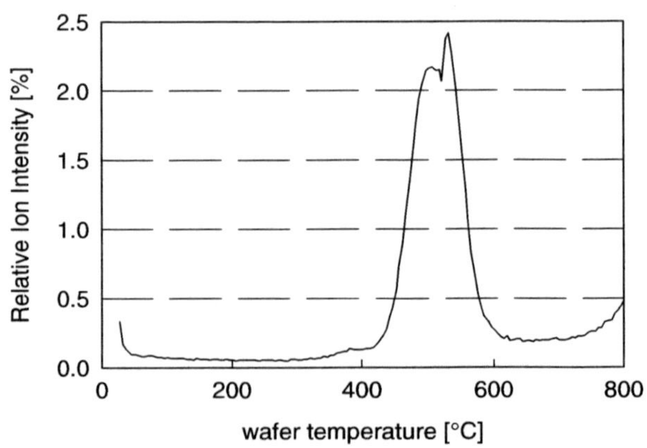

Fig. 2.25. TDS characteristics of hydrogen from the Si(110) surface treated by dilute HF and UPW rinsing

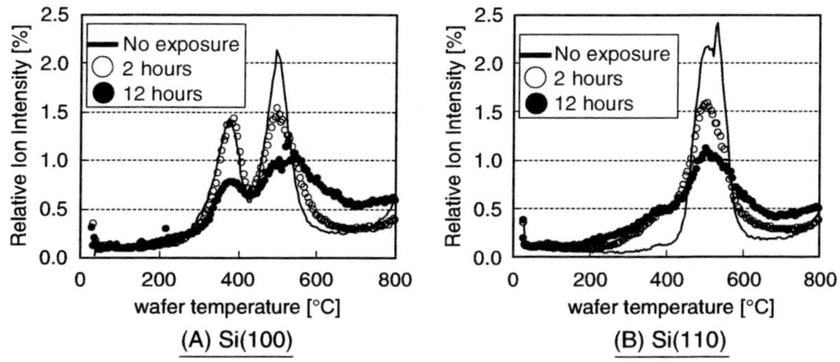

Fig. 2.26. TDS characteristics of (A) Si(100) and (B) Si(110) surfaces just after dilute HF treatment and UPW rinsing and after exposure to air for 2 and 12 h after dilute HF treatment and UPW rinsing

bond of the Si(100) surface is stable and the back bond of silicon cannot be exposed easily to the O_2 or H_2O molecules, when the hydrogen termination of the Si(100) surface is carried out completely, it is very hard to oxidize in. In the case of (B) and (C) surfaces, since a combination of two adjacent silicon atoms appears in the surface, these silicon atoms are easily oxidized by oxygen in the atmosphere. The electrons in the Si–H covalent bonds of the partially-oxidized Si(100) surface oxidized silicon can be attracted toward the oxygen atoms because of their large electronegativity (Oxygen: 3.0). Bonds formed by surface silicon atoms to species other than oxygen (hydrogen atoms and silicon atoms in the case of Si(100) surface, and Si(110) surface, respectively) become weaker, and oxidization of the silicon surface is accelerated further. In the case of the Si(110) surface, the oxidization of the surface is occurs much more readily than for the initially hydrogen-terminated Si(100). Thus, native oxide growth on HF-treated Si(110) cannot be suppressed in the air. It is concluded that wet processes and transfer atmosphere before the

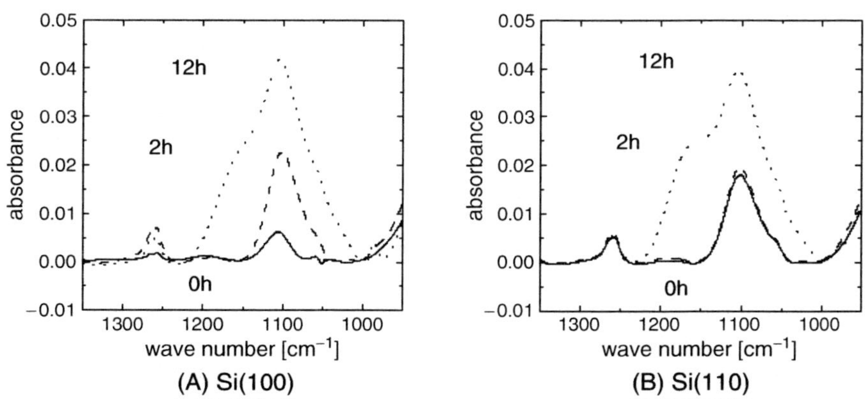

Fig. 2.27. Quantity of Si–O bond evaluated by an ATR/FT-IR spectrometry

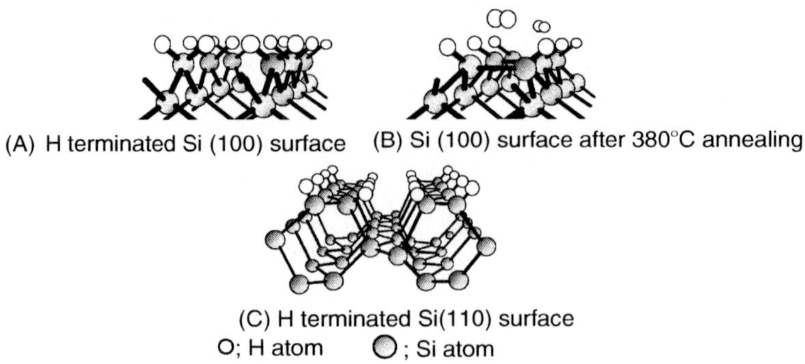

Fig. 2.28. Schematic structure of (**A**) Si(100) surface terminated by two hydrogens a silicon atom, (**B**) Si(100) surface terminated by a hydrogen after desorption at 380°C, and (**C**) Si(110) surface terminated by hydrogens

next process step at which the silicon surface must not be oxidized such as gate oxidation must be carried out in an atmosphere from which oxygen and moisture are removed.

2.4 Conclusions

We have demonstrated the advantages of MOSFET fabrication on the Si(110) surface compared to the Si(100) surface. The current drivability of p-MOSFETs is about three times larger than that on Si(100) and, as a result, balanced CMOS [13] can be realized on the Si(110) surface.

We also demonstrated the relationship between surface or Si/SiO_2 interface roughness and the MOS characteristics of these devices, including their static and noise characteristics. An increase of surface micro-roughness causes the degradation of static characteristics of MOSFETs and an increase of $1/f$ noise. The Si(110) surface is especially easily etched and roughened by OH^- ions in aqueous cleaning solution. A combined five-step room temperature pre-gate oxidation clean which does not employ an alkali solution, followed by radical oxidation can form a high quality Si/SiO_2 interface. As a result, high performance and low noise MOSFETs are realized.

The HF-treated Si(110) surface easily forms a native oxide by air exposure because oxygen can easily reach to the back bond of a hydrogen terminated on Si atom. It is concluded that wet processes and transfer atmosphere before each new process at which the silicon surface must not be oxidized must be carried out in an atmosphere from which oxygen and moisture are removed.

By using a alkali- and oxygen-free cleaning method, radical oxidation and moisture free wafer transfer, high performance, low noise and high reliability ULSI can be realized on the Si(110) surface. Building transistors on this alternative Si surface plane provides very effective improvements in ULSI without shrinking of device dimensions.

Acknowledgements

The authors gratefully acknowledge Professors K. Endo, M. Morita and Dr. K. Arima of Osaka University for STM measurements and discussions. The authors also acknowledge the Ministry of Economy, Trade and Industry and the New Energy and Industrial Technology Development Organization for their financial support.

References

1. J.J. Welser, J.L. Hoyt and J.F. Gibbons, "Electron Mobility Enhancement in Strained-Si N-Type Metal-Oxide-Semiconductor Field-Effect Transistors," IEEE Electron Devices Lett., EDL-15, 1994, pp. 100–102
2. K. Rim, J.J. Welser, J.L. Hoyt and J.F. Gibbons, "Enhanced Hole Mobilities in Surface-channel Strained-Si p-MOSFETs," in IEDM Tech. Dig., 1995, pp. 517–520, December 1995
3. S. Takagi, J.L. Hoyt, J.J. Welser and J.F. Gibbons, "Comparative study of phonon-limited mobility of two-dimensional electrons in strained and unstrained Si metal-oxide-semiconductor field-effect transistors," J. Appl. Phys., Vol. 80, 1996, pp. 1567–1577
4. T. Mizuno, N. Sugiyama, A. Kurobe and S. Takagi, "Advanced SOI P-MOSFETs with Strained-Si Channel on SiGe-on-Insulator Substrate Fabricated by SIMOX Technology," IEEE Trans. Electron Devices, ED-48, 2001, pp. 1612–1618
5. D. Hisamoto, W. Lee, J. Kedzierski, E. Anderson, H. Takeuchi, K. Asano, T. King, J. Bokor, and C. Hu, "A Folded-channel MOSFET for Deep-sub-tenth Micron Era," in IEDM Tech. Dig., 1998, pp. 1032–1034, December 1998
6. F. Yang, H. Chen, F. Chen, Y. Chan, K. Yang, C. Chen, H. Tao, Y. Choi, M. Liang, and C. Hu, "35 nm CMOS FinFETs," Symp. on VLSI Tech., 2002, pp. 104–105, June 2002
7. B. Yu, L. Chang, S. Ahmed, H. Wang, S. Bell, C-Y. Yang, C. Tabery, C. Ho, Q. Xiang, T-J. King, J. Bokor, C. Hu, M-R. Lin, and D. Kyser, "Fin FET scaling to 10 nm Gate Length," in IEDM Tech. Dig., 2002, pp. 251–254, December 2002
8. S. Sugawa, I. Ohshima, H. Ishino, Y. Saito, M. Hirayama, and T. Ohmi, "Advantage of Silicon Nitride Gate Insulator Transistor by using Microwave-Exited High-Density Plasma for applying 100 nm Technology Node," in IEDM Tech. Dig., 2001, pp. 817–820, December 2001
9. T. Mizuno, N. Sugiyama, T. Tezuka, Y. Moriyama, S. Nakaharai, and S. Takagi, "[110]-surface strained-SOI CMOS devices with higher carrier mobility," in VLSI Tech. Dig., 2003, pp. 97–98, June 2003
10. H. Nakamura, T. Ezaki, T. Iwamoto, M. Tago, N. Ikarashi, M. Hane, and T. Yamamoto. "Channel direction impact of (110) surface Si substrate on performance improvement in sub-100nm MOSFETs," in Ext. Abst. SSDM, 2003, pp. 718–719

11. H. Momose, T. Ohguro, K. Kojima, S. Nakamura, and Y. Toyoshima, "1.5-nm Gate Oxide CMOS on (110) Surface-Oriented Si Substrate," IEEE Trans. Electron Devices, vol. 50, No. 4, 2003, pp. 1001–1008
12. H. Momose, T. Ohguro, K. Kojima, S. Nakamura, and Y. Toyoshima, "110 GHz cutoff frequency of ultra-thin gate oxide p-MOSFETs on [110] surface oriented Si substrate," in VLSI Tech. Dig., 2002, pp. 156–157, June 2002
13. A. Teramoto, T. Hamada, H. Akahori, K. Nii, T. Suwa, K. Kotani, M. Hirayama, S. Sugawa and T. Ohmi, "Low noise balanced-CMOS on Si(110) surface for analog/digital mixed signal circuits," IEDM Tech. Dig., pp. 801–804, December 2003
14. A. Teramoto, T. Hamada, M. Yamamoto, P. Gaubert, H. Akahori, K. Nii, M. Hirayama, K. Arima, K. Endo, S. Sugawa, and T. Ohmi, "High Performance CMOS on Si(110) surface," to be submitted to IEEE Trans. Electron Devices
15. T. Sato, Y. Takeishi, H. Hara, and Y. Okamoto, "Mobility Anisotropy of Electrons in Inversion Layers on Oxidized Silicon Surfaces," Phys. Rev. B, Vol. 4, No. 6, pp. 1950–1960, September 1971
16. K. Tanaka, K. Watanabe, H. Ishino, S. Sugawa, A. Teramoto, M. Hirayama, and T. Ohmi, "A Technology for Reducing Flicker Noise for ULSI Applications," Jpn. J. Appl. Phys. Vol. 42, No. 4B, pp. 2106–2109, April 2003
17. W. Kern and D.A. Puotinen, "Cleaning solutions based on hydrogen peroxide for use in silicon semiconductor technology," RCA Review, Vol. 31, No. 2, pp. 187–206, June 1970
18. Tadahiro Ohmi, Koji Kotani, Akinobu Teramoto, and Masayuki Miyashita, "Dependence of Electron Channel Mobility on Si SiO_2 Interface Microroughness", IEEE Electron Device Letters, Vol. 12, No. 12, pp. 652–654, December 1991
19. T. Ohmi, M. Miyashita, M. Itano, T. Imaoka, and I. Kawanabe, "Dependence of thin oxide films quality on surface microroughness," IEEE Trans. Electron Devices, Vol. 137, No. 4, pp. 537–545, March 1992
20. Tadahiro Ohmi, "Total Room Temperature Wet Cleaning for Si Substrate Surface," J. Electrochem. Soc., Vol. 143, No. 9, pp. 2957–2964, September 1996
21. S. Takagi, A. Toriumi, M. Iwase, and H. Tango, "On the Universality of Inversion Layer Mobility in Si MOSFET's: Part II-Effects of Surface Orientation," " IEEE Trans. Electron Devices, vol.41, No.12, pp. 2363–2368, December 1994
22. S. Takagi, A. Toriumi, M. Iwase, and H. Tango, "On the Universality of Inversion Layer Mobility in Si MOSFET's: Part I-Effects of Substrate Impurity Concentration," IEEE Trans. Electron Devices, vol. 41, No. 12, pp. 2357–2362, December 1994
23. P. Gaubert, A. Teramoto, K. Kotani, and T. Ohmi, "Reduction of 1/f noise in Si(110) and (100) surface MOSFET using a new cleaning technology," Proceedings, 20th Annual Meeting of Japanese Association for Science Art and Technology of Fluctuations, pp. 28–30, September 2005
24. H. Akahori, K. Nii, A. Teramoto, S. Sugawa, and T. Ohmi, "Atomic Order Flattening of Hydrogen-Terminated Si(110) Substrate For Next Generation ULSI Devices," Extend Abstracts of the 2003 Int. Conf. SSDM, pp. 458–459, September 2003
25. M. Yamamoto, K. Nii, H. Morinaga, A. Teramoto, and T. Ohmi, "Suppression of Surface Micro-Roughness of Silicon Wafer by Addition of Alcohol into Ultra Pure Water for Rinsing Process," To be published in ECS 208th meeting Proceedings Volume Cleaning Technology in Semiconductor Device Manufacturing IX

26. T. Ohmi, "Surface Chemical Electronics at the Semiconductor Surface," Appl. Surf. Sci., Vol. 121/122, pp. 44–62, November 1997
27. K. Satoa, M. Shikidaa, T. Yamashiroa, M. Tsunekawaa, S. Itob, "Roughening of single-crystal silicon surface etched by KOH water solution Sensors and Actuators," 73, pp. 122–130 (1999)
28. N. Yabumoto, K. Minegishi, K. Saito, M. Morita and T. Ohmi, "An Analysis for Cleaned Silicon Surface with Thermal Desorption Spectroscopy," J. Ruzyllo and R.E. Novak, eds., Semiconductor Cleaning Technology/1989, PU90-5, pp. 265–272, The Electrochemical Society, Pennington, NJ, 1990
29. Y.J. Chabal, G.S. Higashi, and K. Raghavachari, "Infrared spectroscopy of Si(111) and Si(100) surfaces after HF treatment: Hydrogen termination and surface morphology," J. Vac. Sci. Technol. A7(3), pp. 2104–2109, May/Jun 1989

3

Advanced High-Mobility Semiconductor-on-Insulator Materials

B. Ghyselen, I. Cayrefourcq, M. Kennard, F. Letertre,
T. Akatsu, G. Celler, and C. Mazure

Summary. Silicon-on-Insulator (SOI) is today the substrate of choice for several applications. In order to boost further circuit performance, new solutions are being explored. In particular, increasing the charge carrier mobility has been identified as a requirement for the next technology nodes. One possible option is to increase transistor channel mobility through local strain engineering via external stressors, an approach that can be used on bulk silicon as well as standard SOI substrates. Other solutions are based on substrate engineering. The attractiveness of these solutions is largely due to their compatibility with standard CMOS integration processes and architectures and presents the advantage of being independent of transistor geometry. The two approaches can be combined to maximize transistor mobility and on-current. Among the different substrate level approaches, we will focus on three main families: (1) the effect of crystal orientation, (2) strained Si and/or SiGe layers On Insulator, and (3) monocrystalline Ge-On-Insulator substrates.

3.1 Introduction

Silicon-on-insulator (SOI) is today the substrate of choice for several applications, for example high performance and low power ICs. In order to boost further circuit performance, new solutions are being explored at the CMOS process level. In particular, increasing the charge carrier mobility has been identified as a requirement to meet the performance needs of the 65 nm technology nodes and beyond.

One possible option is to increase transistor channel mobility through local strain engineering via stressors like nitride layers or epitaxial SiGe source/drain pockets. This so-called "local strain" or "process-induced strain" approach can be used on bulk silicon as well as standard SOI substrates. Other solutions are to induce a MOSFET carrier mobility increase via substrate engineering, which presents the advantage of being independent of transistor geometry. The attractiveness of wafer level based solutions is largely due to their compatibility with standard CMOS integration processes and architectures. The two approaches can be combined to maximize transistor mobility and on-current.

Among the different substrate level approaches to increased mobilities in SOI architectures, we will focus on three main families:

(1) The effect of crystal orientation on carrier mobility and specific combinations of crystalline orientations of semiconductor layers on Insulator,
(2) Strained Si and/or SiGe layers On Insulator, and
(3) Monocrystalline Ge-On-Insulator substrates.

In each case, we will give an overview of the development status at the substrate level and examples of device performance enhancement.

3.2 Crystalline Orientation Effects

3.2.1 Silicon Crystalline Orientations for Bulk Substrates

The principle of this approach is to take advantage of the fact that the charge carrier mobility is dependent on the current flow direction in the crystal. Historically, (100) oriented bulk substrates with flat or notch location defining the $\langle 110 \rangle$ axis have been the universal standard in the IC industry: see for instance SEMI standards for 200 and 300 mm polished monocrystalline silicon wafers [SEMI M1.9-0699 and SEMI M1.15-0704, respectively]. The reasons behind this choice were to maximize electron mobility, with (100) surfaces having the additional advantage of forming good quality gate oxides.

Figure 3.1 [1] shows a comparison of the effect of different crystalline orientations on both the electron mobility (Fig. 3.1a) and the hole mobility (Fig. 3.1b). It can be seen that while the standard choice for $(100)/\langle 110 \rangle$ wafers is indeed good for electrons, it is not the case for holes. To increase hole mobility, other surface orientations shall be preferred, such as (110) wafers. The gain may be substantial, a factor 2 to 3 depending on the hole current direction flow in the plane. If we stay with (100) wafers, a hole mobility gain in the 10–20% range can be obtained by aligning the PMOS channel along the $\langle 100 \rangle$ directions, which is easily implemented by rotating the substrate (twist) by 45° [2–4]. At the wafering level it corresponds to a change of the notch or flat position, which has no or minor impact on ingot pulling and on the wafer manufacturing. This solution has the advantage that it does not affect the electron mobility, as opposed to other crystalline orientations for which higher hole mobilities are obtained but at the expense of an electron mobility degradation. This advantage has been reported for unstrained silicon [2, 3]. But it has also been demonstrated to be valid for locally strained silicon [4] where the strain induced gain on NMOS was not lost by rotating by 45° the channel flow of the electrons.

For bulk substrates however, changing the crystalline orientations of the active layer means also changing the properties of the handle substrate.

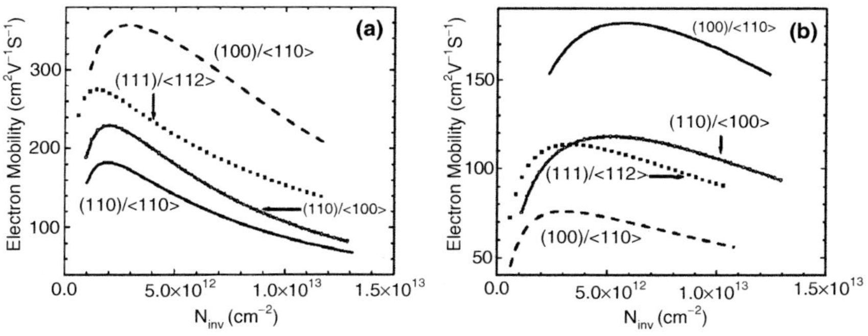

Fig. 3.1. Carrier mobility in the inversion layer with various substrate orientations (gate oxides grown simultaneously in N_2O at 800°C in conventional furnace), for electrons (**a**) and holes (**b**). From [1]

This may have important consequences. For instance resistance to mechanical shocks during handling, die separation and to thermal shocks may be changed, which needs to be taken into account. Rotating the wafers by 45° means also that the cut plane is at 45° angle to the preferred (110) cleavage plane. These limitations are also characteristic of SOI substrates realised from one single bulk substrate (see [6] for a review): SIMOX or other processes based on epitaxy (ZMR, FIPOS, etc.).

3.2.2 Silicon Crystalline Orientations for SOI Substrates

Historically, SOI wafers have duplicated bulk silicon substrates as far as possible. This is the main reason why most of the SOI substrates used today (that could be called in this respect "standard SOI wafers") are composed of a (100)/⟨110⟩ SOI thin film lying on identically oriented (100)/⟨110⟩ handle substrate, as shown in Fig. 3.2a. But SOI wafers do not have to be limited to a duplication of bulk silicon wafers, they can also extend their capabilities. In particular for SOI substrates made using techniques based on wafer bonding and layer transfer (such as BSOI, Smart Cut or ELTRAN, see [6]) two different substrates are used as starting material for the fabrication of one SOI wafer. This has the advantage that the choice of crystalline orientation for the active layer and for the handle substrate can be decoupled and optimized to get the best of both orientations. The choice for the active layer can be guided by optimizing the electrical properties–and especially the charge carrier mobility–while for the handle substrate the historical (100)/⟨110⟩ substrate can be kept for mechanical robustness reasons.

The first simple example of non-standard SOI substrates is based on rotating the donor substrate (active layer) by 45° compared to the handle substrate, as shown in Fig. 3.2b. This is something that can be achieved quite easily. It can be done by rotating one wafer compared to the other at the bonding stage. Or it can be done even more easily by using a donor wafer that has been rotated during notch/flat formation steps. Either way, the standard orientation

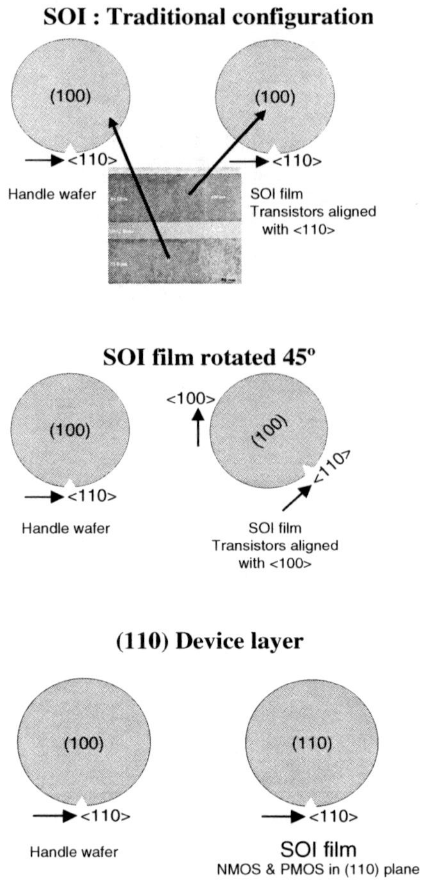

Fig. 3.2. Examples of combinations of crystalline orientations in SOI substrate based on layer transfer techniques: (**a**) traditional configuration (**b**) 45° (100) SOI film rotation. PMOS mobility is increased, NMOS mobility is unchanged [2–4] (**c**) (110) device layer. PMOS is enhanced, NMOS mobility is reduced [1]

is kept for the handle substrates, maintaining the way the SOI wafers will cleave during die separation, while optimizing carrier mobility. The deviation of such a process from standard SOI wafers processes is small and can be implemented easily.

Another example is based on the use of (110) Si wafer for the donor substrate while keeping unchanged the base substrate (regular (100) wafer). There the gain is more significant in terms of hole mobilities (a factor that can be as large as 2–3), but at the expense of a degradation on the electron side. The deviation in wafer fabrication as compared to standard SOI wafers is slightly larger since (110) layers have to be processed. It is well known [7] that some physical and chemical properties (cleaning, oxidation, implantation) are somewhat dependent on crystal orientation. From an SOI substrate manufacturing

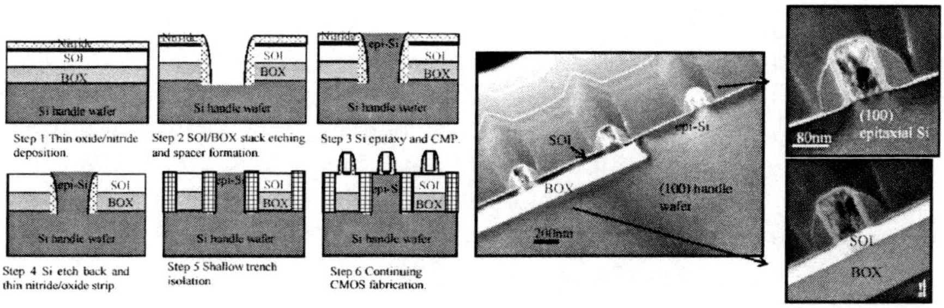

Fig. 3.3. "Hybrid Orientation Technology" (HOT), PFET on (110) surface and nFET on (100) surface. From [1]

point of view, most of it however may look more like engineering work than profound developments. From a device manufacturing point of view, (110) interface states densities and gate insulator properties need to be evaluated. Another practical difference is that (110) bulk material of appropriate quality is needed. Crystal ingot growth as well as wafering steps are involved. For the specific case of Smart Cut, layer splitting within (110) silicon and realisation of (110) SOI layers lying on (100) substrates has already been investigated and demonstrated with some optimizations [8].

When both NMOS and PMOS transistors are built on the same crystalline plane, certain performance compromises are necessary. A concept that has been emerging especially in high performance ICs, is the possibility of combining different layers and addressing separately the improvement of NMOS and PMOS transistors. Figure 3.3 describes schematically the hybrid orientation technology (HOT) approach [1]. Thanks to the flexibility of wafer bonding and layer transfer techniques, it combines the conventional (100) orientation (notch oriented towards $\langle 110 \rangle$ direction) to address NMOS and a (110) orientation for PMOS. In one case the (100) oriented layer is the SOI film (NMOS SOI) while the bulk PMOS is built on a (110) layer grow epitaxially from the handle substrate [1]. The inverse crystal orientation combination results in (110) SOI PMOS and (100) bulk NMOS.

3.2.3 Perspectives for Crystalline Orientations in SOI Substrates

Deviating from the traditional "Full (100)/$\langle 110 \rangle$ oriented silicon substrates" several examples of crystalline orientation evolutions and combination within SOI substrates have been described.

Overall, the advantages of these advanced engineered substrates addressing improved mobilities is that they are still "standard silicon" and "standard SOI" structures. In that respect, such solutions do not represent major complications to the existing processes and can be implemented easily from an SOI substrate manufacturing point of view.

It should be kept in mind that only a few simple examples have been described here. The door is open to other crystalline orientations that would appear more appropriate, and also to more complex substrate structures, with more than one SOI layer to address more than one type of transistor. The optimum crystalline orientation may also depend on whether (and which kind of) strain is applied to the silicon to boost carrier mobilities [5]. As a conclusion, revisiting crystalline orientations should be kept in mind when designing advanced generations of devices and substrates. This is also valid for the next sections where other clues to higher carrier mobilities – like strained Si- and other materials – like Ge- are addressed.

3.3 Strained Si on Insulator Wafers

3.3.1 Introduction to Strained Si on Insulator Wafers

It has been known for a long time that straining semiconductors has an impact on their carrier mobilities. It is true for silicon, for SiGe and Ge, and also for most of III–V materials. Strain can be tensile or compressive, biaxial or uniaxial, parallel or transverse to the current flow. Its effect on carrier mobilities can be positive or negative, and is not equivalent for electrons and holes. The impact of strain on carrier mobilities, although far from being perfectly known and understood, has been documented quite extensively in the literature, especially for the case of silicon: see for instance this book and other references [10–13]. Briefly, conduction or valence band degeneracy splitting, preferential thermal population of electron states with light transport effective mass, reduced interband/intervalley hole/electron-phonon scattering have been used to explain mobility changes [this book + [10–12].

There are several ways to achieve controlled strain levels in silicon. Before going into the specifics of SOI architectures, we will first briefly describe two basic concepts for introducing strain in CMOS transistor channels: the so-called "local strain" and "global strain" approaches. Later we will focus on global strain, considering different manufacturing routes for substrates that combine strained Si layers and SOI architectures. Two types of substrates will be considered depending on whether the final strained SOI wafer contains any SiGe layer. These SOI-based structures are known as SGOI, when an intermediate layer of SiGe is present between strained Si and the buried oxide (BOX), and sSOI when strained Si is placed directly on the BOX. While many techniques look compatible with the realization of SGOI substrates, we will show the unique capability of layer transfer techniques to make Ge-free sSOI. In particular, it will be shown that thin strained films can be transferred to new substrates and that the wafer bonding interface alone can stabilize the built-in strain in adjacent layers.

3.3.2 "Local Strain"

The approach referred to as "local strain" or "process-induced strain", is based upon the uniaxial strain that is introduced during the CMOS manufacturing

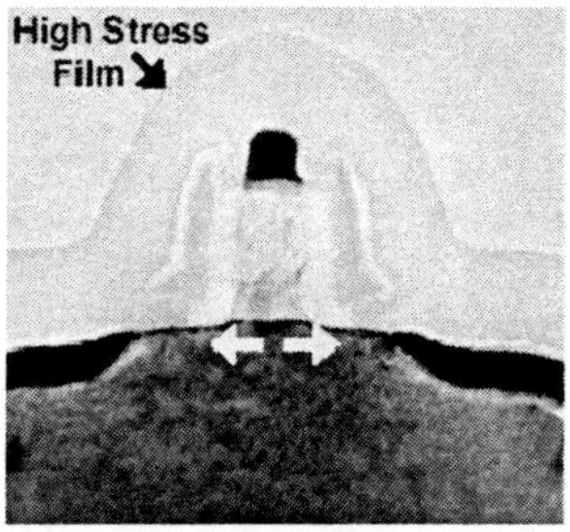

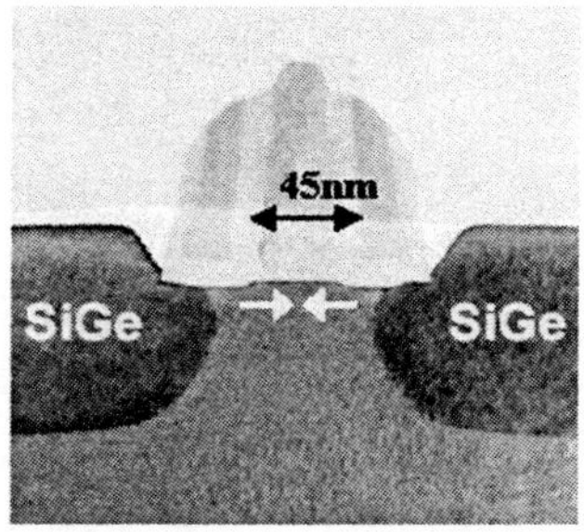

Fig. 3.4. TEM cross-section of NMOS (**a**) device capped with stressed silicon nitride and PMOS (**b**) device showing the SiGe epitaxial source and drain stressors [14]

steps [14–22], such as epitaxial SiGe source/drain pocket formation, nitride spacer deposition, gate encapsulation, shallow trench isolation (STI). Figure 3.4 [14], shows a TEM cross-section of NMOS with tensile strain inducing gate liner and a compressive strain inducing SiGe source/drain stressor.

Uniaxial stressors are generated and introduced during CMOS manufacturing. It has proven very successful and leading edge IC makers have implemented it in 90 nm technology manufacturing. The drawback of uniaxial stressors is that these are device geometry dependent by their nature. The lateral dimensions of the transistors affect the efficiency and the strength of the induced strain. Strain engineering needs to be taken into account at the transistor and IC design level, limiting the technology flexibility to a cell library approach. For each new technology node a redesign of the stressors is needed in order to achieve a mobility increase. Defectivity is another critical parameter. The presence of strong strain gradients associated with the local strain approach may generate crystalline defects that would result in excessive pn junction leakage or generate device failure. Local strain techniques have been industrially implemented in bulk as well as in SOI technologies.

3.3.3 "Global Strain"

Global strain is a further step in strain engineering to further boost device mobility and current drive. It helps overcome the limitations of uniaxial techniques, and its combination with local strain techniques is proposed to maximize performance for the future technology nodes. It has the advantage of being independent of a specific MOSFET geometry, and it applies to large and sub-μm devices. Global strain is not only an alternative but can be combined with process induced strain to enable the optimization of both NMOS and PMOS transistors.

The fabrication of bulk strained silicon is critical. A strained Si layer is grown by hetero-epitaxy on a relaxed $Si_{1-x}Ge_x$ layer, the final strain being an increasing function of the Ge content. Since bulk SiGe substrates with Ge concentrations above 10% do not exist, virtual substrates with relaxed

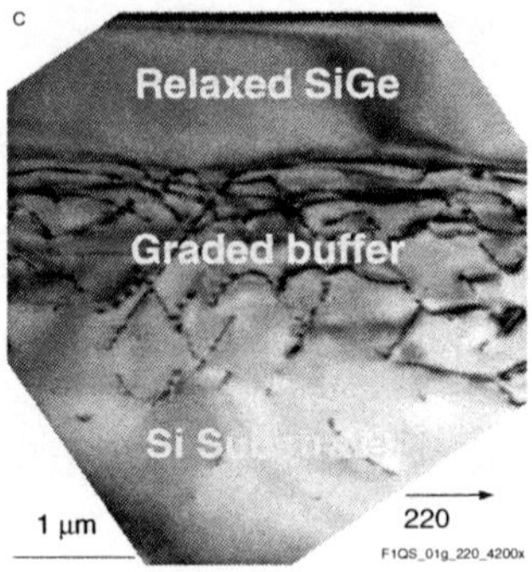

Fig. 3.5. TEM cross-section of a virtual SiGe substrate realized with graded buffer. Only the very surface is eventually useful for the overgrowth of strained Si layer

SiGe are first formed by hetero-epitaxy on bulk Si with the help of different epitaxial techniques. These include forming graded buffers or low temperature buffers that allow for the SiGe lattice to relax to its equilibrium value. This is followed by a low crystal defect epitaxial step in order to fabricate a good quality surface [23–27]. A similar technique is also used in the compound materials field [28]. Figure 3.5 shows as an example a TEM cross-section of a virtual SiGe substrate realized with a graded buffer.

Increasing the Ge content of the SiGe substrate increases the strain within the overlying silicon layer. Figure 3.6 [29] shows experimentally that for electrons a mobility improvement beyond 60% can be achieved for Ge content >18%. For holes, no real gain appears until Ge content reaches 30–40%. For such values the gain in hole mobility becomes very significant: more than 100% improvement is possible.

Technical challenges increase with the Ge content of the films. This is especially true for the epitaxial layers (increasing dislocations densities,

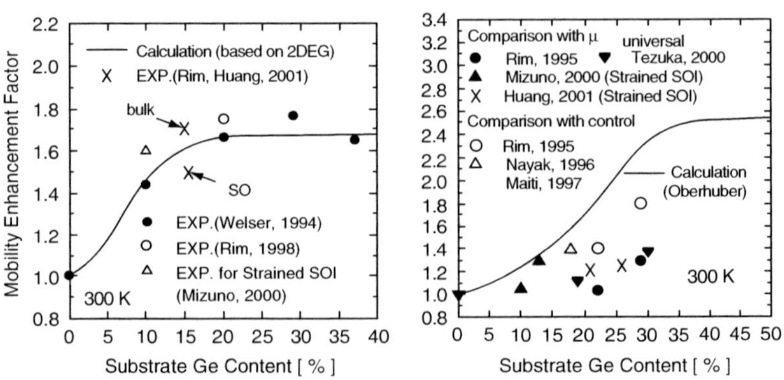

Fig. 3.6. Calculated and measured carrier mobility enhancement factors as published by Tagaki et al. [29]. The left figure is for electrons and the right one for holes

pile-ups, cross-hatch), but also for other process steps such as cleaning or thermal treatments. This is one reason why 20% Ge was chosen for the first generation of strained films: a compromise between the electron mobility improvement and wafer fabrication technology. A second generation of strained substrates, with 35–40% Ge content will address an improved hole mobility.

3.3.4 Two Main Approaches to "Global Strain On Insulator": SGOI Vs. sSOI

SGOI: Strained Si on Insulator via Intermediate Relaxed SiGe Layer On Insulator

Many forms of strained SOI contain at least one thin buried relaxed SiGe layer in addition to the strained Si layer. The role of this layer is to enable the formation of the strain within the Si layer. Usually an SiGeOI substrate is formed first, and then it is used as a template for the final Si epitaxial deposition step. For the latter to be strained, it is important that the SiGe layer On Insulator is relaxed. The relaxation rate, combined with the initial Ge content, determine how strained the silicon layer will be, and ultimately how much the transistor performance will be enhanced. This approach is called "SGOI".

In SGOI, the strain value and the crystalline quality of the overgrown strained Si layer depend on the quality of the underlying SiGeOI layers. The growth of the final strained silicon layer is relatively routine. The main technical challenge is in making appropriate relaxed SiGeOI substrates. In the following sections we will describe different routes that have been demonstrated to realize these structures. Since SGOI involves two semiconductor layers on insulator, the total thickness of the active device region is typically about 30–70 nm, which is suitable for partially depleted (PD) transistors.

sSOI: Strained Si on Insulator Without any Intermediate SiGe Layer

Schematically, the sSOI substrate is a simple evolution of the standard SOI substrate, with a strained Si layer replacing the conventional relaxed Si film. The sSOI substrate does not contain any additional SiGe layer. Compared to SGOI, however, this apparently simple sSOI substrate can be realized only by layer transfer techniques. The main advantages of sSOI are its full compatibility with SOI technology, no Ge contamination risk of the strained Si and elimination of the SiGe/sSi interface which is a source for misfit dislocation nucleation.

Ge is seen by some end-users as an additional and potentially parasitic element to control. Using sSOI substrates becomes therefore a critical advantage for IC makers. The absence of Ge in sSOI eliminates a potential problem of Ge out-diffusion within the transistor active area and increases the flexibility of the device fabrication process flow. sSOI structures are typically very thin and this makes them very suitable for fully depleted technology. It will be shown later that sSOI films can also be made thick enough for PD devices or for multiple gate transistors such as FinFETs.

The two approaches (sSOI and SGOI) can also be compared in terms of their sensitivity to misfit dislocations. In strained epitaxial stacks, relaxation tends to occur thanks to the easy formation of misfit dislocations. These have a very detrimental effect on device behavior when located close to the active area. In the case of SGOI, the danger is that such misfit dislocations may easily appear at the SiGe/strained Si interface.

3.3.5 Different Routes Towards SGOI

Several substrate strategies have been investigated for the formation of relaxed SiGeOI substrates. They include the following:

(a) *SIMOX technology.* Similarly to standard SOI manufacturing by SIMOX, a high dose of oxygen is implanted within a substrate in order to separate the active layer from the rest of the bulk substrate. The main difference is that an additional epitaxial SiGe layer is formed before the oxygen implant [30], as illustrated in Fig. 3.7. Usually in SIMOX, a high temperature (>1,300°C) annealing is performed to promote oxygen diffusion and precipitation into the buried oxide, and also to annihilate crystalline defects created in the top silicon layer. For SGOI formation this final high temperature anneal is still necessary and it limits severely the initial Ge content of the SiGe layer, since its melting point drops with the increase in Ge concentration. It should also be noted that the overall process enables a relaxation of the SiGe layer. Currently it seems there is not much activity to develop further the SIMOX-based process, which may be explained by the emergence of other more promising approaches.

(b) *"Ge condensation effect" [31].* This approach, as illustrated in Fig. 3.8, starts with a "standard" SOI substrate on which an SiGe layer is

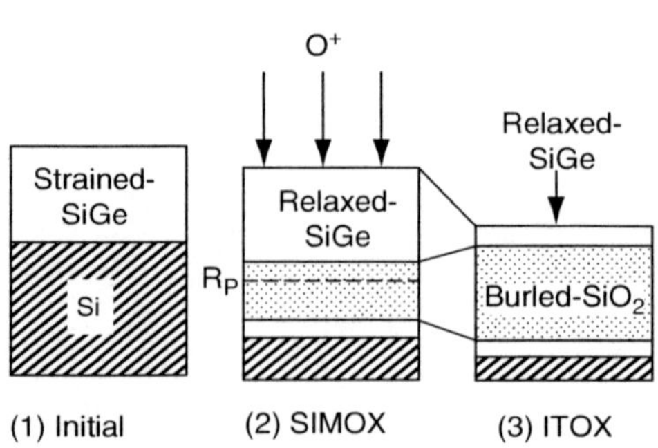

Fig. 3.7. SIMOX process adapted to the realization of SiGeOI substrates (from [30])

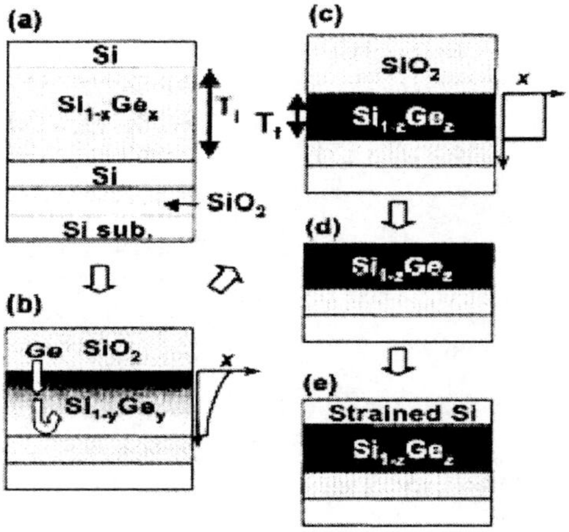

Fig. 3.8. Schematic illustration of the Ge condensation technique (from [31])

deposited. During a subsequent sacrificial oxidation, Ge diffusion (from the SiGe layer) into the SOI layer occurs while the buried oxide blocks diffusion into the substrate below. This leads to a Ge enrichment of the SOI layer and ultimately to a thin relatively uniform, and partially relaxed, SiGe single layer on SiO_2. Subsequently, a strained Si layer is epitaxially grown on such an SiGeOI substrate. This approach has the advantage of starting with standard commercially available SOI wafers. Defectivity (especially dislocation density) and strain relaxation efficiency in the SiGe layer need to be optimized before considering any industrial applications. Since this approach is based on Ge enrichment, it is not compatible with the realization of Ge-free sSOI substrates.

(c) *Layer transfer techniques.* BSOI (Bonded SOI) and Smart Cut^{TM} technologies have been used to realize SiGeOI wafers [32–35]. The main difference compared to the fabrication of standard SOI wafers is that the donor wafers contain relaxed SiGe layers. Figures 3.9 and 3.10 illustrate this concept, with a focus on the Smart Cut^{TM} technology. Donor wafers are produced in the same fashion as the bulk wafers with biaxial strain that were described previously (see Sect. 3.3.3). Buffer layers can be realized by epitaxial growth, for example as graded composition films or as low temperature epi films with a high density of point defects that accommodate the lattice mismatch while enabling relaxation of the SiGe. In other techniques, strained SiGe layers are grown and subsequently relaxed, for instance by the use of H or He implant and anneal [36].

As can be seen in Figs. 3.9 and 3.10, one major advantage of layer transfer techniques is that only the very top, relatively defect-free material is peeled from the donor substrate, without the more defective buffer of the donor substrate. In the case of Smart Cut technology, an additional key advantage

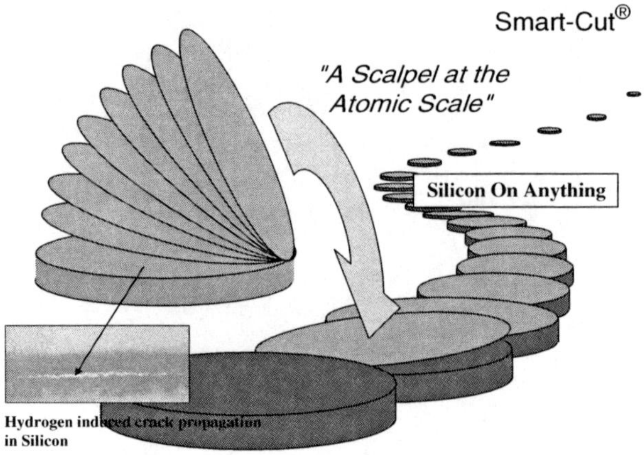

Fig. 3.9. Schematics of the Smart Cut technology, as used today for high volume production of SOI wafers

is the opportunity to save and reclaim the donor substrate, including the epitaxial buffer.

After the relaxed SiGe layer transfer to a new support wafer, the substrate is prepared for strained Si overgrowth. The final preparation can include a surface-smoothing step (CMP) to erase the roughness introduced by the Smart Cut splitting step and to recover a smooth surface that is consistent with device requirements. Other processing can be performed instead of, or in combination with CMP: annealing, sacrificial oxidation, and etching.

To complete SGOI formation, strained Si is grown epitaxially on the relaxed SiGe layer. Epitaxy on SOI-like structures requires some adjustments to accommodate surface emissivity deviations from that of bulk Si. This difference in emissivity can change deposition temperature as has been observed during Si epitaxy on ultra-thin conventional SOI wafers [37]. Details of final strained Si epitaxy on SiGeOI substrates are addressed elsewhere [35].

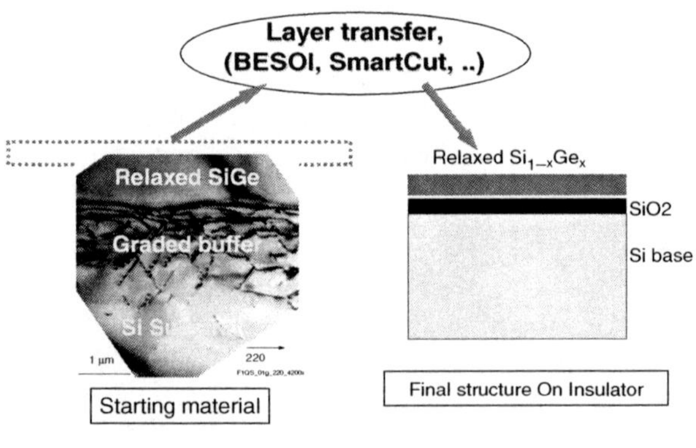

Fig. 3.10. Layer transfer concept applied to the realization of SGOI wafers

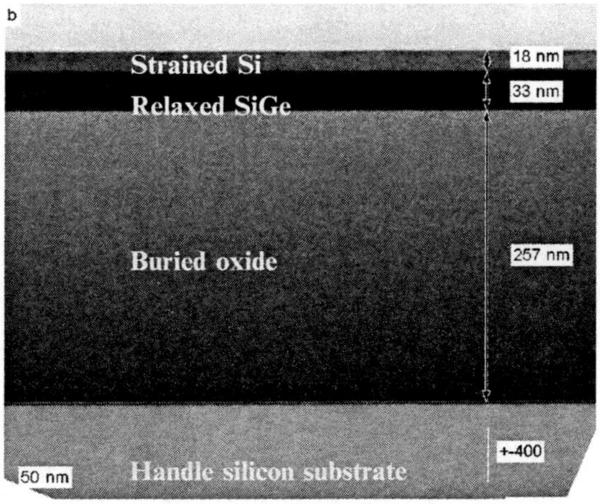

Fig. 3.11. Cross-section TEM showing 18 nm thick strained Si grown onto a 33 nm thick SGOI

Figure 3.11 shows a cross-section TEM image of a finished SGOI substrate after the final strained Si epitaxy.

For all the SGOI methods it is of prime importance that during the final strained Si growth the SiGe template be nearly fully relaxed in order to induce maximum strain within the Si overlayer. This full relaxation should be maintained in spite of the compressive stress force that the strained Si layer is applying to the SiGe layer below it.

The preservation of strain in a thin deposited Si layer has been experimentally verified [32–35]. In the case of Smart Cut, strain within the final Si overlayer was measured by Raman Spectroscopy. A UV mode was used to ensure that the main contribution comes from the Si overlayer and not from the SiGe or the bulk Si handle substrate. While Fig. 3.12 a shows a stress mapping across a quarter of a 200 mm wafer, Fig. 3.12b focuses on the comparison of measured values with a relaxed silicon reference measurement (handle bulk substrate). The distribution of the spectra corresponding to the nine points of Fig. 3.12b is quite narrow compared to the global shift from the reference spectrum. This global shift agrees with values expected from the donor substrate (with 20% SiGe layer with relaxation in the 95–98% range), demonstrating that the final target strain has been achieved within the strained Si.

3.3.6 SGOI Material Concept Validation Through Device Demonstrations

In the last section we showed that strain within silicon could be introduced in SGOI architectures. Further validation of this concept should take into account the device performance. Several transistors demonstrations have been reported for SGOI substrates realized by condensation technique [38] and layer

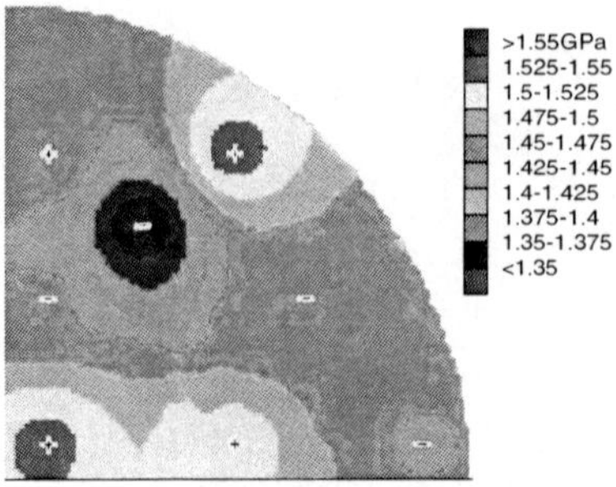

(a)

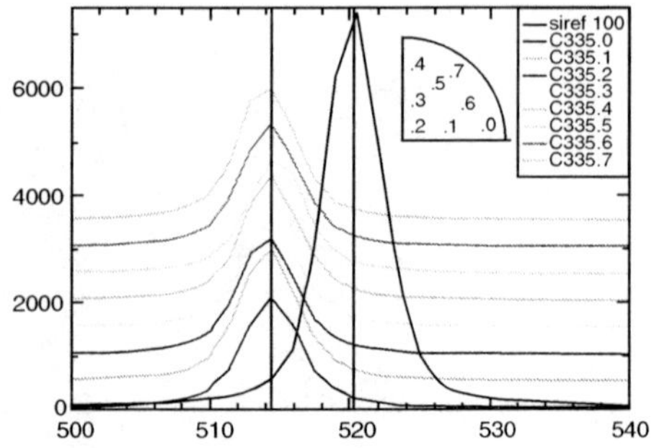

(b)

Fig. 3.12. Stress measurement within the final Si overlayer grown on SiGeOI substrates, as measured by UV Raman spectroscopy (**a**). The measurements have been performed on nine points distributed regularly across a quarter of a 200 mm wafer. For reference, a measurement made on a bulk Si has been added (**b**)

transfer [39–41]. Figure 3.13a shows an example of a partially depleted transistor realized on SGOI made by the Smart Cut technology, with a physical gate length of 33 nm [41]. For long channel transistors improvement in electron mobility is very significant: 43% for an Lg = 1 µm, as shown in Fig. 3.13b. This indicates that the strain was maintained through device processing. Shorter channel results tend to show less or no gain. Some strain relaxation in small structures may be present but other factors appear dominant. Parasitic S/D contact resistance becomes a larger fraction of the total resistance for

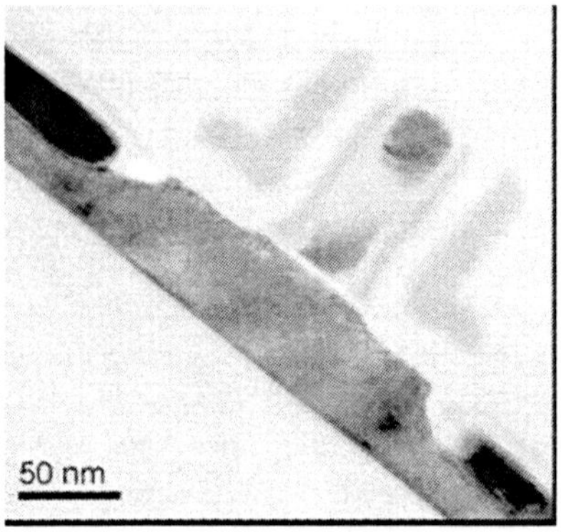

(a)

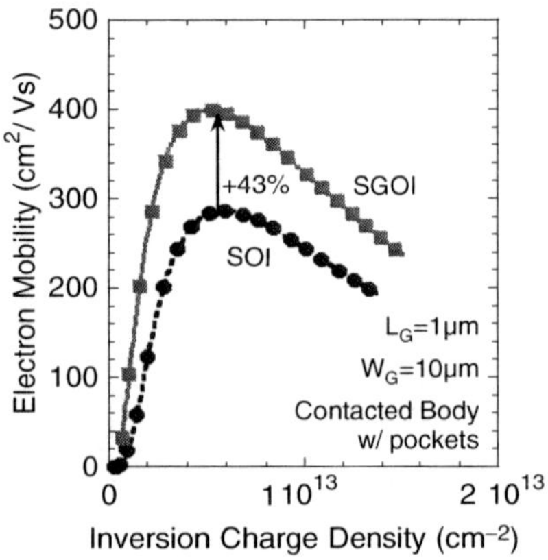

(b)

Fig. 3.13. (**a**) TEM image of an SGOI transistor (physical gate length = 33 nm; strained Si = 12 nm; SiGe = 41 nm). (**b**) Electron mobility vs. inversion charge density for body contacted SGOI and transistors. From [41]

short channel devices and this tends to reduce the enhancement of the drive current Id.

3.3.7 sSOI Substrates: Ge-free Strained Si On Insulator Substrates

sSOI substrates cannot be made by SIMOX and condensation based techniques. Suitable techniques require layer transfer: the Smart Cut technology, and the older BSOI techniques can be used [34, 35, 42–45]. Compared to the SGOI fabrication process, an additional strained Si layer is added on the relaxed SiGe donor substrate before the layer transfer. This layer will become the active layer of the final sSOI substrate. Therefore its defectivity, strain and thickness uniformity, must be precisely controlled. Optimization of these parameters, and the evolution from 200 to 300 mm wafers are addressed elsewhere [45, 46].

In the specific case of Smart Cut, the ions are implanted through the strained silicon layer and into the relaxed SiGe capping layer. This allows the transfer of a double layer (strained Si and some of the relaxed SiGe) such as the one shown in Fig. 3.14a. After the bi-layer transfer, the SiGe film is removed by selective etching (see Fig. 3.14b).

The residual strain inside the Si layers were investigated by TEM and Raman spectroscopy. Detailed results are reported elsewhere [35, 45]. Both techniques have confirmed stress values in the 1.3–1.5 GPa range, close to the predicted value for a donor substrate, consisting of an Si layer on relaxed $Si_{0.8}Ge_{0.2}$. Figure 3.15 shows the Raman spectra of the strained silicon layer on an SiGe donor substrate and of the final strained Si after the total removal

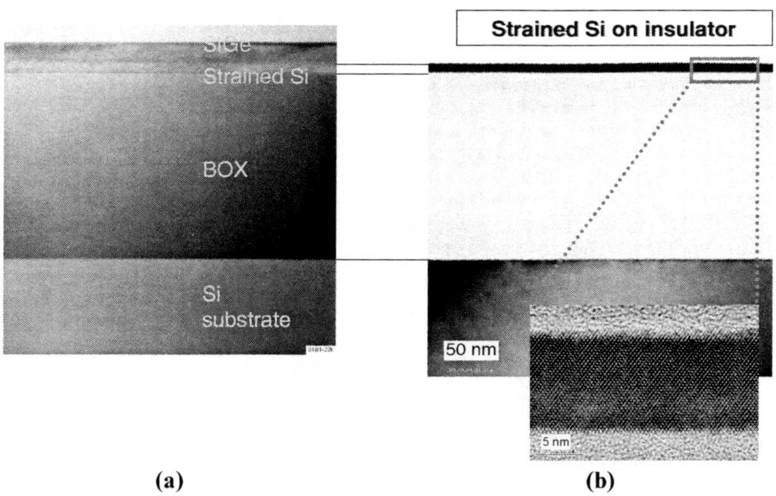

Fig. 3.14. XTEM pictures of 200 mm relaxed SiGe on insulator wafers, before and after the removal of the SiGe overlayer (then generating a sSOI substrate)

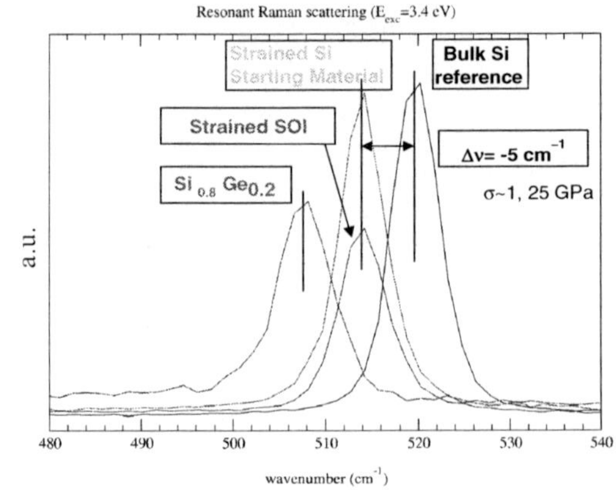

Fig. 3.15. UV-Raman spectra showing that for sSOI substrates the strain in the strained Si layer is maintained through the Smart Cut transfer and the SiGe overlayer removal

of the SiGe layer. This result confirms that the strain is not changed within the Si layer during the layer transfer operation and also that it is maintained after SiGe removal, thanks to the bonding interface alone. In Fig. 3.16, the stress values are mapped accross a quarter of an early 300 mm prototype sSOI substrate.

The effectiveness of the bonding interface in maintaining the strain was further tested by submitting the wafers to "aggressive" thermal annealing. Although rapid thermal annealing is typically used in advanced CMOS

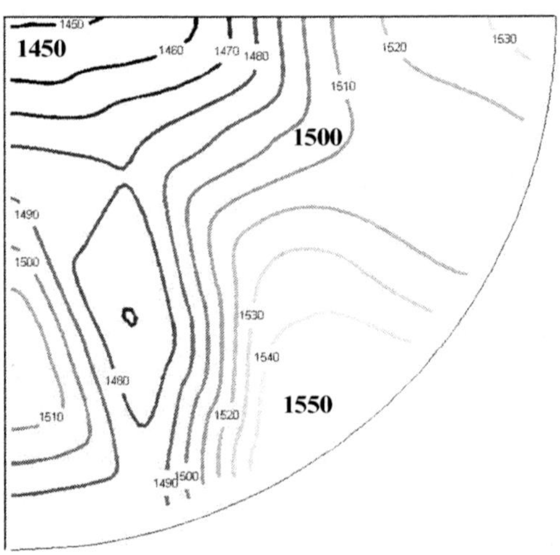

Fig. 3.16. UV macro raman stress measurement showing stress uniformity as measured on early 300 mm sSOI prototype wafers (1.5 GPa and Max–Min ≈ 100 MPa)

manufacturing, the worst-case scenario of furnace anneals was tested on unpatterned wafers, for durations of up to several hours, and temperatures up to 1,100°C. No significant strain relaxation has been detected by UV Raman measurements, indicating that the bonding interface alone can maintain the strain even at high temperatures (see also [42–44] for other similar demonstrations).

3.3.8 Thick sSOI Substrates

The critical thickness of strained silicon is the thickness beyond which strain relaxation occurs through misfit dislocation emission, typically at the Si/SiGe interface [25]. For strain values in the 1.5 GPa range, corresponding to almost 100% relaxed SiGe 20% layers, the critical thickness in silicon is in the 20–25 nm range depending on exact material preparation conditions [25]. Such thickness is suitable for fully depleted SOI devices, but it is not enough for partially depleted technology or for the multiple gate transistors such as FinFETs.

In SGOI, the SiGe/sSi interface is the preferential nucleation site for the misfit dislocations that enable relaxation processes. In sSOI structures, this interface does not exist. This is likely why the maximum sSi thickness before relaxation is higher in sSOI than the critical thickness in SiGe/Si bi-layer system. Lack of relaxation up to 70 nm Si film thickness was confirmed by growing additional epi on 20 nm thick sSOI wafers [45, 46]. Figure 3.17 summarizes corresponding results derived from UV micro-Raman, showing that no relaxation is observed in such thick sSOI structures. Similar results have been obtained by others [47,48]. Again, thermal stability of thick sSOI wafers has been checked successfully with anneals at 1,100°C for 2 h, opening the door to the introduction of strained Si in PD and FinFETs applications.

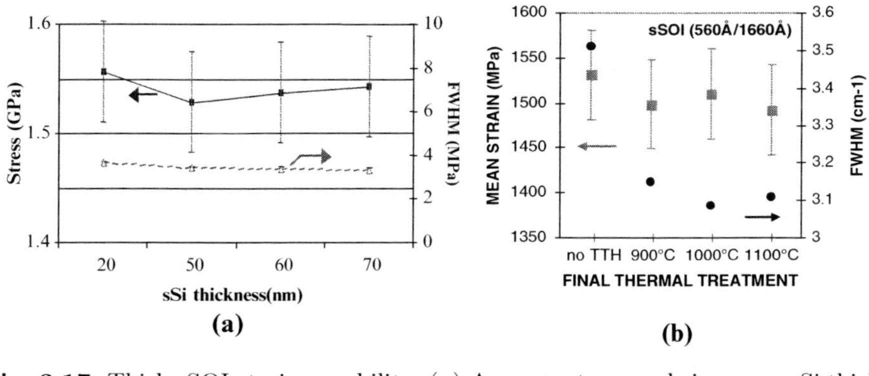

Fig. 3.17. Thick sSOI strain capability: (**a**) Average stress and sigma vs. sSi thickness. 20 nm thick sSOI is the starting wafer. Re-epitaxy on such starting wafers is used here to obtain thicker sSOI films. (**b**) Average stress and sigma vs. final annealing temperature (2 h) for a 56 nm thick sSOI layer

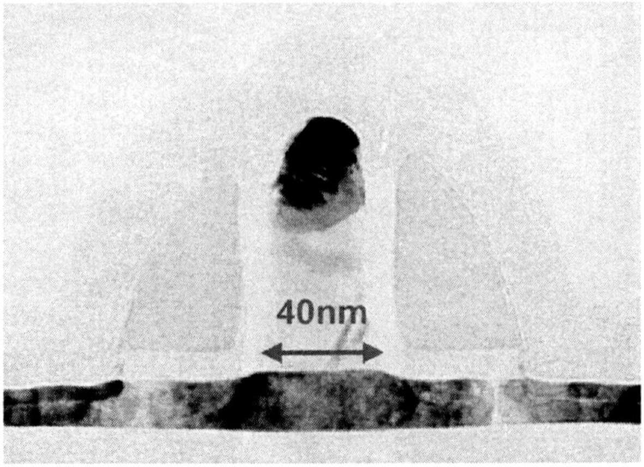

Fig. 3.18. TEM cross-section of a 40 nm device made on 40 nm thick sSOI wafer. From [50]

3.3.9 Device Results on sSOI

Rim et al. [42] showed fully depleted sSOI (FDsSOI) devices in thin sSOI, with an electron mobility enhancement of 125% and a hole mobility enhancement of 21% and demonstrating that strain is preserved through the CMOS process. The nominal strain in Rim's study was around 1.45%, corresponding to a Ge content of the SiGe template of 35%. Several other transistor demonstrations in thin sSOI have also been published [47,49]. For thicker sSOI there is less published data [48,50]. Figure 3.18 shows a cross-section TEM of a device with a 40 nm gate length that was fabricated on 40 nm thick sSOI with very little process tuning compared to the standard SOI process [50]. In this work, transconductance measurements on NMOS long channel devices confirm a 63% performance enhancement that translates into a current increase of 25%. For short channel devices the gain is decreased to 10%, similar to the behavior of SGOI short channel devices. The authors propose that a parasitic S/D series resistance is the predominant cause of lower enhancement, as the devices get smaller [50]. Further optimization of these devices is needed, for example by including raised S/D. However, the results confirm that the strain is preserved in short channel devices down to 40 nm gate length (Fig. 3.19).

3.4 Germanium On Insulator Substrates

3.4.1 Introduction to GeOI Substrates

It is known that compared to silicon the use of germanium would bring significant enhancements to low field carrier mobilities for both electrons and holes. Gains up to 2× for electrons and 4× for holes have been reported for

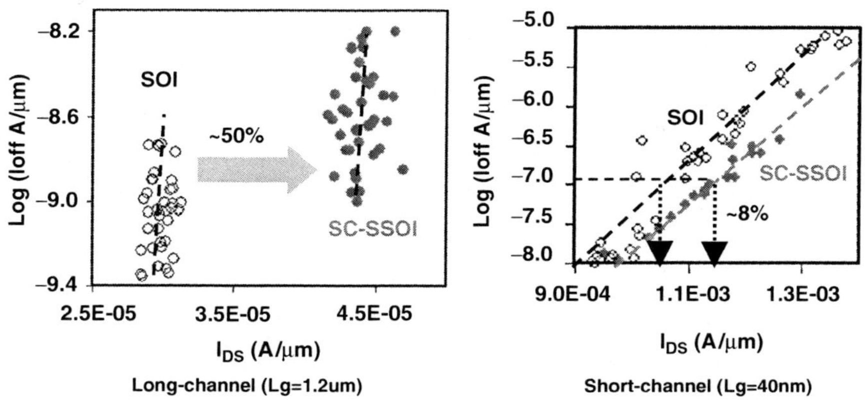

Fig. 3.19. Drive current enhancement of long channel (Lg = 1.2 μm) and short channel (Lg = 40 nm) devices, showing that performance enhancement due to strain is preserved for small channel devices while S/D resistances need to be reduced. SC-SSOI stands for super-critical sSOI, illustrating the use of thick sSOI layers. From [50]

bulk germanium. Ge is also better compatible than silicon with most of high-k materials that have been proposed for gate insulators. This is because, as opposed to the case of silicon, there is no stable phase of Germanium oxide that would form incidentally an additional penalizing layer during the deposition of high-k oxides.

One drawback of germanium is related to its smaller bandgap, which raises concerns about junction leakage. For this reason in particular, it is expected that if germanium comes into mainstream CMOS, it will be in the form of GeOI films. In GeOI, the handle substrate will still be made of silicon. This will provide all the benefits of silicon in terms of mechanical properties, weight, cleanliness, and flatness. Beyond CMOS channel mobility improvement that could allow extending Moore's law, Ge is also an enabling solution for combining different functions on the same substrate. Ge can be used directly as an active material for near-infrared photo-detectors. Thanks to its good lattice parameter matching with GaAs, it is also an excellent template for III–V semiconductor layers that may be monolithically integrated with silicon and used for various optoelectronic components.

In the next section we review the main methods for fabricating GeOI substrates.

3.4.2 GeOI Substrates Manufacturing Routes

Every aspect of Ge technology is today far behind silicon: crystalline ingot growth and bulk wafers formation, growth of epitaxial layers, surface passivation, and so on. Combined with the fact that GeOI is not planned for the

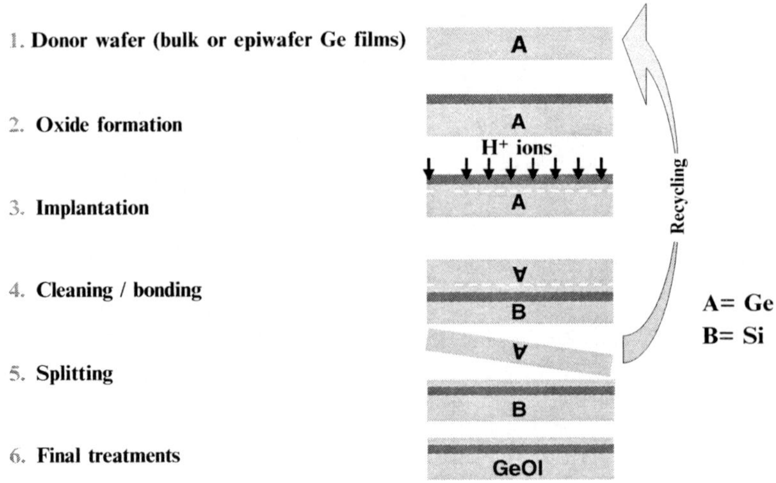

Fig. 3.20. Example of process flow to achieve GeOI using the Smart Cut technology

very next CMOS technology node, it is not surprising that GeOI feasibility demonstrations have been very limited.

For germanium, SIMOX-like techniques are very unlikely. In silicon, implanting a suitable dose of oxygen, followed by high temperature annealing, leads to a formation of a buried SiO_2, therefore an SOI structure is created. This cannot be transposed directly to the case of Ge. On the other hand, it is interesting to note that the condensation technique, initially developed for the realization of SGOI substrates (see SGOI section earlier in this chapter), has been used to demonstrate GeOI formation [51]. The Ge enrichment mechanism, described in Sect. 3.3.5, is extended to its limit to obtain a pure Ge On Insulator structure.

The most frequently reported techniques for producing GeOI substrates are the layer transfer methods: both BSOI as well as the Smart Cut technology. For the sake of the illustration, Fig. 3.20 shows a process flow for the realization of GeOI by the Smart Cut technology [52, 53]. It can be seen in this figure that the process flow is similar to that used to make SOI substrates. Some details are different, however.

Ge Donor Substrate for Layer Transfer Techniques

The first difference is the donor substrate, which needs to contain Ge. The properties of this donor substrate are critical as they largely determine the final quality of the GeOI. Both Bulk Ge and epitaxial Ge layers have been investigated and shown suitable for GeOI structure formation.

Bulk Ge is obtained similarly to bulk Si by CZ crystal pulling and transforming the single crystalline ingots into wafers by slicing, grinding, etching, etc. Ge epitaxy can be performed on bulk Ge to improve defectivity of the active layer. For instance the so-called "Crystal Originated Particles" (COPs), those defects originating from the crystal ingot growth by agglomeration of vacancies can be eliminated. Much more challenging is Ge heteroepitaxy on standard silicon substrates; its purpose is to avoid Ge bulk substrates. This would make easier to fabricate large diameter GeOI wafers which will be needed in the advanced CMOS world (300 mm and beyond). In the latter case, however, the large lattice mismatch between Ge and Si tends to generate high dislocation densities [54]. Currently, advantages and drawbacks of these two approaches are balanced. For some parameters epitaxial germanium may be preferred (low COP densities, compatibility with large diameter wafers) while for others bulk is better (fewer dislocations and improved wafer flatness). Depending on the success of future developments in this field, one of these methods may acquire a clear advantage.

Ge Layer Transfer

Layer transfer techniques to process Ge layers exist, but need further optimization. Layer splitting by the Smart Cut technology has been demonstrated. Compared to Smart Cut in silicon it has been shown that the corresponding process window, as defined by the implantation conditions for splitting, is slightly different [52].

Buried Insulator for GeOI

In SOI substrates, the buried oxide is typically thermally grown on silicon. This provides good bonding properties and ensures that the oxide and its interface with the active SOI layer are of high quality, which is critical for the good electrical behavior of the transistors. This strategy cannot be transferred directly to GeOI. Lack of thermally stable Ge oxides and of proper electrical passivation of a Ge interface are the main impediments. It is expected that solutions developed for gate stacks (nitridation, high-k) may eventually be adopted also for the back gate of GeOI channels, which is defined during the GeOI substrate fabrication. In the meantime, temporary solutions have been explored to demonstrate prototypes of GeOI substrates. These solutions are based on oxide deposition on Ge or on direct bonding of Ge with thermal SiO_2 grown on the Si handle substrate.

Reclaiming Ge Donor Substrates

Germanium substrates are much more expensive than silicon, and the availability of raw material for Ge is very limited. Recycling and conserving Ge material is an important consideration if Ge is ever to move into the mainstream of CMOS technology. Using layer transfer techniques that enable reclaiming of the donor wafers is particularly relevant: the same Ge substrates may be used

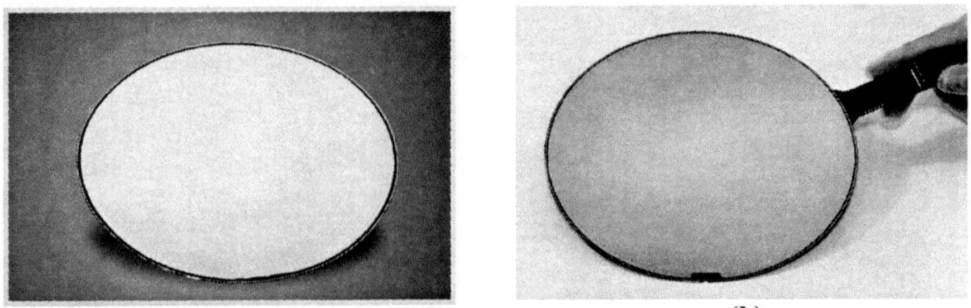

(a) **(b)**

Fig. 3.21. 200 mm GeOI substrates made respectively from (**a**) bulk Ge or (**b**) epitaxial layer Ge donor substrates

as donor wafers to make many GeOI substrates with bulk Si handle wafers. The Smart Cut technology is particularly suitable for multiple reuses of the donor wafers, as it does not sacrifice the donor wafer as in the BSOI approach.

Examples of GeOI Substrates Demonstrations

Figure 3.21 shows the top view of two 200 mm GeOI substrates made by the Smart Cut technology; from a Ge bulk donor wafer (Fig. 3.21a) and from an epi layer grown on an Si substrate (Fig. 3.21b). The quality of such substrates has been characterized. As an example, in Fig. 3.22, we show a TEM cross-section, in which there are no visible dislocations.

3.4.3 Examples of GeOI Substrates Validations at Device Level

GeOI substrates have been used for various device structures and applications. Ge MOSFETs on GeOI are reported in Chap. 4 of this volume (see

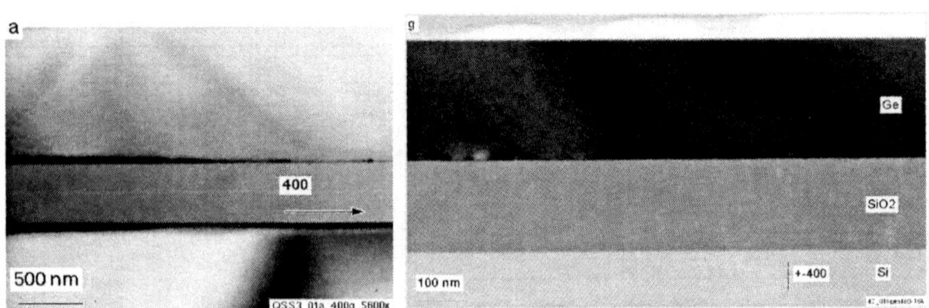

(a) **(b)**

Fig. 3.22. Cross-section TEM of two GeOI substrates made by layer transfer from a bulk Ge donor substrate. In case (**a**), the bonding interface is located within the buried oxide (SiO_2/SiO_2 bonding). In case (**b**), the bonding interface is located on top of the buried oxide at the Ge/SiO_2 interface (Ge/SiO_2 bonding)

also [55, 56]), where it is shown that the device properties are similar, especially in terms of mobilities, to results obtained on bulk Ge substrates. Beyond MOSFETs, such GeOI substrates may also be used for III–V semiconductors growth. As reported by Thomas et al. [57] GaAs HBT structures have been successfully realized, with excellent crystalline properties. The results also show a gain in terms of thermal management as compared to similar structures on bulk GaAs, as can be deduced from the suppression of negative resistance effects. This gain is explained by better thermal conductivity of the silicon handle of the GeOI as compared to bulk GaAs.

3.5 Long Term Perspectives

Three major enhancements of SOI structures have been reviewed here: (a) optimization of crystalline orientation, (b) use of strained layers and (c) use of germanium instead of silicon films. Each of them is attractive for high performance device applications. Since these approaches were proposed only in the last few years, further process optimization may be required before commercial implementation.

Combining some of these approaches in order to further improve the carrier mobility is also possible. A few interesting examples are listed below:

1. Combining non-traditional silicon crystal orientation substrates and strained layers
   - 45° rotated strained silicon on insulator (45° SGOI and 45° sSOI substrates)
   - (110) SGOI; (110) sSOI substrates
   - (100) strained silicon on insulator on (110) bulk (Hybrid oriented and strained substrates)
2. Combining non-traditional crystal orientation and germanium:
   - (111) orientation of GeOI
3. Combining strained layers and germanium. In this case compressive stress in Ge improves mobilities:
   - Compressively strained GeOI, that can be obtained by layer transfer techniques from strained Ge donor substrates. In this case epitaxially grown Ge on silicon (with optional buffer layers) forms appropriate donor substrates
   - SGOI substrates are also excellent templates for growth of compressively strained Ge layers on insulator
4. Combining all three enhancements: non-traditional crystal orientation substrates with strained layers and germanium:
   - Misoriented and compressively strained Ge on insulator
   - Hybrid orientation substrate with compressively strained Ge on (110) handle silicon substrate

There are even more choices if multiple semiconductor layers on insulator are taken into account. Such stacks can be grown by heteroepitaxy. Of particular interest is a concept called a "dual channel transistor" that has already been described in literature [58]. It combines tensile strained silicon from sSOI substrates for the nMOSFETs with an additional compressively strained Ge layer that is grown in areas where pMOSFETs are placed. Multiple layers can also be made by repeating layer transfer operations to form complex SOI-like structures [59].

In addition to the full substrate solutions described above, process-induced strain can be added, to reach complete and flexible optimizations of NMOS and PMOS mobility.

3.6 Conclusions

High mobility semiconductor layers combined with SOI architectures are promising approaches to meet the challenges for CMOS technology outlined in the ITRS roadmap.

We have reviewed three substrate-level solutions: (a) non-traditional silicon crystal orientation substrates, (b) strained silicon layers and (c) germanium. In each case, several scenarios are possible, from simple immediate improvements such as 45° rotated SOI to long term ones such as the introduction of germanium. In the intermediate case of strained silicon, we have described two families of strained SOI substrates that have emerged in the last few years. The first, called SGOI, is characterized by the presence of a buried relaxed SiGe layer within the substrate, which induces and maintains strain within the Si layer that is on top of it. In the second, sSOI substrates are Ge-free and the bonding interface alone maintains the strain. Film thickness in sSOI wafers can be optimized for either fully-depleted or partially depleted devices, while SGOI with its thicker films is primarily suitable for PD applications. We have also given some suggestions on combining multiple enhancements to further improve device performance.

In addition to these substrate solutions, the present industrial success of process-induced strain shall not be forgotten. A reuse of this know-how is indeed to be considered in a catalytic combination of both approaches to further improve both N and P type transistors.

Acknowledgments

The authors want to acknowledge the teams of LETI and SOITEC that made this work possible, as well as collaborations with different partners that include ST, Freescale, ASM, IMEC and UMICORE. This program has been partially supported by the French "Réseau Micro-Nanotechnologies (RMNT)" and the

French "Ministère de l'Industrie", as well as MEDEA+ 2T101 SILONIS program.

References

1. M. Yang, M. Ieong, L. Shi, K. Chan, V. Chan, A. Chou, E. Gusev, K. Jenkins, D. Boyd, Y. Ninomiya, D. Pendleton, Y. Surpris, D. Heenan, J. Ott, K. Guarini, C. D'Emic, M. Cobb, P. Mooney, B. To, N. Rovedo, J. Benedict, R. Mo and H. Ng (2003), High performance CMOS fabricated on hybrid substrate With different crystal orientations, IEDM 2003 Tech. Dig., p. 453
2. H. Sayama, Y. Nishida, H. Oda, T. Oishi, S. Shimizu, T. Kunikiyo, K. Sonoda, Y. Inoue and M. Inuishi, Effect of ⟨100⟩ channel direction for high performance SCE immune PMOSFET with less than 0.15 μm gate length, IEDM 1999 Tech. Dig., p. 657
3. T. Matsumoto, S. Maeda, H. Dang, T. Uchida, K. Ota, Y. Hirano, H. Sayama, T. Iwamatsu, T. Ipposhi, H. Oda, S. Maegawa, Y. Inoue and T. Nishimura (2002), Novel SOI wafer engineering using low stress and high mobility CMOSFET with ⟨100⟩ channel for embedded RF/Analog applications, IEDM 2002 Tech. Dig., paper 27-07
4. T. Komoda, A. Oishi, T. Sanuki, K. Kasai, H. Yoshimura, K. Ohno, M. Iawai, M. Saito, F. Matsuoka, N. Nagashima and T. Noguchi (2004), Mobility improvement for 45 nm node by combination of optimized stress control and channel orientation design, IEDM 2004 Tech. Dig., paper 9.3.1
5. K. Uchida, R. Zednik, C.-H. Lu, H. Jagannathan, J. Mc Vittie, P.C. McIntyre and Y. Nishi (2004), Experimental study of biaxial and uniaxial strain effects on carrier mobility in bulk and ultrathin body SOI MOSFETs, IEDM 2004 Tech. Dig., paper 9.6.1
6. G.K. Celler and S. Cristoloveanu (2003), Frontiers of silicon-on-insulator, J. Appl. Phys., Vol. 93, no. 9, p. 4955
7. S.M. Sze, Physics of Semiconductor Devices, Wiley, New York (1981), Modern Semiconductor Device Physics, Wiley, New York (1998)
8. K.K. Bourdelle, T. Akatsu, N. Sousbie, F. Letertre, D. Delprat, E. Neyret, N. Ben Mohamed, G. Suciu, C. Lagahe-Blanchard, A. Beaumont, A.-M. Charvet, A.-M. Papon, N. Kernevez, C. Maleville and C. Mazure (2004), Smart Cut^{TM} Transfer of 300 mm (110) and (100) Si Layers for Hybrid Orientation Technology, IEEE Intern. SOI Conf. 2004, p. 98
9. H. Irie, K. Kita, K. Kyuno and A. Toriumi (2004), In-plane mobility anisotropy and universality under uni-axis strains in n- and p-MOS inversion layers on (100), (110) and (111) Si layers, IEDM 2004 Tech. Dig., paper 9.5.1
10. M.V. Fischetti and S.E. Laux (1996), Band structure, deformation potentials, and carrier mobility in strained-Si, Ge, and SiGe alloys, J. Appl. Phys. Vol. 80, no. 4, p. 2234
11. J. Welser, J.L. Hoyt, S. Takagi and J.F. Gibbons (1994), Strain dependence of the performance enhancement in strained –Si n-MOSFETs, IEDM 1994 Tech. Dig., p. 373
12. M.D. Giles, M. Armstrong, C. Auth, S.M. Cea, T. Ghani, T. Hoffman, R. Kotlyar, P. Matagne, K. Mistry, R. Nagisetty, B. Obradovic, R. Shaheed, L. Shifren,

M. Stetler, S. Tyagi, X. Wang, C. Weber and K. Zawadzki (2004), Proc. Symp. 2004 VLSI Tech., p. 118
13. C. Mazure and I. Cayrefourcq, Status of device mobility enhancement through strain silicon engineering, IEEE Intern. SOI Conf. 2005, p. 1
14. T. Ghani, M. Armstrong, C. Auth, M. Bost, P. Charvat , G. Glass, T. Hoffmann, K. Johnson, C. Kenyon, J. Klaus, B. McIntyre, K. Mistry, A. Murthy, J. Sandford, M. Silberstein, S. Sivakumar, P. Smith, K. Zawadzki, S. Thompson and M. Bohr (2003), High volume manufacturing logic technology featuring novel 45 nm gate length strained silicon CMOS transistors, IEDM 2003 Tech. Dig., p. 978
15. G. Scott, J. Lutze, M. Rubin, F. Nouri and M. Manley, IEDM 1999 Tech. Dig., p. 827
16. H. Sayama, Y. Nishida, H. Oda, T. Oishi, S. Shimizu, T. Kunikiyo, K. Sonoda, Y. Inoue and M. Inuishi, IEDM 1999 Tech. Dig., p. 657
17. T. Sanuki, A. Oishi, Y. Morimasa, S. Aota, T. Kinoshita, R. Hasumi, Y. Takegawa, K. Isobe, H. Yoshimura, M. Iwai, K. Sunouchi and T. Noguchi, Scalability of strained silicon CMOSFET and high drive current enhancement in the 40 nm gate length technology, IEDM 2003 Tech. Dig., p. 65
18. C. Ge, C. Lin, C. Ko, C. Huang, Y. Huang, B. Chan, B. Perng, C. Sheu, P. Tsai, L. Yao, C. Wu, T. Lee, C. Chen, C. Wang, S. Lin, Y. Yeo and C. Hu, Process-strained Si (PSS) CMOS technology featuring 3D strain engineering, IEDM 2003 Tech. Dig., p. 73
19. T. Komoda, A. Oishi, T. Sanuki, K. Kasai, H. Yoshimura, K. Ohno, M. Iwai, M. Saito, F. Matsuoka, N. Nagashima and T. Noguchi, Mobility improvement for 45 nm node by combination of optimized stressand channel orientation design, IEDM 2004 Tech. Dig., p 217.
20. A. Steegen, M. Stucchi, A. Lauwers and K. Maex, Silicide induced pattern density and orientation dependent transconductance in MOS transistors, IEDM 1999 Tech. Dig., p. 497
21. V. Chan, R. Rengarajan, N. Rovedo, W. Jin, T. Hook, P. Nguyen, J. Chen, E. Nowak, X. Chen, D. Lea, A. Chakravarti, V. Ku, S. Yang, A. Steegen, C. Baiocco, P. Shafer, H. Ng, S. Huang and C. Wann, High speed 45 nm gate length CMOSFETs integrated into a 90 nm bulk technology incorporating strain engineering, IEDM 2003 Tech. Dig., p. 77
22. H.S. Yang, R. Malik, S. Narasimha, Y. Li, R. Divakaruni, P. Agnello, S. Allen, A. Antreasyan, J.C. Arnold, K. Bandy, M. Belyanski, A. Bonnoit, G. Bronner, V. Chan, X. Chen, Z. Chen, D. Chidambarrao, A. Chou, W. Clark, S.W. Crowder, B. Engel, H. Harifuchi, S.F. Huang, R. Jagannathan, F.F. Jamin, Y. Kohyama, H. Kuroda, C.W. Lai, H.K. Lee, W-H. Lee, E.H. Lim, W. Lai, A. Malikkarjunan, K. Matsumoto, A. McKnight, J. Nayak, H.Y. Ng, S. Panda, R. Rengarajan, M. Steigerwalt, S. Subbanna, K. Subramanian, J. Sudijono, G. Sudo, S.-P. Sun, B. Tessier, Y. Toyoshima, P. Tran, R. Wise, R. Wong, I.Y. Yang, C. H. Wann, L.T. Su, M. Horstmann, Th. Feudel, A. Wei, K. Frohberg, G. Burbach, M. Gerhardt, M. Lenski, R. Stephan, K. Wieczorek, M. Schaller, H. Salz, J. Hohage, H. Ruelke, J. Klais, P. Huebler, S. Luning, R. van Bentum, G. Grasshoff, C. Schawn, E. Ehrichs, S. Goad, J. Buller, S. Krishnan, D. Greenlaw, M. Raab and N. Kepler (2004), Dual stress liner for high performance sub-45 nm gate length SOI CMOS manufacturing, IEDM 2004 Tech. Dig., p. 75

23. E.A. Fitzgerald, Y.-H. Xie, M.L. Green, D. Brasen, A.R. Kortan, J. Michel, Y.-J. Mii and B.E. Weir (1991), Totally Relaxed GexSi1-x Layers with low threading dislocation densities grown on si substrates, Appl. Phys. Lett. 59, p. 811
24. J.M. Hartmann, Y. Bogumilowicz, P. Holliger, F. Laugier, R. Truche, G. Rolland, M.N. Séméria, V. Renard, E.B. Olshanetsky, O. Estibal, Z.D. Kvon, J.C. Portal, L. Vincent, F. Cristiano and A. Claverie (2004), Reduced pressure chemical vapour deposition of SiGe virtual substrates for high mobility devices, Semicond. Sci. Technol. 19, p. 311
25. F. Schäffler (1997), High mobility Si and Ge structures, Semiconductor Sci. Techn. 12, p. 1515
26. K.K. Linder, F.C. Zhang, J.-S. Rieh, P. Bhattacharya and D. Houghton (1997), Reduction of dislocation density in mismatched SiGe/Si using a low-temperature Si buffer layer, Appl. Phys. Lett. 70, p. 3224
27. M. Bauer, K. Lyutovich, M. Oehme, E. Kasper, H.-J. Herzog and F. Ernst (2000), Relaxed SiGe buffers with thicknesses below 0.1 µm, Thin Solid Films 369 (2000), p. 152
28. M.T. Bulsara, V. Yang, A. Thilderkvist and E.A. Fitzgerald (1998), Graded InxGa1-xAs/GaA 1.3 µm wavelength light emitting diode structures grown with molecular beam epitaxy, J. App. Phys. 83, 592
29. S. Takagi, T. Tezuka, N. Sugiyama, T. Mizuno and A. Kurobe (2002), Structure and characteristics of strained-Si-on-insulator (Strained-SOI) MOSFETs, Mat. Res. Soc. Symp. Proc. Vol. 686, p. A1.3.1
30. T. Mizuno, N. Sugiyama, T. Tezuka and S.-I. Takagi (2002), Relaxed SiGe-on-insulator substrates without thick SiGe buffer layer, Appl. Phys. Lett. 80, p. 601
31. T. Tezuka, N. Sugiyama and S. Takagi (2001), Fabrication of strained Si on an ultra-thin SiGe-On-Insulator virtual substrate with a high Ge fraction, Appl. Phys. Lett. Vol. 79, p. 1798
32. G. Tarashi, T.A. Langdo, M.T. Currie, E.A. Fitzgerald and D.A. Antoniadis (2002), Relaxed SiGe-on-Insulator fabricated via wafer bonding and etch back, J. Vac. Sci. Technol. B 20, p. 725
33. L.J. Huang, J.O. Chu, D.F. Canaperi, C.P. D'Emic, R.M. Anderson, S.J. Koester and H.-S. Philip Wong (2001), SiGe-On-Insulator prepared by wafer bonding and layer transfer for high-performance field-effect transistors, Appl. Phys. Lett. 78, p. 1267 (2001)
34. B. Ghyselen, C. Aulnette, B. Osternaud, T. Akatsu, C. Mazure, C. Lagahe-Blanchard, S. Pocas, J.-M. Hartmann, P. Leduc, T. Ernst, H. Moriceau, Y. Campidelli, O. Kermarrec, P. Besson, Y. Morand, M. Rivoire, D. Bensahel and V. Paillard (2003), Strained silicon on insulator wafers, Proc. 3rd Intern. Conference on SiGe(C) Epitaxy and heterostructures, Santa Fe, 9–12th March 2003
35. B. Ghyselen, J.-M. Hartmann, T. Ernst, C. Aulnette, B. Osternaud, Y. Bogumilowicz, A. Abbadie, P. Besson, O. Rayssac, A. Tiberj, N. Daval, I. Cayrefourq, F. Fournel, H. Moriceau, C. Di Nardo, F. Andrieu, V. Paillard, M. Cabié, L. Vincent, E. Snoeck, F. Cristiano, A. Rocher, A. Ponchet, A. Claverie, P. Boucaud, M.-N. Semeria, D. Bensahel, N. Kernevez and C. Mazure (2004), Engineering strained silicon on insulator wafers with the Smart Cut^{TM} technology, Solid State Electronics (Special issue on strained Si and heterostructures and Devices), 48 (2004), pp. 1285–1296

36. B. Holländer, St. Lenk, S. Mantl, H. Trinkaus, D. Kirch, M. Luysberg, T. Hackbarth, H.-J. Herzog and P.F.P. Fichtner (2001), Strain relaxation of pseudomorphic Si1-xGex/Si(100) heterostructures after hydrogen or helium ion implantation for virtual substrate fabrication, Nucl. Instr. Meth. Phys. Res. B 175–177, p. 357
37. J.M. Hartmann, A. Abbadie, M. Vinet, L. Clavelier, P. Hollinger, D. Lafond, M.N. Séméria and P. Gentile (2003), Growth kinetics of Si on fullsheet, patterned and silicon-on-insulator substrates, J. Cryst. Growth 257, p. 19
38. T. Mizuno, N. Sugiyama, T. Tezuka, T. Numata and S. Takagi (2003), High performance strained-SOI CMOS devices using thin film SiGe-on-insulator technology, IEEE Trans. Electron Devices, vol. 50, p. 988
39. K. Rim, L. Shi, K. Chan, J. Chu, D. Boyd, K. Jenkins, J. Ott, D. Lacey, P. Mooney, M. Cobb, N. Klymko, F. Jamin, S. Koester and T. Kanarsky (2003), Strained Si MOSFETs on bulk and SiGe-on-Insulator (SGOI) substrates, in Proc. ISTDM Conf., 2003, p. 9
40. M. Sadaka, A.V.-Y. Thean, A. Barr, D. Tekleab, S. Kalpat, T. White, T. Nguyen, R. Mora, P. Beckage, D. Jawarani, S. Zollner, M. Kottke, R. Liu, M. Canonico, Q-H Xie, X-D Wang, S. Parsons, D. Eades, M. Zavala, B.-Y. Nguyen, C. Mazure and J. Mogab (2004), Fabrication and operation of Sub-50 nm Strained-Si on Si 1-x Ge, on Insulator (SGOI) CMOSFETs, IEEE Intern. SOI Conf. 2004, p. 209
41. F. Andrieu, T. Ernst, O. Faynot, O. Rozeau, Y. Bogumilowicz, J.-M. Hartmann, L. Brévard, A. Toffoli, D. Lafond, H. Dansas, B. Ghyselen, F. Fournel, G. Ghibaudo and S. Deleonibus (2005), In-depth study of strained SGOI nMOSFETs down to 30 nm gate length, Proc. of ESSDERC 2005, Grenoble (FR), p. 297
42. K. Rim, K. Chan, L. Shi, D. Boyd, J. Ott, N. Klymko, F. Cardone, L. Tai, S. Koester, M. Cobb, D. Canaperi, B. To, E. Duch, I. Babich, R. Carruthers, P. Saunders, G. Walker, M. Steen and M. Ieong (2003), Fabrication and mobility characteristics of ultra-thin strained Si directly on insulator (SSDOI) MOSFETs, IEDM Tech. Dig. 2003, p. 49
43. T.A. Langdo, A. Lochtefeld, M.T. Currie, R. Hammond, V.K. Yang, J.A. Carlin, C.J. Vineis, G. Barithwaite, H. Badawi, M.T. Bulsara and E.A. Fitzgerald (2002), Preparation of Novel SiGe-free strained Si on insulator substrates, IEEE Intern. SOI Conf. 2002, p. 211
44. T.S. Drake, C.Ni Chleirigh, M.L. Lee, A.J. Pitera, E.A. Fitzgerald, D.A. Antoniadis, D.H. Anjum, J. Li, R. Hull, N. Klymko and J.L. Hoyt (2003), Fabrication of ultra-thin strained silicon on insulator, J. Electron. Mat. 32, p. 972
45. I. Cayrefourcq, M. Kennard, F. Metral, C. Mazuré, A. Thean, M. Sadaka, T. White and B.Y. Nguyen (2005), Mobility enhancement through substrate engineering, Silicon-on-Insulator Technology and Devices XII, ECS Proc. Vol 2005-03, edited by G.K. Celler et al. (The Electrochemical Society, Pennington, NJ, USA, 2005)
46. M. Kennard, C. Figuet, Y.-M. Le Vaillant, F. Triolet, C. Villeneuve, C. Aulnette, C. Berne, L. Portigliati, E. Guiot, N. Daval, R. Larderet, C. Le Moingne, S. Bisson, F. Métral, Z. Chahra, I. Cayrefourcq, C. Mazure, N. Cody, C. Arena and C. Werkhoven (2005), The critical role of epitaxy in the fabrication of enhanced mobility substrates, Proc. 4th International Conference on Silicon Epitaxy and heterostructures (ICSI-4), Awaji island-Hyogo, Japan, 23–26th May 2005

47. T.A. Langdo, M.T. Currie, Z.-Y. Cheng, J.G. Fiorenza, M. Erdtmann, G. Braithwaite, C.W. Leitz, C.J. Vineis, J.A. Carlin, A. Lochtefeld, M.T. Bulsara, I. Lauer, D.A. Antoniadis and M. Somerville (2004), Strained Si on insulator technology: from materials to devices, Solid State Electronics (special issue on strained Si and heterostructures and Devices), 48 (2004), pp. 1357–1367
48. I. Lauer, T.A. Langdo, Z.-Y. Cheng, J.G. Fiorenza, G. Braithwaite, M.T. Currie, C.W. Leitz, A. Lochtefeld, H. Badawi, M.T. Bulsara, M. Somerville and D.A. Antoniadis (2004), Fully depleted n-MOSFETs on Supercritical Thickness Strained SOI, IEEE Elect. Dev. Lett., Vol. 25, No. 2, p. 83
49. F. Andrieu, T. Ernst, O. Faynot, Y. Bogumilowicz, J.-M. Hartmann, J. Eymery, D. Lafond, Y.-M. Levaillant, C. Dupré, R. Powers, F. Fournel, C. Fenouillet-Beranger, A. Vandooren, B. Ghyselen, C. Mazure, N. Kernevez, G. Ghibaudo and S. Deleonibus (2005), Co-integrated dual strained channels on fully depleted sSDOI, IEEE Intern. SOI Conf. 2005, "late news"
50. A.V.Y. Thean, T. White, M. Sadaka, L. McCormick, M. Ramon, R. Mora, P. Beckage, M. Canonico, X.-D. Wang, S. Zollner, S. Murphy, V. Van Der Pas, M. Zavala, R. Noble, O. Zia, L-G. Kang, V. Kolagunta, N. Cave, J. Cheek, M. Mendicino, B.Y. Nguyen, M. Orlowski, S. Venkatesan, J. Mogab, C.H. Chang, Y.H. Chiu, H.C. Tuan, Y.C. See, M.S. Liang, Y.C. Sun, I. Cayrefourcq, F. Metral, M. Kennard and C. Mazure (2005), Performance of super-critical strained-Si directly on insulator (SC-SSOI) CMOS based on high-performance PD-SOI technology, Proc. Symp. 2005 VLSI Tech, p. 134
51. S. Nakaharai, T. Tezuka, N. Sugiyama, Y. Moriyama and S.-I. Takagi (2003), Characterization of 7-nm-thick strained Ge-on-insulator layer fabricated by Ge-condensation technique, Appl. Phys. Lett., vol. 83, no. 17, p. 3516
52. C. Deguet, C. Morales, J. Dechamp, J.M. Hartmann, A.M. Charvet, H. Moriceau, F. Chieux, A. Beaumont, L. Clavelier, V. Loup, N. Kernevez, G. Raskin, C. Richtarch, F. Allibert, F. Letertre and C. Mazure (2004), Germanium-On-Insulator (GeOI) structures realized by the Smart Cut ™ technology, IEEE Intern. SOI Conf. 2004, p. 96
53. T. Akatsu, C. Deguet, C. Richtarch, F. Letertre, L. Sanchez, N. Kernevez, L. Clavelier, C. Le Royer, J.M. Hartmann, V. Loup, M. Meuris, B. De Jaeger and G. Raskin (2005), 200 mm Germanium On Insulator (GeOI) by Smart Cut technology and recent GeOI pMOSFETs achievements, IEEE Intern. SOI Conf. 2005, p. 137
54. J.M. Hartmann, J.-F. Damlencourt, Y. Bogumilowicz, P. Hollinger, G. Rolland and T. Billon (2005), Reduced pressure-chemical vapor deposition of intrinsic and doped Ge layers on Si(001) for microelectronics and optoelectronics purposes, J. Cryst. Growth, 274, p. 90
55. B. De Jaeger, R. Bonzom, F. Leys, O. Richard, J. Van Steenbergen, G. Winderickx, E. Van Moorhem, G. Raskin, F. Letertre, T. Billon, M. Meuris and M. Heyns (2005), Optimisation of a thin epitaxial Si layer as Ge passivation layer to demonstrate deep sub-micron n- and p-FETs on Ge-On-Insulator substrates, M. Eng., 80 (2005), p. 26
56. L. Clavelier, C. Le Royer, C. Tabone, J.M. Hartmann, C. Deguet, V. Loup, C. Ducruet, C. Vizioz, M. Pala, T. Billon, F. Letertre, C. Arvet, Y. Campidelli, V. Cosnier and Y. Morand (2005), Fully Depleted Germanium p-MOSFETs with High-K and metal gate fabricated on 200 mm GeOI substrates, Proceedings Silicon Nanoelectronics Workshop, June 2005, Kyoto, Japan

57. S.G. Thomas, E.S. Johnson, C. Tracy, P. Maniar, X. Li, B. Roof, Q. Hartmann and D.A. Ahmari (2005), Fabrication and characterization of InGaP/GaAs heterojunction bipolar transistors on GOI substrates, IEEE Electr. Dev. Lett., Vol. 26, no. 7, J 2005, p. 438
58. M.L. Lee and E.A. Fitzgerald (2003), Optimized strained Si/strained Ge dual channel heterostructures for high mobility p- and n-MOSFETs, IEDM 2003 Tech. Dig., p. 429
59. B. Ghyselen (2003), Advanced SOI structures based on wafer bonding: A short review, Wafer Bonding VII : Science, Technology and Applications, ECS Proceedings (PV 2003-19, ISBN 1-56677-402-0), Pennington, NJ (2003)

4

Passivation and Characterization of Germanium Surfaces

S.R. Amy and Y.J. Chabal

Summary. This chapter reviews the chemical passivation of germanium surfaces achieved in a variety of environments (ultra high vacuum or UHV, gas and liquid phases). The UHV studies make it possible to better control the chemical nature of surface termination and to study the resulting structure and bonding geometries. Gas phase or vapor processing helps span a larger pressure range, sometimes important to achieve thin film growth (e.g., nitridation). Finally, wet chemical treatments are attractive for large industrial batch processing. They remain, however, the most difficult methods to fully control germanium passivation. For that reason, emphasis is given to understanding how various experimental characterization techniques can best study germanium surfaces in all these environments, and to summarizing the best current processing conditions.

4.1 Introduction

As a group IV semiconductor, germanium (Ge) is expected to display many of the same properties as silicon (Si). Yet, despite its better bulk electrical performances (higher hole and low-field mobilities, and narrower bandgap) [1], Ge has not been used much in the past because its oxide is much less stable than SiO_2. As a result, the electrical properties of the Ge/GeO_x interface are much worse than for Si/SiO_2, where x is used to emphasize that GeO_2 is not necessarily the dominant oxide form for germanium. The nature of germanium oxide is such that wet chemical procedures for cleaning Ge (so critical for device fabrication) have been very challenging to establish. Yet, with renewed interest for high mobility substrates as Si is reaching some of its fundamental limits, the need to understand and control the passivation of Ge surfaces is even greater.

Key to developing an understanding of surface passivation are surface characterization techniques such as X-ray photoelectron spectroscopy (XPS), Fourier-transform infrared (FTIR) spectroscopy, and scanning tunneling microscopy (STM). The value of these techniques for Ge surfaces is briefly described in the next section.

The chapter is organized as follows: after a brief description of relevant experimental techniques, the methods for cleaning Ge in an ultra-high vacuum (UHV) environment are discussed to help understand the fundamental surface chemistry of Ge surfaces. The nature of Ge oxide is then considered as the growth and removal of such oxide is central to proper control of passivation. The rest of the chapter deals with surface passivation (i.e., chemical functionalization) using various elements or compounds, including hydrogen, nitride, oxynitride, sulfur, and chlorine, which are all relevant to subsequent processing (e.g., high-κ dielectrics growth) on Ge. For each case, the work done in UHV is presented first, followed by the description of wet chemical methods.

4.2 Experimental Methodology

4.2.1 X-ray and UV Photoemission Spectroscopy

X-ray photoemission spectroscopy (XPS) also known as electron spectroscopy for chemical analysis (ESCA) is most useful to identify elements, and their chemical environment, in the surface region. Using a source of high-intensity monochromatic X-ray from a rotating anode, or a synchrotron, XPS is performed by detecting photo-emitted electrons and measuring their kinetic energy. The short mean free path of electrons in solids (generally ~2–100 Å) makes XPS a very surface sensitive technique and limits its application to films less than 100 Å. Some depth resolution is achievable by varying the detection angle of the emitted electrons, using grazing emission for highest surface sensitivity and normal emission for highest bulk contribution. The sensitivity varies with elements, but on average it is ~1 atom%. For the studies reviewed in this chapter, the relevant core levels are the Ge2p and Ge3d, the O1s, the C1s, subject to sensitivity factors given in Table 4.1.

UV photoemission is ideal to study the valence band region, which is more sensitive to surface states and to details of surface chemical bonding. This technique is convenient because it does not rely on synchrotron radiation (a simple UV lamp is sufficient).

Table 4.1. Binding energies and sensitivity factors of Ge2p, Ge3d, O1s and C1s

Element	Binding energy (eV)	Sensitivity
Ge2p	1248.1/1217	
Ge3d	29.2/29.8	6.1
O1s	543.1	0.66
C1s	284.2	0.25

4.2.2 Fourier Transform Infrared Spectroscopy

Infrared (IR) spectroscopy is ideal to probe the chemical nature of surfaces. By identifying molecular vibrations, it makes it possible to infer the chemical bonding of a variety of species at the surface, including oxides, nitrides, and halogens. An advantage of IR spectroscopy is its high sensitivity to hydrogen, which is much harder to detect by XPS or Auger Electron Spectroscopy (AES) for instance. In contrast to XPS, IR spectroscopy requires the use of a reference surface to eliminate contribution from the bulk. This requirement underscores the fact that IR spectroscopy, like all other optical methods, is not intrinsically surface sensitive. The biggest challenge consists therefore in using the most appropriate reference surface to lower the bulk and other unwanted contributions as much as possible. Another challenge is to enhance sensitivity, which is sometimes done by using Multiple Internal Reflection (MIR) spectroscopy (Fig. 4.1a), in which the IR beam is reflected many times on the surface and therefore traverses long distance (3–10 cm) in the bulk material.

Among the most common electronic semiconductors (e.g., Si, InP, GaAs, SiC), Ge is the best optical substrate with relatively low frequency phonon absorption. It is therefore possible to measure surface modes as low as 700 cm^{-1} using the MIR geometry, and as low as ~500 cm^{-1} using direct transmission (Fig. 4.1b). This makes it possible to detect all oxide and nitride modes using MIR. Figure 4.1 illustrates the IR beam pathway in the two most common geometries used to perform IR spectroscopy. MIR is best to study Ge in liquid environments, and direct transmission is most convenient for in situ studies of passivation and growth.

Thin films at surfaces such as oxides and nitrides are characterized by extended phonon modes reflecting the interaction between the individual building blocks (e.g., SiO_2, Si_3N_4). Usually, IR spectroscopy is only sensitive to transverse optical (TO) phonon modes because the electric field of the probing radiation is normal to the direction of propagation (i.e., to its wavevector). However, when probing a thin film at a surface with non-normal incidence, there are components of the electric field (and wavevector) both perpendicular and parallel to the surface, so that both the transverse optical (TO) and longitudinal optical (LO) phonons can be excited. The longitudinal optical phonon is characterized by a dipole perpendicular to the surface and its frequency is blue shifted due to Coulomb interactions.

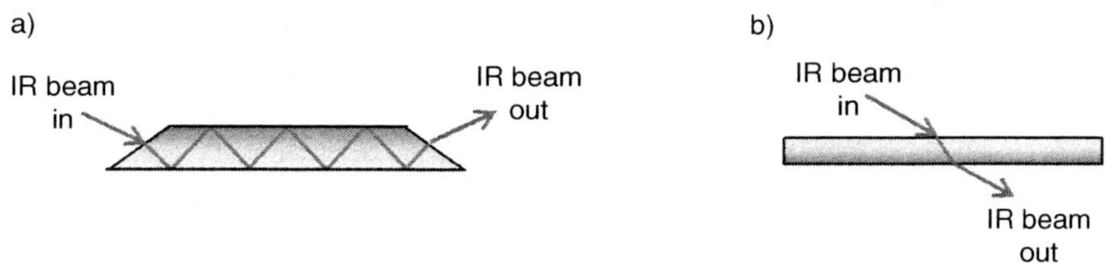

Fig. 4.1. Schematic representation of (**a**) MIR and (**b**) transmission geometries

4.2.3 Scanning Tunneling Microscopy

STM is a relatively new technique developed by Binning and Rohrer in the early 1980s that images electrically conductive surfaces with atomic resolution. When a sharp tip (tungsten, for example) is positioned close to a conductive surface (so close that the wavefunctions of the closest atom from the tip and the atom from the surface overlap) and a potential difference is applied between the tip and the surface, a tunneling current can be established. By varying either the potential or the tip/surface distance, the tunneling current varies reflecting details of the local density of states and surface topography. However, the biggest limitation for STM is the need of a conductive sample. Moreover, although the STM does not need vacuum to operate, it appears that for stability purposes as well as surface contamination issues, a UHV environment is preferable. Low temperature measurements are also often performed to increase both stability and spectral resolution when performing scanning tunneling spectroscopy (STS). Most of the experiments on germanium substrates described in this chapter are performed in a UHV chamber ($\sim 10^{-10}$ Torr), with typical bias (-1.0 V) and tunneling current (0.5 nA).

4.3 Clean Ge Surfaces

To appreciate the issues associated with Ge cleaning, it is important to consider first cleaning in an ultra-high vacuum environment. Several methods have been used initially based mostly on ion sputtering, thermal annealing or wet chemistry. Ion sputtering with Ne^+ or Ag^+ leads to near-surface damage and disorder, and therefore requires postannealing. Yet, even if the removal of the oxide is complete after ion sputtering, it appears that the surface is strongly roughened and cannot be recovered even after annealing [2–4]. Furthermore, STM images on those clean sputtered/annealed surfaces show the presence of carbon contamination as two-dimensional islands and the formation of irregular step edges with a wide distribution of terraces (Fig. 4.2) [4].

Other methods have added an ex situ wet chemical pretreatment to the cleaning sequence, followed by a repetition of Ar^+ sputtering and annealing in UHV. After such a process, a homoepitaxial Ge layer is grown at 300–350°C. Such a method leads to clean germanium surfaces with almost no defect sites aside from the steps (arising from miscut) as observed with scanning tunneling microscopy [5]. Because of the ion-sputtering-induced roughening of the surface, other cleaning methods have been investigated leading to Ge surface passivation methods described below.

4.4 Oxidation of Ge Surfaces

Germanium oxide has long been known as more complex and less stable than silicon oxide. Therefore, much of the early work was dedicated to

Fig. 4.2. STM image of a Ge(100) surface prepared by sputter/anneal cycles (reprinted with permission from [4])

characterizing Ge oxides. In the 1930s, crystals of germanium dioxide were examined by Laubergayer et al. [6] who identified the existence of three different phases of GeO_2, one amorphous and two crystalline (hexagonal and tetragonal). Both the amorphous and hexagonal phases are water soluble, but the tetragonal one is not. In all cases, the GeO bond is believed to be covalent as suggested by the elementary dipole moment (0.04e for GeO; for reference SiO is 0.31e). All soluble and insoluble oxide structures have been studied spectroscopically, using Raman Spectroscopy and Fourier Transform Infrared (FTIR) spectroscopy, and both the longitudinal $[\mathrm{Im}(-1/\varepsilon)]$ and transversal $[\mathrm{Im}(\varepsilon)]$ optical modes of GeO_2 at 857 and 930 cm^{-1} (Fig. 4.3) have been identified [7, 8].

The oxidation of Ge(100) and Ge(111) surfaces under UHV conditions has also been studied using XPS to identify the various oxidation states [9–11]. The XPS data reveal the formation of a +2 oxidation state upon annealing. In the work from Schmeisser et al. [10], the stability of Ge oxides is discussed and compared to that of oxides of other elements close to Ge. Prabhakaran et al. [12, 13] were the first to study wet oxidation processes using both XPS and UPS. They compared the formation and removal of oxides on Ge(100) and Ge(111) using wet chemical treatments and in situ oxidation (i.e., by introducing oxygen gas directly into the UHV chamber). Focusing on the Ge3d, Ge2p and O1s core levels, they showed that a cleaning procedure involving a sequential treatment in deionized water (H_2O), hydrofluoric acid

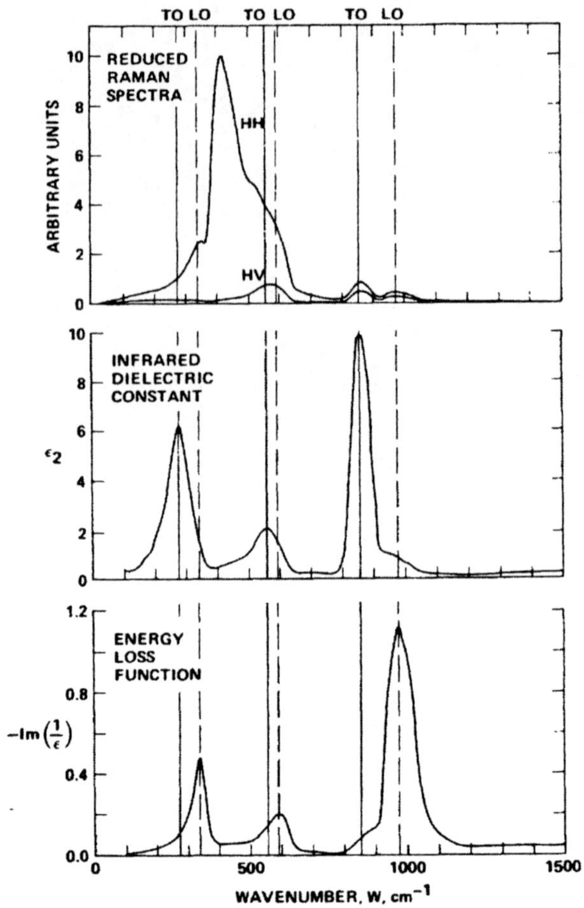

Fig. 4.3. Comparison of reduced Raman spectra of bulk $\nu - GeO_2$ with the experimentally determined transverse energy loss function, ε_2, and the longitudinal energy loss function $\mathrm{Im}(-1/\varepsilon)$ (reprinted with permission from [8])

(HF) and hydrogen peroxide ($H_2O_2 : H_2O$) leads to the formation of both GeO (in a non-bridged configuration) and GeO_2 (Fig. 4.4).

This comprehensive study shows that in situ oxidation in UHV results in mostly GeO and some related suboxides, wet chemical oxidation leads to the formation of a mixture of GeO/GeO_2 and suboxides, and native oxidation in room air produces primarily GeO_2. Most of the Ge suboxides appear to be thermally more stable than both GeO_2, requiring annealing temperatures above 450°C for removal, and are not water soluble. These findings have been confirmed by Zhang et al. [14] who performed a water dip to remove the 1–2 nm thick native oxide, leaving behind 1–2 Å thick film of GeO and C, the latter presumed to originated from the contamination during wet chemical treatment. Zhang et al. [14] suggested that the dominant phase

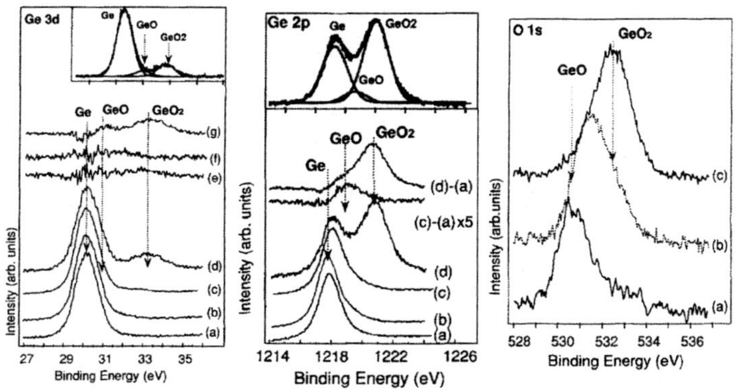

Fig. 4.4. Ge3d (*left panel*), and Ge2p (*middle panel*) core level spectra of (**a**) clean Ge(111), (**b**) clean Ge(111) exposed to 100 L O_2 at 250°C, (**c**) ex situ oxide prepared by dipping the wafer in a mixture of H_2O_2 and H_2O at room temperature, and (**d**) clean Ge(111) exposed to clean room air for 6 h. The difference spectra (**e**) = (**b**−**a**), (**f**) = (**c** − **a**), and (**g**) = (**d** − **a**) are plotted for the Ge3d core level (*left panel*), and the fittings used for curves (**d**) are shown as *inset* at the top of the left and middle panels. Data for O1s core level spectra (*right panel*) are for (**a**) clean Ge(111) exposed to 100 L O_2 at 250°C, (**b**) Ge(111) dipped in a mixture of H_2O_2 and H_2O at room temperature, and (**c**) Ge(111) exposed to clean air for 6 h (reprinted with permission from [12])

of GeO_2 formed in this manner is hexagonal. This GeO_2 can be thermally removed in UHV, leaving behind a clean, reconstructed Ge(100)2×1 surface, characterized by the sharp surface state emission of surface dimers. Further investigations have shown that carbon contamination plays a crucial role as the Ge–C bond is very strong and difficult to break [14]. In fact, carbon contamination is present on the surface (observed through C1s core level in XPS) after any wet chemical treatment, and can only be removed after a 30 min UV-ozone irradiation (Fig. 4.5).

The ozone irradiation treatment leads to the formation of ~2 nm-thick GeO_2 including a 1.5 Å suboxide layer at the interface that is strongly dependent on the UV-ozone exposure time [14]. This oxide thickness is higher than reoxidation of a clean Ge sample in air for 24 h. Recent FTIR measurements performed in our laboratory show that a 30 min-ozone treatment using a UV light produces an oxide similar to a 10 s H_2O_2-grown oxide, i.e., GeO_2 is formed as evidenced by the appearance of GeO_2 LO and TO modes (Fig. 4.6). An analysis using Gaussian fits suggests that Ge=O and CH_x are not present in a UV-ozone oxide. In general, the growth of UV-ozone oxides on H/Ge(100) surface seems to be slower than chemical oxidation (Fig. 4.6) suggesting that a denser oxide is formed during UV-ozone treatment. In contrast, the oxide formed after 12 h reoxidation in air is dominated by suboxides and CH_x species, with no dioxide (Fig. 4.6a).

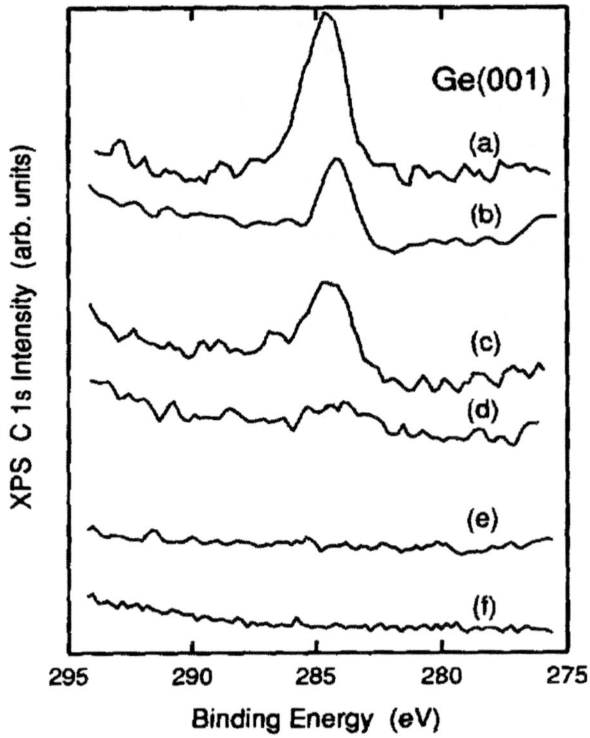

Fig. 4.5. C1s core level spectra from Ge(100) wafers after undergoing (**a**) a degreasing step involving successive rinses of trichloroethylene, acetone and methanol (process 1), (**b**) process 1 followed by a thermal anneal to 620 K, (**c**) process 1 followed by an ultrasound rinse in deionized water for 15 min (process 2), (**d**) process 2 followed by a thermal anneal to 620 K, (**e**) process 2 followed by a 30 min UV/ozone exposure (process 3) and (**f**) process 3 followed by a thermal anneal to 620 K (reprinted with permission from [14])

A recent FTIR and XPS study has examined the removal of oxides prepared by various chemical treatments on epi-ready Ge(100) [15]. The IR spectrum (Fig. 4.7a) obtained after rinsing a native oxide (epi-ready and subsequent air exposure) in deionized (DI) water shows the loss (negative absorption) of a broad absorbance feature in the 750–1050 cm^{-1} range, with a more intense negative component at 785 cm^{-1}. This region generally corresponds to germanium oxide, but the assignment is difficult because this band is wide and relatively featureless. The removal of this oxide in water suggests that it consists mostly of GeO_2 in either an hexagonal (solubility: 4.5 $g\,l^{-1}$) or amorphous (solubility: 5.2 $g\,l^{-1}$) phase (Fig. 4.7) [15].

Further treatment in H_2O_2 leads to the growth of two components at 830 and 980 cm^{-1} indicating the regrowth of an oxide (although the resulting oxide is thinner than the original native oxide since the overall integrated area is still negative). This new H_2O_2-grown oxide is also soluble in water suggesting that

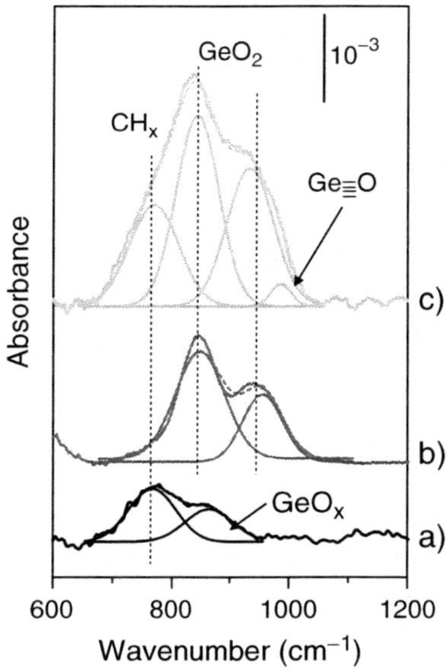

Fig. 4.6. Transmission absorption IR spectra of germanium oxides formed (**a**) after 90 min in air, (**b**) after 30 min UV/ozone exposure and (**c**) after 10 s in H_2O_2 (30%). The reference spectrum is H/Ge(100) in all cases (see Sect. 4.55). Gaussian lines are used for the fitting (unpublished results)

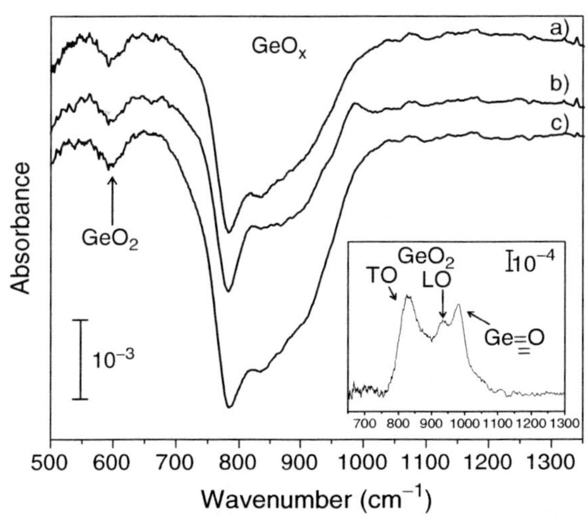

Fig. 4.7. Transmission absorption IR spectra of Ge(100) after treatment in: (**a**) deionized H_2O, (**b**) deionized H_2O followed by 10 s in H_2O_2. All the spectra are referenced to an as-received Ge(100) wafer. The *inset* shows the oxide region after H_2O_2 oxidation, using the H_2O-treated Ge(100) surface as reference (reprinted with permission from [15])

it is primarily composed of amorphous or polycrystalline GeO_2 [15]. By referencing this H_2O_2-grown oxide to the surface previously treated in water (inset in Fig. 4.7 corresponding to the difference of spectra b and a) it is apparent that this newly formed oxide is characterized by three distinct components at 830, 940 and 980 cm^{-1}. The intensity dependence of these components on the photon incidence angle (varied from 60° to 10° with respect to the normal) indicates that the pair at 830 and 940 cm^{-1} exhibits the polarization dependence expected for the transverse (TO, parallel) and longitudinal (LO, perpendicular) optical phonon modes of GeO_2 (involving primarily the Ge–O–Ge asymmetric stretch mode) [7,8]. The TO and LO modes associated with the Ge–O–Ge *symmetric* stretch are observable at 583 cm^{-1} (weak shoulder) and 595 cm^{-1} respectively. The assignment of the third absorbance feature at 980 cm^{-1} is challenging. Its relatively high frequency points to the existence of germanium doubly-bonded to oxygen (Ge=O). Its dependence on the incidence angle indicates that it is mostly but not completely perpendicular to the surface (Fig. 4.8).

XPS data(15) confirm the presence of GeO and GeO_2 and are in complete agreement with the data from Prabhakaran et al. [12] and Zhang et al. [14]. Using a simple attenuation layer model and assuming a 15 Å mean free path for the photoelectrons, the total native oxide thickness is estimated to be ~15 ± 5 Å. The GeO_2 component completely disappears after DI H_2O rinse and only GeO remains on the surface [15]. Longer rinse time in DI water does

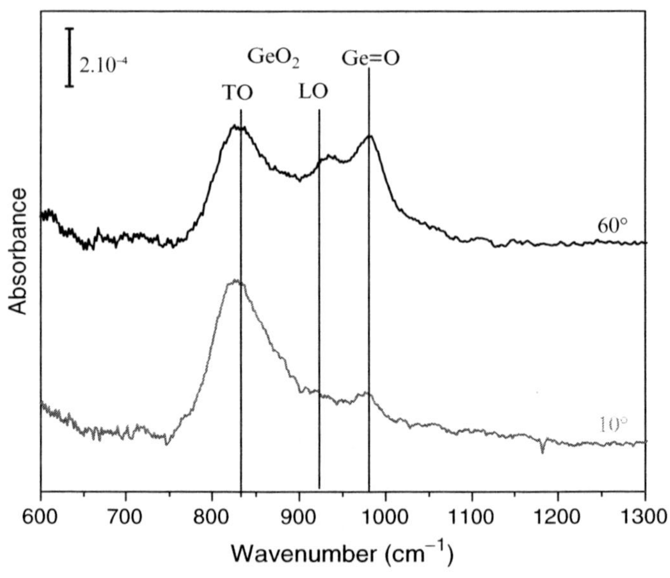

Fig. 4.8. Transmission absorption IR spectra of Ge(100) after H_2O, H_2O_2 chemical treatment for two incidence angles: 10° and 60° with respect to the normal. The reference spectra are H_2O-treated Ge(100) (unpublished results)

not remove GeO [16]. Hydrogen peroxide grows a mixture of GeO and GeO_2 oxides very similar to the initial oxides present on the as-received substrates. Using the same approach as for the as-received samples, we estimate the H_2O_2 oxide to be about 10 Å thick [14].

4.5 Hydrogenation of Germanium Surfaces

4.5.1 Hydrogenation in Ultra High Vacuum

Hydrogenation of clean germanium surfaces was first performed using atomic hydrogen. In 1963, Becker and Gobeli [17] pioneered the use of multiple internal reflection to examine hydrogen chemisorption on silicon. Later, UHV/FTIR studies of hydrogen adsorption on both Si(100)-(2×1) and Ge(100)-(2×1) gave clear evidence for the formation of both mono-, di- and tri-hydrides on Si(100) but primarily monohydride on Ge(100) [18], as later confirmed by STM studies. Specifically, surface IR spectra as a function of the hydrogen coverage show the presence of two absorbance components at 1,979 and 1,991 cm^{-1} oriented parallel and perpendicular to the surface, respectively, corresponding to the asymmetric and symmetric stretch modes of the monohydride structure (Fig. 4.9). Only at high exposures can spectral components associated with dihydride be observed, but they remain weaker than for comparable exposures on Si(100). At very low coverage, the hydrogen atoms are randomly distributed and there is evidence that most of the dangling bonds are initially unpaired [18].

Later STM studied confirmed that, the (2×1) reconstruction is preserved [11, 18–22] as a stable monohydride phase after moderate atomic H exposures at room temperature (Fig. 4.10) [19, 20, 22]. Prolonged exposures at room temperature leads to the development of line defects, involving the formation of GeH_2 rows with local H/Ge(100)-(3×1) arrangement, while a local (2×1) reconstruction still dominates. The GeH_2 structure is unstable and can easily be reduced to monohydride because of collision-induced H_2 desorption or by dissociation (Fig. 4.11) [22].

At substrate temperatures higher than 400 K, hydrogenation of Ge(100)-(2×1) results in etching. Single atoms and dimer vacancies are formed, followed by line defects along the germanium dimer row, which leads to the formation of V-groove-shaped etch pits (Fig. 4.12) [20]. GeH_2 formation is an essential first step in the etching process because it initiates Ge-Ge bond breaking. Etching proceed with release of GeH_4 gas, which has been detected directly with mass spectrometry [21].

4.5.2 Wet Chemical Treatment of Flat Single Crystal Germanium Surfaces

Several groups have studied the effect of hydrofluoric (HF) acid treatment on chemically grown GeO_2 surfaces [15, 23–28]. Dependence on HF etching time

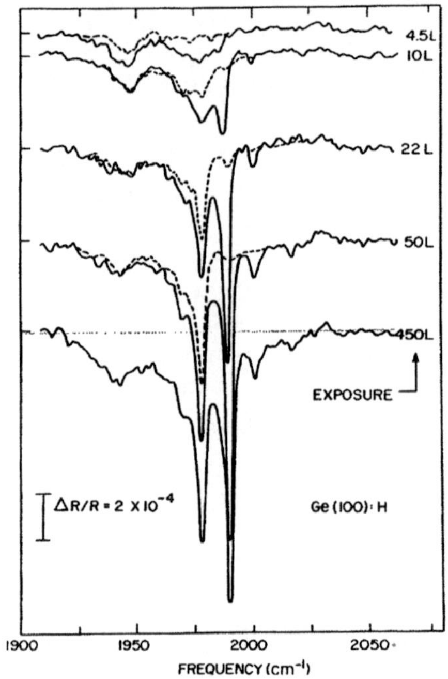

Fig. 4.9. Multiple internal reflection infrared spectra as a function of H coverage on a flat Ge(100)-(2 × 1) surface at room temperature. The exposures are quoted in terms of molecular hydrogen exposures (the actual atomic H exposure are exactly proportional and roughly two orders of magnitude lower than the molecular exposures) (reprinted with permission from [18])

and concentration (ranging from 2 to 50%) has been investigated. Few studies report complete H-passivation, although most agree that HF does remove wet-chemical oxides. The main difficulty of such studies is the lack of direct evidence for H termination.

A common procedure consists of multiple cycles consisting of immersion in HF (10 s) and DI water (20 s) repeated five times. Studies of the resulting surfaces include XPS, second harmonic generation (SHG) and ellipsometry for Ge(111) and Ge(100). The XPS studies of Deegan et al. [24] report no evidence for the presence of oxide after the above treatment, i.e., no shoulders are observed in the Ge3d and Ge2p core level spectra at higher kinetic energy (Fig. 4.13). However, similar experiments performed by Bodlaky et al. [28] reveal the presence of a non-negligible sub-oxide contribution (~5 Å) on the Ge2p core level spectrum, consistent with the observation of the O1s peak (Fig. 4.14).

Using SHG, Bodlaky et al. [28] report that the measured rotational anisotropy for H–Ge(111) surface is inconsistent with what is expected for this surface. Namely, since the Ge–O bond is more polar that the Ge–H bond,

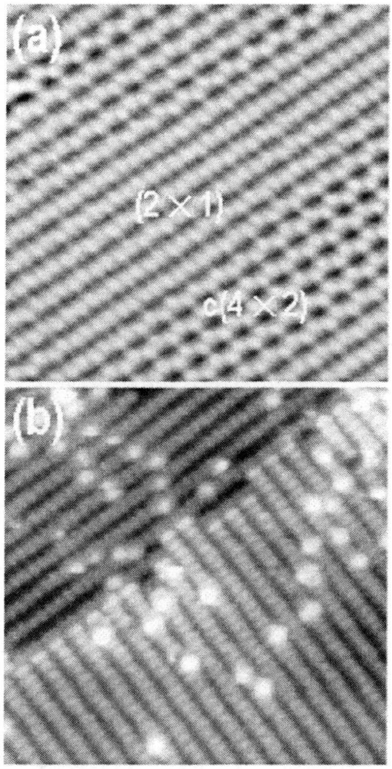

Fig. 4.10. Filled-state STM images ($15 \times 15\,\mathrm{nm}^2$) of (**a**) the adsorbate-free Ge(100)-(2×1) surface and (**b**) the nearly monohydride-saturated surface (reprinted with permission from [22])

the SHG-rotational anisotropy pattern for H/Ge and GeO_2/Ge should be different. The observed similarity of both surfaces, H/Ge and GeO_2/Ge, suggests that the hydrogen monolayer is either inhomogeneous or defective, i.e., must coexist with an oxide layer. The HF-etched surface is also highly unstable,

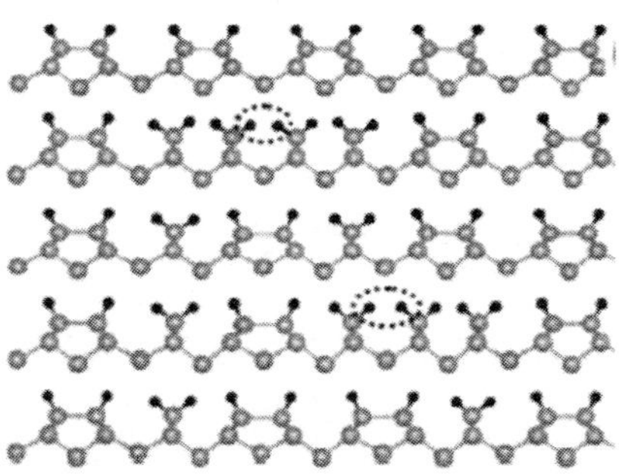

Fig. 4.11. Schematic diagram for the Ge 2×1 : H to 3×1 : H transition and antiphase monohydride dimer row formation (reprinted with permission from [22])

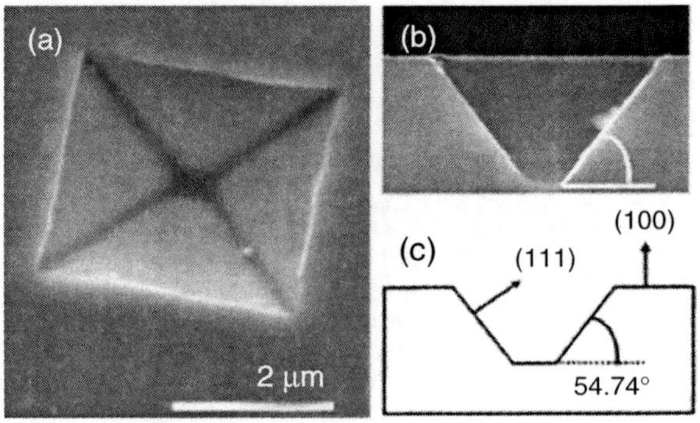

Fig. 4.12. FE-SEM images of (**a**) a typical etch pit of V-groove shape formed on a Ge(100) surface exposed to $H_2(g)$ of $\sim 1 \times 10^6$ L at $T = 400$ K, (**b**) the cross-sectional view of (**a**) and (**c**) a schematic diagram of (**b**) showing that the etch pits are bounded by a set of four (111) planes (reprinted with permission from [20])

exhibiting a rapid oxidation in air. Specifically, H-terminated Ge(111) surfaces oxidize quickly without the typical induction period observed for H/Si surfaces that undergo a very slow initial oxidation. The authors note that a thick oxide layer (∼5 Å) is formed on H/Ge surfaces and suggest that some

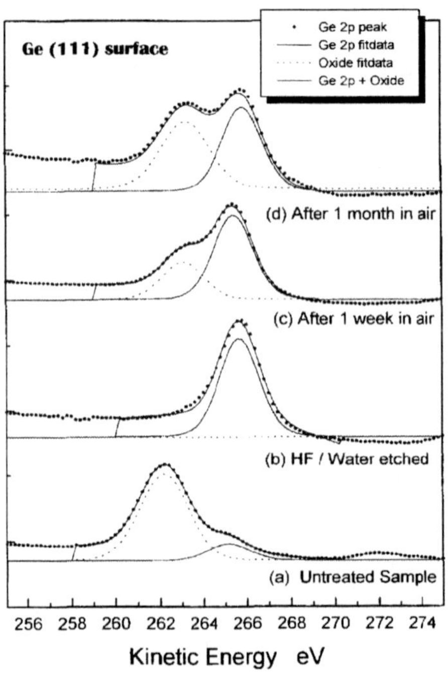

Fig. 4.13. $Ge2p_{3/2}$ core level of Ge(111) (**a**) as received, (**b**) after HF/water etched, (**c**) etched sample after 1 week in air, (**d**) etched sample after 1 month in air (reprinted with permission from [24])

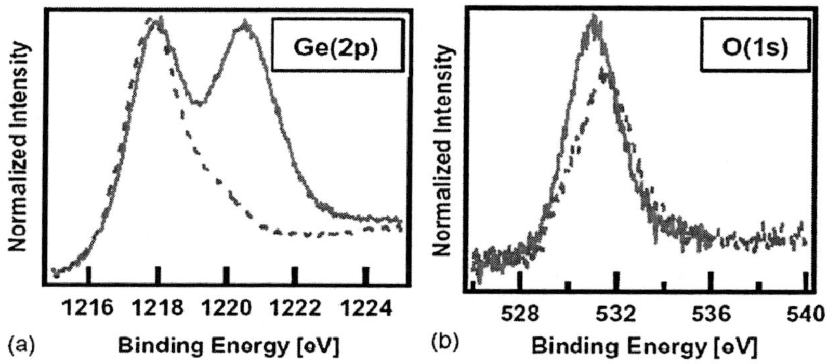

Fig. 4.14. Ge2p$_{3/2}$ and O1s core level spectra of Ge(111) as received (*solid line*), and after HF/water etched (*dashed line*) (reprinted with permission from [23])

oxide already exists on the HF-etched Ge surface. Ellipsometric measurement of the H/Ge(111) surface reveal that the oxide growth in air follows a logarithmic law [29]. Recent FTIR studies of H/Ge stability in air give strong evidence that hydrocarbon contamination of the surface initially dominates. Within the first hour in air, the growth of a CH_x mode is faster than that of suboxides (GeO_x) (Fig. 4.6 a). After 90 min in air, there is ~10% of the initial hydrogen coverage left, with relatively little oxide [16].

Another procedure to achieve H-passivation of Ge substrates is to immerse a chemically oxidized Ge directly in aqueous HF (10%) for 10 min. Choi et al. [23] first showed that such a procedure did lead to hydrogen-termination using FTIR. A hydride termination was clearly derived from the presence of a broad absorbance features centered at 2,010 cm^{-1} and corresponding to GeH_x stretching modes. The width of the band probably arises from a relatively large surface roughness estimated to be ~3–4 nm after this treatment. For reaction times lower than 10 min in aqueous HF, the authors find that germanium surfaces are not fully hydrogenated. For longer reaction times in HF, the hydrogen coverage surprisingly decreases, suggesting a complicated process (Fig. 4.15). More dilute aqueous HF solutions (<10%) also result in incomplete hydrogen-termination (weak GeH_x stretch modes) while higher concentration solution (e.g., 25% HF) leads to surface roughening (8–10 nm as determined by AFM), with deep pitting of the germanium substrate [23]. These observations have led the authors to believe that etching can take place just as it does for Si using higher pH solutions, and that kinetics rather than thermodynamics dominate during the H-passivation process. A recent study on the etching effect of different aqueous hydrohalogenic acid solutions on oxidized germanium surfaces clearly shows that high HF concentration (49%) for a short time (5 min) do etch GeO_2 and leave suboxides on the surface that prevent full hydrogen termination of the germanium surface [25]. Surprisingly the authors observe with XPS that longer exposures in HF or other hydrohalogenic acids do not help in achieving full H passivation.

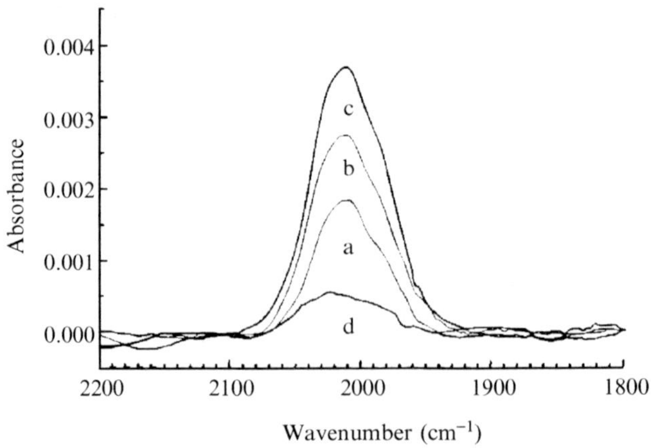

Fig. 4.15. ATR-FTIR spectra of the $\nu(GeH_x)$ stretching region for the hydrogen-terminated Ge(100) surface prepared by etching with an aqueous 10% HF solution: after (**a**) 2 min, (**b**) 5 min, (**c**) 10 min, (**d**) 15 min (reprinted with permission from [23])

A recent FTIR study on chemically oxidized samples ($H_2O + H_2O_2$) provides a direct measure of the hydrogen coverage after HF etching [15,23–28]. Using a transmission geometry instead of ATR to access the full spectral range (400–4,000 cm^{-1}), all the relevant Ge–H modes (bend, scissor, and stretch modes) are observed. This study shows that etching in 10% HF does lead to the complete removal of the H_2O_2-grown germanium oxide (GeO_2) as well as germanium suboxide (GeO) [15], and to full H-termination as mono- and di-hydrides. Figure 4.16 shows the scissor (834 cm^{-1}) and rocking (640 cm^{-1}) modes of GeH_2 together with a broad stretch mode around 2,000 cm^{-1} attributed to Ge-H_x. This broad absorbance feature is composed of unresolved modes centered at 1,987, 2,020 and possibly 2,060 cm^{-1}, corresponding to the germanium mono, di- and tri-hydrides, respectively. An upper limit for the trihydrides can be set at 10% of the total passivated surface, less than what is observed on Si. Although a flat Ge(100) surface could in principle support both mono- and di-hydrides, with a local 3×1 reconstruction, the wet etched surface is clearly not atomically flat. From the IR spectrum shown in Fig. 4.16, the hydrogen-coverage has been estimated to be around 80% of a monolayer. This partial coverage reflects the poor stability of the HF-etched surface in air as it can be increased by faster transport into the N_2-purged spectrometer immediately after etching.

Complementary XPS information confirms that all oxides and sub-oxides evident in the Ge3d core level have been removed by HF etching. This is consistent with the IRAS observation of complete H-termination right after HF etching. The lack of a well-defined contribution in the O1s core level confirms the absence of oxide (<0.1 monolayer). However, a careful decomposition of the Ge3d peak reveals the presence of a small component on the higher

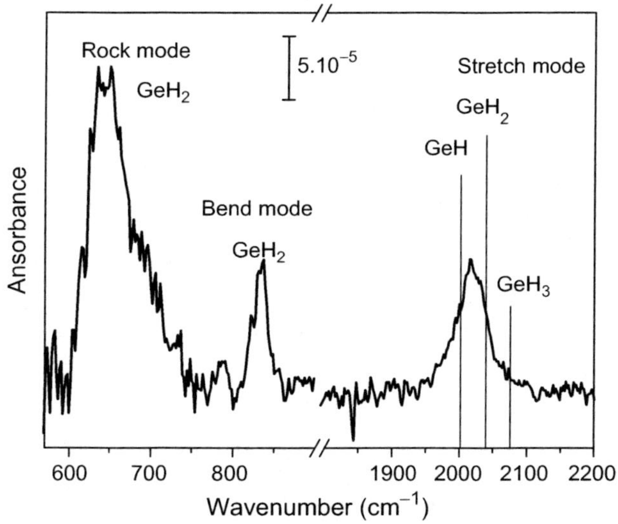

Fig. 4.16. Transmission absorption IR spectrum of H/Ge(100) after HF wet chemical treatment for 10 min, using the ultra-thin native epiready Ge wafer as reference. The positions for mono-, di- and tri-hydride stretch modes are indicated with vertical lines (reprinted with permission from [15])

binding energy side (+1.6 eV with respect to bulk Ge3d). This component cannot be attributed to GeO but it is likely due to the presence of carbon. Since the chemical treatment is performed in air, ambient exposure to hydrocarbons resulting in partial carbon contamination after etching and prior to IRAS or XPS measurements cannot be avoided. In fact, the carbon level detected by XPS is often higher on HF-etched surfaces than on the as-received oxidized substrates or on wet-chemically oxidized surfaces. Figure 4.17 shows, for example, that the as-received and GeO- or GeO_2-terminated surfaces (after H_2O or H_2O_2, respectively) show less carbon contamination. The nature of this carbon contamination is of great interest. There is no doubt that, just as in the case of hydrophobic H-terminated Si, hydrocarbon molecules can be physisorbed on the surface. In fact, the presence of such hydrocarbons is observed with FTIR with multiple components at 790, 1,440 and 2,800–2,950 cm^{-1}, corresponding to deformation and stretching mode of CH_x, respectively [16]. This observation suggests that some hydrocarbon molecules react with the germanium surface removing H in the process. Direct measurement of Ge–C bonds would greatly help in understanding this contamination process.

4.5.3 Electrochemistry on Flat Single Crystal Germanium Surfaces

Electrochemistry has also been used for hydrogenation and hydroxylation of germanium electrodes [30–34]. Typically, polished flat (111) and (100) germanium substrates are first cleaned with sulfochromic acid/hydrochloric

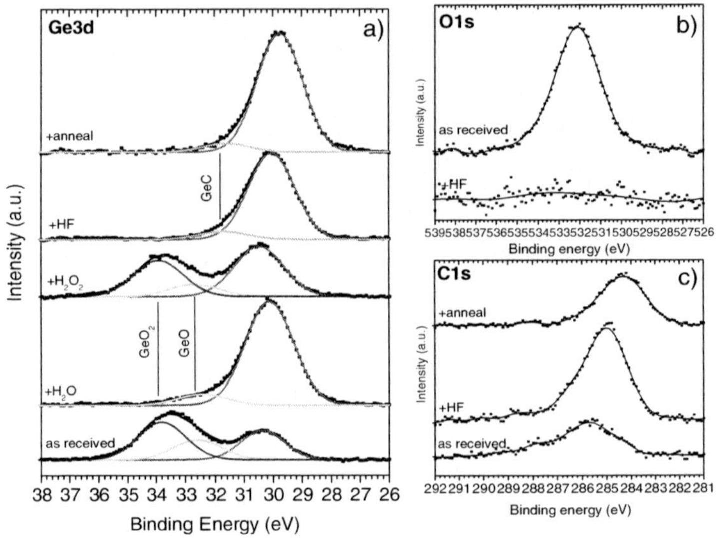

Fig. 4.17. XPS data for (**a**) Ge3d, (**b**) O1s and (**c**) C1s for as-received, H_2O, H_2O_2, HF, and annealed surfaces. On the Ge3d core level spectra, the *vertical lines* indicate the center position of the GeO, GeO_2 and GeC reactive components obtained after peak decomposition (reprinted with permission from [15])

acid cycle and then immersed in an acidic ($HClO_4$) solution. The initial state of the germanium surface is believed to be hydroxyl-terminated. Application of a negative potential ($-0.5/-0.7$ V) on this starting surface clearly shows the formation of germanium hydride and the removal of the hydroxyl in IR spectra while anodization of the hydrogen-terminated germanium surface allows the substrate to recover its initial state (Fig. 4.18) [28, 31, 32]. The observed change in surface chemistry (GeH $\rightarrow$ GeOH) is believed to be associated with a band-edge shift. In contrast to silicon surfaces, the change is due to free carriers (surface ionic charge) equilibration with ionic charges in the electrolytes rather than surface dipoles [32].

In situ FTIR and voltametry measurements on Ge(111) and Ge(100) give similar results for both surfaces after hydrogenation indicating that the H-terminated Ge surfaces are atomically rough. Formation of GeH and GeH_2 is observed in both surfaces but in contrast to silicon surfaces, tri-hydrides are always absent. As the absorbance feature of GeH_x is broad and unresolved, a Gaussian fit is used to distinguish the presence of two components centered at 1,970 cm^{-1} and 2,020 cm^{-1} with 20–30 cm^{-1} half-widths, which correspond to GeH and GeH_2 stretching modes, respectively (Fig. 4.19). Interestingly, while the 2,020 cm^{-1} component remains unchanged, the position of GeH varies from 1,950 to 1,990 cm^{-1} as the applied potential is more negative. Its initial position can be ascribed to an oxidized environment of Ge atoms whereas the position at 1,950 cm^{-1} can be interpreted as the penetration of hydrogen atoms into the Ge lattice breaking the weak Ge–Ge bond.

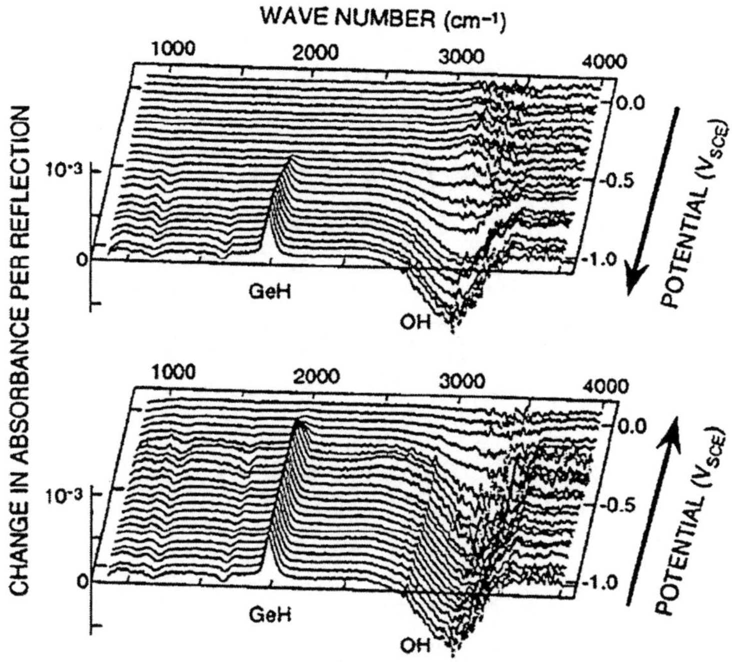

Fig. 4.18. IR spectra as a function of potential for Ge(100) surface (p-polarization); *top figure*: cathodic scan and *bottom figure*: anodic scan. The reference is taken at the upper potential bound (reprinted with permission from [31])

Electrochemical incorporation of hydrogen in Ge(111) has been reported earlier after a prolonged cathodic treatment [34]. FTIR experiments as a function of polarization clearly support the penetration of the hydrogen into the Ge bulk as the position of the observed band red-shifts while the intensity of the s- and p-polarization remains unchanged: a thin disordered highly hydrogenated layer is thus formed close to the surface [34].

The combined voltametry and FTIR data also suggest the possible existence of (111) microfacetting for Ge(100) as an explanation for the observed intensity changes of the GeH_x stretch mode. The voltametric peaks (β and α) appear to be associated with hydrogenation of microfacets and steps generating GeH and GeH_2.

4.5.4 Electrochemistry on Porous Germanium Substrate

Electrochemical hydrogenation of porous material is natural since it represents an intrinsic step of their formation. Porous materials have been studied extensively because of their photoluminescence properties important for optoelectronics [35, 36]. Two procedures have been reported. The first one involves an anodic etch of the c-Ge(100) substrate in a 50 wt.% HF solution

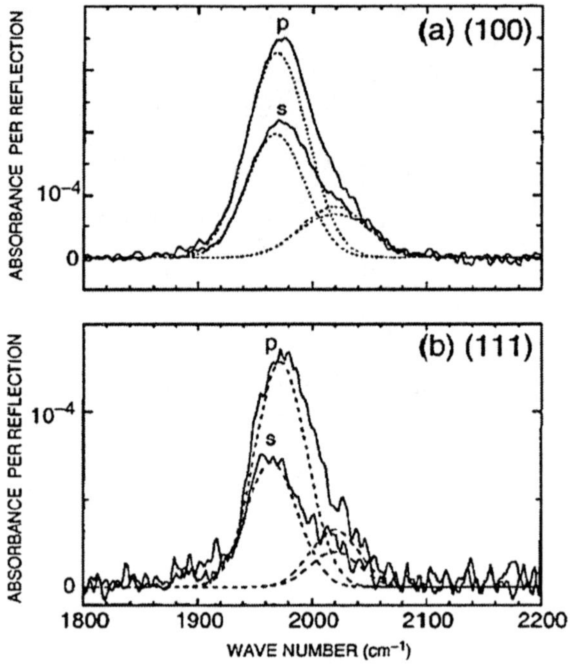

Fig. 4.19. Analysis of the νGeH_x spectrum for Ge (**a**) (100) and (**b**) (111), for s- and p-polarization at $-0.825\,V_{sce}$. The data exhibit evidence for two contributions, with distinct polarization dependences. The *dotted curves* show the fitting curves (reprinted with permission from [31])

in presence of an halogen lamp (500 W). The second process is based on a bipolar electrochemical etching (BEE) technique, whereby the single crystalline Ge(100) substrate is first cleaned in an ethanoic HCl solution and then anodized for 5 min, followed by a cathodization step for 1 min. Both methods lead to similar H-termination, characterized by a broad stretch mode at ~2,030 cm^{-1} corresponding to Ge–H_x (Fig. 4.20) and by GeH_2 deformation modes observed at 580 and 830 cm^{-1} [35, 36]. Little characterization of germanium prepared with the anodic etching method has been performed. For H-terminated porous germanium prepared by BEE, the starting surface is probably hydroxyl-terminated and becomes hydrogen terminated when a positive potential is applied. The Ge–Ge bonds are protonated upon cathodization leading eventually to the formation of GeH_4 and the dissolution of the germanium bulk. This results in both hydrogenation and formation of pores. The observed etching process is in fact close to what is observed when Ge(100)-2 × 1 surfaces are exposed to $H_{2(g)}$ at 400 K [35, 36]. FTIR measurements of this porous substrate reveal a broad Ge–H stretching band with two components at 2,044 and 2,015 cm^{-1}, i.e., higher in frequency than usually measured. Choi and Buriak [35] suggest that all forms of hydrides

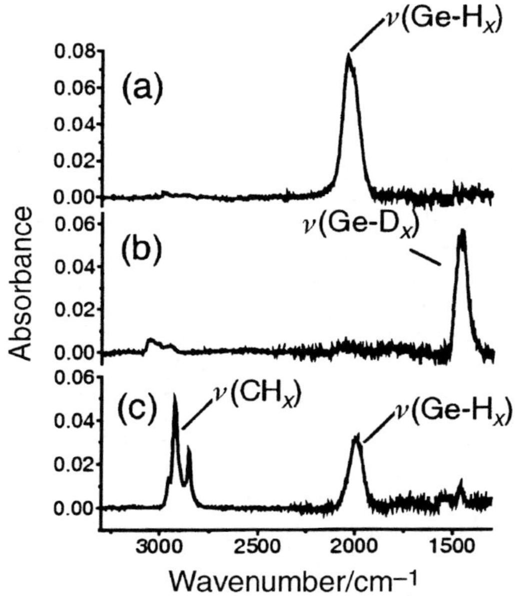

Fig. 4.20. Transmission FTIR spectra of porous germanium: (**a**) etched by BEE using ethanoic HCl, (**b**) using deuterated ethanol and DCl, and (**c**) thermally hydrogermylated using 1-dodecene (reprinted with permission from [35])

(mono- and di-hydrides) are formed because of the surface roughness as in the case of porous silicon.

4.6 Nitridation and Oxynitridation of Germanium Surfaces

Nitridation of germanium surfaces has been investigated mostly because a thin nitride (or oxynitride) layer is believed to effectively prevent Ge oxidation and Ge diffusion into high-κ dielectrics [37–44]. Nitride layers are grown primarily by exposure to ammonia gas or plasma (or even HN_3 or N_2H_4 [42]), and also by N implantation. Again, much of the early work was done in UHV, initially at relatively low temperatures (exposure below room temperature) to investigate the chemisorption of NH_3 on clean Ge surfaces. At ~200 K in UHV, there is evidence for molecular adsorption upon exposure of NH_3 on clean Ge(100)-(2 × 1) as determined by XPS, UPS as well as optical spectroscopy measurements. The modes related to NH_3 have clearly been identified with HREELS at 3,270 (symmetric stretching mode), 1,570 (asymmetric bending mode) and 1,150 cm^{-1} (symmetric bending mode) (Fig. 4.21) [38, 42].

Under these conditions, three adsorption regimes can be distinguished: monolayer, second layer and condensed multilayer [44]. Interestingly, as the temperature is increased to ~500 K, ammonia is completely removed from

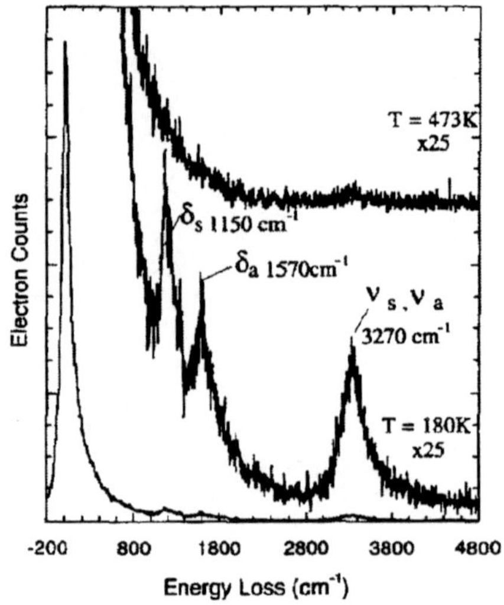

Fig. 4.21. HREELS spectra of NH_3 on Ge(100)-(2×1). The lower curve shows Ge(100) exposed to 10 L of NH_3 at $T = 180$ K. The upper curve was taken after annealing the sample at 473 K for 2 min and then cooling the sample back down to 180 K (reprinted with permission of [42])

the surface (Fig. 4.22) [38]. Since there is no observable spectroscopic features related to NH_2 throughout the whole temperature range, it was concluded that NH_3 does not dissociate on Ge(100). This is in contrast to silicon where NH_3 does dissociate into NH_2 and H, although it mostly recombines to desorb as NH_3 at higher temperatures. On Ge(100), NH_3 adsorbs initially on the electrophilic surface dimer atom (atom lower in the dimer) and produces a (2×2) reconstruction because it adsorbs in a zig-zag fashion, causing a flip of the dimer tilt direction of every second dimer row. In this HREELS study [38], germanium nitride could only be formed after electron irradiation at 110 K, as evidenced by the appearance of a component at $\sim$800 cm^{-1}.

First-principles pseudopotential calculations have confirmed that NH_3 dissociation into NH_2 + H on Ge(100) is not energetically stable [43]. NH_3 gas treatment performed at either room or high temperatures (up to 870 K) does not form any nitride layer. Using relatively low pressures, XPS measurements confirm that less than 6.3×10^{13} nitrogen atoms/cm^2 are observed on Ge(100), yielding an upper limit for the sticking (reaction) of 3×10^{-6} [45]. This value is consistent with earlier studies [46] of NH_3 adsorption at 180 K, which had derived a sticking coefficient $<10^{-4}$.

Nitridation requires higher pressures. Early work was performed on a thin evaporated (porous) germanium films. In the presence of a low pressures of argon, ammonia was observed to dissociate on the Ge film even at room temperature [40]. Infrared experiments on a Ge clean film show the presence of the Ge–NH_2 deformation mode in the 1,550–1,600 cm^{-1} region and Ge–H stretch mode at 1,920 cm^{-1} consistent with dissociation into NH_2 and

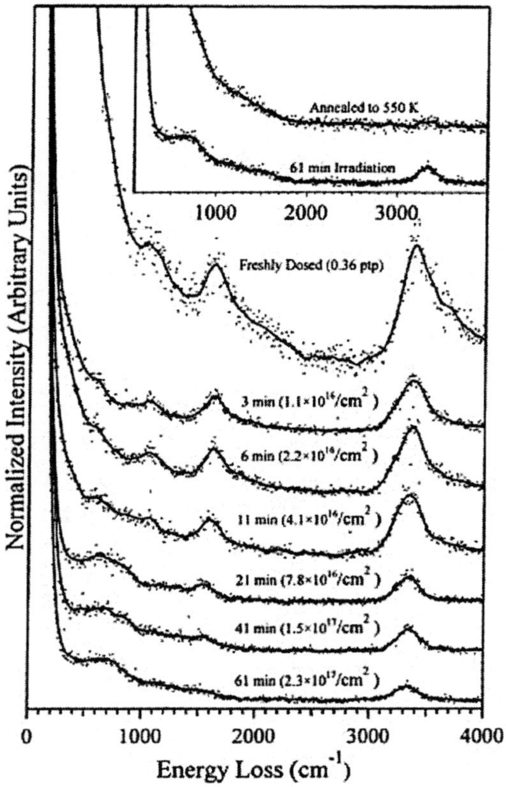

Fig. 4.22. HREELS spectra for ammonia adsorbed on Ge(100)-(2 × 1) at 110 K after several exposure times (reprinted with permission from [38])

H. The weak Ge–H feature disappears after 17 h, indicating that GeH is not as stable (Fig. 4.23) The interpretation for the observation of NH_2, however, is different from simple NH_3 dissociation. The authors [40] suggest that at low exposure or in absence of NH_3 gas, the bonded hydrogen atoms recombine forming $H_{2(g)}$ which is released. At higher exposure of ammonia gas, $NH_{3(g)}$ reacts with the Ge–H forming Ge–NH_2 on the surface and releasing $H_{2(g)}$: this explains why Ge–H stretching mode is weaker and observed only briefly. The mode observed at 1,450 cm^{-1} after many hours of NH_3 exposure is attributed to adsorbed $NH_4{}^+$. To explain the presence of this mode, the authors [40] suggest either a ionic reaction between $NH_{3(g)}$ and Ge–H forming so Ge–$NH_4{}^+$ (less probable due to the area of Ge–NH_2) or the possible reaction of $NH_{3(g)}$ with water. Various studies have actually demonstrated that oxygen incorporation during the thermal nitridation is unavoidable and lead to the formation of an oxynitride on top of the germanium substrate [39, 47].

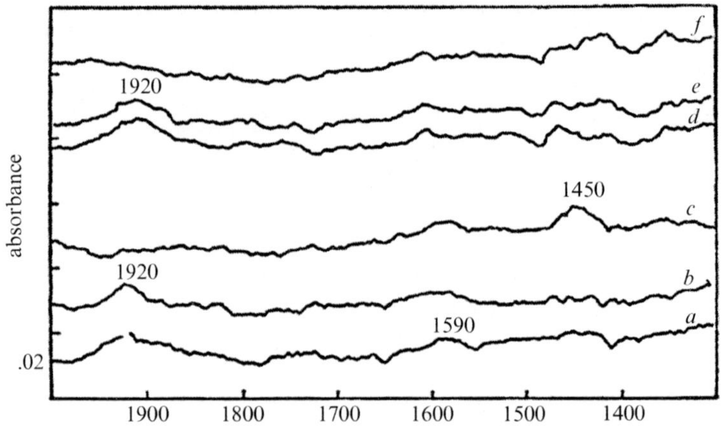

Fig. 4.23. IR spectra of ammonia adsorbed on germanium at 300 K, 67 Pa (**a**) after 30 min, (**b**) after 4 h, (**c**) after 17 h. (**d**)–(**f**) obtained from separate runs at 665 Pa (**d**) after 30 min, (**e**) after 2 h30 and (**f**) pumped for 18 h (reprinted with permission from [40])

Using ellipsometry and optical micrography, Hua et al. [39] have shown that the growth of a smooth and uniform nitride film takes place at 870–930 K in 2–4 h on a hydrophobic germanium surface. They also observed an increase of the refractive index during the growth they attributed to the presence of oxide [39]. The growth mechanism is dependent on the diffusion of the nitrogen and oxide into and through the film. However, when nitrogen gas is used instead of ammonia, there is no growth of a nitride layer indicating that N_2 does not dissociate. Finally, nitridation without oxygen incorporation or carbon contamination has been recently achieved using a plasma-enhanced nitridation technique [41]. Auger spectra show that a two-step cleaning procedure, involving out-gassing at low temperature and thermal annealing at high temperature (870 K) in UHV leads to the formation of a good quality nitride layer (1.6 nm thick) with a stoichiometry close to Ge_3N_4. Cross-sectional HR-TEM images show a sharp interface between the germanium substrate and the crystalline Ge_3N_4 film (Fig. 4.24).

Oxynitridation has been investigated more recently as an alternative to GeO_2 to passivate Ge surfaces. Layers consisting of Ge_2N_2O are insoluble in water and their interface with Ge appears to yield good electrical properties [48–50]. Oxynitride layers are obtained following a two-step oxidation/nitridation process: after an ex situ chemical treatment, the hydrophobic germanium(100) surface is introduced in a furnace at 820 K and maintained for 90 min in an atmosphere of O_2/N_2 (1 : 4), followed by NH_3 gas flow at 870 K for 30–45 min. TEM pictures of such a films (~22 nm thick) display a sharp interface between the crystalline germanium and the amorphous oxynitride layer comparable to a Si/SiO_2 interface (Fig. 4.25) [48]. However the authors [48] mention that the electron beam stimulates the

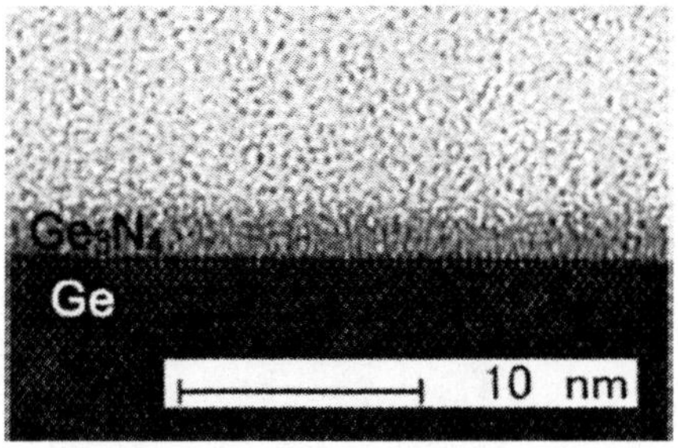

Fig. 4.24. A cross-sectional HR-TEM image of Ge_3N_4/Ge substrate (reprinted with permission from [41])

decomposition and reaction of the oxynitride layer with the germanium substrate, leading to roughening. Oxynitrides have been used as an intermediate layer for the growth of high-κ dielectrics because they also prevent diffusion of metal species [48, 49].

4.7 Sulfur Passivation of Germanium Surfaces

Sulfur has also been investigated as an alternative element to passivate germanium surfaces [28, 51–55]. Exposure of a clean Ge(100)-(2 × 1) surface (obtained after various cycles of Ar^+ ion sputtering and annealing at 870 K under UHV) to a molecular beam of elementary sulfur leads to the formation of an S/Ge(100)-(1 × 1) surface, with ideal Ge–S–Ge (monosulfide) termination [55]. XPS data show that there is only one chemical environment for the sulfur surface atoms (Fig. 4.26). The bivalent sulfur atoms are therefore believed to bond in the bridge position between the topmost germanium atoms of the

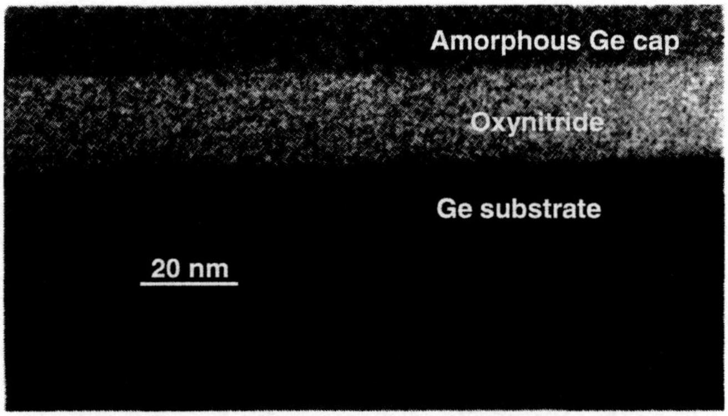

Fig. 4.25. High resolution TEM image of an oxynitride film grown on a Ge(100) wafer obtained after a two stage-oxidation/nitridation process. The oxynitride layer is capped with amorphous Ge (reprinted with permission from [48])

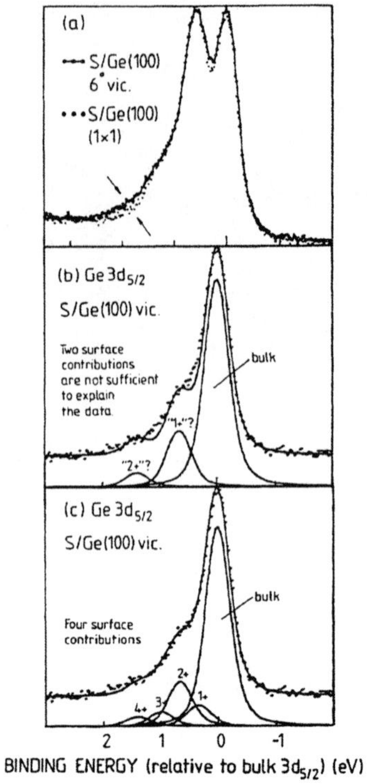

Fig. 4.26. Ge3d core level photoemission spectra for clean and sulfur-covered Ge(100) surfaces. (**a**) spin orbit splitting and secondary electron background, (**b**) determination of bulk emission, (**c**) surface-bulk deconvolution for S/Ge(100)(1×1) (reprinted with permission [55])

Ge(100) surface: a coverage of one sulfur atom per Ge surface atom is expected and no sulfur island is present (Fig. 4.27). Local density approximation (LDA) calculations confirm that the dimer bonds are broken with the S adatoms saturating the dangling bonds of the top layer Ge atoms restoring the (1×1) reconstruction with an optimal S–Ge bond length of 2.36 Å. In that configuration, the total surface energy is lowered by 0.71 eV [52]. The S-terminated surface is inert and stable in a UHV environment with no observed contamination after 2 days.

Cohen [46], Kuhr [56] and Nelen [51] have used H_2S to passivate Ge(100)-(2×1). They all find that H_2S dissociates on a clean Ge(100) into SH, S and H at room temperature with an S saturation coverage of 0.25 ML. This coverage can be increased to 0.5 ML by exposing on a substrate at 553 K. This higher coverage is due to hydrogen desorption generating extra sites for HS and S adsorption.

There has been little work done using wet chemistry. Anderson et al. [54] have used $(NH_4)_2S$ in a wet chemical treatment similar to that used for GaAs

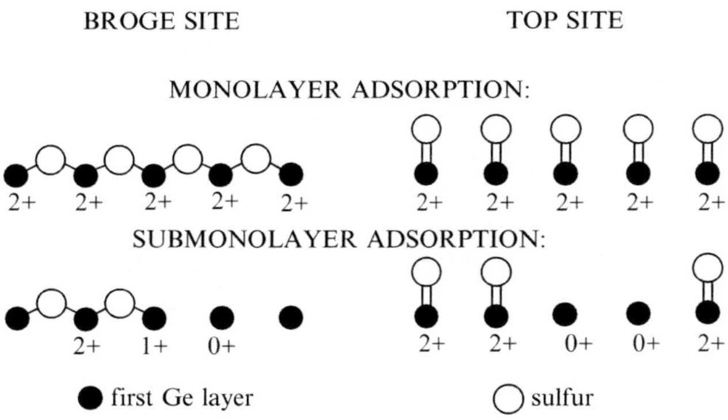

Fig. 4.27. Bridge-site and top-site bonding can be distinguished by the appearance of the oxidation state 1+ in the case of low-coverage adsorption (reprinted with permission from [55])

and InP. After a first HF(1%) etching for 1min, the Ge(100) substrate is immersed in an aqueous $(NH_4)_2S$ solution for 20 min at 340 K and rinsed thoroughly in methanol or water before drying. The surface was then introduced in UHV for LEED and XPS analysis. The surface symmetry is S/Ge(100)-(1×1) surface similar to what Weser [55] observed on S/Ge(100) obtained by dosing atomic sulfur on Ge(100)2×1. However, while the germanium is also passivated with one layer of bridge-bonded sulfur atoms, the wet chemical preparation leads to trace amounts of oxygen and carbon contamination. In a vacuum environment, sulfur starts to be released from the surface at 460 K, and is completely removed by 650–750 K, although a Ge(1×1) pattern is still observed in LEED. Since carbon contamination is observed on the surface after such an anneal, it is possible that C contamination prevents the restoration of a clean Ge(100)-(2×1) reconstructed surface. An attractive aspect of the S passivation is the fact that the Ge(100) surface remains very flat, even after annealing, with an RMS roughness ranging between 3 and 12 Å [54]. The S-passivated surface is believed to be stable in atmosphere, preserving the (1×1) reconstruction with no trace of oxide. This stability is confirmed by Bodlaki et al. [28] who observed with second harmonic generation that no oxidation of an S/Ge(111) surface occurred for up to 2 months.

4.8 Chlorine Passivation of Germanium Surfaces

Chlorination of germanium can be achieved using Cl_2 gas, both in UHV and at atmospheric pressure, as well as by wet chemistry [57–61]. For instance, the clean Ge(111)-(2×8) surface was exposed to chlorine gas under ultra high vacuum by Citrin et al. [57] who found that the (2×8) reconstruction

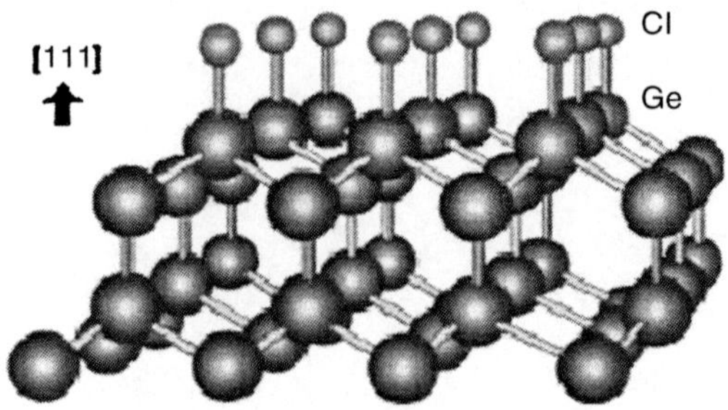

Fig. 4.28. Schematic view of an ideally chlorine-terminated germanium(111) surface (reprint with permission from [60])

is maintained as the Cl is adsorbed in an atop configuration. Using SEXAFS, they verified that the reconstruction and Cl coverage are maintained upon annealing to ∼670 K for 5 min, suggesting a good stability of chlorine on germanium surfaces.

Lu et al. [60,61] have shown using XPS that immersion of Ge(111) covered with native oxide in dilute aqueous HCl (10%) for 10 min results in a chlorine-passivated germanium surface. Combining XPS to X-ray absorption near edge spectroscopy (XANES), the authors argue that a flat monochloride structure is formed and remains stable in air (Fig. 4.28).

The above studies also show that the orientation of the chemical Ge–Cl bond is strictly perpendicular to the surface: the chlorine is bonded to the germanium substrate in the atop position. These results are in good agreement with the UHV study performed on the Ge(111)-(2×8) discussed above [57]. A finer analysis using X-ray absorption fine structure (EXAFS) finds a Ge–Cl bond of 2.17 nm which is confirmed by theoretical calculations [61]. Multiple scattering clusters and DV-Xα methods show that the Cl–Cl distance has to be 4 Å to match the Ge(111) lattice constant in order to built a stable Cl/Ge(111) adsorption structure [58]. The observation of Ge–Cl instead of Ge–H upon HCl treatment is explained by Lu et al. [60, 61] by invoking thermodynamics considerations rather than kinetic barriers: the bond energy of Ge–Cl is higher than Ge–H and is therefore energetically favorable [60]. Unpublished results using FTIR on Ge(100) using the same wet chemical treatment (10% HCl for 10 min) confirm that the germanium oxide and suboxides are completely removed with no evidence of Ge–H. Unfortunately, the Ge–Cl stretch mode (350–400 cm^{-1}) cannot be observed directly because it is below the current detector sensitivity.

The stability of flat Cl/Ge(111) surfaces has been studied by Bodlaki et al. [28] using SHG and XPS. Their work confirms the absence of oxide as well as suboxides and shows that the surface is unstable as reoxidation occurs after 1 h in air. (Fig. 4.29): the Ge2p core level starts exhibiting a shoulder at higher binding energy characteristic of GeO_x while the intensity of the Cl2p core level is decreasing. As observed for H/Ge, the oxide

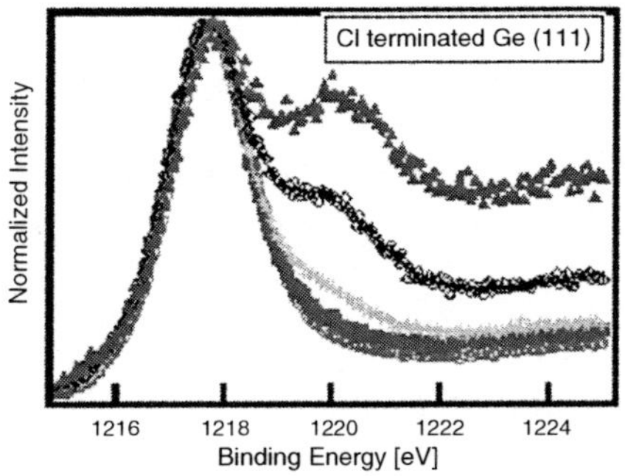

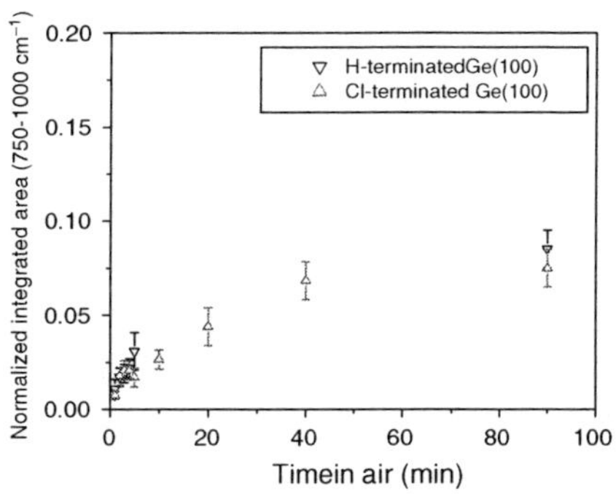

Fig. 4.29. (**a**) Ge2p core level spectra of the oxidation of Cl-terminated Ge(111) immediately after preparation (○), after 1 h (■), after 1 day (+), after 1 week (◇) after 1 month (▲) air exposure (reprint with permission from [28]) (**b**) Variation of the integrated area for the increase of GeO_x/CH_x modes as a function of the passivated Ge(100) surface exposed to air for H- and Cl-terminated Ge(100) surfaces (reprint with permission from [16])

growth is logarithmic but seems to be faster after an initial period. A recent FTIR study confirms that the Cl/Ge(100) surface is undergoing reoxidation/hydrocarbon insertion, similarly to what is observed for H/Ge(100) (Fig. 4.11). As for H/Ge(100), the Cl-terminated surface gets contaminated in air: after 1 min in air, the stretching mode related to CH_x (770 cm^{-1}) is clearly observed suggesting that hydrocarbons play a key role in the unstability Cl/Ge(100) [16]. However, in contrast to H-terminated Ge(100), the regrowth of germanium oxide is dominated by the formation of GeO_2 (instead of GeO_x), as evidenced by the appearance of two modes at 910 cm^{-1} and 830 cm^{-1} observed after 5 min in air. These results suggest that the reoxidation/hydrocarbon contamination depends on the chemical nature of the starting surface.

4.9 Organic Molecules as Passivating Agent of Germanium Surfaces

Passivation of germanium surfaces can be achieved by organic molecules as well [62, 63]. Under UHV conditions, part of the chemistry that has been developed takes advantage of the partial π bond of Ge dimers's double bond on clean Ge(100)-(2 × 1). This specific bond is actually known to mimics the double bond of alkene molecules, which allows pericyclic reactions such as [2 + 2] cycloaddition or Diels–Alder reaction (Fig. 4.30) [64]. Short alkene chains such as ethylene or cycloalkene (cyclopentene and cyclohexene) molecules undergo a [2+2] cycloaddition reaction at room temperature by bonding on Ge(100)-(2 × 1) by breaking the Ge=Ge dimer row, thus forming a Ge–C

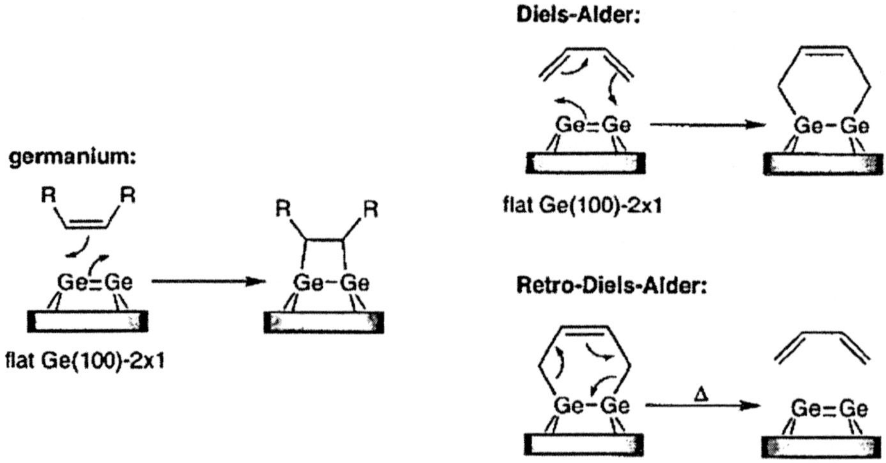

Fig. 4.30. Schematic view of (**a**) the [2 + 2] cycloaddition reaction and (**b**) the [4 + 2] Diels–Alder reaction and retro Diels–Alder reaction (reprint with permission from [63])

bond [65, 66]. In all cases, no significant molecular dissociation is observed as suggested by the absence of GeH stretching mode at ~1,980 cm^{-1}, indicating that chemisorption involves a bridge-type bonding. For cycloalkene molecules, at low coverage (~0.1 monolayer), the sticking probability is relatively small, e.g., ~10% of the Si(100)-(2 × 1) surface.

Thermal adsorption/desorption of ethylene and cycloalkenes have also been investigated. Lee et al. have demonstrated that desorption of the cycloalkene appears ~360 K [66]. However, in the case of cyclohexene, the observation of desorption of molecular hydrogen at 590 K, and of 1,3-cyclohexadiene and benzene indicates that the adsorption/desorption process is not reversible. For cyclopentene, not such effect is observed. These results suggest that competitive mechanisms, alkene π bond formation and hydrogen elimination need to be considered. In contrast, upon annealing the desorption of ethylene is observed at 390 K but this process is more complex as an undetermined structure is formed and stays stable until 425 K [65]. Other molecules such as 1,3-butadiene and 2,3-dimethyl-1,3-butadiene have been shown to react on Ge(100)-(2 × 1) at room temperature [67]. FTIR measurements show a clear signature for Diels–Alder adducts with the formation of two Ge–C bonds (Fig. 4.30a): no terminal vinylic CH_2 stretch at 3,090 cm^{-1} are observed confirming the chemisorption of the molecule. Upon annealing, the thermal reaction pathway is different from what is observed on silicon. The observation of 1,3-butadiene at 570 K suggests that a retro Diels–Alder reaction (Fig. 4.30b) is taking place [67].

Passivation with organic molecules using wet chemical treatments has also been considered, starting from either H-terminated or Cl-terminated

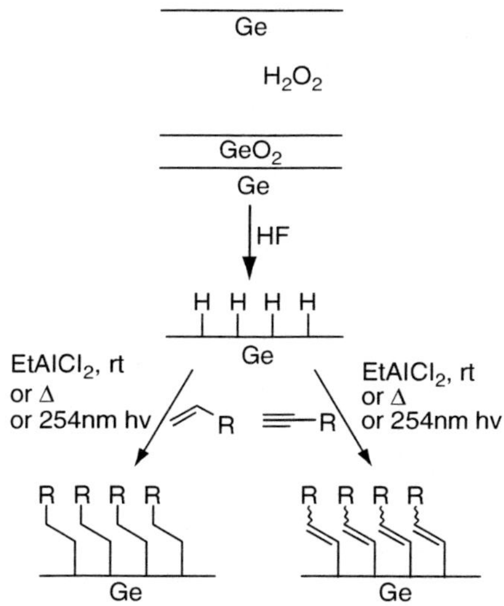

Fig. 4.31. Schematic view of the hydrogermylation process (reprint with permission from [63])

Ge surfaces. On flat Ge(100), germanium quantum dots and Ge nanowires, hydrogermylation has been studied using alkynes and alkenes chains with different chain lengths [23, 68, 69]. Choi et al. have demonstrated that hydrogermylation can take place forming a Ge–C bond using an hexane solution of $EtAlCl_2$ at room temperature, irradiating the sample in solution with UV light, or heating the solution at 200–220°C (Fig. 4.31).

The authors identify with FTIR modes related either to alkynes or alkenes molecules. The presence of a C=C stretching mode at 1,594 cm^{-1}, for example, suggests that the alkynes molecules are bound through a vinyl group. However, there is no evidence for the Ge–C stretching mode at 620 cm^{-1} [70, 71]. The chemical stability of a 1-hexadecyl layer has also been demonstrated by observing only a 20% decrease of the CH_x stretching modes intensity after washing in 25% HF. Further chemical cleaning using sonication in chloroform for 5 min, immersion in boiling water followed by boiling chloroform cannot alter the organic layer, thus confirming that these alkyl monolayers are stable. Thermally activated hydrogermylation has also been investigated on Ge nanowires [69]. The functionalization is performed on H-passivated Ge nanowires with alkenes, alkyne and diene molecules that remain in the reactor after the growth. The hydrogermylation reaction of dienes molecules on GeH nanowires is supposed to involve a [2 + 2] or [4 + 2] cycloaddition mechanism similar to what has already been observed under UHV with unsaturated bond reacting with Ge=Ge dimers. The functionalized Ge nanowires exhibit a sharp interface that remains stable. This is in contrast to the surface on non-passivated Ge nanowires that have been removed from the reactor after the growth. Interestingly, there is no oxidation on functionalized Ge nanowires after exposure to air or water confirming the good stability of the functionalization.

More recently, various groups have looked at the alkylation of hydrogen-terminated germanium surfaces through alkanethiol chains [72–74]. The dependence on concentration of alkanethiol solution dissolved in propanol as well as chains length has been investigated using FTIR spectroscopy, XPS and AFM. The results demonstrate that the thiol-terminated chains bond on the germanium surface via the sulfur group forming a hydrophobic surface. Within a first-order Langmuir isotherm model, the adsorption kinetics of the thiol chains on H-terminated Ge surfaces exhibit a two-step mechanism: fast adsorption with strongly entangled chains during the first stage followed by a slower adsorption leading to a denser and more ordered alkyl monolayer [72]. Similar kinetics have been previously observed on gold or silicon substrates. In general, the longer the alkyl chains, the fastest the reaction: this chains length dependence is attributed to the increasing chain-to-chain interaction. The stability of these alkyl chains has been checked as a function of the temperature: the monolayer is found to be stable until 450 K. Above 550 K, the monolayer is completely removed as the molecule desorbs primarily by Ge–S bond cleavage [73]. Upon chemical treatment, the alkyl layer shows that the chemical Ge–S bond is not as strong as the Ge–C one.

Organometallic reagents have been used on Cl-terminated Ge surfaces for alkylation [75]. Following the reaction process developed for Cl-terminated silicon surfaces, Grignard reagents of various alkyl length chains have been investigated starting with an atomically flat Cl/Ge(111) surface obtained by a 10 min dip in diluted HCl solution. After 6 h to 7 days of reaction (depending of the chain length), FTIR and XPS data show the formation of a densely packed alkyl monolayer on the Ge surface. This layer appears to be stable in air after 5 days or after various chemical treatments, such as boiling water, boiling chloroform or diluted HCl suggesting that the Ge–C bond is strong.

4.10 Conclusions

In general, cleaning and passivation of germanium surfaces is less well developed and more complex than for silicon surfaces, except in an ultra-high vacuum environment. Part of the problem comes from the poor stability of germanium oxides, with many of them soluble in water, and part comes from the poor stability of the Ge–H surface in air. Some of these problems may turn out to be advantages for high-κ dielectrics growth. For instance, the poor oxide stability may make it easier to thermally remove GeO_x in an ALD reactor prior to growth and to minimize or eliminate an interfacial GeO_2 at the Ge/high-κ

dielectrics interface. However, the preparation methods, particularly the wet chemistry methods have to be optimized in a manner that is often radically different from what is done for Si. For instance, the standard SC-2 clean, based on HCl and peroxide etching, is not appropriate for Ge since its oxide is soluble in HCl. Even hydrogen passivation is much more delicate than for Si because the chemical history (cleaning steps) and the concentration of HF used are critical to the degree of H-passivation. Moreover, the stability in air of the H-terminated Ge surfaces is much lower than that of H-terminated Si surfaces, making the HF last step much less relevant for the microelectronic industry.

A key to further understanding of Ge passivation is the ability to probe the chemical nature of the surface as a function of processing parameters. A powerful combination is the use of ex situ FTIR spectroscopy to characterize the chemical bonding (particularly for H-containing species) at the surface after wet chemistry and ex situ XPS to quantify the concentration of important chemical elements such as C, O, N, Cl, and S. For XPS, care has to be taken during the introduction of Ge samples in the UHV system, as this step often leads to contamination (e.g. by hydrocarbons). The great advantage of FTIR is that it does not require a load-lock procedure and samples can therefore be placed in position rapidly without contamination. In this respect, FTIR is much more convenient than HREELS to monitor the surface chemistry, except for modes that are lower than $400\,\mathrm{cm}^{-1}$. FTIR is also readily amenable to study in situ surface chemistry during gas phase oxidation, nitridation or high-κ dielectrics growth on Ge. The transmission geometry discussed in Sect. 4.2.2 makes it possible to detect most films containing O, N, and C. Another important parameter to control is the surface morphology prior and during high-κ dielectrics growth. This is best done with imaging techniques, such as TEM, STM and AFM that are often used ex situ, but could also benefit from in situ optical methods such as ellipsometry. Ultimately, a correlation of such chemical and structural measurements with the electrical properties of the surfaces and interfaces is needed.

Considering the relatively low research activity performed on Ge surface passivation compared to what has been done on Si, there are reasons to believe that progress will be made in the future. However, the passivation of Ge and other high mobility substrates (GaAs and InP) may remain a challenge for many years to come and a real limitation as to their use for mainstream microelectronics, i.e., as a true replacement for Si-based technologies.

Summary

	Ultra high vacuum		
Techniques used	XPS	FTIR/HREELS	STM
Oxide-passivation (hexagonal phase)	**Ge(100)/Ge(111)** [10] Formation of +1, +2, +3, +4 oxidation states **Ge3d:** $E_b + 0.85$ eV/Ge-O bond Upon annealing, +2 oxidation state	**GeO_2**: [8] LO: 930 cm^{-1} TO: 857 cm^{-1}	
H-passivation		**Ge(100)-2 × 1** **GeH** [18] // comp: 1,979 cm^{-1} ⊥ comp: 1,991 cm^{-1}	**Ge(100)-2 × 1** [20, 22] Monohydride at RT 2 × 1 reconstruction Stable *Prolonged* H_2 *exposure* Dihydride at RT 3 × 1 reconstruction Unstable $T > 400\,K$: Etching of Ge Formation of V-groove shaped etch pits
Nitridation/oxynitridation	No nitridation observed up to 870 K NH_3 sticking coefficient on Ge: 3.10^{-6} [45]	**Ge(100)-2 × 1** [38, 42] *Ammonia gas* $T <$ RT: **NH_3 (molecular ads.):** 3,270 cm^{-1} (sym stretch) 1,570 cm^{-1} (asym bend) 1,150 cm^{-1} (sym bend) $T \sim 500$ K: No NH_3 dissociation, no formation of nitride layer	

S-passivation	**Ge(100)-2 × 1** [55] *Elementary beam of sulfur:* 1 × 1 reconstruction monosulfide **Ge3d:** $E_b + 0.665\,eV$ Inert in UHV H_2S *gas*: [51] Dissociation into SH, S and H Coverage: 0.25 ML at RT, 0.5 ML at 553 K	
Cl-passivation	**Ge(111)-2 × 8** [57] Cl_2 *gas*: Cl adsorbed in an atop configuration 2 × 8 reconstruction	
Organic molecules		**Ge(100)-2 × 1** [65,66] *Alkene chains:* [2 + 2] cycloaddition, formation of GeC *1,3 butadiene, 2,3-dimethyl 1,3 butadiene* Diels Alder adduct formed at RT Formation of GeC bond Upon annealing (570 K), retro Diels–Alder reaction

	Wet chemical process		
Techniques used	XPS	FTIR	Other
Oxide-passivation (hexagonal phase)	**Ge(100)/Ge(111)** [12,14] +2 *oxidation state*: **Ge3d**: $E_b + 1$ eV **Ge2p:** $E_b + 1.4$ eV +4 *oxidation state*: **Ge3d:** $E_b + 3$ eV **Ge2p:** $E_b + 3$ eV	**Ge(100)** [15,16] **GeO_2**: LO: 940 cm^{-1} TO: 830 cm^{-1} **Ge=O:** 980 cm^{-1} **GeO_x**: 850–930 cm^{-1}	
H-passivation	**Ge(100)/Ge(111)** [24] Cyclic H_2O/HF: No oxidation visible **Ge(100)** [12,14] $H_2O/H_2O_2/HF$: No oxidation visible *GeC* [15] **Ge3d:** E_b: +1.6 eV	**Ge(100)** [15,16,63] $H_2O/H_2O_2/HF$: **GeH:** 1,987 cm^{-1} **GeH_2**: 2,020 cm^{-1} **GeH_3**: 2,060 cm^{-1} **CH_x**: 790, 1,440, 2,800–2,950 cm^{-1} Stability: Air: unstable (few min) **GeO_x**: 880 cm^{-1} **CH_x**: 770 cm^{-1} N_2: stable (few hours)	*SHG* [28] **Ge(100)** Cyclic H_2O/HF: oxidation visible ~5 Å

		Ge(100)/Ge(111) *Electrochemistry:* **GeH:** 1,970 cm^{-1} (→ 1,950 cm^{-1} if potential too negative: insertion of H in Ge bulk) $\mathbf{GeH_2}$: 2,020 cm^{-1}	
Nitridation/oxynitridation		***Ge(100)*** [40] *Ammonia gas*: $\mathbf{Ge\text{-}NH_2}$ 1,550–1,600 cm^{-1} **GeH:** 1,920 cm^{-1} Ammonia plasma $\mathbf{Ge_3N_4}$: **730/910** cm^{-1} (β-phase) $\mathbf{Ge_3N_4}$: 780 cm^{-1} (α-phase)	*Ellipsometry/optical micrography* [39] *Ammonia gas*: Smooth and uniform nitride layer formed at 870–930 K but oxide present → oxynitride *Nitrogen gas* No nitridation
S-passivation	**Ge(100)** [54] $(NH_4)_2S$: 1 × 1 reconstruction Monosulfide with traces of oxygen and carbon **Ge3d:** $E_b + 0.665$ eV Sulfur released at 460 K, removed at 650–750 K		*SHG* [28] **Ge(111)** $(NH_4)_2S$: Monosulfide No oxidation Stable in air

	Wet chemical process		
Techniques used	XPS	FTIR	Other
Cl-passivation	**Ge(111)** [27,60] *HCl:* Flat monochloride **Ge2p:** $E_b + 1.87\,eV$ Stability: Air: unstable (few min) $+2 \rightarrow +4$ oxidation state	**Ge(100)** [16] *HCl:* No oxidation, no GeH Stability: Air: unstable (few min) $\mathbf{GeO_2}$: 830 and 940 cm^{-1} $\mathbf{CH_x}$: 770 cm^{-1}	*EXAF* [61] Ge–Cl: 2.17 nm
Organic molecules		**H/Ge(100)** [23,68–72] *Alkynes or alkenes solution:* At RT with EtAlCl2 or UV irradiation or annealing Alkylation of Ge Stability: Stable to boiling water, chloroform, boiling chloroform 20% loss of the alkyl layer in HF *Thiol-terminated alkyl chains:* Bonded through sulfur atom Stability: Stable until 450 K T > 550 K, alkyl chains removed **Cl/Ge(100)** [75] *Grignard reagent:* Densely packed layer bonded through Ge–C Stability: Stable in air and in various chemical treatment	

Acknowlegment

The authors acknowledge the support of the National Science Foundation (CHE-0415652).

References

1. S.M. Sze, *Physics of Semiconductor Devices*, edited by Wiley, New York, (1987).
2. J.A. Dillon Jr., H.E. Farnsworth, J. Appl. Phys., 28, 174–184, (1957)
3. H.E. Farnsworth, R.E. Schlier, T.H. George, R.M. Burger, J. Appl. Phys., 29, 1150–1161, (1958)
4. L.H. Chan, E.I. Altman, Y. Liang, J. Vac. Sci. Technol. A, 19, 976–981, (2001)
5. T. Fukuda, T. Ogino, Phys. Rev. B, 56, 13190–13193, (1997)
6. A.W. Laubengayer, D.S. Morton, J. Am. Chem. Soc., 54, 2303–2320, (1932)
7. J.F. Scott, Phys. Rev. B, 1, 3488–3493, (1970)
8. F.L. Galeener, G. Lucovsky, Phys. Rev. Lett., 37, 1474–1478, (1976)
9. C.M. Garner, I. Lindau, J.N. Miller, P. Pianetta, W.E. Spicer, J. Vac. Sci. Technol., 14, 372–375, (1977)
10. D. Schmeisser, R.D. Schnell, A. Bogen, F.J. Himpsel, D. Rieger, G. Landgren, J.F. Morar, Surf. Sci., 172, 455–465, (1986)
11. L. Surnev, M. Tikhov, Surf. Sci., 138, 40–50, (1984)
12. K. Prabhakaran, T. Ogino, Surf. Sci., 325, 263–271, (1995)
13. K. Prabhakaran, F. Maeda, Y. Watanabe, T. Ogino, Appl. Phys. Lett., 76, 2244–2246, (2000)
14. X.-J. Zhang, G. Xue, A. Agarwal, R. Tsu, M.-A. Hasan, J. E. Greene, A. Rockett, J. Vac. Sci. Technol. A, 11, 2553–2561, (1993)
15. S. Rivillon, Y.J. Chabal, F. Amy, A. Kahn, Appl. Phys. Lett., 87, 253101, (2005)
16. S. Rivillon Amy, Y.J. Chabal, F. Amy, A. Kahn, C. Krugg, P. Kirsch, *Wet Chemical Cleaning of Germanium Surfaces for Growth of High-κ Dielectrics*, MRS, 2006
17. G.E. Becker, G.W. Gobeli, J. Chem. Phys., 38, 1, (1963)
18. Y.J. Chabal, Surf. Sci., 168, 594–608, (1986)
19. J.Y. Lee, J.Y. Maeng, A. Kim, Y.E. Cho, S. Kim, J. Chem. Phys., 118, 1929–1936, (2003)
20. J.Y. Lee, S.J. Jung, J.Y. Maeng, Y.E. Cho, S. Kim, S. K. Jo, Appl. Phys. Lett., 84, 5028–5030, (2004)
21. Y.-J. Zheng, P.F. Ma, J.R. Engstrom, J. Appl. Phys., 90, 3614–3622, (2001)
22. J.Y. Maeng, J.Y. Lee, Y.E. Cho, S. Kim, Appl. Phys. Lett., 81, 3555–3557, (2002)
23. K. Choi, J.M. Buriak, Langmuir, 16, 7737–7741, (2000)
24. T. Deegan, G. Hughes, Appl. Surf. Sci., 123/124, 66–70, (1998)
25. B. Onsia, T. Conard, S. De Gendt, M. Heyns, I. Hoflijk, P. Mertens, P. Meuris, G. Raskin, S. Sioncke, I. Teerlinck, A. Theuwis, J. Van Steenbergen, C. Vinckier, Solid State Phenom., 103–104, 23–26, (2005)
26. H. Okumura, T. Akane, S. Matsumoto, Appl. Surf. Sci., 125, 125–128, (1998)
27. Y.F. Lu, Z.H. Mai, W.D. Song, W.K. Chim, Appl. Phys. A, 70, 403–406, (2000)
28. D. Bodlaki, H. Yamamoto, D.H. Waldeck, E. Borguet, Surf. Sci., 543, 63–74, (2003)

29. R.J. Archer, J. Electrochem. Soc., 104, 619–625, (1957)
30. F. Maroun, F. Ozanam, J.-N. Chazalviel, Chem. Phys. Lett., 292, 493–499, (1998)
31. F. Maroun, F. Ozanam, J.-N. Chazalviel, Surf. Sci., 427–428, 184–189, (1999)
32. F. Maroun, F. Ozanam, J.-N. Chazalviel, J. Phys. Chem. B, 103, 5280–5288, (1999)
33. F. Maroun, J.-N. Chazalviel, F. Ozanam, D. Lincot, J. Electroanal. Chem., 549, 161–163, (2003)
34. K.C. Mandal, F. Ozanam, J.-N. Chazalviel, Appl. Phys. Lett., 57, 2788–2790, (1990)
35. H.C. Choi, J.M. Buriak, Chem. Comm., 1669–1670, (2000)
36. S. Miyazaki, K. Sakamoto, K. Shiba, M. Hirose, Thin Solid Films, 255, 99–102, (1995)
37. L. Chun-rong, S. Zhao-qi, X. Jing-zhou, X. Zheng, J. Phys. D: Appl. Phys., 24, 2215–2216, (1991)
38. M. Sanders, J.H. Craig Jr., App. Surf. Sci., 171, 283–287, (2001)
39. Q. Hua, J. Rosenberg, J. Ye, E.S. Yang, J. Appl. Phys., 53, 8969–8973, (1982)
40. A. Metcalfe, S.U. Shankar, J.C.S. Faraday I, 76, 489–496, (1980)
41. T. Maeda, T. Yasuda, M. Nishizawa, N. Miyata, Y. Morita, S. Takagi, Appl. Phys. Lett., 85, 3181–3183, (2004)
42. C. Tindall, J.C. Hemminger, Surf. Sci., 330, 67–74, (1995)
43. R. Miotto, R.H. Miwa, A.C. Ferraz, Phys. Rev. B, 68, 115436, (2003)
44. W. Ranke, Surf. Sci., 342, 281–292, (1995)
45. D. Aubel, M. Diani, J.L. Bischoff, D. Bolmont, L. Kubler, J. Vac. Sci. Technol. B, 12, 2699–2704, (1994)
46. S.M. Cohen, Y.L. Yang, E. Rouchouze, T. Jin, J. D'Evelyn, J. Vac. Sci. Technol. A, 10, 2166, (1992)
47. Y. Igarashi, K. Kurumada, T. Niimi, Jap. J. Appl. Phys., 7, 300–301, (1968)
48. D.C. Paine, J.J. Rosenberg, S.C. Martin, D. Luo, M. Kawasaki, Appl. Phys. Lett., 57, 1443–1445, (1990)
49. H. Kim, P.C. McIntyre, C.O. Chui, K.C. Saraswat, M.-H. Cho, Appl. Phys. Lett., 85, 2902–2904, (2004)
50. A. Dimoulas, G. Mavrou, G. Vellianitis, E. Evangelou, N. Boukos, M. Houssa, M. Caymax, Appl. Surf. Sci., 86, 032908, (2005)
51. L.M. Nelen, K. Fuller, C.M. Greenlief, Appl. Surf. Sci., 150, 65–72, (1999)
52. P. Kruger, J. Pollmann, Phys. Rev. Lett., 64, 1808–1811, (1990)
53. M. Rohlfing, P. Kruger, J. Pollmann, Phys. Rev. B, 54, 13759–13766, (1996)
54. G.W. Anderson, M.C. Hanf, P.R. Norton, Z.H. Lu, M.J. Graham, Appl. Phys. Lett., 66, 1123–1125, (1995)
55. T. Weser, A. Bogen, B. Konrad, R.D. Schnell, C.A. Schug, W. Steinmann, Phys. Rev. B, 35, 8184–8188, (1987)
56. H.J. Kuhr, W. Ranke, Surf. Sci., 189/190, 420–425, (1987)
57. P.H. Citrin, J.E. Rowe, P. Eisenberger, Phys. Rev. B, 28, 2299–2301, (1983)
58. S. Cao, J.C. Tang, S.-L. Shen, J. Phys.: Condens. Matter, 15, 5261–5268, (2003)
59. M.W.C. Dharma-wardana, M.Z. Zgierski, D. Ritchie, J.G. Ping, H. Ruda, Phys. Rev. B, 59, 15766–15771, (1999)
60. Z.H. Lu, Appl. Phys. Lett., 68, 520–522, (1996)
61. Z.H. Lu, T. Tyliszczak, A.P. Hitchcock, M.W.C. Dharma-wardana, Surf. Sci., 442, L948–L952, (1999)

62. P.W. Loscutoff, S.F. Bent, Annu. Rev. Phys. Chem., 57, 467–495, (2006)
63. J.M. Buriak, Chem. Rev., 102, 1271–1308, (2002)
64. M.A. Filler, S.F. Bent, Prog. Surf. Sci., 73, 1–56, (2003)
65. P. Lal, A.V. Teplyakov, Y. Noah, M.J. Kong, G.T. Wang, S.F. Bent, J. Chem. Phys., 110, 10545–10553, (1999)
66. S.W. Lee, J.S. Hovis, S.K. Coulter, R.J. Hamers, C.M. Greenlief, Surf. Sci., 462, 6–18, (2000)
67. A.V. Teplyakov, P. Lal, Y. Noah, S.F. Bent, J. Am. Chem. Soc., 120, 7377–7378, (1998)
68. E. Fok, M. Shih, A. Meldrum, J.G.C. VEinot, Chem. Comm., 4, 386–387, (2004)
69. T. Hanrath, B.A. Korgel, J. Am. Chem. Soc., 126, 15466–15472, (2004)
70. A. Mahajan, B.K. Kellerman, J.M. Heitzinger, S. Banerjee, A. Tasch, J.M. White, J.G. Ekerdt, J. Vac. Sci. Technol. A, 13, 1461–1468, (1995)
71. J. Vilcarromero, F.C. Marques, F.L. Freire Jr., J. Appl. Phys., 84, 174–180, (1998)
72. M.R. Kosuri, R. Cone, Q. Li, S.M. Han, Langmuir, 20, 835–840, (2004)
73. S.M. Han, W.R. Ashurst, C. Carraro, R. Maboudian, J. Am. Chem. Soc., 123, 2422–2425, (2001)
74. D. Wang, Y.-L. Chang, Z. Liu, H. Dai, J. Am. Chem. Soc., 127, 11871–11875, (2005)
75. J. He, Z.-H. Lu, S.A. Mitchell, D.D.M. Wayner, J. Am. Chem. Soc., 120, 2660–2661, (1998)

5

Interface Engineering for High-k Ge MOSFETs

S.J. Lee, C. Zhu, and D.L. Kwong

Summary. Recent progress and current understanding of high-k/Ge interface and its impact on device performances are summarized. After reviewing the properties of Ge oxide and high-k/Ge interface, several reported Ge surface passivation techniques and improved characteristics of high-k/Ge system using surface passivations are discussed.

5.1 Introduction

A success of Si CMOS technology for decades is partly due to the existence of stable oxide and high-quality SiO_2/Si interface, which is in direct contact with CMOS channel region and permits low-defect charge density ($\sim 10^{10}\,cm^{-2}$) and interface state density ($\sim 10^{10}\,cm^{-2}eV^{-1}$), providing high performances of CMOSFET and reliability characteristics.

Renewed interest in Ge as a material of choice for future electronics is mainly due to its attractive properties, including the lower effective mass and higher mobility of carriers for increasing drive current capability of MOSFETs, and smaller optical band gap for broader absorption wavelength spectrum. In addition, the recent technological progress in high-k gate dielectric has increased the possibility of high-k/Ge system to be implemented for future gate stack. However, several years' studies on high-k/Si system indicate that tremendous efforts should be paid on the high-k interface engineering in order to achieve successful scaling of EOT with low-leakage current, good subthreshold characteristics, high carrier mobility, and acceptable reliability.

In this chapter, a review of current work and literature in the area of high-k/Ge interface will be given. First, the physical and chemical properties of Ge oxide will be discussed, followed by the review of recent experimental reports on the high-k/Ge interface and its difference against high-k/Si interface. In addition, various reported Ge surface passivation techniques, including surface nitridation, Si passivation, plasma-PH_3 passivation, and AlN passivation, will be discussed with their physical and electrical properties.

5.2 Germanium Oxide and High-k/Ge Interface

Since the first MOSFET was introduced [1], Ge had been considered as one of the most important semiconductors. A number of studies had been previously carried out to investigate the Ge MOSFETs using various gate dielectrics; pyrolytic SiO_2 [2], CVD SiO_2 [3], high-pressure oxidation of germanium [4], germanium oxynitride [5–7], Ge_3N_4 [8]. Although most of these previous works demonstrated higher carrier mobilities than Si-MOSFETs, the poor interface quality in Ge MOSFETs had been observed with high density of interface traps or midgap bulk semiconductor traps, which result in higher leakage current and poor subthreshold characteristics. Lack of sufficiently stable oxide of Ge, unlike Si, had been considered as one of the main reasons explaining these results.

Ge is known to oxidize in various environments and form an oxide layer consisting of a mixture of mainly monoxide (GeO) and dioxide (GeO_2) species [9,10]. Experimental thermal decomposition studies of Ge and Si oxides showed that annealing causes the transformation of GeO_2 to GeO on the surface and finally desorbs from the surface at $\sim$425°C [11]. It was also reported that thermal desorption of Ge oxides and oxidation of Ge takes place successively, which results in the loss of Ge from the surface [12]. Moreover, in contrast to chemical stability of SiO_2 that can be etched only in hydrofluoric acid, both GeO and GeO_2 are amphoteric, dissolving both in dilute acidic and alkaline solutions, even in warm water [9]. Due to these significantly different physical and chemical properties of Ge oxide, Si has been the main material in CMOS devices for decades.

Recent progresses in development of high-k dielectric have reopened the possibility of realizing CMOS devices in Ge. Studies on high-k dielectric in Si devices showed that the unavoidable and uncontrollable formation of a thin low-k interfacial oxide layer between high-k and substrate obstructs the further scaling of EOT and affects the device performances [13, 14]. However, above mentioned relatively unstable nature of Ge oxide implies that the removal of interfacial oxide can be readily achieved more easily than high-k/Si system. This potential advantage of high-k/Ge system has been demonstrated by several experimental results [15–22]. TEM pictures for MOCVD-HfO_2 layers deposited on different starting surfaces, in Fig. 5.1, clearly indicate that a thinner interfacial layer is formed on Ge than on Si during the HfO_2 deposition [15]. Similar results have been observed on chemical oxide passivated Ge substrate by using various high-k deposition techniques; ALD-ZrO_2 [16], ALD-HfO_2 [17–19], PVD-ZrO_2 [20–22], and PVD-HfO_2 [23]. Detailed investigations by XPS and MEIS indicate that the chemical composition of interfacial layer is GeO_2 and its thickness is about 0.3 nm [19]. Consistent results from different groups regardless of high-k deposition methods suggest that this can be explained by the intrinsic properties related to thermodynamic instability of Ge oxide, inhomogeneous dissociation of Ge oxide. Obtained EOT of $\sim$5 Å from PVD-ZrO_2 [20] also supports the absence of GeO_2 ($k = 3.0 \sim 3.8$ [24])

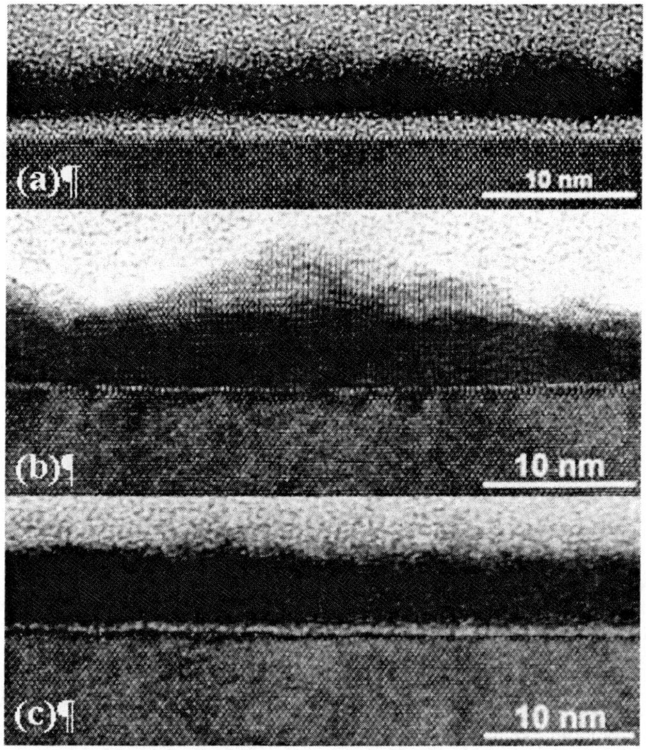

Fig. 5.1. TEM pictures of a HfO_2 layer deposited on (**a**) Si HF-last, (**b**) Ge HF-last, and (**c**) Ge HF last + NH_3 anneal starting surface. Note the significant thinner interfacial layer for the Ge substrate [15]

at the interface. From the electrical measurements of high-k on Ge with different surface cleaning prior to high-k deposition, it is also reported that better performances (EOT, interface states, and C–V hysteresis) can be achieved by eliminating the poor quality and unstable oxide interlayer [20]. Interesting comparison between ZrO_2 and HfO_2 on Ge was performed by Kamata et al. [25]. Results show that ZrO_2 demonstrated the thinner interfacial layer, higher dielectric constant with intermixing with Ge, and improved thermal stability than HfO_2. As a result, a ZrO_2/Ge showed better CET-Jg characteristics, as shown in Fig. 5.2. This observation implies that different material properties and surface chemistries at interface between Ge and Si may create new issues and requirements for high-k/Ge system, which have not been studied for high-k/Si system.

In spite of this EOT scaling advantage of high-k/Ge system over high-k/Si, other drawbacks have also been reported. First, unlike the situation observed on similarly passivated Si substrate, the greater roughness was observed at the interface between ALD-HfO_2 film and the interfacial layer [17], and at the top HfO_2 surface grown on the HF-last Ge [15]. In addition, the

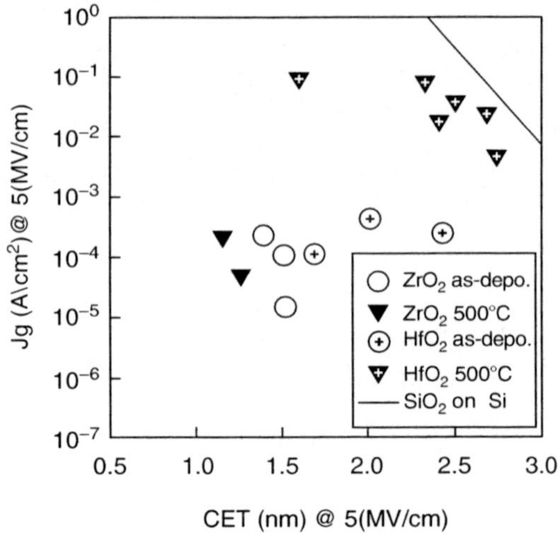

Fig. 5.2. CET-Jg characteristics of ZrO_2 and HfO_2 before and after 500°C annealing in N_2 [25]

local epitaxial growth of high-k on Ge was observed from MOCVD-HfO_2 [15], ALD-ZrO_2 [16], and ALD-HfO_2 [18, 19]. Figure 5.3 shows the cross-sectional and plan-view image of ALD-ZrO_2 grown on HF-vapor-cleaned Ge substrate, exhibiting local epitaxial growth of ZrO_2 on Ge without a distinct interfacial layer. These results, which are not typically observed on Si substrate, can be also attributed to the unstable nature of GeO_2 which is converted to GeO in contact with Ge and finally sublimes at low temperature [11], inducing a partial reaction between HfO_2 and Ge oxide [17]. In [19], the thickness-dependent local epitaxial crystallization was explored, showing that partial crystallization happens for HfO_2 thicker that 9 nm. It is believed that the presence of a small amount of O or other impurities hinder the epitaxial alignment between high-k film and substrate. The case of the high-k/Si system, and the effects of Ge surface passivation with incorporation of N, discussed in following section, are also consistent with this explanation [15, 17, 18]. This epitaxial growth of high-k on Ge induces the large areal density of interfacial dislocation ($\sim 7 \times 10^{12}\,\text{cm}^{-2}$ [16]) due to the lattice mismatch or intrinsic differences in bonding coordination across the chemically abrupt high-k/Ge interface.

The interface quality of high-k/Ge and its impacts on electrical performances of devices are of great concern. It has been reported that C–V measurements of MOS capacitors with high-k on Ge substrate generally exhibit abnormal C–V characteristics [16, 18, 20]. Figure 5.4 shows the C–V curves of MOS capacitor made with Pt/HfO_2/Ge stack, measured at different frequencies. In contrast to Si devices, the low-frequency behavior of the high frequency C–V is clearly observed with a high ac inversion capacitance at high frequencies, which indicates the existence of high density of interface or bulk traps in high-k/Ge system. A large hysteresis (>100 mV) in C–V when swept both

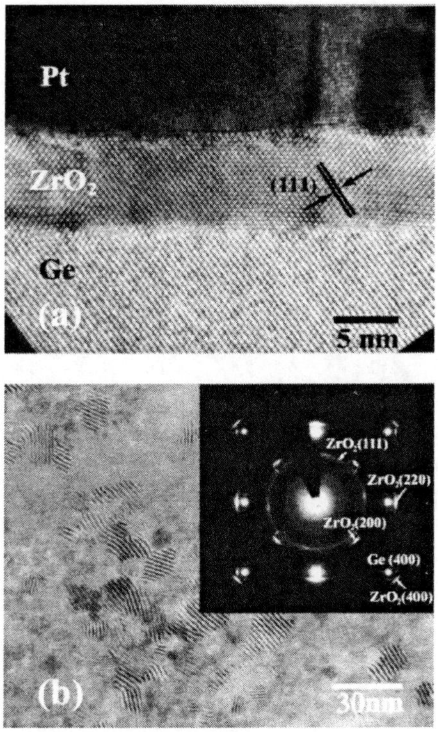

Fig. 5.3. (**a**) Cross-sectional high-resolution-TEM micrograph of Pt/55 Å ZrO_2/Ge (100) along the ⟨110⟩ zone axis and (**b**) bright-field plan-view image of 55 Å ZrO_2/Ge (100). *Inset* of figure shows selected area electron diffraction pattern and corresponding epitaxial relationship [16]

from accumulation to inversion and from inversion to accumulation directions has been observed [17, 18, 20], also indicating significant density of interface states. The calculated density of interface states of high-k/Ge systems is in the range of $10^{12} \sim 10^{13}\,eV^{-1}\,cm^{-2}$ [16,26,27]. Although the microscopic reason and mechanism for such traps are not understood yet, it is believed to be attributed to the poor quality of bulk Ge starting material or to insufficient surface passivation which creates high density of interface traps or midgap bulk semiconductor traps.

The degradation of interface quality of ALD-HfO_2/Ge devices was explained by the diffusion of Hf into the interfacial oxide/Ge substrate [17]. Figure 5.5 shows the MEIS spectra and fitted curves for HfO_2 on chemical oxide and nitride passivated Ge substrate. A distinct shoulder detected for HfO_2 on chemical oxide indicates the Hf diffusion into underlying layer, which causes the poor interface quality. C–V characteristics similar to MIM structures were also reported for ALD-HfO_2 on chemical oxide passivated Ge, implying a formation of Hf–germanide bonds at the interface [18]. On the other hand, in [15], the indiffusion of Ge into the MOCVD-HfO_2 film was observed

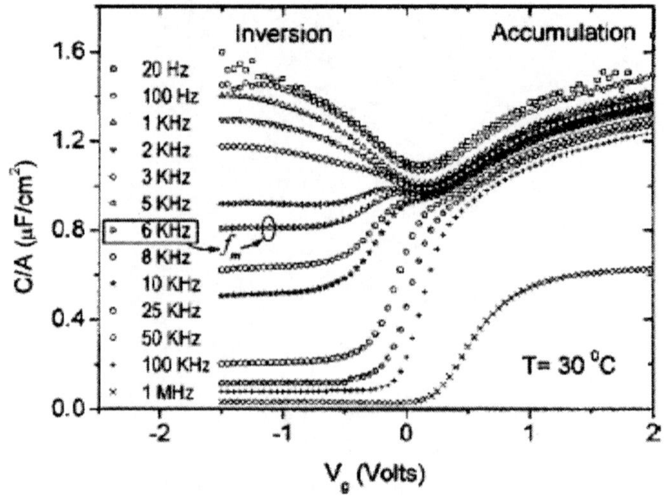

Fig. 5.4. Room temperature C–V characteristics of an n type Pt–HfO_2/Ge MIS measured at different frequencies from 20 to 1 MHz. The transition frequency fm ~6 kHz defined as the frequency at which the inversion capacitance is midway between low and high values, marks the transition from a low- to high-frequency behavior [26]

(Fig. 5.6) and considered as a source of degradation of electrical performances. TOF SIMS profile in Fig. 5.6 shows that severe indiffusion of Ge is observed in HfO_2 deposited at 485°C and after PDA at 550°C. However, a different result was reported for ALD-HfO_2 in [19], where no significant concentration of Ge in the ALD-HfO_2 deposited at 300°C was detected by TOF SIMS. It seems that the interdiffusion and electrical properties of high-k/Ge system are sensitive to high-k deposition and PDA temperature. Thermal stability of ALD-HfO_2/Ge stack was investigated by annealing in vacuum [18]. Figure 5.7 shows MEIS backscattering spectra in Hf, Ge, and O for as-deposited and annealed HfO_2 on wet chemical treated Ge substrate, indicating the HfO_2/Ge stack is stable up to 780°C in terms of HfO_2 film decomposition, interface reaction, or interface regrowth.

The abnormal low-frequency behavior of high-frequency inversion C–V for high-k/Ge was also explained by the high-intrinsic carrier concentration n_i of Ge [26–28]. Due to lower energy band gap and higher intrinsic carrier concentration, Ge is expected to show the higher density of minority carriers than Si. It was shown that the ac conductance in inversion is thermally activated varying proportional to n_i or n_i^2 depending on the temperature range and the minority carrier response time in Ge is orders of magnitude shorter than in Si depending inversely proportional to n_i around room temperature, which could explain the observed low-frequency behavior [26–28].

Energy band alignment at the high-k/Ge interface is of another concern, since the sufficient barriers for electron and hole are needed to suppress the tunneling leakage current. By using XPS, the valence band offset of ZrO_2/Si,

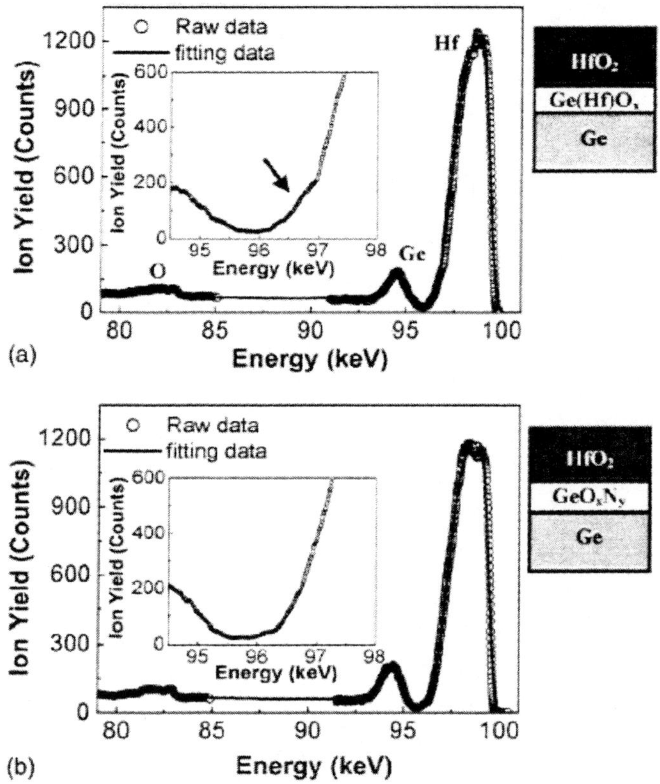

Fig. 5.5. MEIS spectra and fitted curves, and the corresponding model structure of (**a**) HfO_2 on Ge substrate with chemical oxide and (**b**) HfO_2 on Ge substrate nitrided at 600°C for 1 min in NH_3 ambient after cyclic HF stripping [17]

$ZrO_2/Si_{0.75}Ge_{0.25}$, and ZrO_2/Ge interfaces are determined to be 2.95, 3.13, 3.36 eV, respectively, while the conduction band offsets are found to be the same value of 1.76 eV [29]. The HfO_2/Ge interface band diagram was determined using internal photoemission of electrons and holes from Ge into the HfO_2 [30]. Results show that the offsets of conduction and valence band at the interface are ~2.0 and ~3.0 eV, respectively, and PDA at 650°C results in ~1 eV reduction of valence band offset attributed to the growth of GeO_2 interlayer. These results suggest that the barrier height for high-k/Ge is comparable to that of high-k/Si, enough to obtain low-leakage current using high-k.

The excellent leakage current characteristics of ZrO_2/Ge were obtained in spite of a large number of interface defects and grain boundaries existing in ZrO_2 [16]. In addition, leakage current comparison with different substrates in [17] shows, although there are significant differences in C–V and interface qualities, comparable leakage currents were observed from ALD-HfO_2 films deposited on chemical oxide terminated Si and Ge, and nitrided Ge (Fig. 5.8). However, different results were reported from MOCVD-HfO_2 in reference [15],

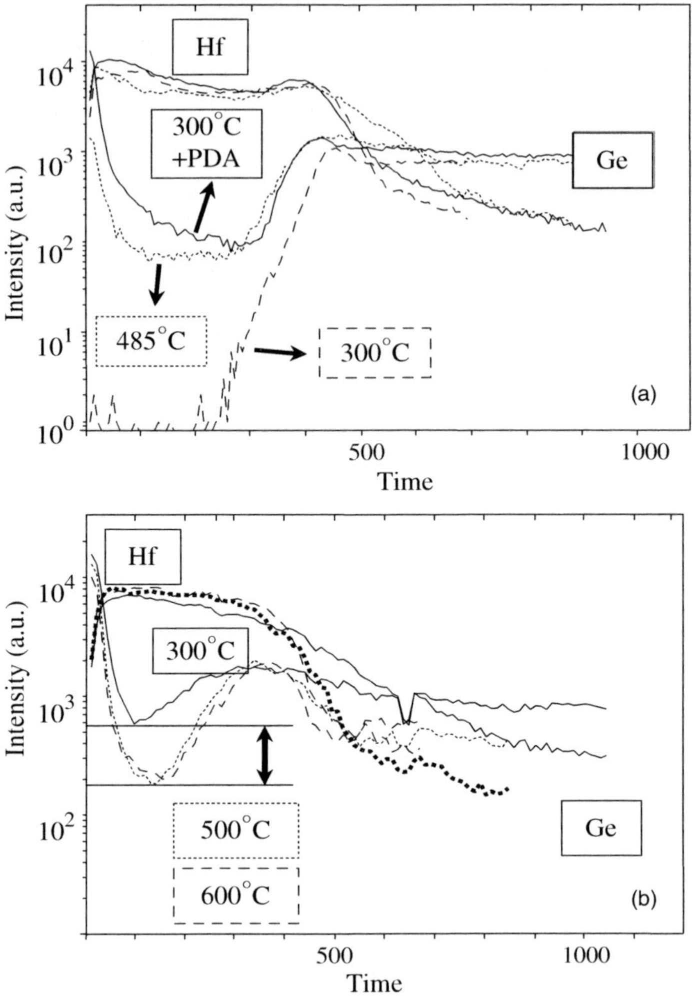

Fig. 5.6. TOF SIMS depth profiles for (**a**) HfO_2 layer deposited at 300 (with or without PDA) and 485°C on an HF-last starting surface, (**b**) HfO_2 layers deposited on an HF-last NH_3 anneal (prior to deposition) starting surface. Ge indiffusion is clearly temperature related and can be reduced by a surface pretreatment [15]

where the MOCVD-HfO_2 on HF-last Ge showed several orders of magnitude higher leakage current than on nitrided Ge.

5.3 Surface Nitridation

The direct deposition of high-k on DHF-cleaned germanium substrates has been demonstrated with poor interface properties, which possibly results from

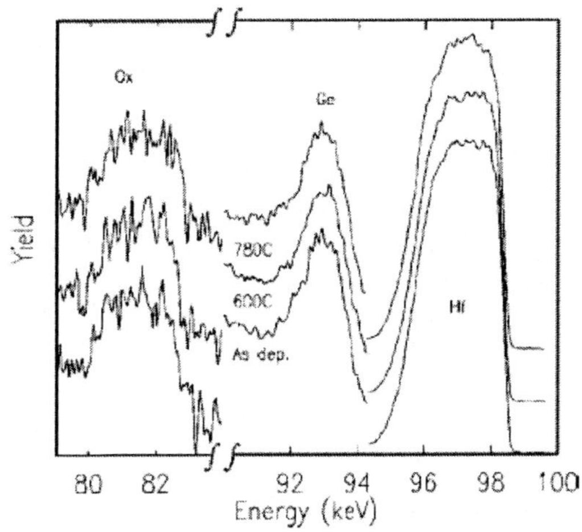

Fig. 5.7. MEIS backscattering spectra in the hafnium, germanium, and oxygen regions for as-deposited (at 300°C) 5 nm HfO_2 films and stacks after in situ annealing in vacuum at 600 and 780°C. Prior to HfO_2 deposition, Ge surface was prepared by wet chemical treatment [18]

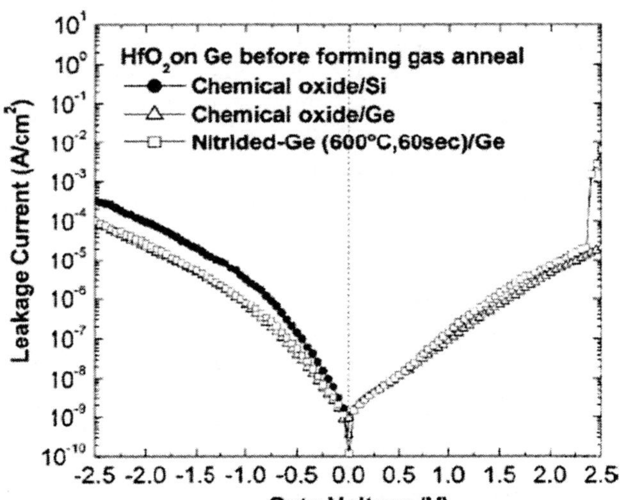

Fig. 5.8. *J–V* characteristics of $Pt/HfO_2/Si$ and $Pt/HfO_2/Ge$ capacitors with different surface passivations (the physical thickness of HfO_2 is identical; 4.4–4.6 nm) [17]

Ge out diffusion from substrate into high-k dielectric layer and reaction of Ge with high-k. Surface treatment prior to high-k deposition is believed to improve the interface quality by inserting an interlayer between Ge and high-k to minimize the Ge out diffusion and the reaction between Ge and high-k. Surface nitridation, which is one of the most common methods in forming high-k on Si substrate, has been first proposed and investigated to form high-k gate stack on germanium substrate by introducing a thin nitride layer on top of Ge.

Different high-k gate dielectrics such as HfO_2 and ZrO_2 on nitrided germanium substrate using different deposition methods including ALD and MOCVD have been investigated [17, 18, 31–35].

5.3.1 Physical Characterization

To study the surface nitridation effect on Ge substrate before HfO_2 deposition, two samples were prepared for XPS characterization, with one just after DHF cleaning (Sample A) and another one after DHF cleaning and thermal annealing in NH_3 ambient (Sample B) [32]. The Ge 2p3 core-level XPS spectra at 90° photoelectron take-off angle are characterized and shown in Fig. 5.9. The line with closed boxes is the signal of the Ge sample (Sample A) right after the cleaning. The main peak located at 1,217.8 eV is attributed to metallic Ge 2p3 spectrum from the substrate. The shoulder, ranging from 1,219 to 1,221 eV, is attributed to $GeO_x (x \leq 2)$ bonds, which is believed to be introduced during sample transportation. The circle-dotted line is the signal of the Ge sample (Sample B) annealed in NH_3 (600°C, 10 min, base pressure of 5×10^{-6} Torr) after the same cleaning procedure. From the figure, it is observed that the small shoulder in Sample A has evolved to a broad peak (~1,219.5 eV). Considering all the possible bonds that the Ge atoms may have, it is reasonable to infer that

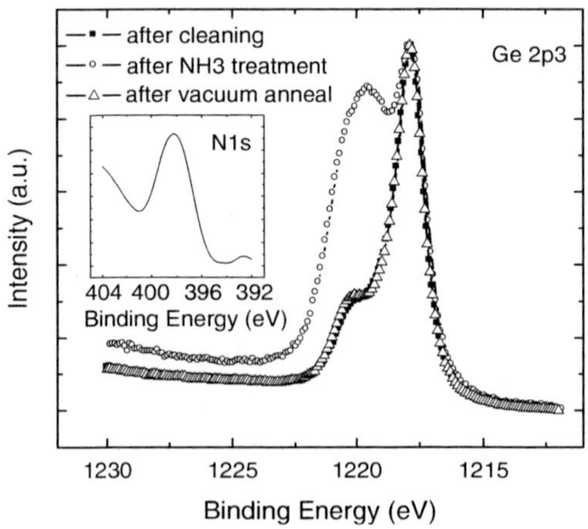

Fig. 5.9. XPS analysis of the NH_3 annealing effect on the Ge substrates with the Ge 2p3 spectra [32]

the broad peak consists of two types of Ge bonds: the Ge–O (1,220.1 eV) and Ge–N (1,218.9 eV). Detailed curve fitting verifies that the whole Ge $2p3$ signal consists of three peaks, which are identified as metallic Ge (1,217.8 eV), Ge–N bond (1,218.9 eV) and Ge–O bond (1,220.1 eV), respectively. The nitrogen existence is also confirmed by the N $1s$ spectrum (inset of Fig. 5.9). The concentrations of oxygen and nitrogen were quantified by integrating each peak area and subtracting the background, and a layer of $GeO_{0.83}N_{0.17}$ is found on top of the Ge substrate after surface nitridation. To clarify the oxygen source in the $GeO_{0.83}N_{0.17}$ during the NH_3 annealing, one more sample (Sample C) was put in the NH_3 treatment chamber under standby condition (same temperature without NH_3 flow) for 10 min after the same cleaning procedure. The XPS result is also shown in Fig. 5.9 with the empty-triangle line. It clearly shows that, comparing to the sample after cleaning (Sample A), there is no substantial oxidation in the Sample C. This means that the oxygen detected in the $GeO_{0.83}N_{0.17}$ of Sample B (after NH_3 surface annealing) is likely introduced by NH_3 gas source, even though the NH_3 gas purity is 99.999% (with the main impurities of O_2, H_2O, and H_2). The high oxygen concentration in GeON after surface nitridation implies that germanium is easier to be oxidized than nitrified. Similar results of high oxygen concentration in interfacial GeO_xN_y layer after surface nitridation was also reported in [35].

XPS spectra of Ge $2p3$ and Hf $4f$ for HfO_2 deposited on germanium with and without surface NH_3 annealing (600°C, 30 s) were characterized to understand the effect of surface nitridation on the following deposited HfO_2 [32], as shown in Fig. 5.10. A 5-nm layer of HfO_2 was deposited on both the DHF cleaned and surface nitrided Ge substrates using MOCVD, with Hf *tert*-butoxide as the metal-organic precursor in an $N_2 + O_2$ ambient at 400°C with a base pressure of 3×10^{-3} Torr. For comparison, XPS spectrum of Hf $4f$ peak for HfO_2 deposited on Si with postdeposition annealing is also characterized.

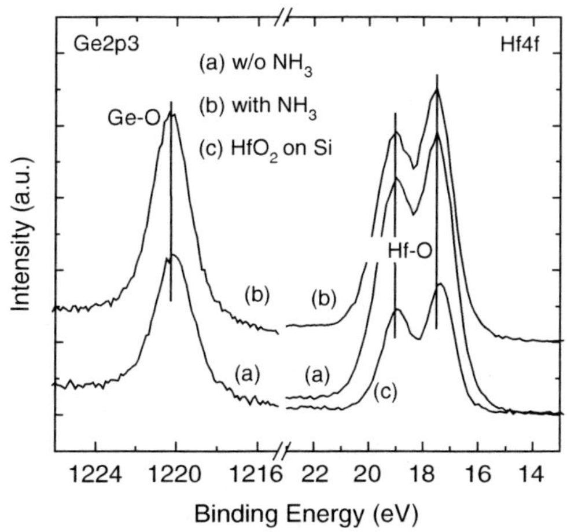

Fig. 5.10. XPS spectra of Hf $4f$ and Ge $2p3$ for HfO_2 deposited on Ge substrates [32]

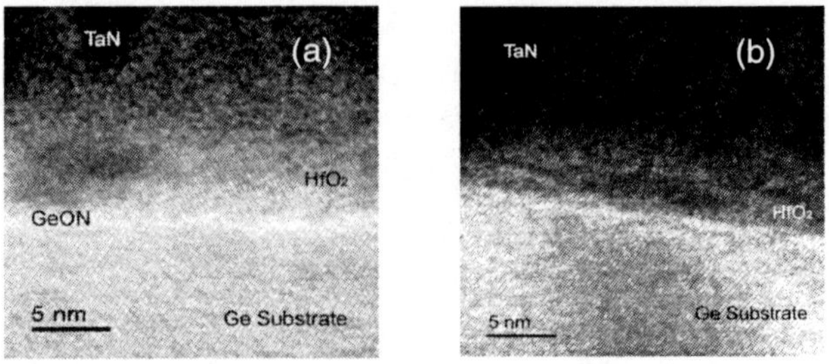

Fig. 5.11. HR-XTEM of the MOS structure of TaN/HfO_2/GeON/Ge (**a**) with and (**b**) without surface nitridation [32]

All the XPS are characterized at a 10° take-off angle. No substantial binding energy difference is observed in Hf $4f$ peaks between the Si and Ge substrates. On the other hand, Ge $2p3$ peaks at 1,220.1 eV were detected and are identified as Ge–O bonds for both of the Ge samples with and without surface nitridation. Since there are no metallic Ge bonds found in these signals, it is believed the Ge signals are from the as-deposited HfO_2. This implies that Ge is incorporated in the HfO_2 films. This Ge incorporation phenomenon may result from the Ge diffusion from substrate during CVD deposition process (~400°C). Further, since no Ge–Hf bond is detected in both samples, the dielectric is of good electrical insulating property in terms of chemical states. Kim et al. also analyzed the role of the interfacial GeON layer [17] formed by NH_3 nitridation of HF cleaned Ge substrate using the MEIS analysis. It was observed that the nitride interfacial layer also blocks Hf out diffusion into the Ge substrate.

HR-XTEM image of the Ge MOS structure with NH_3 annealing was performed [32], as shown in Fig. 5.11a. The dielectric thickness measured from the TEM picture is ~51 Å and the dielectric constant of dielectric is ~18.9. From Fig. 5.11a, it is also noticed that, unlike the interfacial layer on Si substrate, the interfacial layer on Ge substrate is crystallized. The crystallized interfacial layer may be related to its high oxygen concentration, since GeO_2 formed by gaseous O_2 on Ge substrate was found to be polycrystalline. It is also noted from the TEM picture that the dielectric film is crystallized, which implies that the crystallization temperature of the HfO_2 film with Ge is lower than 600°C. For comparison, the TEM image of MOS stack without NH_3 anneal is also presented in Fig. 5.11b. Nonuniform interfacial layer is observed between the dielectric and the substrate. Gusev et al. also characterized TEM pictures of the interfacial nitride layer [18]. It was suggested that the insertion of nitride layer prevents the possible occurring of the epitaxy in the case of direct deposition of HfO_2 on germanium substrate.

5.3.2 MOS Capacitor Characteristics

Figure 5.12 shows the C–V characteristics of HfO_2 Ge MOS capacitor (area = 100 × 100 μm, sweeping from inversion to accumulation) with surface

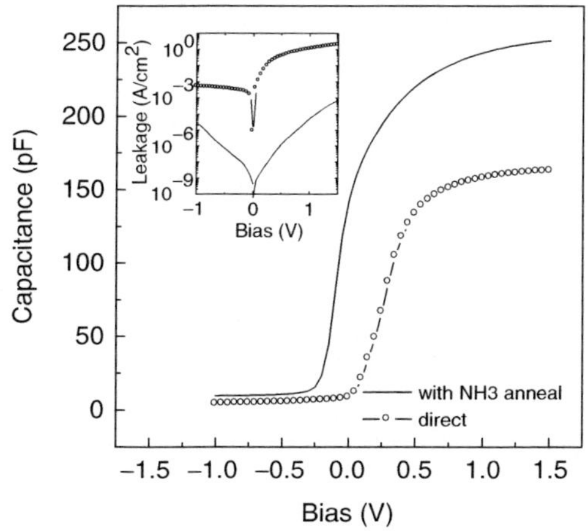

Fig. 5.12. *C–V* curves measured at frequency of 100 kHz and *J–V* curves plotted as *inset* [32]

nitridation, and its *J–V* curve is plotted as inset [32]. For comparison, both *C–V* and *J–V* characteristics of HfO_2 Ge MOS capacitor without surface nitridation (same dimension) are also included. By fitting the *C–V* data while taking into account the quantum confinement effects, it is calculated that a small EOT of 10.5 Å and a low-leakage current of 5.02×10^{-5} A cm^{-2} at $V_g = 1$ V can be achieved for the MOS capacitor with surface nitridation. While for the MOS capacitor without surface nitridation, an EOT of 16.8 Å is obtained with a leakage current of 1.01 A cm^{-2} at $V_g = 1\,V$. Thus, though the Ge incorporation in HfO_2 films is observed for both samples with and without surface nitridation and both samples show similar chemical states, the NH_3 annealing is very effective in improving the electrical properties of Ge MOS capacitors. Considering together the fact that there is no significant chemical bonding difference in the dielectrics of both samples, the large difference of the leakage currents could be likely attributed to the interfacial layer. Therefore, the large leakage current of the Ge MOS capacitor implies the interfacial layer (GeO_x) is of poor quality when there is no surface nitridation. The negative shift of the flatband voltage of the surface nitridation device may be due to significant positive charge ($\sim 5.3 \times 10^{12}$ cm^{-2}) introduced during the NH_3 annealing.

5.3.3 MOSFET Performance

Figure 5.13 shows the I_d–V_d characteristics and hole mobility of p-channel MOSFETs with surface nitridation and HfO_2 gate dielectric. The mobility is extracted using split CV method. It can be seen that higher hole mobility over the universal mobility curve of Si/SiO_2 system has been successfully demonstrated in MOSFET made on Ge substrate.

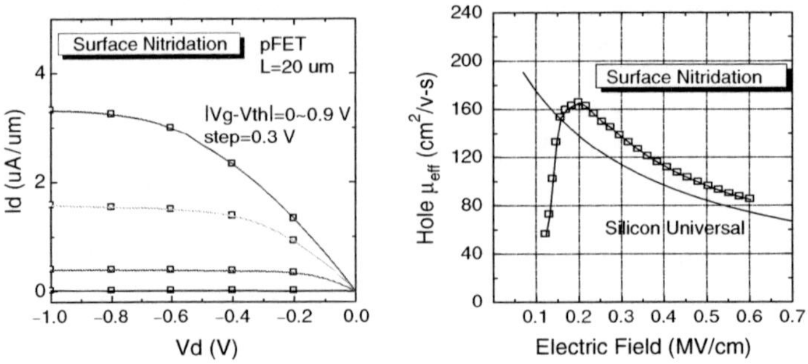

Fig. 5.13. (**a**) I_d–V_d characteristics and (**b**) hole mobility of p-MOSFET with HfO_2 gate dielectric and surface nitridation [36]

Though a higher hole mobility of Ge over the Si counterparts has been successfully demonstrated in MOSFETs with surface nitridation, a reasonably high-electron mobility has not been demonstrated in n-channel MOSFETs on Ge substrate using surface nitridation treatment. This is possibly because of the poor thermal stability of surface nitridation, which degrades the interface properties since a higher source/drain annealing temperature is needed to activate phosphorus than boron. Thus, a novel surface passivation with better thermal stability may be needed to provide well performed n-channel MOSFETs.

5.4 Surface Silicon Passivation

In Sect. 5.3, surface nitridation (SN) prior to high-k deposition on germanium substrate is proposed and investigated by introducing an interlayer of ultrathin nitride layer. However, the easier oxidation over nitridation of the Ge substrate makes the nitride layer containing a high concentration of oxygen, which may degrade the blocking properties and causing poor interface properties (thermal stability). Recently, a surface passivation using SiH_4 annealing prior to high-k deposition has been proposed and investigated for the Ge MOSFETs [36–38]. The silane surface passivation introduces an ultrathin layer of Si which contributes to the interfacial layer during the subsequent high-thermal processing steps. This passivation is found to be able to provide both high hole and electron mobility in p- and n-channel MOSFETs, respectively, due to the good interface properties.

5.4.1 Physical Characterization

An effective surface silicon passivation (SP) on germanium should meet the following criteria (1) Si must completely cover the germanium surface and the germanium surface should be free of germanium oxide and (2) the silicon

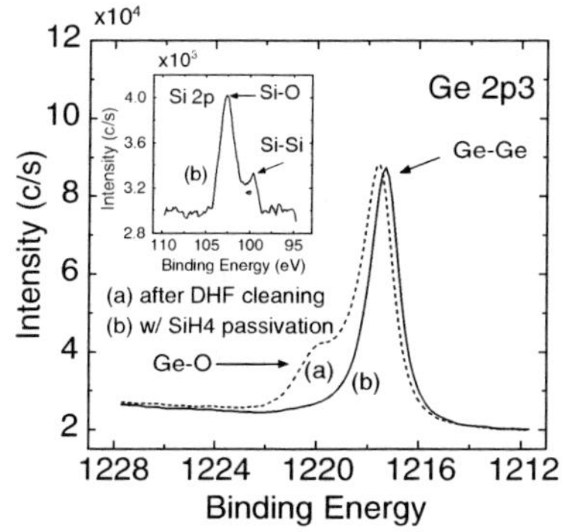

Fig. 5.14. The Ge 2p3 spectra of XPS on the passivation process on Ge surface by SiH_4. The samples before and after passivation were compared. Inset shows the Si 2p spectra of the sample after SiH_4 passivation [37]

passivation layer should be thin enough and consumed during the subsequent high-k deposition so that the MOSFET channel is still kept in germanium.

Figure 5.14 shows the Ge 2p3 XPS spectra of the sample just after DHF cleaning and the sample after the SiH_4 surface passivation. The inset is the Si 2p XPS spectrum of the sample after the SiH_4 surface passivation. As can be seen, the Ge–O peak observed on the DHF as-cleaned Ge surface disappeared after the SiH_4 passivation. Moreover, the Si-passivated Ge surface is stable even the sample is exposed to the air. The existence of silicon is confirmed by the Si 2p spectrum of the same sample, as shown in the inset. This Si signal includes two types of bonds: one is the elemental Si, and the other is SiO_x (which is likely introduced during sample transportation). Based on the above analysis, the germanium surface is completely covered by elemental Si and free of germanium oxide after the SiH_4 passivation. It is also noticed in Fig. 5.14 that (1) the intensity of the Si–O peak (~102.6 eV) is much lower (~65 times) than that of the Ge–Ge peak (~1217.4 eV) for the sample after passivation and (2) the Ge–Ge peak (~1217.4 eV) of the sample after passivation shows a negligible intensity reduction compared to the as-cleaned sample. Both imply that the amount of the top silicon is so little that the thickness is much less than the inelastic mean free path (IMFP) of Ge 2p3 electrons traveling in silicon. The small value of the IMFP (~8.7 Å) indicates the silicon layer could be only a few monolayers. AFM analysis was performed to examine the surface roughness. It was found that comparable surface roughness between the DHF-cleaned sample (RMS = 0.157 nm) and the Si-passivated sample (RMS = 0.166 nm) were observed.

The effect of the SiH_4 surface passivation on the subsequent HfO_2 deposition was studied by angle-resolved XPS analysis and shown in Fig. 5.15. The sample after HfO_2 deposition without Si-passivation is included for comparison. Both samples are characterized at take-off angles of 10° and 90°, respectively. As can be seen, only the sample with Si passivation at the 90°

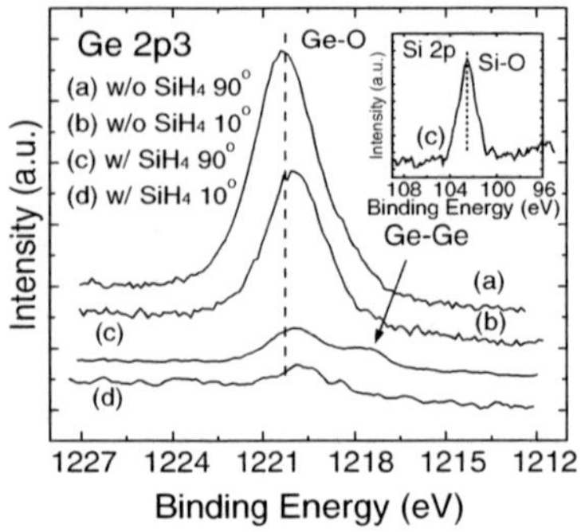

Fig. 5.15. The Ge 2*p*3 spectra of angle-resolved XPS on the samples with and without passivation after HfO_2 MOCVD [37]

take-off angle shows a two-peak spectrum [curve (c)], representing the elemental germanium and the germanium oxide states. The detection of element Ge bonds indicates that the HfO_2 film on this sample is thinner than the sample without Si passivation [curve (a)]. Moreover, the Ge–O peak area of the Si passivated sample [curve (c)] is much smaller than that of the sample without passivation [curve (a)]. This means the passivation is very effective in suppressing formation of germanium oxide. The small take-off angle (10°) in the analysis was also used to examine the dielectric film. Obviously, the amount of Ge in the sample with Si passivation at 10° take-off angle is much less than that of the sample without Si passivation. Thus, the significant Ge out diffusion during HfO_2 MOCVD is greatly suppressed by Si passivation. On the other hand, only Si at its oxide state is detected (the inset of Fig. 5.15), indicating that there is no Hf–silicide formation at the interface.

5.4.2 MOS Capacitors and Thermal Stability

Figure 5.16 shows a comparison between the two surface treatments. The gate leakage currents of the p-MOS capacitors at $|V_g - V_{fb}| = 1V$ under accumulation were plotted as a function of the EOT. The theoretical gate leakage currents of SiO_2/Si and HfO_2/Si systems contributed by direct tunneling mechanism are also plotted with solid lines. As is shown in Fig. 5.16, direct CVD HfO_2 on germanium results in a large gate leakage current that is higher than thermal SiO_2 on Si at the same moderate EOTs. The formation of unstable germanium oxide during the HfO_2 CVD is the main cause of this large leakage. The large gate leakage current can be reduced by orders of magnitude by using either surface nitridation or Si passivation, as shown in Fig. 5.16 with square and triangle dots. Although the two surface passivation techniques both result in higher leakage currents than HfO_2/Si system, it is noticed that the gate leakage currents can still meet the requirements of both high performance (HP) and low operation power (LOP) MOSFETs after the

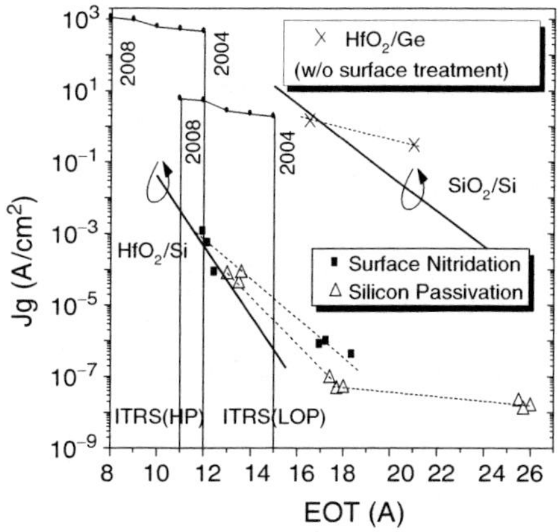

Fig. 5.16. Gate leakage current as a function of EOT of Ge MOS capacitors with different surface passivation

year 2008 according to ITRS2003 (see Fig. 5.16). Further scaling of EOT can be achieved by thinning down the high-k layer.

Since for n-MOS devices, a postmetal annealing temperature higher than 500°C is more favorable for activating the source/drain dopants and for repairing the source/drain implant damage, the thermal stability of the postmetal annealing (higher than 500°C) on gate stack in terms of leakage current and EOT is analyzed with SP processed n-MOS capacitors (Fig. 5.17). The leakage current is taken at -1 V gate bias. From the device without PMA to the device processed with the highest thermal budget (600°C, 30 s), the leakage current increases by around one order of magnitude, and by a 2.8 Å increase of EOT. Thus, a higher temperature PMA on the gate stack generally degrades the I_g-EOT characteristic. The increase of EOT might be due to additional growth

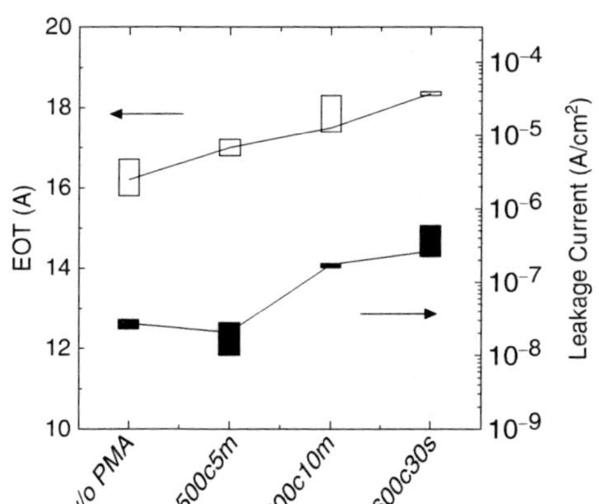

Fig. 5.17. Dependences of EOT and gate leakage current on annealing condition

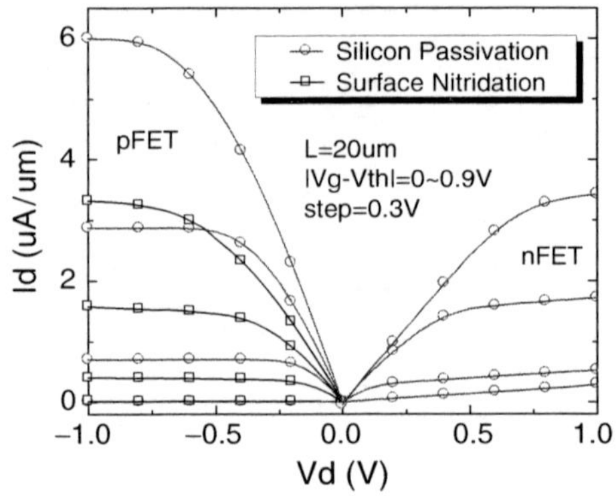

Fig. 5.18. I_d–V_d characteristics of p- and n-FETs with silane surface passviation and surface nitridation [36]

of the interfacial layer. The increase of leakage current might be related to the Ge out diffusion from the substrate into the oxide during the thermal anneal.

5.4.3 MOSFET Performance

To obtain high performance Ge p- and n-FETs with silicon passivation, two different silicon passivation conditions (thin silicon and thick silicon) were employed in the device fabrication. The starting Ge wafers (100) were subjected to different surface treatments after DHF precleaning. The silicon passivation (SP) was carried out by annealing in SiH_4. The ultrathin silicon layer will be oxidized during the subsequent MOCVD HfO_2 deposition and contributed to the interfacial layer. For comparison, both Ge devices with surface nitridation (conducted by annealing in NH_3) and Si devices were fabricated. After HfO_2 deposition and postdeposition anneal (PDA), PVD TaN was deposited and patterned as metal gate. Boron and phosphorus were implanted and activated as source/drain. Finally, forming gas annealing was performed at 350°C.

Charge pumping measurement was conducted and the interface trap densities are $1.6 \times 10^{12}\,\mathrm{cm^{-2}\,eV^{-1}}$, $3.2 \times 10^{11}\,\mathrm{cm^{-2}\,eV^{-1}}$, and $4.5 \times 10^{11}\,\mathrm{cm^{-2}\,eV^{-1}}$ for SN Ge p-FET, SP Ge p-FET, and SP Ge n-FET, respectively. Figure 5.18 shows the $I_d - V_d$ characteristics of the transistors. The SP p-FET shows a ~76% improved drain current over the SN p-FET under the same gate overdriver. Figure 5.19 shows the I_d–V_g characteristics of the transistors. The Ge transistors with SP technique exhibit larger on/off ratios than the transistors with SN technique. A record-low subthreshold swing ($72\,\mathrm{mV\,dec^{-1}}$) for p-FET is also achieved by SP technique. Split-CV measurement was used to extract effective hole and electron mobility, respectively, for both n- and p-FETs. Figure 5.20 shows the effective hole and electron mobility of the transistors. The SP devices exhibit ~82% higher hole mobility than the SiO_2/Si universal

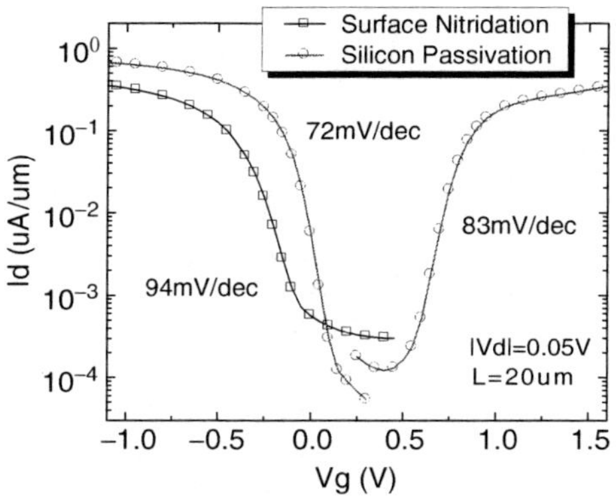

Fig. 5.19. I_d–V_g curves of p- and n-FETs with silane surface passivation and surface nitridation [36]

curve at high field of 0.6 MV cm^{-1}, and ~61% higher peak electron mobility than the HfO_2/Si device.

Further improvement in surface passivation may be needed to provide better mobility, particularly for the electron mobility. Bai et al. recently proposed a new surface passivation of using Si interlayer passivation and the subsequent nitridation of the Si interlayer [39]. An improved interface properties has been demonstrated with this surface passivation technique.

5.4.4 BTI and Charge Trapping

BTI characteristics were measured in MOSFETs biased under inversion. A voltage (V_{stress}) is applied to the gate of a device, while the source/drain and substrate are grounded. Measurements were conducted during the stress intervals. The measurement time between two consecutive stresses was ensured to be minimal in order to reduce the possible detrapping of the oxide trapped charge.

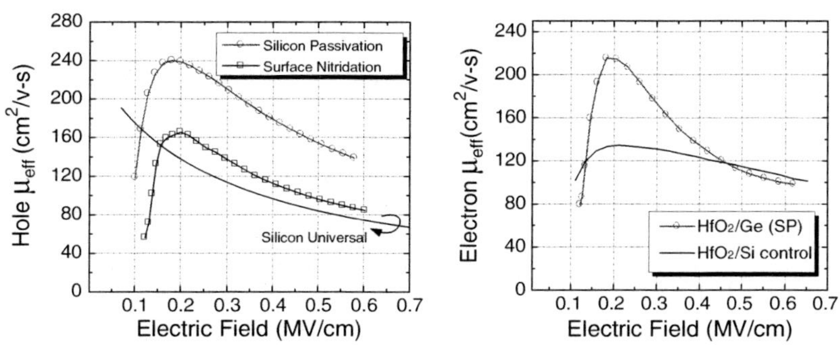

Fig. 5.20. (**a**) effective hole mobility and (**b**) effective electron mobility as a function of electrical field of p- and n-FETs made with silane surface passivation and surface nitridation [36]

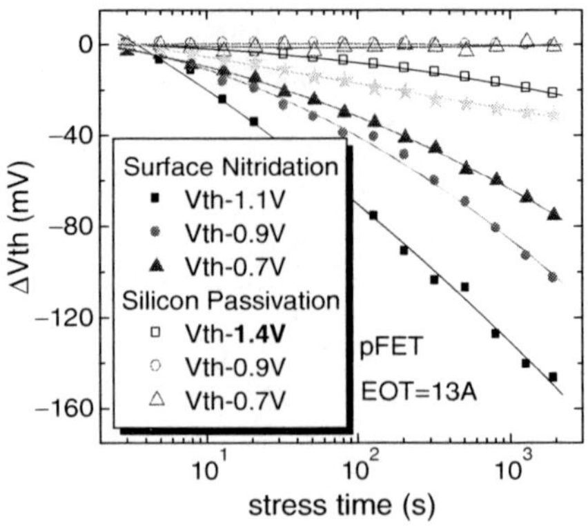

Fig. 5.21. NBTI degradation of V_{th} shift in p-FETs with silicon passivation and surface nitridation [36]

Figure 5.21 shows the NBTI degradation of V_{th} shift with stressing for Ge p-FETs as well as Si p-FETs. The V_{th} shift of the SP Ge p-FET is smaller than the silicon control, which might be attributed to the larger valence band offset of HfO_2/Ge and hence less hole trapping in the high-k dielectric than the silicon counterpart. Thus, the SP Ge p-FET shows better NBTI characteristics than the silicon control. Figure 5.22 shows the PBTI degradation of V_{th} shift with stressing for Ge n-FETs with SP as well as Si n-FETs. The SP Ge n-FET shows a larger V_{th} shift than the silicon control, which suggests a severer charge trapping in the gate oxide. This may relate to the lower processing temperature during the device fabrication, which results in a larger amount of preexisting electron traps in the dielectric. Thus, for high-k Ge MOSFETs, PBTI degradation is more significant than NBTI degradation.

Charge trapping characterization was conducted on Ge MOSFETs under accumulation stress with a same stressing scheme as used in BTI

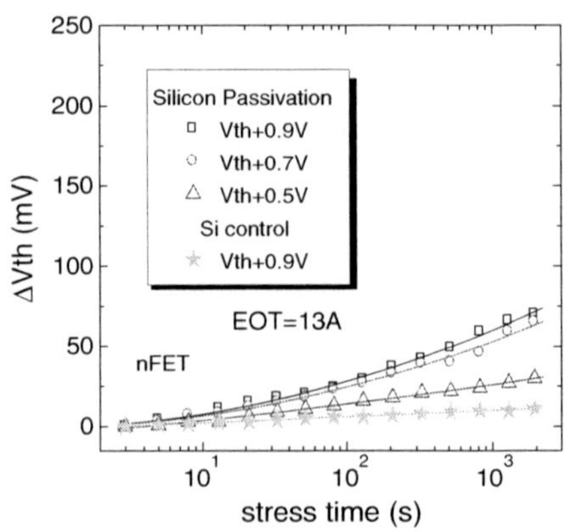

Fig. 5.22. PBTI degradation of V_{th} in n-FETs with silicon passivation and surface nitridation [36]

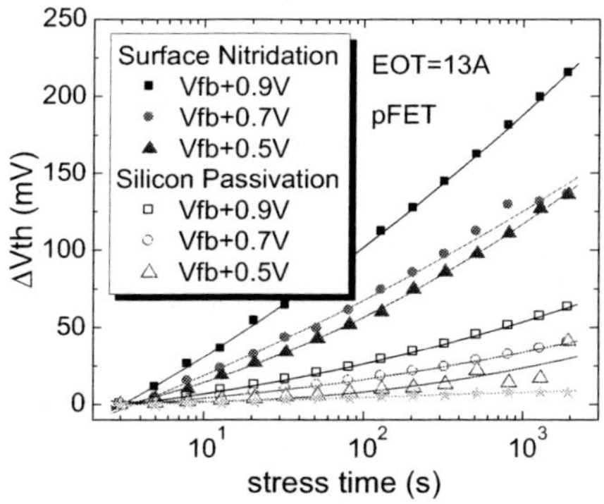

Fig. 5.23. V_{th} degradation of p-FETs due to charge trapping with silicon passivation and surface nitridation [36]

characterization. Figure 5.23 shows the V_{th} shift stressed at various voltages for all the p-FETs. SN Ge p-FET shows the highest V_{th} shift. The larger positive V_{th} shift in SP Ge p-FET than Si control is due to more electron trapping in HfO_2 on Ge substrate. This is consistent with the PBTI in n-FETs. Figure 5.24 shows the V_{th} shift with stressing for both the Ge and Si n-FETs. It is noticed that the V_{th} shift in Ge n-FET is less than that in Si control. These also imply less hole trapping occurred in HfO_2 on Ge substrate due to the larger valence band offset (hole barrier) compared to the HfO_2 on Si substrate.

5.5 Plasma-PH_3 and AlN Surface Passivation

There have been noticeable efforts to study the reconstruction of semiconductor surfaces by removal of dangling bonds and surface states with proper

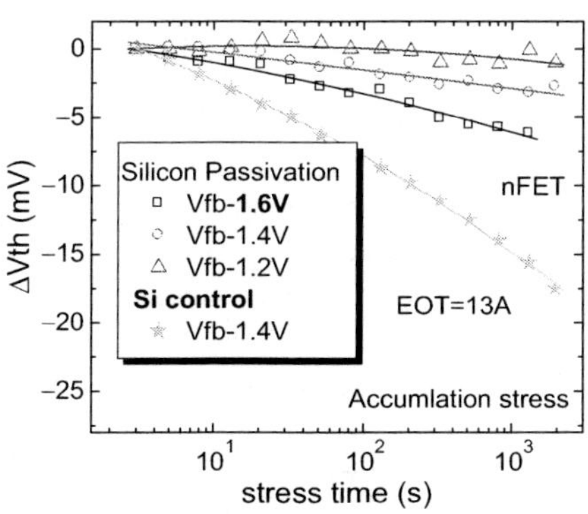

Fig. 5.24. V_{th} degradation of n-FETs due to charge trapping with silicon passivation and surface nitridation [36]

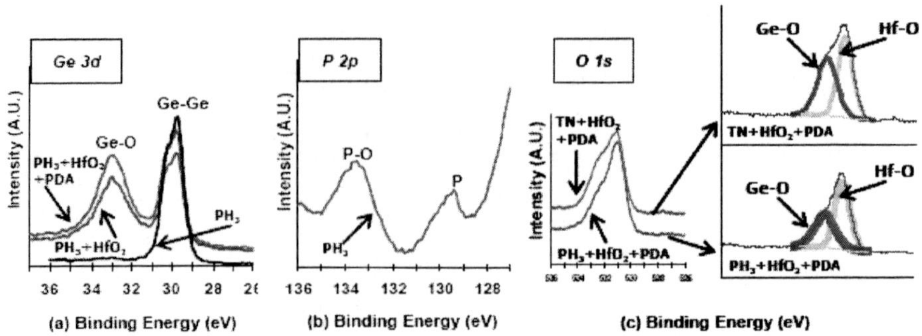

Fig. 5.25. XPS analysis of HfO_2 on plasma-PH_3 treated and thermal nitride Ge substrate; (**a**) Ge 3*d* spectra after plasma-PH_3 treatment, HfO_2 deposition and PDA, (**b**) P 2*p* spectra on Ge after passivation by plasma-PH_3 and HfO_2 deposition, and (**c**) O 1*s* spectra on Ge after thermal nitridation and plasma-PH_3 treatment with HfO_2 [43]

surface treatments in order to restore the ideal bulk-terminated geometry on semiconductor surfaces. On the Ge surfaces, several adsorbates have been investigated to passivate the Ge surfaces; a monolayer of As by MBE [40], chemisorption of S by dissociation of Ag_2S in UHV [41], adsorption of Cl and Br by dissociation of AgCl and AgBr [42].

Plasma-PH_3 surface treatment was employed to passivate the Ge substrate prior to HfO_2 deposition [43]. In this experiment, the HF-last n- and p-type Ge wafers received plasma treatment in PH_3 ambient at 400°C, followed by MOCVD HfO_2 deposition and PDA at 500°C, all the processes performed in in situ multicluster CVD system without breaking vacuum. Figure 5.25 shows the XPS analysis of Ge, P, and O spectra for HfO_2 on plasma-PH_3 and thermal nitrided Ge substrate before and after PDA. First, the Ge 3d spectra indicate that the Ge oxide is formed mainly during the CVD process with a little increase after PDA. The successful incorporation of P at the surface is detected by the P–O peak in P 3*p* spectra. By comparing the O 1*s* spectra and its deconvolution for both plasma-PH_3 treated and thermal nitrided samples, it was found that Ge oxide formation at the surface is suppressed by plasma-PH_3 passivation. Moreover, no significant out diffusion of Ge in HfO_2 is confirmed by EDS and SIMS analysis. As a result, the improved C–V characteristics without interface state-related behavior were demonstrated by plasma-PH_3 treated HfO_2/Ge for both n- and p-MOS devices (Fig. 5.26). Thermal stability study at various postannealing temperatures also indicates that a plasma-PH_3 treated HfO_2/Ge stack maintains stable EOT and leakage current up to 600°C anneal.

The predeposition of a layer that is thermodynamically more favorable to form oxide than Ge, prior to high-k deposition on Ge, can be expected to be effective to suppress the Ge oxide formation at the interface. In fact, the ZrO_2 formation in [20] was performed by the sputtering of Zr precursor films on

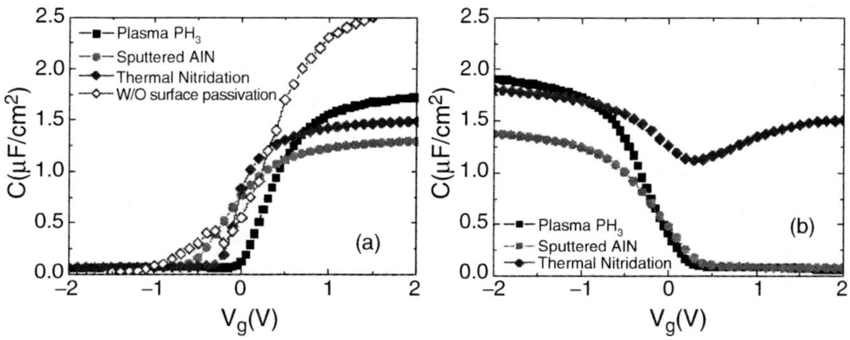

Fig. 5.26. Measure high frequency *C–V* of (**a**) p-MOS and n-MOS capacitors of TaN/HfO_2/Ge stack with different Ge surface passivations [43]

the pretreated Ge surface, followed by oxidation with UV ozone technique, demonstrating free of any significant interfacial layer and EOT of ~5 Å. The sputtered HfO_2 films were deposited on Ge both by direct HfO_2 deposition on Ge and by preformation of 1.5 nm Hf metal layer on Ge followed by HfO_2 deposition [44]. Results showed that the thinner interfacial layer and thus thinner EOT were obtained from HfO_2 deposited on Hf bottom layer. In [45], a thin AlN layer was deposited on HF-last Ge substrate prior to ALD-HfO_2 deposition. XPS analysis showed that a thin AlN layer is oxidized during HfO_2 and PDA, forming AlON layer, which is more effective to reduce the interfacial layer than thermal nitrided passivation. Excellent *C–V* and thermal stability in terms of EOT and leakage current were also demonstrated.

5.6 Conclusion

A review of ongoing researches on high-k/Ge interface engineering is presented. Recent results show that the high-k/Ge system revealed different phenomena, compared to high-k/Si system, with regards to interfacial layer formation/control, high-k layer growth, and electrical properties. In order to prevent the degradation of electrical performances of high-k/Ge devices, several Ge passivation techniques prior to high-k deposition have been studied using nitridation, Si passivation, plasma-PH_3 passivation, and AlN passivation. Experimental results show that they are effective to suppress the formation of unstable Ge oxide and the diffusion of Ge into upper high-k layer, resulting in improved *C–V* and mobilities.

References

1. J. Bardeen and W.H. Brattain, Phys. Rev., p. 230 (1948)
2. L.L. Chang and H.N. Yu, Proc. IEEE, Vol. 53, p. 316 (1965)

3. K.L. Wang and P.V. Gray, TED, Vol. ED-22, p. 353 (1975)
4. E.E. Crisman, et al., J. Elec. Soc., Vol. 129, 8, p. 1845 (1982)
5. J.J. Rosenberg. et al., EDL, Vol. 9, 12, p. 639 (1988)
6. S.C. Martin, et al., EDL, Vol. 10, 7, p. 325 (1989)
7. T.N. Jackson, et al., EDL, Vol. 12, 11, p. 605 (1991)
8. A.V. Rzhanov, et al., Thin Solid Films, Vol. 58, p. 37 (1979)
9. K. Prabhakaran, et al., Surf. Sci., Vol. 325, p. 263 (1995)
10. N. Tabet, et al., Surf. Rev. Lett., Vol. 6, p. 1053 (1999)
11. K. Prabhakaran, et al., Thin Solid Films, Vol. 369, p. 289 (2000)
12. J. Oh, et al., J. Elect. Mat., Vol. 33, 4, p. 364 (2004)
13. S.J. Lee, et al., JAP, Vol. 92, 5, p. 2807 (2002)
14. G.D. Wilk, et al., JAP, Vol. 89, p. 5243 (2001)
15. S.V. Elshocht, et al., APL, Vol. 85, 17, p. 3824 (2004)
16. H. Kim, et al., APL, Vol. 83, 13, p. 2647 (2003)
17. H. Kim, et al., APL, Vol. 85, 14, p. 2902 (2004)
18. E.P. Gusev, et al., APL, Vol. 85, 12, p. 2334 (2004)
19. A. Delabie, et al., JAP, Vol. 97, 064104 (2005)
20. C.O. Chui, et al., EDL, Vol. 23, 8, p. 473 (2002)
21. C.O. Chui, et al., IEDM 2002
22. C.O. Chui, et al., VLSI 2002
23. K. Kita, et al., APL, Vol. 85, 1, p. 52 (2004)
24. V. Cracium, et al., APL, Vol. 75, p. 1261 (1999)
25. Y. Kamata, et al., JJAP, Vol. 44, 4B, p. 2323 (2005)
26. A. Dimoulas, et al., APL, Vol. 86, 223507 (2005)
27. A. Dimoulas, et al., APL, Vol. 86, 032908 (2005)
28. A. Dimoulas, et al., ME, Vol. 80, p. 34 (2005)
29. S.J. Wang, et al., APL, Vol. 85, 19, p. 4418 (2004)
30. V.V. Afanas'ev, et al., APL, Vol. 84, 13, p. 2319 (2004)
31. W.P. Bai, et al., VLSI Symp. Tech. Dig., **2003**, Vol. 121 (2003)
32. N. Wu, et al., Appl. Phys. Lett., Vol. 84, 3741 (2004)
33. H. Shang, et al., IEEE Electron Dev. Lett., Vol. 25, 135 (2004)
34. C.O. Chui, et al., IEDM Tech. Dig., **2003**, 18.3.1 (2003)
35. C.O. Chui, et al., IEEE Electron Dev. Lett., Vol. 25, 274 (2004)
36. N. Wu, et al., IEDM Tech. Dig., **2005**, 22.5 (2005)
37. N. Wu, et al., Appl. Phys. Lett., Vol. 85, 4127 (2004)
38. N. Wu, et al., IEEE Electron Dev. Lett., Vol. 25, 631 (2004)
39. W.P. Bai, et al., IEEE Electron Dev. Lett., Vol. 26, 378 (2005)
40. R.D. Bringans, et al., Phys. Rev. Lett., Vol. 55, p. 533 (1985)
41. T. Weser, et al., Phys. Rev. B, Vol. 35, p. 8184 (1987)
42. R.D. Schnell, et al., Phys. Rev. B, Vol. 32, p. 8053 (1985)
43. S.J. Whang, et al., IEDM 2004
44. K. Kita, et al., SSDM 2003, p. 292 (2003)
45. F. Gao, et al., APL, Vol. 86, p. 1 (2005)

6

Effect of Surface Nitridation on the Electrical Characteristics of Germanium High-κ/Metal Gate Metal-Oxide-Semiconductor Devices

D.Q. Kelly, J.J.-H. Chen, S. Guha, and S.K. Banerjee

Summary. We discuss the effect of surface nitridation on germanium high-κ/metal gate metal-oxide-semiconductor capacitors. Results from various groups concerning Ge surface cleaning, surface annealing in ammonia, and the effect of nitridation on C–V and leakage current characteristics are presented and discussed. Electrical data obtained by our group for devices incorporating alumina (Al_2O_3) and hafnia (HfO_2) dielectrics is presented and explained. Capacitance–voltage and leakage current data for capacitors with and without surface nitridation are compared, and the effect of various postdeposition anneals is given. It was found that surface nitridation improves the electrical characteristics of both Al_2O_3 and HfO_2 devices as evidenced by reduced C–V hysteresis and lower leakage. Flat band voltage shifts are presented as a consequence of surface nitridation that can be remedied by postdeposition annealing.

6.1 Introduction

Germanium MOSFETs have attracted attention in recent years due to their enhanced carrier transport and the advances that have been made in high-κ gate dielectric technology [1–9]. Its higher bulk mobility for both electrons (2X) and holes (4X) relative to Si, as well as its demonstrated compatibility with high-κ dielectrics, makes Ge a potential candidate for improving MOSFET channel mobility while allowing for continued aggressive scaling of the gate dielectric. Despite its demonstrated ability to improve drive current capacity in MOSFETs, many engineering hurdles must still be overcome before Ge can be integrated into future CMOS technology nodes.

Although a significant knowledge gap still remains in materials-related issues for Ge-based CMOS, Ge is not unfamiliar to the semiconductor industry, at least historically. The first transistors were made using Ge crystals. Silicon totally eclipsed Ge, however, due to the advantageous properties of its native silicon dioxide (SiO_2). Unlike Si, Ge does not possess a stable natural oxide that can passivate the surface, act as etch protection, or be used as a high-quality gate insulator [10, 11]. The thermodynamic instability of germanium dioxide (GeO_2) was shown by Prabhakaran et al. [12], who reported that low-temperature annealing of GeO_2 resulted in the transformation of GeO_2 to GeO on the surface, which thermally desorbed at around 420°C. Thermal desorption at such low temperatures makes GeO_2 impractical as a gate dielectric for CMOS processes that require high-temperature steps. Another major drawback of GeO_2 is that it is soluble in water [10–12], which rules out almost all wet chemical processes that would attempt to use the Ge native oxide as an etch barrier.

The native dielectrics of Ge, including GeO_2, germanium nitride (Ge_3N_4) [13–15], and germanium oxynitride (GeO_xN_y) [13,16], have all been explored as potential gate insulators for CMOS. In the 1990s, creative methods were employed for growing better-quality GeO_2 on Ge for MOS applications [17,18]. But since GeO_2 is volatile at relatively low temperatures and is water-soluble, alternative fabrication schemes that cover one of the native dielectrics (or the bare Ge surface) with a non-native insulator have proven more promising. One of the earliest examples of the non-native dielectric approach was reported in 1971 by Iwauchi and Tanaka, who demonstrated Ge MOSFETs with Al_2O_3 dielectrics using reactive DC sputtering of Al [19]. In 1986, Chang et al. obtained good interface properties through the use of thick films of an aluminium–phosphorus oxide mixture deposited over native GeO_2 [20]. Even SiO_2 has been explored as a possible non-native insulator on Ge [21, 22], but while SiO_2 has proven extremely successful on Si, its use on Ge has been proven to be much less effective. The main reason for this is the poor quality of the SiO_2/Ge interface. A solution to the SiO_2/Ge interface problem was proposed by Vitkavage et al., who examined the possibility of depositing a thin layer of Si onto the Ge substrate as a buffer layer prior to SiO_2 deposition [21]. The Si interlayer seemed to improve the interface with the Ge substrates, which was evidenced by a drastic reduction in the density of interface states [21].

The initial studies investigating non-native insulators on Ge concentrated on achieving good interface and bulk characteristics, and not necessarily on reducing the equivalent oxide thickness (EOT). As a result, most of the films were relatively thick, and would be unlikely to offer an EOT of less than 10 Å to advance beyond the sub-2 nm regime. For example, the 2005 ITRS roadmap predicts the EOT for high-performance logic to be 9 Å by the year 2008 and 5 Å by 2012 [23]. Fortunately, the natural progression of these lines of research had led to the use of high-κ dielectrics, which do offer the potential of scaling below 10 Å. The demonstrated compatibility of various high-κ materials with Ge has ignited a flurry of interest in Ge-channel MOSFETs. The current research is focused mainly on MOSFET feasibility for bulk Ge [1–6, 9], Ge-on-insulator (GOI) [8, 24], and strained Ge channels on Si [7, 25, 26], with particular attention to understanding the Ge/high-κ interface and reducing the EOT.

Much of the groundwork for high-κ/metal gate Ge-based MOSFETs has already been established. In 2002, Shang et al. reported a Ge p-MOSFET with a thin GeO_xN_y and low-temperature oxide (LTO) gate stack on a bulk Ge substrate, demonstrating 40% hole mobility enhancement over Si and an EOT of 80 Å [5]. In the same year, Chui et al. reported mobility-enhanced Ge p-MOSFETs with ultrathin (6–10 Å) ZrO_2 dielectrics fabricated by UV ozone oxidation of sputtered Zr [3]. In the following year (2003), Bai et al. reported low-leakage MOS capacitors on Ge substrates with an EOT of 13 Å formed by rapid-thermal treatment of the Ge surface in NH_3 followed by in situ rapid-thermal CVD of HfO_2 [9]. Later that year, the same gate process was used by Ritenour et al. to fabricate strained Ge-channel p-MOSFETs on Si substrates (using relaxed $Si_{1-x}Ge_x$ buffer layers) with twice the low-field mobility of Si and an EOT of 16 Å [7].

An important finding in the recent literature on Ge MOSFETs is the effectiveness of using thin GeO_xN_y as a surface passivation layer for subsequent high-κ deposition [9, 27–31]. For example, the study by Bai et al. showed that the formation of a GeO_xN_y passivation layer by rapid-thermal annealing in NH_3 prior to HfO_2 deposition reduced the leakage current in MOS capacitors from 150 to 6 mA cm^{-2} and lowered the EOT by a factor of 2 [9]. Van Elshocht et al. attributed the improved electrical performance and lower EOTs of nitrided devices to reduced updiffusion of Ge atoms into the HfO_2 layer, prevention of HfO_2 epitaxy, and significantly reduced interfacial layer thickness [27].

In this chapter, we will discuss results obtained by our group that examine the effect of surface nitridation on the electrical characteristics of Ge MOS devices incorporating Al_2O_3 and HfO_2 dielectrics. A detailed study of Ge wafer cleaning conditions, dielectric deposition conditions, and anneal conditions was performed, and their effect on the electrical properties of metal-gated Al_2O_3/Ge and HfO_2/Ge capacitors was evaluated. Given the high level of interest within the research community in this and related topics, we note that a number of groups have published reports that complement our results. With this in mind, the specific results obtained by our group will be presented in the larger context of results obtained by other groups. We will show that surface nitridation prior to Al_2O_3 or HfO_2 deposition is important in reducing hysteresis, interfacial layer formation, and leakage current. We will also observe that surface nitridation introduces positive trapped charges and/or dipoles at the Ge/high-κ interface, resulting in significant flat band voltage shifts that can be mitigated by postdeposition anneals.

6.2 Germanium Surface Cleaning

Germanium surface preparation prior to high-κ deposition is vital for obtaining high-quality, low-leakage gate stacks. Various methods have been proposed for removing the native oxide, eliminating surface contaminants, and providing a passivation layer for the Ge surface [10, 11, 32–37]. Some cleaning and passivation methods are more suitable for MOS fabrication than others.

A common yet less appropriate method for Ge surface cleaning is to use 500–1,000 eV sputtered Ar^+ ions followed by thermal annealing. Although ion sputtering is effective in removing surface contaminants, it creates significant ion-induced defects which cannot be removed completely by annealing alone [38]. Alternative methods involve wet chemical etching followed by the formation of a native oxide passivation layer. The passivating oxide is then removed in some cases by annealing under ultrahigh vacuum, leaving a surface free of contaminants [32–37]. If annealing in vacuum is not appropriate for a particular MOS process, the native oxide is removed by other means, such as HF vapor etching [39].

For MOS devices it is clear that the cleaning of the Ge surface prior to surface passivation and high-κ deposition is important for achieving high-quality interfaces with low interfacial layer thicknesses. The techniques that involve wet chemical treatment followed by thermal annealing include the work of Prabhakarana et al. [32] and Akane et al. [33, 34]. Prabhakarana et al. used a cyclical procedure of dipping the Ge substrate in a mixture of H_2O_2 and H_2O and removing the oxide by dipping in HF, followed by the formation of a final native oxide passivation layer that was removed by thermal decomposition under ultrahigh vacuum annealing [32]. Using this procedure, the authors were unable to detect any C or O impurities with X-ray photoelectron spectroscopy (XPS) or Auger electron spectroscopy (AES). Akane et al. used an aqueous ammonia solution to remove the Ge native oxide, followed by a rinse in diluted H_2SO_4 and an H_2O_2 treatment to form a passive oxide layer [33]. Extremely smooth oxide surfaces were found using atomic force microscopy (AFM). The passive oxide layer was again removed in ultrahigh vacuum by thermal annealing, and the surfaces were analyzed *in situ* using reflection high-energy electron diffraction (RHEED) to reveal well-ordered 2×1 diffraction patterns. A clean, smooth surface was inferred by subsequent high-quality crystal growth. The same group also showed the effectiveness of cyclical H_2O_2 and HCl treatments followed by the formation of a protective oxide using $NH_4OH/H_2O_2/H_2O$ and thermal desorption in ultrahigh vacuum [34]. In this case, a clean surface was inferred by the lack of C contamination on the surface, as indicated by AES.

Zhang et al. tried UV ozonation of Ge to form a native GeO_2 layer instead of using wet chemical treatments [35]. Germanium wafers were degreased in successive rinses of trichloroethylene, acetone, and methanol followed by ultrasonic rinsing in deionized (DI) water. The wafers were then exposed to UV in laboratory air for various times, which resulted in passive oxide layers with thicknesses on the order of a few nanometers. Annealing in ultrahigh vacuum was then carried out, followed by characterization with XPS, AES, and electron energy loss spectroscopy (EELS). For the samples that were cleaned with the optimized UV exposure process, no C or O impurities were found with XPS or AES, and EELS spectra revealed no peaks corresponding to Ge–O bonds.

Cyclical treatments using HF and DI water have also been proposed as an effective cleaning technique. This process was introduced by Deegan et al. [36]. The process involved rinsing Ge samples in running purified DI water for approximately 20 s, dipping in 50% HF solution for 10 s, and rinsing again. This procedure was repeated a total of five times before drying the samples in filtered N_2 gas and immediately inserting them into a vacuum chamber for thermal desorption of the native oxide and XPS analysis. The XPS data showed a drastic reduction in Ge–O bonding. Due to the simplicity of this cleaning technique, it has gained some popularity in the recent Ge MOSFET literature.

The cleaning techniques described thus far have all involved a thermal annealing step in ultrahigh vacuum for native oxide removal. For some MOS processes that include high-κ deposition, an annealing step in ultrahigh vacuum may not be desirable or even feasible. In fact, the native GeO_2 in some cases may even be kept so it can be used as a starting layer for the formation of GeO_xN_y [29]. Instead of removing the native oxide in ultrahigh vacuum, Chui et al. used an HF vapor etch prior to ZrO_2 deposition [39]. MOS capacitor results were compared with samples that were treated in DI water for native oxide removal as well as samples that retained the native oxide. The devices that underwent HF vapor treatments showed the highest electrical quality, as evidenced by having the lowest hysteresis and frequency dispersion in the capacitance–voltage characteristics. These devices also showed a shift of only 1.24 mV in the flat band voltage after 200 s of constant-current stressing, which indicates a good-quality gate stack.

The work performed by our group uses a recipe developed by Dr. Huiling Shang from the IBM T.J. Watson Research Center. The Ge wafers were prepared by solvent cleaning in successive rinses of hot isopropanol, acetone, and methanol (10 min each), followed by a 15-min DI water rinse to remove the native oxide. To eliminate surface contamination, Ge surface etching was then performed using a cyclical rinse of H_2O_2 (30%), HCl/H_2O (1 : 4), and DI water. The H_2O_2 oxidized the Ge surface and the HCl/H_2O etched away the resulting GeO_2. The HCl-based etch has been shown by Okumura et al [34] to reduce the roughness of the Ge surface, as opposed to the use of HF as the etchant. After repeating this etch three times, a protective layer of GeO_2 was formed by dipping the wafer in a solution of $NH_4OH/H_2O_2/H_2O$ (0.5 : 1 : 10) prior to inserting the wafers into the load lock chamber of an ultrahigh vacuum reactive atomic-beam deposition system. The system is equipped with a boron nitride resistive heater that was used to heat the cleaned Ge wafers to 650°C for 30 min to thermally desorb the native oxide layer. XPS studies showed that a 650°C anneal in ultrahigh vacuum for 30 min resulted in complete removal of the residual surface oxygen, as seen in Fig. 6.1. The O1s 531 eV core-level peak is no longer observable after the 650°C anneal. The inset in Fig. 6.1 also shows that the Ge surface is carbon free [28].

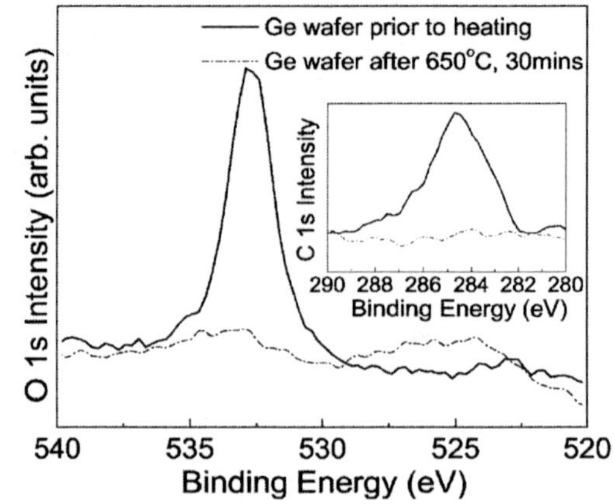

Fig. 6.1. XPS scan measured on a Ge wafer before and after heating for 30 min at 650°C, showing a drastic reduction in the O1*s* peak intensity [28]. The C1*s* profile is shown in the *inset*, and confirms that the surface is also free of C contamination

6.3 Surface Pretreatment with NH_3 (Surface Nitridation)

Nitridation of the Ge surface using NH_3 annealing has been shown to be an important step for achieving high-quality interfaces with the high-κ dielectric [9, 27–31, 40]. An extensive review of nitrogen as an impurity in Ge was provided by Chambouleyrona and Zanatta [41]. Van Elshocht et al. showed that compared to HF-last cleaning without surface nitridation, the NH_3 pretreated samples had much smoother interfaces with HfO_2 deposited by metal-organic chemical vapor deposition (MOCVD) [27]. Using time-of-flight secondary ion mass spectrometry (TOF-SIMS) depth profiles, they showed the temperature dependence of Ge interdiffusion with the HfO_2. The results are shown in Fig. 6.2, with intensity plotted as a function of sputtering time.

Figure 6.2a shows TOF-SIMS Ge and Hf depth profiles for samples that did not undergo NH_3 annealing (no surface nitridation) for HfO_2 deposited under the following three conditions: 300°C with a postdeposition anneal (PDA–550°C, N_2, 200s), HfO_2 deposited at 300°C without a PDA, and HfO_2 deposited at 485°C without a PDA. As seen in Fig. 6.2a, the sample that underwent a PDA at 550°C, as well as the sample grown at 485°C, show significant Ge diffusion into the HfO_2. This clearly shows the temperature dependence of Ge indiffusion, and it also indicates that both growth temperature and postdeposition annealing have an effect. TOF-SIMS results for samples that underwent surface nitridation prior to HfO_2 deposition are shown in Fig. 6.2b. All of the surface-nitrided samples had a HfO_2 layer that was grown at 485°C, and the NH_3 anneals were done at 300°C, 500°C, and 600°C. The authors claim that the N content in the GeO_xN_y layer was 0% for the sample annealed at 300°C, 4% for 500°C, and 10% for 600°C. As seen in

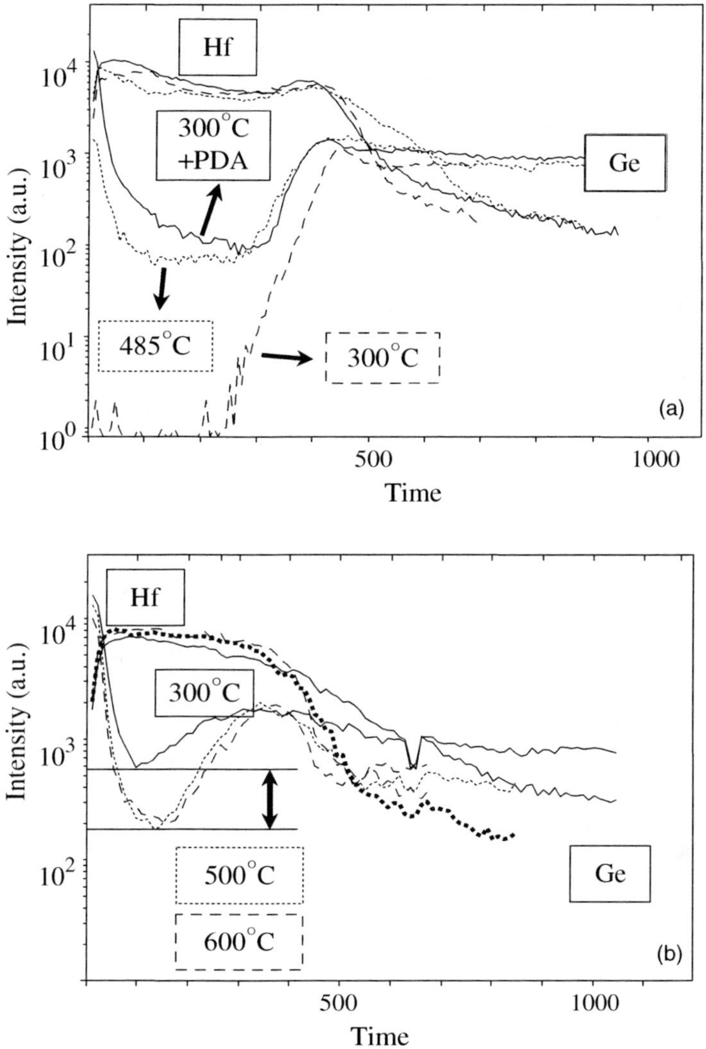

Fig. 6.2. TOF-SIMS profiles of Hf and Ge atoms for HfO_2 layers grown on Ge by MOCVD [27]. (**a**) Depth profiles for HfO_2 grown on Ge at 300°C (with and without PDA) and 485°C with no subsequent NH_3 annealing. (**b**) Depth profiles for HfO_2 grown on Ge at 485°C after annealing in NH_3 at 300°C, 500°C, and 600°C. Reused with permission from Van Elshocht et al. Applied Physics Letters, 85, 3824 (2004). Copyright 2004, American Institute of Physics

Fig. 6.2b, the 300°C NH_3 anneal did not prevent Ge diffusion into the HfO_2 at all. The samples annealed at 500°C and 600°C, however, show suppressed Ge diffusion. Since the TOF-SIMS data for the 500°C and 600°C anneals in Fig. 6.2b look the same, the authors claim that a minimum amount of N is required to prevent Ge diffusion, which they observed to saturate at around 4%. The authors conclude that the prevention of Ge diffusion into the HfO_2 might be the reason for the improved electrical characteristics for surface-nitrided Ge/high-κ MOS devices reported in the literature [27]. They also point out that while samples without nitridation (Fig. 6.2a) have lower Ge diffusion, the high-κ oxides are more prone to crystallization, which causes much higher leakage.

It should be noted here that Seo et al. observed interfacial layer formation prior to molecular beam deposition of HfO_2 caused increased traces of Ge atoms to be detected in the oxide [42]. This result contradicts the results described earlier, but they attributed it to higher instability of the interfacial layers grown with an atomic O or N beam. In this case, the Ge atoms originate from the interfacial layers rather than the substrate. They also used lower growth temperatures (225°C) where diffusion from the substrate is limited.

Van Elshocht et al. also examined the effect of surface nitridation on film quality with cross-sectional transmission electron microscopy (XTEM) [27]. Figure 6.3 shows images for HfO_2 deposited on Si, Ge, and Ge with NH_3 annealing. From the XTEM images we see that the HfO_2 films that were grown directly on Ge without surface nitridation showed much higher interface roughness compared to surface-nitrided devices (compare Fig. 6.3b, c). In addition, comparisons of XTEM images with HfO_2 grown on Si wafers (Fig. 6.3a) revealed that the interfacial layer is much thinner on Ge. The authors conclude that both Ge diffusion and crystallization increase the interface (and/or surface) roughness, and that these can be mitigated using surface nitridation.

Lu et al. studied Ge diffusion and its impact on the electrical properties of HfO_2/Ge capacitors with TaN gates [30]. Figure 6.4 shows SIMS results for three Ge samples that underwent different surface preparation techniques. The ion intensity ratio of Ge to Hf is plotted as a function of sputtering depth for samples with Ge native oxide (no surface preparation), HF-last cleaning, and surface nitridation. It is immediately apparent that the surface-nitrided sample shows reduced Ge diffusion into the HfO_2 layer. This is consistent with the results of Van Elshocht et al. [27]. The authors of [30] proposed two mechanisms that could lead to Ge diffusion into the high-κ dielectric, both of which are related to the existence and amount of GeO_2. The first proposed mechanism was the following reaction taking place at the interface: $Ge + GeO_2 \rightarrow 2GeO$. The resulting volatile GeO (see also Kamata et al. [43]) was thought to diffuse through the oxide and into the high-κ dielectric during growth and postdeposition annealing, with the rate of diffusion enhanced by impurities in the HfO_2. The other proposed mechanism was the desorption of Ge-enriched volatile Hf–Ge–O formed at the HfO_2/GeO_2 interface. One Hf atom would replace a Ge atom from the oxide since Hf is more electropositive

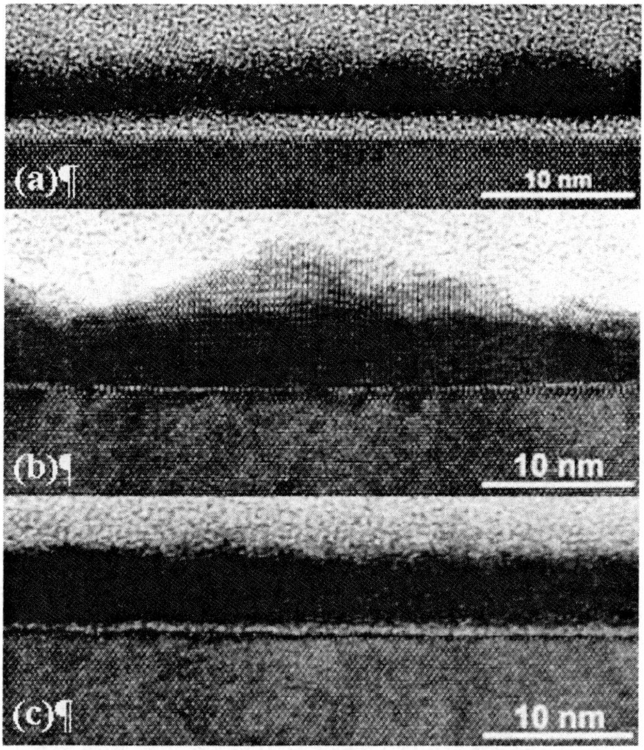

Fig. 6.3. XTEM images of HfO_2 deposited on (**a**) Si, (**b**) Ge, and (**c**) Ge with NH_3 annealing for surface nitridation [27]. The surface-nitrided sample shows a much smoother interface, and both Ge samples have reduced interfacial layer thicknesses. Reused with permission from Van Elshocht et al. Applied Physics Letters, 85, 3824 (2004). Copyright 2004, American Institute of Physics

[44]. Since surface nitridation consumes all of the GeO_2, both of the proposed mechanisms for Ge diffusion are suppressed.

6.4 Effect of Surface Nitridation on the Electrical Characteristics of Germanium MOS Capacitors

Several groups have observed the positive effect of surface nitridation on the electrical characteristics of Ge MOS capacitors [9,27–31,40]. The consensus in the research literature is that nitridation (typically achieved through NH_3 annealing) decreases leakage current and improves the C–V characteristics. Gusev et al. compared C–V and leakage characteristics for HfO_2/Ge capacitors with Ge surface nitridation at with different NH_3 annealing temperatures [40]. Figure 6.5 shows C–V characteristics and leakage current data (inset) for Ge samples annealed in NH_3 for 1 min at 550°C, 600°C, and 650°C. As seen in

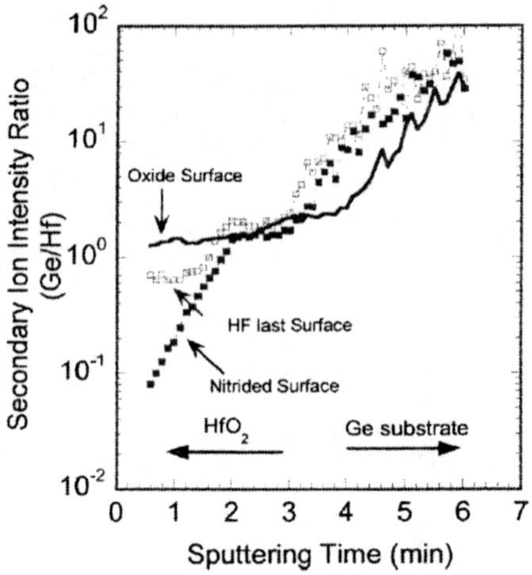

Fig. 6.4. SIMS profiles showing the ratio of Ge and Hf ion intensities in Ge samples with native GeO_2 (no surface preparation), HF-last surface preparation, and surface nitridation [30]. The surface-nitrided sample shows significantly reduced diffusion of Ge atoms into the HfO_2. Reused with permission from Lu et al. Applied Physics Letters, 87, 051922 (2005). Copyright 2005, American Institute of Physics

Fig. 6.5, the EOT decreases with increasing annealing temperature. Although not shown in [40], the *C–V* characteristics for samples that did not undergo NH_3 annealing were said to resemble the characteristics for MIM structures, which was attributed to the formation of Hf–germanide bonds at the interface.

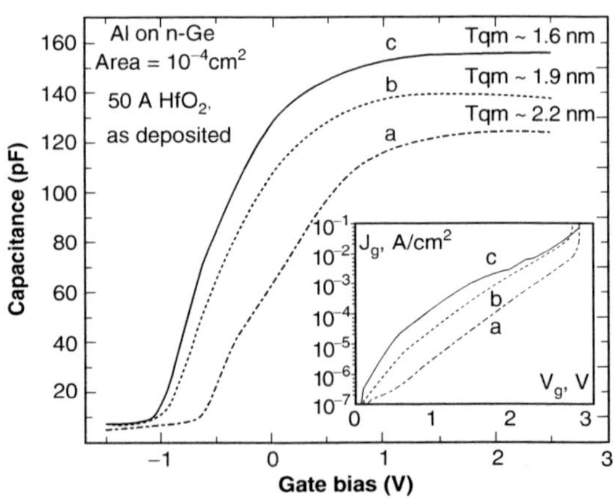

Fig. 6.5. Capacitance–voltage and leakage (*inset*) characteristics of HfO_2/Ge capacitors fabricated on Ge samples annealed in NH_3 for 1 min at (**a**) 550°C, (**b**) 600°C, and (**c**) 650°C. Reused with permission from Gusev et al. Applied Physics Letters, 85, 2334 (2004). Copyright 2004, American Institute of Physics

The study performed by Lu et al. [30] summarized the effect of different surface preparations on the electrical characteristics of HfO_2/Ge capacitors. The data is shown in Figs. 6.6 and 6.7. As seen from this data, the lowest leakage currents, EOTs, flat band shifts, and interface state densities were obtained on samples that underwent surface nitridation.

Our group recently studied the effect that surface nitridation has on MOS capacitor characteristics for two different high-κ dielectrics, Al_2O_3 and HfO_2 [28]. The rest of this section will focus on the results obtained by our group. The layer deposition and capacitor fabrication process will be described first, followed by the electrical results.

6.4.1 Al_2O_3/Ge and HfO_2/Ge Capacitor Fabrication

Al_2O_3 is a possible high-κ candidate for the replacement of SiO_2 for Si CMOS transistors, since it is an inert, high-temperature material, with good thermal stability with Si, expected to withstand Si processing conditions. Furthermore, it has a large bandgap of ~9 eV, conduction and valence band alignments similar to that of SiO_2, and it is a good barrier to ionic transport. Recently, high-quality Al_2O_3 ultrathin films have been grown on Si using various deposition techniques, such as reactive sputtering [45], low-temperature atomic layer chemical vapor deposition (ALCVD) [46] and low-pressure chemical vapor deposition (LPCVD) [47]. In particular, Guha et al. [48] reported the growth of high-quality ultrathin Al_2O_3 films with EOTs as low as 16 Å by reactive atomic-beam deposition. Gate leakage currents were 5 orders of magnitude lower than those of an SiO_2 film of corresponding electrical thickness. Using this deposition technique, the thick interfacial reaction layers previously reported in other Al_2O_3 studies were not observed. Device-quality Al_2O_3 interfaces with interface trap densities in the range of $10^{10}\,cm^{-2}\,eV^{-1}$ were

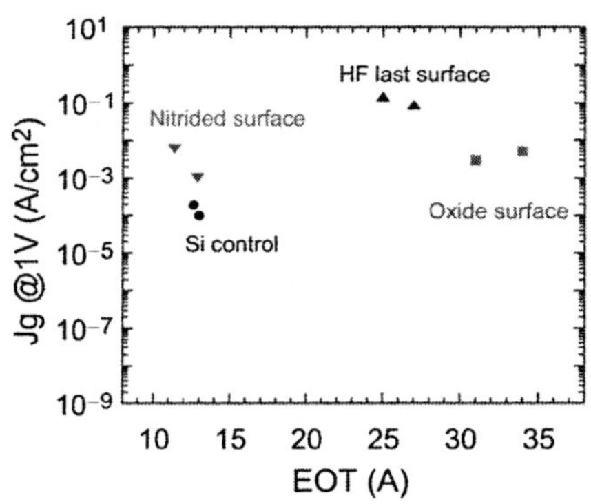

Fig. 6.6. Leakage current and EOT data for HfO_2/Ge and Si control capacitors with different surface preparations. Reused with permission from N. Lu et al. Applied Physics Letters, 87, 051922 (2005). Copyright 2005, American Institute of Physics

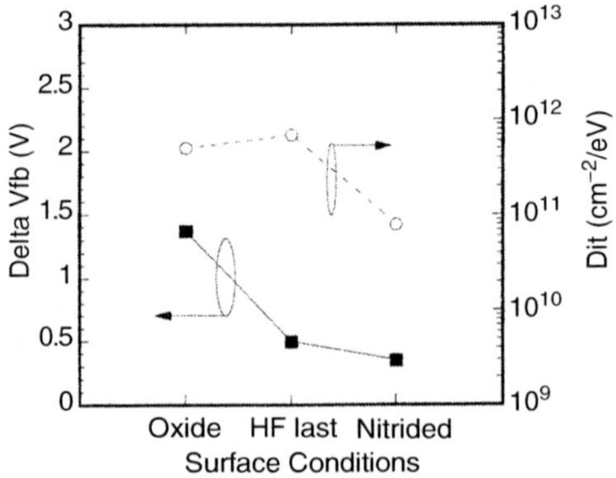

Fig. 6.7. Flat band voltage shifts and interface state densities for HfO_2/Ge capacitors with different surface preparations. Reused with permission from N. Lu et al. Applied Physics Letters, 87, 051922 (2005). Copyright 2005, American Institute of Physics

obtained. We made use of the same reactive atomic-beam deposition technique to deposit ultrathin Al_2O_3 films on bulk Ge.

Following the oxide desorption described earlier, Al_2O_3 was deposited on the Ge wafers in situ using an atomic-beam deposition system. The system consisted of a modified molecular beam epitaxy (MBE) chamber equipped with two radio frequency (RF) discharge sources, used to excite either O_2 or N_2 molecules to produce beams of atomic oxygen or nitrogen directed at the sample surface. Al_2O_3 was grown at temperatures between 250°C and 300°C through simultaneous exposure of Al from a resistively heated high-temperature effusion source and reactive atomic oxygen from the RF discharge source [49]. The Al_2O_3 films were deposited on either as-cleaned or surface-nitrided Ge substrates. Surface nitridation of the Ge took place between 350°C and 600°C by exposure to an atomic nitrogen beam from an RF source at 350 W for 30 s. The N_2 flow rate is maintained at 3 sccm. The Al_2O_3 films were then treated to various postdeposition anneals (under forming gas or O_2) before Al electrodes were evaporated through a shadow mask to define capacitors with areas ranging from 10^{-5} to 10^{-3} cm^2.

HfO_2 has received considerable attention as the leading high-κ dielectric candidate for sub-100 nm technology due to its high dielectric constant (~21–25), reasonably large bandgap (6 eV), and superior thermal stability with conventional polysilicon gates [50]. Kang et al. report that even after a high-temperature anneal treatment at 1,000°C, no detectable interfacial layer formation occurred between the HfO_2 and polysilicon [51]. The thermal stability of HfO_2 separates it from other high-κ candidates that have similar dielectric constants and bandgaps but lower thermal stability, such as ZrO_2. (In fact, ZrO_2 has been shown to react with polysilicon to form silicides at temperatures above 900°C, which degrades the electrical properties [52, 53]. It is also unclear whether the HfO_2/Ge and ZrO_2/Ge systems show the same thermal stability characteristics.) Furthermore, it has been shown that it is possible to deposit ultrathin HfO_2 films (EOT ~ 10 Å) on Si that have leakage current levels more than 4 orders of magnitude lower than that of an SiO_2 film with equivalent thickness [54].

Using the same reactive atomic-beam deposition technique that we used to deposit Al_2O_3, we have deposited ultrathin HfO_2 films on Ge. Similar to our work, Dimoulas et al. have examined HfO_2 deposited using an atomic-beam technique, but focused instead on frequency dispersion effects in much lower EOT gate stacks and Pt gates [55]. For our work, two different gate metals, Al and W, were evaluated for the gate electrode. The surface preparation steps for these wafers prior to HfO_2 deposition were identical to those applied to the Ge wafers with Al_2O_3 deposition. Following oxide desorption, as was the case with the Al_2O_3 samples, surface nitridation of the Ge was performed at temperatures between 350°C and 600°C by exposure to an atomic N beam from a RF discharge source, at 350 W for 30 s. For HfO_2 growth, the Ge substrate was heated to temperatures between 50°C and 300°C and exposed simultaneously to Hf from an electron-beam evaporation source and reactive atomic oxygen from the RF discharge source. The growth temperature was kept low to minimize interaction between Ge and Hf.

After HfO_2 deposition, the Ge wafer was drawn back into the load lock of the system and subjected to UV ozone oxidation to further improve the stoichiometry of the film. Room-temperature UV ozone oxidation has been shown to be successful in producing good-quality oxide films, without significantly increasing the interfacial layer thickness [56]. The load lock is equipped with a Hg vapor lamp, operated at about 50 mW cm^{-2}, which emits UV radiation at wavelengths of 185 and 254 nm. In the presence of the UV radiation, the load lock is flooded with O_2 gas at a pressure of 600 Torr for 60 min. The energies corresponding to these particular wavelengths of UV light are very close to the bond energies of O_2 (~5 eV) and O_3 (~1.1 eV) [57] and interact with the oxygen gas to produce oxygen radicals and ozone, both of which react with the deposited HfO_2 film. MOS capacitors were fabricated using either Al or W metal. HfO_2 capacitors with Al gates were treated to postdeposition anneals before Al metal evaporation, which was performed through a shadow mask in the same manner as the Al_2O_3 capacitors. The capacitors with W gates were formed by in situ e-beam evaporation of blanket W films after the deposition of the HfO_2, followed by standard photolithography and wet etch for defining the capacitor electrodes. Subsequent postdeposition anneals were then carried out.

The deposition of HfO_2 directly on Ge, without surface nitridation, resulted in the formation of an interfacial GeO_x layer, although this result cannot be generalized (Fig. 6.3b). This was verified by medium-energy ion scattering (MEIS) analysis on as-deposited HfO_2/Ge samples, and the depth profiles are shown in Fig. 6.8 [28]. Figure 6.8a shows that for the non-surface nitrided sample, the O depth profile extends well beyond that of the Hf. This indicates the presence of an interfacial oxide. A quantitative modeling of the MEIS spectrum gave 6 Å of interfacial GeO_2 and 33 Å of HfO_2. HfO_2 deposited

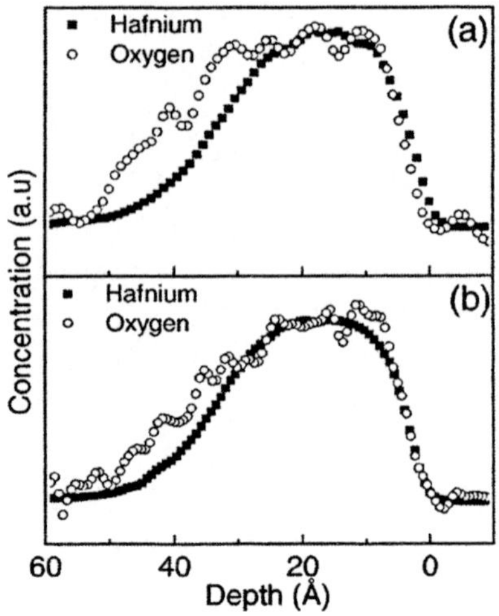

Fig. 6.8. MEIS depth profiles of Hf and O distributions for HfO_2 films deposited on Ge. (**a**) HfO_2 deposited on non-nitrided Ge [28]. The O atoms extend deeper than the Hf, due to an interfacial GeO_2 layer. (**b**) Hf and O depth distributions for HfO_2 deposited on nitrided Ge. The interfacial GeO_2 layer is still present, but its extent is reduced compared to (**a**)

on a surface-nitrided Ge was also analyzed with MEIS (Fig. 6.8b), and it was shown that the O atoms did not extend as far beyond the Hf. The surface-nitrided sample was modeled to have a thinner interfacial GeO_2 layer of 4 Å.

6.4.2 Electrical Characterization of $Al/Al_2O_3/Ge$ Capacitors

High-frequency C–V measurements were carried out on the $Al/Al_2O_3/Ge$ capacitors from 10 kHz to 1 MHz using an HP4284A LCR meter. Figure 6.9

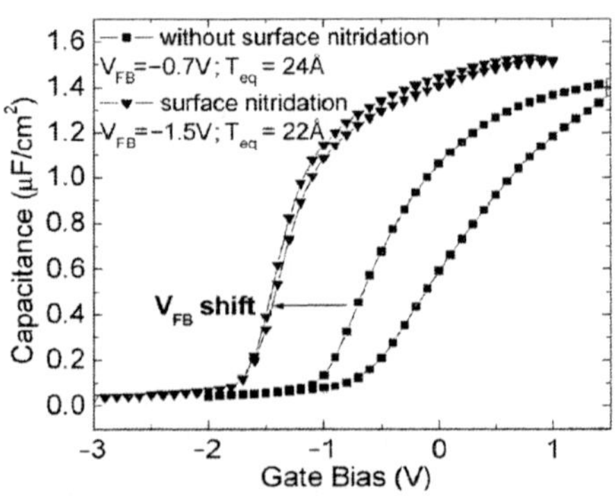

Fig. 6.9. Comparison of C–V characteristics for $Al/Al_2O_3/Ge$ capacitors with and without surface nitridation [28]. The surface-nitrided samples had significant reduction in hysteresis, but also exhibited a large shift in the flat band voltage

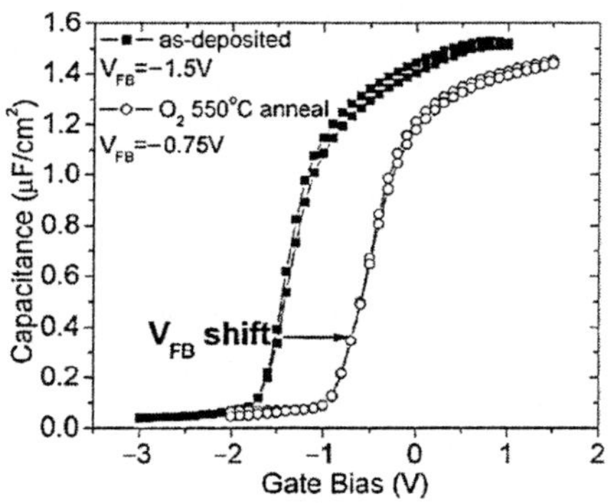

Fig. 6.10. Comparison of C–V characteristics for $Al/Al_2O_3/Ge$ surface-nitrided capacitors with and without a PDA in O_2 for 30 min at 550°C [28]. The surface-nitrided samples had significant reduction in hysteresis [14], but also exhibited a large shift in the flat band voltage

shows high-frequency (1 MHz) C–V measurements taken for two $Al/Al_2O_3/Ge$ capacitors, with and without surface nitridation. The surface nitridation, carried out at 350°C, significantly reduced hysteresis from 500 mV to 20 mV, implying a reduction in charge trapping at the interface. Surface nitridation also reduced the interface state density (D_{it}) from $2 \times 10^{12}\,eV^{-1}\,cm^{-2}$ to $4 \times 10^{11}\,eV^{-1}\,cm^{-2}$, which indicates a higher quality interface. As seen in Fig. 6.9, however, the flat band voltage (V_{FB}) exhibits a large negative shift for the surface-nitrided samples. The ΔV_{FB} is approximately −0.8 V. The explanation could be that positive charge was somehow introduced at the interface during the nitridation process. It might also be possible that a dielectric polarization layer formed as a result of the nitridation. The V_{FB} shift can be partially corrected, however, by performing a PDA in O_2 ambient. Figure 6.10 compares high-frequency C–V characteristics for a $Al/Al_2O_3/Ge$ capacitor that underwent a PDA in O_2 for 30 min at 550°C to a device that did not undergo an O_2 anneal. We notice immediately that the ΔV_{FB} is reversed, pushing the flat band voltage toward its ideal position.

The EOT of the non-nitrided sample dielectric was 24 Å, while the EOT for nitrided gate stack was reduced to 22 Å. Both Al_2O_3 films were deposited under identical growth conditions except for the starting Ge surface. This reduction in equivalent film thickness is due to the presence of the thin nitride layer on the nitrogen-passivated Ge surface, which provides a sufficient barrier to interfacial reaction during dielectric deposition. Similar reduction has been also observed for other high-κ dielectrics deposited on Si [15] and Ge [9] substrates.

Surface nitridation was also observed to reduce the leakage current density, as shown in Fig. 6.11. The leakage current density for the non-nitrided $Al/Al_2O_3/Ge$ capacitor at 1 V is $7\times10^{-3}\,A\,cm^{-2}$. In order to make a meaningful comparison between the leakage current densities between the capacitors

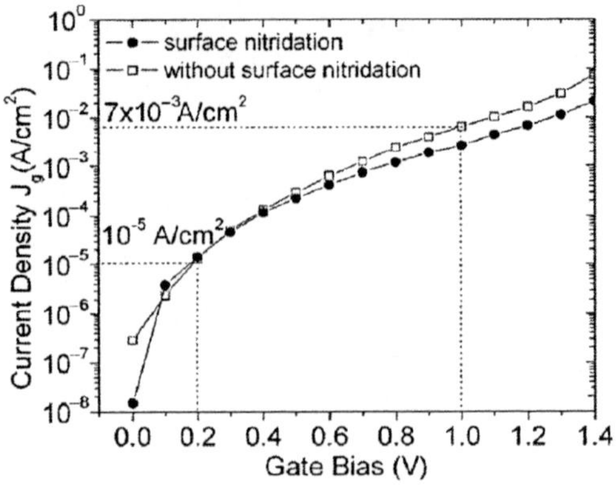

Fig. 6.11. Comparison of leakage current density characteristics for $Al/Al_2O_3/Ge$ capacitors with and without surface nitridation [28]. Surface nitridation reduces the leakage current by nearly three orders of magnitude at equivalent $V - V_{FB}$, denoted by the *dashed lines*

with and without surface nitridation, we have to take into account the difference in the V_{FB}. (Many reports in the literature show leakage currents at applied biases of $\pm 1\,V$, but sometimes this can be misleading. It is always important to take into account the differences in V_{FB} for measuring leakage currents, since bulk and interface charges can compensate the applied bias.) Since the nitrided sample exhibited a ΔV_{FB} of -0.8 V, we should compare the leakage current measured at a bias of 0.2 V. The corresponding leakage current density for the $Al/Al_2O_3/Ge$ sample with surface nitridation (at 0.2 V) was about 1×10^{-5} A cm^{-2}, representing a reduction of nearly three orders of magnitude.

The O_2 anneal at 550°C for 30 min, in addition to recovering the ΔV_{FB}, also improved the leakage current. The EOT increased after the O_2 anneal (Fig. 6.10), which explains the lower leakage. Figure 6.12 shows the effect of two types of PDA on the leakage current density characteristics of nitrided $Al/Al_2O_3/Ge$ capacitors. The O_2 anneal improves the leakage current density by an order of magnitude compared to the as-deposited stack. The current density measured at 1 V is 2×10^{-6} A cm^{-2}. This leakage current is three orders of magnitude lower than SiO_2 with equivalent thickness [58]. An anneal in forming gas at 550°C for 30 min, by comparison, makes the leakage current worse than the as-deposited case.

Charge trapping measurements were carried out on the $Al/Al_2O_3/Ge$ devices using a square pulse technique described in [59]. This technique allows for the measurement of both the trapped charge density (N_{ox}) and the trapping probability (P_n). We define trapping probability as $P_n = q\Delta N_{ox}/\Delta Q_{inj}$, where ΔN_{ox} is the change in the trapped charge density and ΔQ_{inj} is the change in the injected charge. The charge trapping measurements done under substrate injection conditions at a stress bias of 1 V. Figure 6.13a shows the N_{ox} as a function of injected charge for the Al_2O_3 capacitor stack on non-nitrided and surface-nitrided Ge, including the case where the nitrided sample underwent a

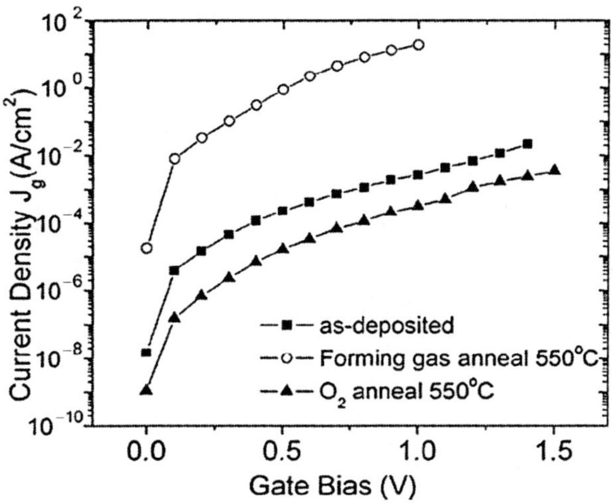

Fig. 6.12. Comparison of leakage current density characteristics for surface-nitrided Al/Al_2O_3/Ge with different PDA conditions [28]. The O_2 anneal (550°C, 30 min) reduces the leakage current by an order of magnitude, while a similar anneal in forming gas increases the leakage current

550°C O_2 anneal. Surface nitridation reduced N_{ox} by about an order of magnitude compared to the non-nitrided case. The O_2 anneal reduced N_{ox} even further ($2–4 \times 10^{10}\,cm^{-2}$). Figure 6.13b is a plot of P_n as a function of the injected charge for both nitrided and non-nitrided samples. The nitrided sample has a P_n that is reduced by an order of magnitude. These results indicate that charge traps could be located in the interfacial layer, which is minimized by the surface-nitridation treatment.

6.4.3 Electrical Characterization of Al/HfO_2/Ge and W/HfO_2/Ge Capacitors

Capacitors made with HfO_2 on non-nitrided Ge surfaces showed significant hysteresis that was slightly higher than the Al_2O_3 films. An example is given in the high-frequency (1 MHz) C–V characteristics shown in Fig. 6.14 for an Al/HfO_2/Ge capacitor, comparing the characteristics of HfO_2 films under two different PDA conditions to that of an as-deposited film. The hysteresis for the non-nitrided, as-deposited HfO_2 film is ~0.7 V. PDAs in forming gas at 450°C (30 min) resulted in a slight decrease in the hysteresis (0.5 V), and a large decrease in the EOT from 25 to 18 Å. Leakage current density characteristics are shown in Fig. 6.15. The current density for both the as-deposited HfO_2 films and the films annealed in forming gas at 450°C were about 1 mA cm^{-2} at 1 V. The leakage was increased by two orders of magnitude for the sample annealed in forming gas at 550°C. It is quite possible that the Ge diffusion effect observed by Van Elshocht et al. [27] became significant at this higher temperature, resulting in an interdiffused interface resulting in higher leakage.

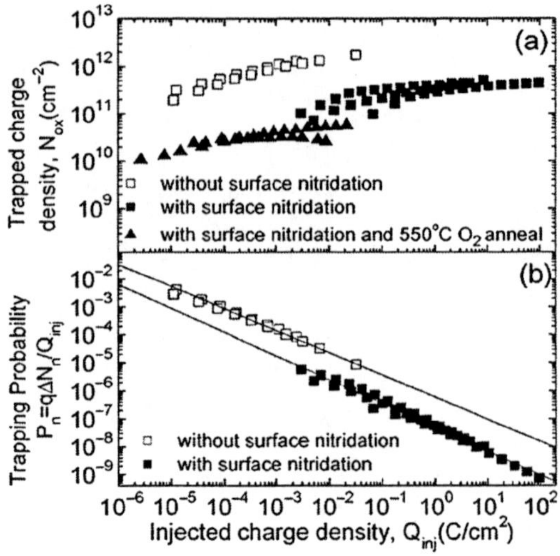

Fig. 6.13. (**a**) Trapped charge density ($N_{\rm ox}$) as a function of injected charge ($Q_{\rm inj}$) for $Al/Al_2O_3/Ge$ capacitors [28]. $N_{\rm ox}$ is reduced by surface nitridation, and is reduced further by a 550°C, 30-min O_2 anneal. (**b**) Effect of surface nitridation on trapping probability ($P_{\rm n}$)

Surface nitridation of the $Al/HfO_2/Ge$ capacitors (prior to HfO_2 deposition) was carried out at 500°C. The effect on the C–V characteristics can be seen in Fig. 6.16. Figure 6.16 shows C–V characteristics for $Al/HfO_2/Ge$ capacitors with and without surface nitridation, with either 450°C or 550°C forming gas PDAs. For the 450°C forming gas PDA, we observe a reduction in the EOT from 18 to 14 Å. The nitridation again introduced a negative

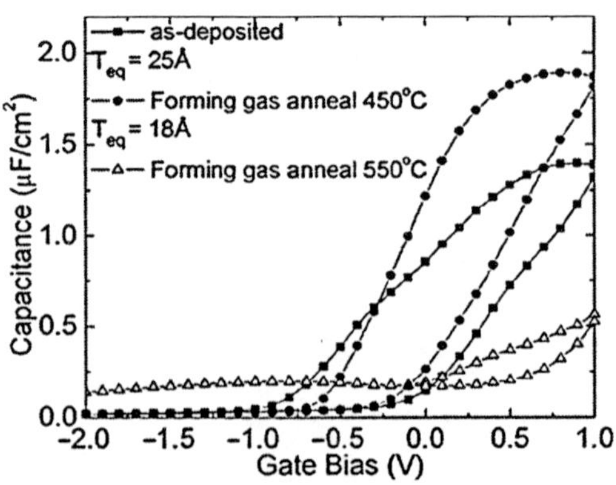

Fig. 6.14. C–V characteristics for non-nitrided $Al/HfO_2/Ge$ capacitors under different PDA conditions [28]. The forming gas anneal at 450°C slightly reduced the hysteresis and decreased the EOT. The 550°C severely degraded the C–V characteristics

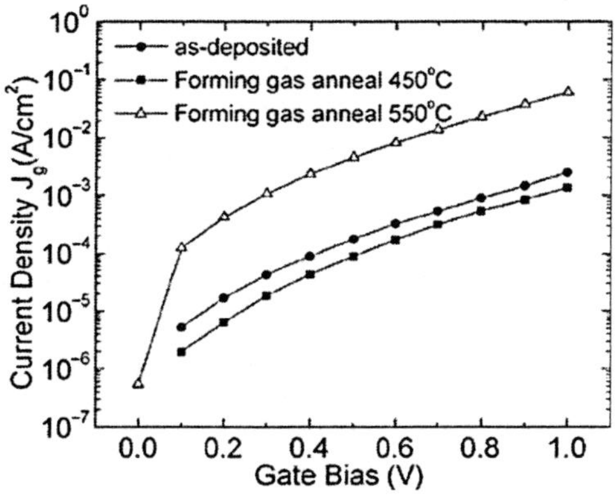

Fig. 6.15. Leakage characteristics for $Al/HfO_2/Ge$ capacitors under different PDA conditions [28]. The as-deposited HfO_2 film and the film annealed at 450°C in forming gas had similar gate leakage, but the sample annealed at 550°C had two orders or magnitude higher leakage

ΔV_{FB} of -0.7 V (from -0.5 V). The hysteresis on the 450°C PDA sample was reduced from 500 to 80 mV with nitridation. These results are similar to the trends observed in our surface-nitrided $Al/Al_2O_3/Ge$ capacitors. Observing the C–V characteristic for the sample annealed at 550°C, we see that the nitridation process afforded an increase in the PDA temperature. This is advantageous because at the higher temperature of 550°C, we observe a reversal of the ΔV_{FB} back toward 0 V, indicating charge compensation or reduction. The C–V hysteresis was reduced to only 10 mV.

Figure 6.17 shows leakage current curves for nitrided and non-nitrided $Al/HfO_2/Ge$ capacitors with 450°C or 550°C forming gas anneals. The nitrided capacitors exhibited two orders of magnitude lower leakage than the

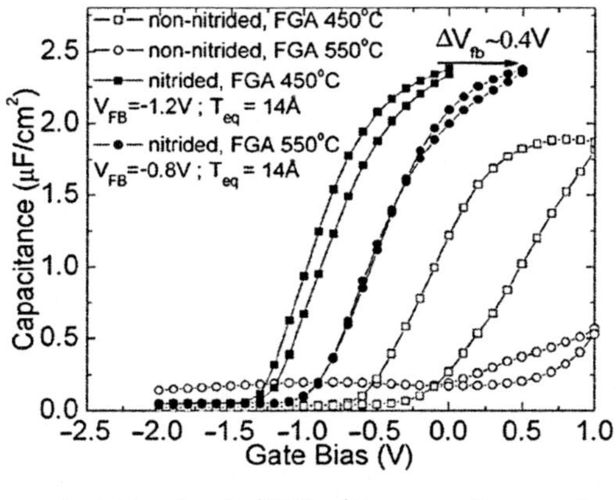

Fig. 6.16. C–V characteristics for $Al/HfO_2/Ge$ capacitors under two different forming gas PDA conditions, with and without surface nitridation [28]. Surface nitridation improved hysteresis and reduced the EOT

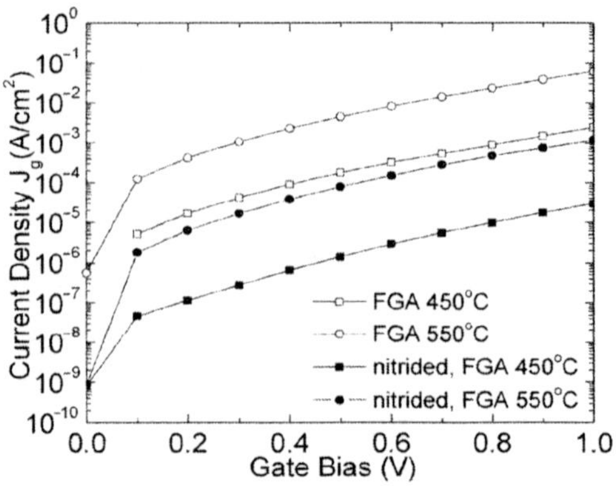

Fig. 6.17. Leakage characteristics for Al/HfO_2/Ge capacitors under two different forming gas PDA conditions, with and without surface nitridation [28]. Surface nitridation reduced the leakage current by two orders of magnitude

non-nitrided samples. Although the forming gas PDA at 550°C is raised by slightly more than an order of magnitude over the sample with a 450°C PDA, the leakage current remained low. The current density was $1 \times 10^{-5}\,\mathrm{A\,cm^{-2}}$ at ($V_{FB} + 1\,\mathrm{V}$), which is six orders of magnitude less than SiO_2 of the same equivalent thickness of 14 Å [58].

The surface-nitrided Al/HfO_2/Ge capacitor, after undergoing forming gas PDA at 550°C, gave the most promising results (EOT $\sim$ 14 Å, hysteresis $\sim$10 mV, and low leakage current density of $1 \times 10^{-5}\,\mathrm{A\,cm^{-2}}$) but some issues remain that need to be addressed. These capacitors still exhibit some frequency dispersion behavior in the C–V depletion region due to a large concentration of interface states. Using the combined high–low frequency capacitance method, the interface state density, D_{it}, was calculated to be around $8 \times 10^{12}\,\mathrm{eV^{-1}\,cm^{-2}}$ near the midgap.

A few conclusions may be drawn regarding the role of surface nitridation for both the Al_2O_3 and the HfO_2 films. Nitrided films are consistently thinner than non-nitrided ones, suggesting reduced interfacial GeO formation. This is consistent with what is observed in the case of nitridation of Si surfaces prior to SiO_2 gate dielectric formation. Nitridation also substantially reduces the hysteresis observed in the C–V plots. The reason for this is not clear, though it may be tied to the reduction in GeO formation. We also see that surface nitridation produces a negative shift in V_{FB} that can be partially recovered by annealing. Finally, the thin nitride layer at the interface reduces leakage current across the film.

The results shown so far have been for Al electrodes on the HfO_2. However, Al electrodes are known to form an Al_2O_3-containing interface layer [59]. This interfacial layer inevitably increases the EOT. We explored an alternative metal, W, in order to examine whether a capacitor with even smaller EOT could be achieved. HfO_2 layers grown on Ge were capped with an in situ W layer (without exposing the HfO_2 to atmosphere). Figure 6.18 shows the

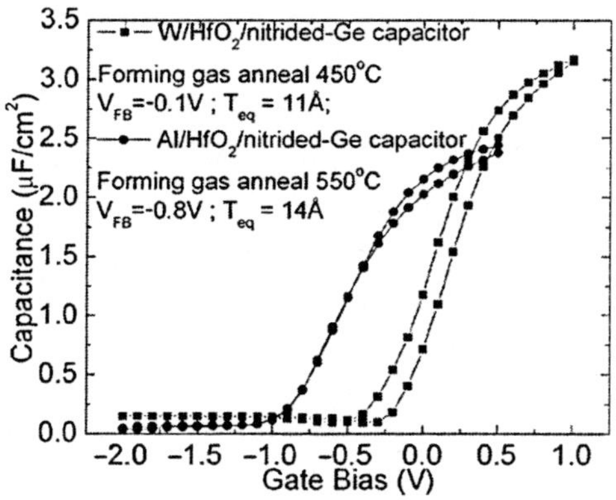

Fig. 6.18. *C–V* characteristics for W/HfO_2/Ge surface-nitrided capacitors under two different forming gas PDA conditions [28]. The device annealed at 450°C had an EOT of 11 Å, but the hysteresis was better for the sample annealed at 550°C

high-frequency (1 MHz) *C–V* measurements taken for W/HfO_2/Ge capacitors under two different PDA conditions, including surface nitridation. The surface nitridation condition (500°C, 30 s, 350 W) and film deposition condition were identical to that of our Al/HfO_2/Ge capacitors. The gate dielectric of the device annealed at 450°C had an EOT of 11 Å, representing a reduction of 3 Å compared to the Al/HfO_2/Ge devices. Correcting for quantum-mechanical effects, which are significant in the case of very thin films, we estimate an EOT of 5.3 Å [39]. We postulate that a lower permittivity Al/HfO_2 interfacial layer had formed in our Al/HfO_2/Ge devices, while no interfacial layer formed between W and HfO_2. This afforded the W/HfO_2/Ge capacitors a smaller EOT. It is unlikely that the in situ (vs. non in situ) capping of the HfO_2 film reduced the interfacial GeO formation, since W has significant O_2 solubility.

With the aim of reducing *C–V* hysteresis (as before), several PDAs of different conditions were carried out. The *C–V* characteristic for the W/HfO_2/Ge capacitor in Fig. 6.18 showed that the best-case hysteresis of 120 mV was achieved after a 550°C forming gas PDA. The D_{it} of the W/HfO_2/Ge capacitor was extracted by the high-frequency/low-frequency method to be around $6 \times 10^{12}\,eV^{-1}\,cm^{-2}$ near midgap. This was similar to the value obtained for the Al/HfO_2/Ge device. From Fig. 6.19, we see that the leakage current density of the W/HfO_2/Ge capacitor was 0.1 mA cm^{-2} (measured at $V_{FB} + 1\,V$), which remains six orders of magnitude less than that for SiO_2 of the same EOT of 11 Å [5].

6.5 Conclusions

High-κ dielectrics have opened the door to Ge-based MOS devices, but many materials-related issues still remain to be resolved. Surface passivation remains

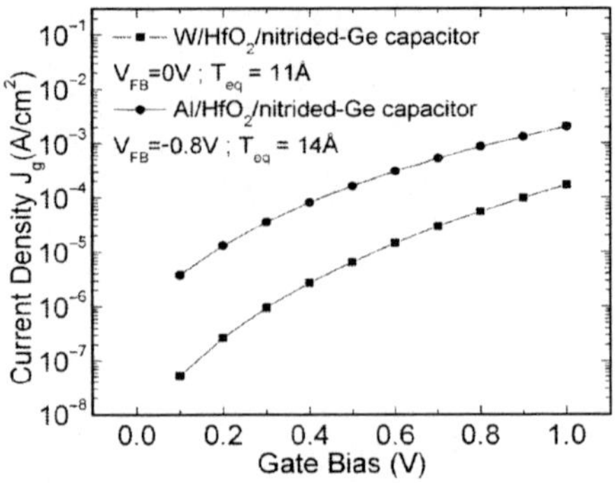

Fig. 6.19. Leakage characteristics for $W/HfO_2/Ge$ surface-nitrided capacitors under two different forming gas PDA conditions [28]

one of the key concerns within this research. In this chapter we have presented a detailed study of the effect of surface nitridation and postdeposition annealing on the electrical characteristics of Ge MOS capacitors incorporating Al_2O_3 and HfO_2 dielectrics. It was found that surface nitridation tends to improve leakage currents and C–V hysteresis, but it also introduces positive charges or dipole layers that cause the flat band voltage to exhibit a negative shift. We have seen that the flat band voltage shift can be mitigated by postdeposition anneals following high-κ deposition.

Acknowledgments

This work was supported in part by SRC, AMRC, and the TARP programs. We also acknowledge Joe Donnelly, Sachin Joshi, and Sagnik Dey for valuable discussions.

References

1. C.O. Chui, S. Ramanthan, B.B. Triplett, P.C. McIntyre, and K. C. Saraswat, "Ultrathin high-κ gate dielectric technology for germanium MOS applications," De. Res. Conf. Proc., pp. 191–192, 2002
2. C.O. Chui, H. Kim, P.C. McIntyre, and K.C. Saraswat, "A germanium NMOSFET process integrating metal gate and improved hi-κ dielectrics," IEDM Tech. Dig., pp. 437–440, 2003
3. C.O. Chui, H. Kim, D. Chi, B.B. Triplett, P.C. McIntyre, and K.C. Saraswat, "A sub-400°C germanium MOSFET technology with high-k dielectric and metal gate," IEDM Tech. Dig., 2002, p. 437
4. H. Shang, H. Okorn-Schimdt, J. Ott, P. Kozlowski, S. Steen, E.C. Jones, H.-S.P. Wong, and W. Hanesch, "Electrical characterization of germanium p-channel MOSFETs," IEEE Electron Dev. Lett., vol. 24, no. 4, pp. 242–244, April 2003

5. H. Shang, H.O. Schmidt, K.K. Chan, M. Copel, J. Ott, P. Kozlowski, S.E.Steen, S.A. Cordes, H.S.P. Wong, E.C. Jones, and W.E. Haensch, "High mobility p-channel germanium MOSFETs with a thin Ge oxynitride gate dielectric," IEDM Tech. Dig., p. 441, 2002
6. N. Wu, Q. Zhang, C. Zhu, D.S.H. Chan, A. Du, N. Balasubramanian, M.F. Li, A. Chin, J.K.O. Sin, and D.-L. Kwong, "A TaN–HfO_2–Ge pMOSFET with novel SiH_4 surface passivation," IEEE Electron Dev. Lett., vol. 25, no. 9, pp. 631–633, September 2004
7. A. Ritenour, S. Yu, M.L. Lee, N. Lu, W. Bai, A. Pitera, E.A. Fitzgerald, D.L. Kwong, and D.A. Antoniadis, "Epitaxial strained germanium p-MOSFETs with HfO_2 gate dielectric and TaN gate electrode," IEDM Tech. Dig., pp. 433–436, 2003
8. C.H. Huang, M.Y. Yang, A. Chin, W.J. Chen, C.X. Zhu, B.J. Cho, M.F. Li, and D.L. Kwong, "Very low defects and high performance Ge-on-insulator p-MOSFETs with Al_2O_3 gate dielectrics," Symp. VLSI. Tech., p. 119, 2003
9. W.P. Bai, N. Lu, J. Liu, A. Ramirez, D.L. Kwong, D. Wristers, A. Ritenour, L. Lee, and D. Antoniadis, "Ge MOS characteristics with CVD HfO_2 gate dielectrics and TaN gate electrode," VLSI Tech. Dig., pp. 121–122, 2003
10. K. Prabhakaran and T. Ogino, "Oxidation of Ge(100) and Ge(111) surfaces: an UPS and XPS study," Surf. Sci., vol. 325, pp. 263–271, 1995
11. J.S. Hovis, R.J. Hamers, and C.M. Greenlief, "Preparation of clean and atomically flat germanium (001) surfaces," Surf. Sci., vol. 444, pp. L815–L819, 1999
12. K. Prabhakaran, F. Maeda, Y. Watanabe, and T. Ogino, "Distinctly different thermal decomposition pathways of ultrathin oxide layer on Ge and Si surfaces," Appl. Phys. Lett., vol. 76, no. 16, p. 2244, 2000
13. D.J. Hymes and J.J. Rosenberg, "Growth and materials characterization of native germanium oxynitride thin films on germanium," J. Electrochem. Soc., vol. 135, p. 961, 1988
14. T. Maeda, T. Yasuda, M. Nishizawa, N. Miyata, Y. Morita, and S. Takagi, "Ge metal–insulator–semiconductor structures with Ge_3N_4 dielectrics by direct nitridation of Ge substrates," Appl. Phys. Lett., vol. 85, no. 15, p. 3181, 2004
15. A.B. Young, J.J. Rosenberg, and I. Szendro, "Preparation of germanium nitride films by low pressure chemical vapor deposition," J. Electrochem. Soc., vol. 134, p. 2867, 1987
16. C.O. Chui, F. Ito, and K.C. Saraswat, "Scalability and electrical properties of germanium oxynitride MOS dielectrics," IEEE Electron Dev. Lett., vol. 25, no. 9, pp. 613–615, September 2004
17. Y. Wang, Y.Z. Hu, and E.A. Irene, "Electron cyclotron resonance plasma and thermal oxidation mechanism of germanium," J. Vac. Sci. Tech. A, vol. 12, p. 1309, 1994
18. V. Craciun, I.W. Boyd, B. Hutton, and D. Williams, "Characteristics of dielectric layers grown on Ge by low temperature vacuum ultraviolet-assisted oxidation," Appl. Phys. Lett., vol. 75, p. 1261, 1999
19. S. Iwauchi and T. Tanaka, "Interface properties of Al_2O_3-Ge structure and characteristics of Al_2O_3-Ge MOS transistors," Jap. J. Appl. Phys, vol. 10, no. 2, pp. 260–265, 1971

20. R.P.H. Chang and A. Fiory, "Inversion layerse on germanium with low-temperature-deposited aluminum-phosphorus oxide dielectric films," Appl. Phys. Lett., vol. 49, p. 1534, 1986
21. D.J. Vitkavage, G.G. Fountain, R.A. Rudder, S.V. Hattangandy, and R.J. Markunas, "Gating of germanium surfaces using pseudomorphic silicon interlayers," Appl. Phys. Lett., vol. 53, p. 692, 1988
22. R.S. Johnson, H. Niimi, and G. Lucovsky, "New approach for the fabrication of device-quality $Ge/GeO_2/SiO_2$ interfaces using low-temperature remote plasma processing," J. Vac. Sci. Technol. A, vol. 18, p. 1230, 2000
23. The International Technology Roadmap for Semiconductors, 2005 Edition, Semiconductor Industry Association, San Jose, CA, 2005, from http://public.itrs.net/
24. D.S. Yu, C.H. Huang, A. Chin, C. Zhu, M.F. Li, B.J. Cho, and D.-L. Kwong, "Al_2O_3–Ge-on-insulator n- and p-MOSFETs with fully NiSi and NiGe dual gates," IEEE Electron Dev. Lett., vol. 25, no. 3, pp. 138–140, March 2004
25. H. Shang, J.O. Chu, X. Wang, P.M. Mooney, K. Lee, J. Ott, K. Rim, K. Chan, K. Guarini, and M. Ieong, "Channel design and mobility enhancement in strained germanium buried-channel MOSFETs," VLSI Tech. Dig., pp. 204–205, 2004
26. H. Shang, J.O. Chu, S. Bedell, E.P. Gusev, P. Jamison, Y. Zhang, J.A. Ott, M. Copel, D. Sadana, K.W. Guarini, and M. Ieong, "Selectively formed high mobility strained Ge PMOSFETs for high performance CMOS," IEDM Tech. Dig., pp. 157–160, 2004
27. S. Van Elshocht, B. Brijs, M. Caymax, T. Conard, T. Chiarella, S. De Gendt, B. De Jaeger, S. Kubicek, M. Meuris, B. Onsia, O. Richard, I. Teerlinck, J. Van Steenbergen, C. Zhao, and M. Heyns, "Deposition of HfO_2 on germanium and the impact of surface pretreatments," Appl. Phys. Lett., vol. 85, no. 17, p. 3824, 2004
28. J.J.-H. Chen, N.A. Bojarczuk, Jr., H. Shang, M. Copel, J.B. Hannon, J. Karasinski, E. Preisler, S.K. Banerjee, and S. Guha, "Ultrathin Al_2O_3 and HfO_2 gate dielectrics on surface-nitrided Ge," IEEE Trans. Elec. Dev., vol. 51, no. 9, pp. 1441–1447, September 2004
29. C.O. Chui, H. Kim, P.C. McIntyre, and K.C. Saraswat, "Atomic layer deposition of high-κ dielectric for germanium MOS applications – substrate surface preparation," IEEE Electron Dev. Lett., vol. 25, no. 5, pp. 274–276, May 2004
30. N. Lu, W. Bai, A. Ramirez, C. Mouli, A. Ritenour, M.L. Lee, D. Antoniadis, and D.L. Kwong, "Ge diffusion in Ge metal oxide semiconductor with chemical vapour deposition HfO_2 dielectric," Appl. Phys. Lett., vol. 87, no. 5, art. 051922, 2005
31. N. Wu, Q. Zhang, C. Zhu, C.C. Yeo, S.J. Whang, D.S.H. Chan, M.F. Li, B.J. Cho, A. Chin, D.-L. Kwong, A.Y. Du, C.H. Tung, and N. Balasubramanian, "Effect of surface NH_3 anneal on the physical and electrical properties of HfO_2 films on Ge substrate," vol. 84, no. 19, pp. 3741–3743, May 2004
32. K. Prabhakarana, T. Ogino, R. Hull, J.C. Bean, and L.J. Peticolas, "An efficient method for cleaning Ge (100) surface," Surf. Sci., vol. 316, pp. L1031–L1033, 1994
33. T. Akane, J. Tanaka, H. Okumura, and S. Matsumoto, "Preparation of high-quality Ge substrate for MBE," Appl. Surf. Sci., vol. 108, p. 303, 1997
34. H. Okumura, T. Akane, and S. Matsumoto, "Carbon contamination free Ge (100) surface cleaning for MBE," Appl. Surf. Sci., vol. 125, p. 125, 1998

35. X.J. Zhang, G. Xue, A. Agarwal, R. Tzu, M.-A. Hasan, J.E. Greene, and A. Rockett, "Thermal desorption of ultraviolet-ozone oxidized Ge (001) for substrate cleaning," J. Vac. Sci. Technol. A, vol. 11, p. 2553, 1993
36. T. Deegan and G. Hughes, "An X-ray photoelectron spectroscopy study of the HF etching of native oxides on Ge (111) and Ge (100) surfaces," Appl. Surf. Sci., vol. 123–124, pp. 66–70, 1998
37. B. Onsia, T. Conard, S. De Gendt, M. Heyns, I. Hoflijk, P. Mertens, M. Meuris, G. Raskin, S. Sioncke, and I. Teerlinck, "A study of the influence of typical wet chemical treatments on the germanium wafer surface," Solid State Phenom., vol. 103/104, pp. 27–30, 2005
38. J.C. Bean, in *Layered Structures, Epitaxy and Interfaces*, Eds. J.M. Gibson and L.R. Dawson (Materials Research Society, Pittsburg), vol. 37, p. 245, 1985
39. C.O. Chui, S. Ramanathan, B.B. Triplett, P.C. McIntyre, and K. C. Saraswat, "Germanium MOS capacitors incorporating ultrathin high-κ gate dielectric," IEEE Electron Dev. Lett., vol. 23, no. 8, pp. 473–475, August 2002
40. E.P. Gusev, H. Shang, M. Copel, M. Gribelyuk, C. D'Emic, P. Kozlowski, and T. Zabel, "Microstructure and thermal stability of HfO_2 gate dielectric deposited on Ge(100)," Appl. Phys. Lett., vol. 85, no. 12, pp. 2334–2336, September 2004
41. I. Chambouleyrona and A.R. Zanatta, "Nitrogen in germanium," J. Appl. Phys., vol. 84, no. 1, p. 1, 1998
42. J.W. Seo, Ch. Dieker, J.-P. Locquet, G. Mavrou, and A. Dimoulas, "HfO_2 high-k dielectrics grown on (100) Ge with ultrathin passivation layers: Structure and interfacial stability," Appl. Phys. Lett., vol. 87, 221906, 2005
43. Y. Kamata, Y. Kamimuta, T. Ino, and A. Nishiyama, "Direct comparison of ZrO_2 and HfO_2 on Ge substrate in terms of the realization of ultrathin high-κ gate stacks," Jap. J. Appl. Phys., vol. 44, no. 4B, pp. 2323–2329, 2005
44. S. Van Elshocht, M. Caymax, S. De Gendt, T. Conard, J. Pétry, L. Daté, D. Pique, and M.M. Heyns, "Composition and growth kinetics of the interfacial layer for MOCVD HfO_2 layers on Si substrates," J. Electrochem. Soc., vol. 151, no. 4, pp. F77–F80, 2004
45. L. Manchanda, W.H. Lee, J.E. Bower, F.H. Baumann, W.L. Brown, C. Opila, P.J. Silverman, T.W. Sorsch, and G.R. Weber, "Gate quality doped high-κ films for CMOS beyond 100nm: 3–10 nm Al_2O_3 with low leakage and low interface states," IEDM Tech. Dig., p. 605, 1998
46. D.A. Buchanan, E.P. Gusev, E. Cartier, H. Okorn-Schmidt, K. Rim, M.A. Gribelyuk, A. Mocuta, A. Ajmera, M. Copel, S. Guha, N. Bojarczuk, A. Callegari, C. D'Emic, P. Kozlowski, K. Chan, R.J. Fleming, P.C. Jamison, J. Brown, and R. Arndt, "80 nm poly-silicon gated n-FETs with ultra-thin Al_2O_3 gate dielectric for ULSI applications," IEDM Tech. Dig., p. 223, 2000
47. T.M. Klein, D. Niu, W.S. Epling, W. Li, D.M. Maher, C. Hobbs, R. I Hegde, I.J.R. Baumvol, and G.N. Parsons, "Evidence of aluminum silicate formation during chemical vapor deposition of amorphous Al_2O_3 thin films on Si (100)," Appl. Phys. Lett., vol. 75, p. 4001, 1999
48. S. Guha, E. Cartier, N.A. Bojarczuk, J. Bruley, L. Cignac, and J. Karasinski, "High-quality aluminum oxide gate dielectrics by ultra-high-vacuum reactive atomic-beam deposition," J. Appl. Phys., vol. 90, p. 512, 2001
49. S. Guha, E. Cartier, M.A. Gribelyuk, N.A. Bojarczuk, and M.C. Copel, "Atomic beam deposition of lanthanum- and yttrium-based oxide thin films for gate dielectrics," Appl. Phys. Lett., vol. 77, p. 2710, 2000

50. S.J. Lee, H.F. Luan, C.H. Lee, T.S. Jeon, W.P. Bai, Y. Senzaki, D. Roberts, and D.L. Kwong, "Performance and reliability of ultrathin CVD HfO_2 gate dielectrics with dual poly-Si gate electrodes," VLSI Symp. Tech. Dig., p. 133, 2001
51. L. Kang, K. Onishi, Y. Jeon, B.H. Lee, C.S. Kang, W.J. Qi, R. Nieh, S. Gopalan, R. Choi, and J.C. Lee, "MOSFET devices with poly silicon on single-layer HfO_2 high-*k* dielectrics," IEDM Tech. Dig., p. 35, 2000
52. M. Copel, M. Gribelyuk, and E. Gusev, "Structure and stability of ulthathin zirconium oxide layers on Si (001)," Appl. Phys. Lett., vol. 76, p. 436, 2000
53. C. Hobbs, L. Dip, K. Reid, D. Gilmer, R. Hegde, T. Ma, B. Taylor, B. Cheng, S. Samavedam, H. Tseng, D. Weddington, F. Huang, D. Farber, M. Schippers, M. Rendon, L. Prabhu, R. Rai, S. Bagchi, J. Conner, S. Backer, F. Dumbuya, J. Locke, D. Workman, and P. Tobin, "Sub-quarter micron Si-gate CMOS with ZrO_2 gate dielectric," Symp. VLSI Technol. Syst. Appl., pp. 204–207, 2001
54. S.J. Lee, H.F. Luan, W.P. Bai, C.H. Lee, T.S. Jeon, Y. Senzaki, D. Roberts, and D.L. Kwong, "High quality ultrathin CVD HfO_2 gate stack with poly-Si gate electrode," IEDM Tech. Dig., p. 31, 2000
55. A. Dimoulas, G. Mavrou, G. Vellianitis, E. Evangelou, N. Boukos, M. Houssa and M. Caymax, "HfO_2 high-*k* gate dielectrics on Ge (100) by atomic oxygen beam deposition," Appl. Phys. Lett., vol. 86, 032908, 2005
56. S. Ramanathan and P.C. McIntyre, "Ultrathin zirconia/SiO_2 dielectric stacks grown by ultraviolet-ozone oxidation," Appl. Phys. Lett., vol. 80, p. 3793, 2002
57. H. Okabe, *Photochemistry of Small Molecules*, Wiley, New York, 1978
58. E. Gusev, D.A. Buchanan, E. Cartier, A. Kumar, D. DiMaria, S. Guha, A. Callegari, S. Zafar, P.C. Jamison, D.A. Neumayer, M. Copel, M.A. Gribelyuk, H. Okorn-Schmidt, C. D'emic, P. Kozlowski, K. Chan, N. Bojarczuk, L. Ragnarsson, P. Ronsheim, K. Rim, R.J. Fleming, A. Mocuta, and A. Ajmera, "Ultrathin high-*k* gate stacks for advanced CMOS devices," IEDM Tech. Dig., pp. 451–454, 2001
59. S. Zafar, A. Callegari, V. Narayanan, and S. Guha, "Impact of moisture on charge trapping and flatband voltage in Al_2O_3 gate dielectric films," Appl. Phys. Lett., vol. 81, pp. 2608–2610, September 2002

7

Modeling of Growth of High-k Oxides on Semiconductors

C.J. Först, C.A. Ashman, K. Schwarz, and P.E. Blöchl

Summary. The replacement of SiO_2 by so-called high-k oxides is one of the major challenges for the semiconductor industry to date. Based on electronic structure calculations and ab initio molecular dynamics simulations, we are able to provide a consistent picture of the growth process of a class of epitaxial oxides around SrO and $SrTiO_3$. To the best of our knowledge this is the only theoretical study which considers the whole growth process starting from the clean Si substrate. Knowledge of the initial growth steps such as metal adsorption on the Si surface has proven to be vital for the understanding of the growth process. The knowledge of the interfacial binding principles has also allowed us to propose a way to engineer the band-offsets between the oxide and the silicon substrate. The results obtained for the Si substrate are directly transferable to Ge.

7.1 Introduction

The replacement of SiO_2 as a gate dielectric in microelectronic devices is one of the key challenges of the semiconductor industry in the next years [1]. Due to their larger dielectric constants, so-called high-k oxides can be deposited with greater physical thicknesses which reduces the quantum-mechanical leakage currents through insulating oxide layers. The growth of these, mostly transition metal containing oxides on the technologically relevant Si(001) surface has proven to be highly challenging. While in a first phase amorphous oxides, based on ZrO_2 or HfO_2 in connection with an ultra-thin SiO_2 layer, will be employed, industry requires solutions for crystalline oxides with an epitaxial interface to the silicon substrate as soon as 2013 [1]. However, there is only one clear demonstration of an atomically well-defined interface between silicon and a high-k oxide so far, namely $SrTiO_3$. Despite the clear experimental evidence of an atomically well-defined interface [2], the interfacial stoichiometry and structure have remained elusive until recently. Key process parameters as well as the electronic properties at the interface are still under debate. Using ab initio molecular dynamics simulations we have been able to unravel the growth process of $SrTiO_3$ on Si(001) and propose a way to engineer the band offsets with respect to silicon [3, 4]. The detailed understanding of the chemistry of group II–IVa transition metals on the silicon(001) surface has proven to be vital in order to rationalize the mechanisms of interface formation. Recent results show that the basic binding principles can also be applied to investigate the interfacial chemistry between silicon and $LaAlO_3$ [5]. This chapter summarizes the results obtained by the authors in this context. In depth information can be found in our previous publications [3–7]. Other theoretical studies on $SrTiO_3$ can be found in [8–10]. We will concentrate on our work since it provides a consistent picture of the growth process from the clean silicon surface to the first few monolayers of the oxide film.

7.2 Computational Approach

Simulations close the gap between analytical theory as well as experiment and constitute a third field of modern materials research. They allow the exploration of models which cannot be solved analytically and therefore contain fewer approximations. Atomistic simulations give information about the position and momentum of individual atoms in molecules, on surfaces or in bulk materials and allow the calculation of the electronic structure. Computational materials scientists can furthermore perform computer experiments under any imaginable condition or stoichiometry, which enables them to explore situations that are not accessible experimentally. For example, knowledge about transitions states is crucial for the understanding of chemical processes but, due to their short life-times, they may not be easily detectable via experiment. However, it is possible to accurately determine transition states computationally. Computational simulation is a rather new and novel field in materials science. The first approaches towards simulation were done during World War II in the context of the Manhattan project. The limited computer capabilities in previous decades and also the nonavailability of reliable models and efficient numerical techniques was responsible that it took several decades until computer simulations became "everyday business."

The basis of every simulation is a reliable model that is a simplified description of the real system. Such models can provide us for example with the total energy as a function of atomic configurations, and maybe also other parameters such as volume, temperature or electronic structure. The proper physical framework in microscopic systems is quantum mechanics. So-called ab initio or first principles models are purely based on the theorems of quantum mechanics and do not require any experimental input parameters other than the chemical composition and charge state of the system. Since for systems of interest it is impossible to solve the Schrödinger equations exactly, approximations have to be made.

The results presented here are based on density functional theory [11, 12], an efficient method to calculate the ground state energies from first principles which has become state-of-the-art in computational materials science. We use the projector augmented wave method [13, 14] to represent wave-functions and densities. The trajectories of the atoms are computed using the ab initio molecular dynamics scheme [15]. All calculations were done on a five layer-model of the (001) surface of the silicon substrate where the bottom layer was saturated with hydrogen and frozen at bulk positions. Further details about the computational procedure can be found in our previous publications in this field [3–7].

7.3 The Chemistry of the Substrate

Silicon crystallizes in the diamond structure. Each atom is tetrahedrally coordinated and forms four covalent bonds. Cleaving the crystal along a (001) plane creates a surface with a quadratic (1×1) array of silicon atoms (compare Fig. 7.1a). Each surface atom exhibits two partially occupied dangling bonds pointing out of the surface. This situation is energetically highly unfavorable and leads to the (4×2) buckled dimer row reconstruction of the silicon (001) surface. We will shortly review its atomic and electronic structure, since it is a key to understand the oxide growth: In a first reconstruction step, two neighboring silicon atoms approach each other and form the so-called dimer bond which is parallel to the surface. The result is the (2×1) dimer row reconstructed silicon surface which leaves only one singly occupied dangling bond state per surface silicon atom (compare Fig. 7.1b). The bonding and anti-bonding states of the dimer bonds move out of the gap into the valence and conduction band, respectively.

The half-filled dangling bond band is then, in a second step, split into two subands, one fully occupied the other empty. The reason for this lift in degeneracy is a geometrical distortion, namely the tilt of the dimer bond. One silicon atom moves up and adopts a quasi-tetrahedral, sp^3-like bonding arrangement, the other one moves down and ends up in a planar, sp^2-like bonding environment. This process is called "buckling." Along one dimer row, the buckling is strongly correlated in an anti-parallel manner. Across the rows, the correlation is comparably weak. This second step results in the final $c(4 \times 2)$ reconstruction of the silicon surface (compare Fig. 7.1c). The fully occupied sub-band is formed by the sp^3-orbitals of the upper silicon atoms and the empty subband with p_z character by the lower silicon atoms. The upper silicon

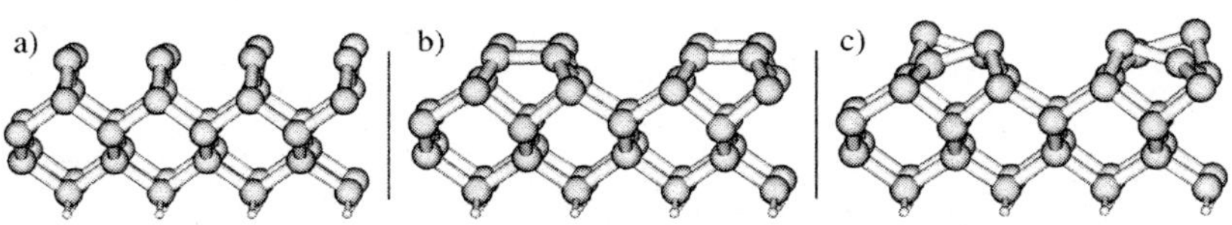

Fig. 7.1. The reconstruction of the Si(001) surface. Panel (**a**): the unreconstructed (1×1) surface; panel (**b**): the (2×1) reconstruction; panel (**c**): the (4×2) reconstruction

atoms are thus formally charged "−1" and the lower silicon atoms "+1." It is important to note that, despite this reconstruction processes, the silicon surface is still considered to be a highly reactive surface. The presence of oxygen or transition metals leads to oxidation and silicidation even at ambient conditions.

7.4 Metal Adsorption on Si(001)

In order to maintain the integrity of the substrate, it is important to avoid the oxidation of silicon. Therefore it is mandatory to start the growth process with the deposition of the metal, an approach followed by most experimental groups. The critical part during growth of an oxide with molecular beam epitaxy (MBE) is the deposition of the metal and the oxidation of the adsorbed metal layers. Therefore a detailed understanding of the adsorption of the metal layer is important to guide the growth process. We have investigated the deposition of metal ions selected from the three most relevant groups in the context of high-k oxides on the silicon(001) surface [16, 17]. These are the divalent earth-alkali metals and the three- and four-valent transition metals exemplified by strontium (Sr)[3], lanthanum (La)[7], and zirconium (Zr)[6].

We simulated a wide range of adsorbed metal layers and scanned the relevant phase space, which is a considerable task. The search was guided by chemical insight, geometric classification and ab initio simulations that provide an unbiased exploration of a small region of the phase space. While this approach cannot exclude, with certainty, that a particular low-energy structure has been missed, the use of a mixture of various search strategies and a sufficiently wide search space seems to yield fairly reliable results.

Figure 7.2 shows the results for Sr adsorption on silicon. The symbols mark the adsorption energies normalized per (1×1) silicon surface cell as function of coverage in monolayers (represented by appropriate supercells). The adsorption energies are obtained relative to a reservoir of bulk silicon and the most stable silicide, in this case $SrSi_2$. Only the lower envelope is physically relevant. At coverages where the line segments meet we predict stable phases. The slopes of the two line segments correspond to the chemical potentials of the phase boundaries. Figure 7.3 shows the phase diagram as a function of the Sr chemical potential derived from Fig. 7.2. The chemical potential can be related to experimental parameters like partial pressure and temperature. The straight line segments indicate coexistence of the two adjacent stable phases. As the coverage increases within one of the line segments, the surface area of the low-coverage phase is converted into that of the high-coverage phase. Thus we are able to predict the sequence of stable phases at zero temperature. From simple thermodynamic considerations we expect that the qualitative features will not change at finite, but not too high temperatures.

One of the main problems during growth of high-k oxides is the formation of silicide grains. Bulk thermodynamic stability of the relevant metals has

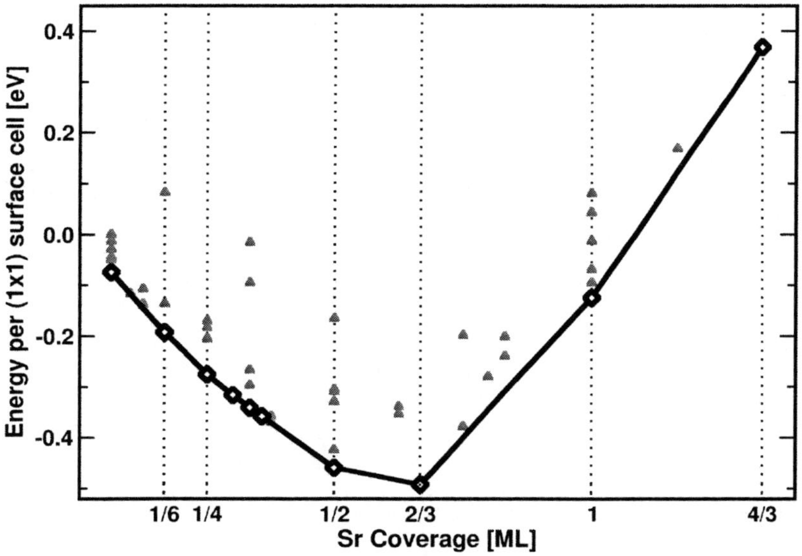

Fig. 7.2. Surface energy vs. coverage for Sr. The *open diamonds* represent thermodynamically accessible structures, the *triangles* correspond to metastable structures

been investigated by Schlom and co-workers [16, 17] but may be misleading for film growth in the monolayer range, since it neglects the effects of epitaxial strain and does not account for the binding of the metal ad-atom layer to the substrate. Going one step further is to compare the energies of the adsorbed metal layer with that of a bulk silicide. Since the adsorption energies have

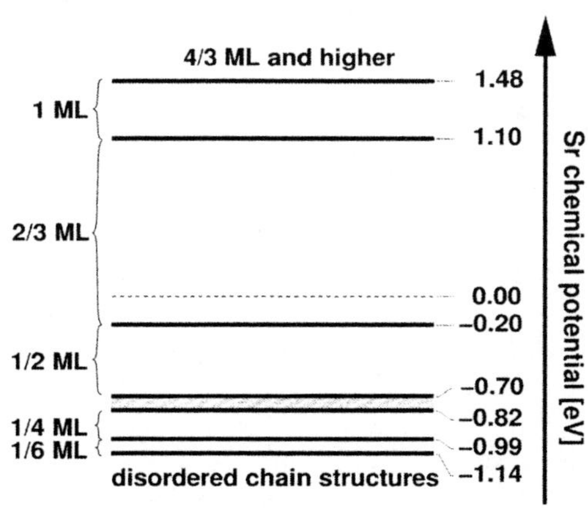

Fig. 7.3. Adsorption phase diagram of Sr on the Si(001) surface as a function of the Sr chemical potential for coverages up to 4/3 ML. For chemical potentials above zero, bulk silicide formation is thermodynamically favorable. The *shaded* region between the 1/4 and 1/2 ML phases indicates disordered surface structures

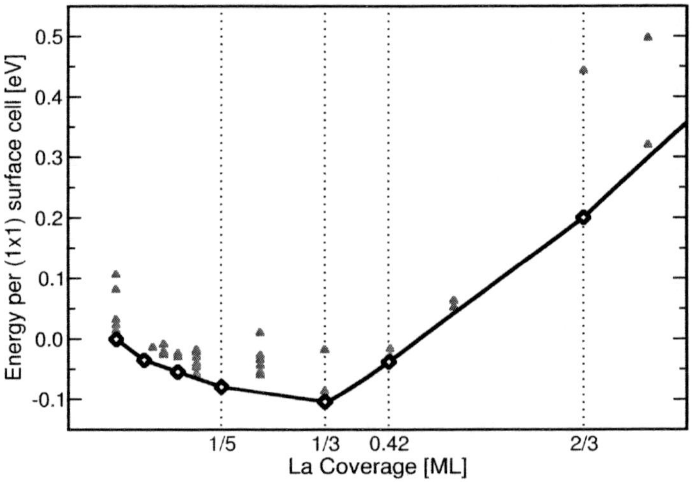

Fig. 7.4. Surface energy vs. coverage for La. The *open diamonds* represent thermodynamically accessible structures, the *triangles* correspond to metastable structures

been calculated with respect to the silicide as a particle reservoir, we obtain the stability limit, where the lower envelope in Fig. 7.2 changes its slope to positive values. Choosing another reservoir would change the slopes of the straight lines, however, the thermodynamically stable reconstructions (i.e., the nodes of the lower envelope) remain the same.

In the case of Sr (compare Fig. 7.2 and 7.3) we calculate coverages up to 2/3 monolayer (ML) as thermodynamically stable [3], whereas for the La surface we find reconstructions only to be stable up to the coverage of 1/3 ML [7] (compare Fig. 7.4). In case of Zr and Hf, all surface reconstructions are unstable with respect to silicide formation, which excludes the possibility of metal pre-deposition [6]. In the case of Zr, we find that the adsorption structures become more stable with increasing coverage indicating that the silicide grows spontaneously. In addition we directly observed the onset of silicide formation in the form of Zr atoms migrating below the Si surface. Zr-silicide formation has also been identified experimentally [18].

The stable surface reconstructions found for Sr and La are driven by the atomic and electronic topology of the Si(001) surface. Here we will summarize the basic principle with the example of Sr. La behaves conceptually similar at low coverages and, due to a change in oxidation state from 3+ to 2+ above 1/3 ML, even identical to Sr at high coverages [7].

The chemistry of Si and Sr is probably best understood in terms of the Zintl–Klemm concept [19–21]. It explains in a simple ionic picture the relationship between stoichiometry and structure for a wide range of compounds between electronegative elements, essentially groups IV–VII, and electropositive elements, mainly groups I and II. The electropositive element, in our example Sr, donates its electrons to the electronegative element, which will

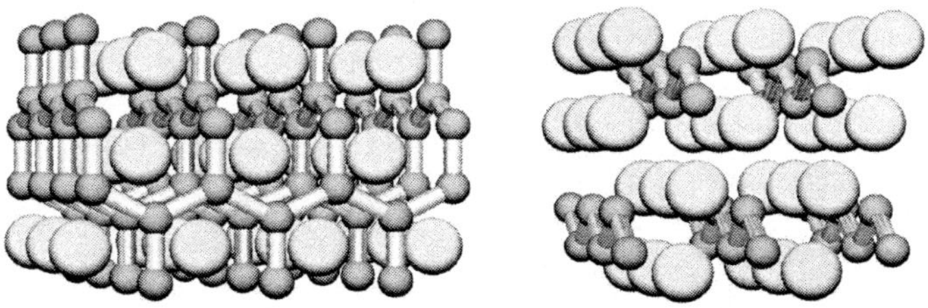

Fig. 7.5. Two typical bulk Sr silicides: $SrSi_2$ (*left*) and SrSi (*right*). Si atoms are visualized by the small spheres connected by sticks and Sr by the big spheres

be Si in this context. Each electron of the electropositive element can saturate one partially filled dangling bond of the electronegative element. As a result the electronegative element forms structures which have a smaller number of covalent bonds. Si^0 forms four bonds, Si^{1-} forms three bonds, Si^{2-} forms two bonds, and so on. The bonding behavior corresponds to a shift to the right in the periodic table of elements for each electron Si receives. Si^-, Si^{2-}, Si^{3-} and Si^{4-} are thus found to be isosteric to P, S, Cl and Ar.

For the example of two bulk Sr silicides shown in Fig. 7.5 we can demonstrate that this simple concept is indeed a useful tool to rationalize the chemistry between Sr and Si. In the left panel we observe that each Si atom forms three covalent bonds. This leaves one valency per Si atom which has to be saturated via electron donation from Sr. Since Sr can donate two valence electrons, the final stoichiometry, according to Zintl–Klemm, must be $SrSi_2$ ($Sr^{2+}Si_2^-$), which is indeed correct. In the example given in the right panel of Fig. 7.5, each Si atom forms two covalent bonds which correspond to two valences that need to be saturated via electron donation or, in other words, formal Si^{2-} ions. The stoichiometry is indeed SrSi ($Sr^{2+}Si^{2-}$). Furthermore, SrSi with its chain structures is a nice example showing that Si^{2-} is isosteric to sulfur which is known for forming chain structures.

Translating the Zintl–Klemm concept to the silicon(001) surface implies that each Sr ad-atom will saturate the dangling bonds of one silicon dimer with its two valence electrons. A saturated dimer looses its buckling, since all dangling bonds are filled and both Si atoms prefer the tetrahedral sp^3 configuration. The atomic structure around an isolated Sr-ad-atom is schematically shown in Fig. 7.6. The preferred adsorption site is in the center of four dimers. One of the neighboring Si dimers is unbuckled due to the electron donation from the ad-atom. The other three dimers are orientated such that the upper, and thus negatively charged Si atom points towards the Sr^{2+} ion. The energy penalty for placing a lower and thus positively charged silicon atom next to an Sr ad-atom is roughly 0.4 eV. Therefore the ad-atoms pin the dimer buckling as observed experimentally [22].

The filled dimer offers a preferred adsorption site in the next valley as indicated by the open circle in Fig. 7.6. As a result, diagonal and zig-zag chain structures turn out to be the thermodynamically stable reconstructions

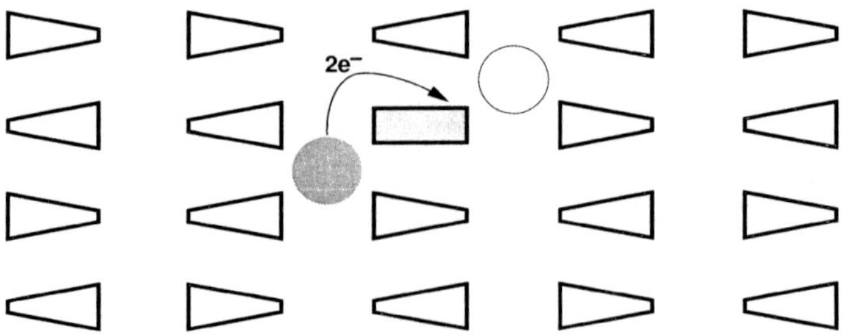

Fig. 7.6. Schematic representation of the isolated Sr ad-atom at the preferred adsorption site position. The *filled circle* represents the Sr ad-atom, the *rectangle* represents a filled and therefore unbuckled Si dimer. The *triangles* represent buckled dimers. The *flat side* of a buckled dimer indicates the upper Si atom with a filled dangling bond, whereas the pointed side indicates the lower Si atom with the empty dangling bond. The charge transfer from the Sr ad-atom to one of the surrounding dimers is indicated by the *arrow*, the preferred adsorption site in the neighboring valley by the *open circle*

at low coverages. At 1/6 ML these chains condense (compare left panel of Fig. 7.7). It is not possible to stack them with a separation of only two instead of three dimer spacings because that would imply that a lower and thus positively charged Si atom points towards an Sr ion which involves a large energy penalty as discussed above. This is the reason for a distinct phase at a coverage of 1/6th ML (compare phase diagram in Fig. 7.3). The next stable phase is made of double chains at 1/4 ML as shown in the right panel of Fig. 7.7. Chain structures at low coverages have also been observed experimentally [23–25].

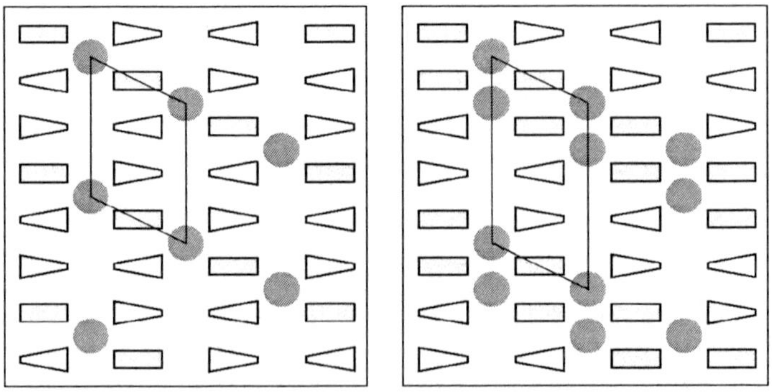

Fig. 7.7. Chain structures of Sr ad-atoms at low coverages. The symbols are explained in Fig. 7.6. The surface unit cells are outlined. *Left*: Condensed diagonal chains of Sr at a coverage at 1/6 ML. *Right*: Diagonal double chains of Sr at a coverage of 1/4 ML

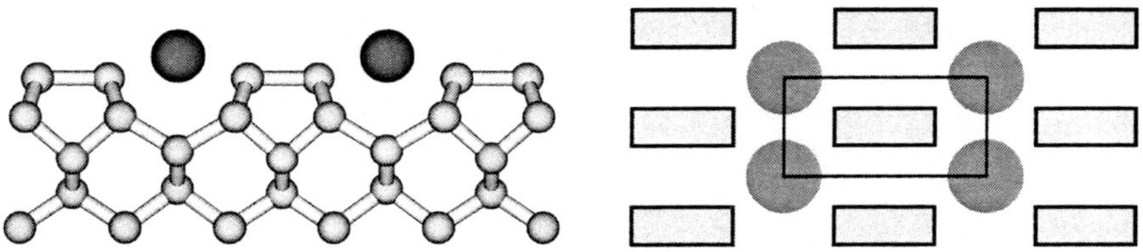

Fig. 7.8. The (2×1) reconstruction at a coverage of 1/2 monolayer. *Left panel*: ball-stick model; *right panel*: schematic top-view

The phases at 1/6 ML and at 1/4 ML correspond to those observed by McKee et al. by their (1×3) and (1×2) RHEED pattern [18]. The presence of these phases has been one main argument for proposing a solid-state transformation to a silicide phase as expected from the bulk phase diagram. However, our calculations clearly show that the structural model proposed by McKee [26] is higher in energy than the structures discussed above. Therefore, we rule out the model of a transformation into a silicide layer of the form proposed by McKee et al. [2] for clean substrates.

At one half monolayer, the Sr ad-atoms occupy all favorable positions in the center of four dimers (compare Fig. 7.8). We labeled this coverage as the "canonical coverage" because all dangling bond states are saturated. Consequently we find that there are no surface states deep in the gap of silicon. This surface is isoelectronic to a hydrogen terminated silicon surface and thus is expected to be chemically comparably inert. Experimentally it has been found that the surface is exceptionally inert against oxidation at 1/2 ML of Sr [27].

Above 1/2 ML, the electrons donated by Sr ad-atoms enter the dimer antibonding states leading to a partial breakup of the dimer bonds. At 2/3 ML we observe a (3×1) reconstruction with alternating rows of Si dimers and isolated Si atoms as depicted in Fig. 7.9. All reconstructions above this coverage are thermodynamically unstable against bulk silicide formation.

Lanthanum adsorption follows similar principles with two main differences [7]: Firstly, La donates three electrons instead of two. This changes the phase-diagram considerably, despite very similar building principles (Fig. 7.10 shows the double-stepped La rows at 1/6 ML as compared to the single-stepped Sr rows shown in Fig. 7.7). Secondly, La deviates from the Zintl–Klemm concept in that it changes its charge state from being a formal La^{3+} ion at lower coverages to a La^{2+} ion at coverages above 1/3 ML. However, the structures above 1/3 ML are no more thermodyamically stable against the formation of La silicide. Note however, that these predictions do not exclude the formation of metastable layers beyond a coverage of 1/3 ML.

These studies of the adsorption of two-, three-, and four-valent metals on Si(001) have led to a unified picture of the processes and have added new insight that allows to explain a number of experimental data. The results furthermore have led to the discovery of the interface structure between silicon and $SrTiO_3$.

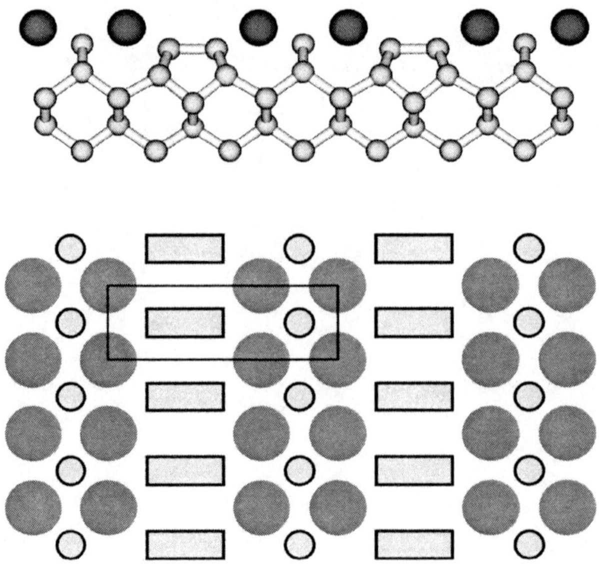

Fig. 7.9. The (3 × 1) reconstruction at an Sr coverage of 2/3 ML. *Top panel*: ball-stick model; *bottom panel*: schematic top-view

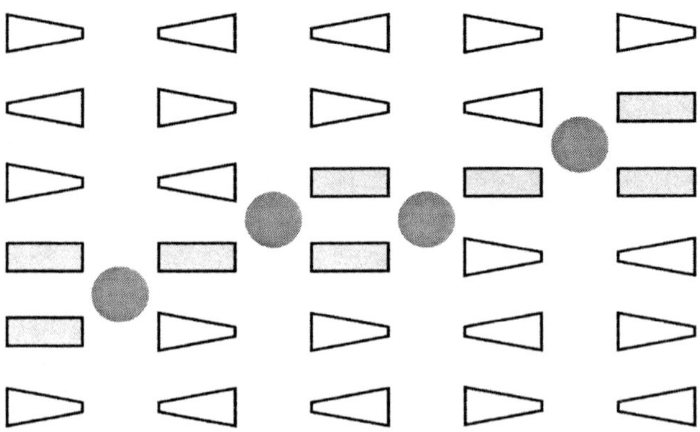

Fig. 7.10. Single La chain on the silicon substrate. Due to the different electron count (La donates three valence electrons to the substrate) we predict double-stepped chains

7.5 Interface of $SrTiO_3$ and Si(001)

The chemical bonding in silicon and $SrTiO_3$ is fundamentally different. While silicon is a covalently bonded material, $SrTiO_3$ is an ionic crystal with some covalent character in the Ti–O bonds. More specifically, $SrTiO_3$ crystallizes in the perovskite structure with Ti being octahedrally coordinated by oxygen and Sr placed in a cubic oxygen cage. The (001) planes of $SrTiO_3$ are

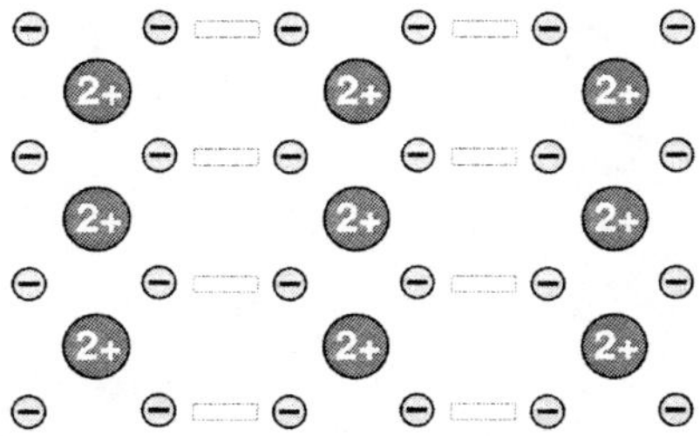

Fig. 7.11. Charge pattern of the silicon surface covered by 1/2 ML of Sr. The Sr ions have a formal charge of 2+, Si of 1− due to the filled dangling bond. The empty *rectangles* denote the dimer bonds, that are included to relate this figure with the lower left panel of Fig. 7.12

alternating SrO and TiO_2 planes that are electronically saturated ($Ti^{4+}O_2^{2-}$ and $Sr^{2+}O^{2-}$) and thus are unable to form covalent bonds. The SrO terminated surface does not exhibit states in the band gap. An electronically saturated Si–$SrTiO_3$ stack must thus exhibit an interfacial layer which provides a covalent bonding environment towards the silicon substrate and in addition an ionic template compatible with that of $SrTiO_3$. The only Sr covered surface meeting these requirements is the reconstruction at 1/2 ML. It is the only one which saturates all silicon dangling bonds and does not have surface states in the band-gap of silicon. The quasi-ionic interaction of Sr with Si furthermore prepares an ionic template with a formal charge distribution as visualized in Fig. 7.11. The resulting two-dimensional ionic layer is compatible with the NaCl-type charge pattern of an SrO-terminated $SrTiO_3$ crystal.

We started from this Sr-passivated substrate and simulated the deposition of one layer of SrO. During a heating cycle to 600 K this single oxide layer reconstructs significantly. However, after placing two or more layers of SrO or $SrTiO_3$ on top of the reconstructed SrO layer, the oxide layers crystallize into the perfect bulk structure. Thus we obtain an atomically abrupt interface between the silicon substrate and the high-k oxide. This interface structure, denoted A and shown in Fig. 7.12, corresponds to the Sr-passivated silicon surface matched to the nonpolar SrO layer of the oxide. It emerged as the only possible candidate for an electronically saturated interface structure. Furthermore it is compatible with all unambiguous features of the Z-contrast image of McKee et al. [2] Similar structures can also be formed by directly growing $SrTiO_3$ with the TiO_2 layer in direct contact with the Sr-covered Si surface.

7.6 Band Offset Engineering

An electrically inactive interface is only one of the requirements a high-k oxide has to meet. A common problem with high-k oxides is a conduction-band offset between silicon and the oxide which is too small. This offset acts as an electron injection barrier, which prevents electrons from being drawn

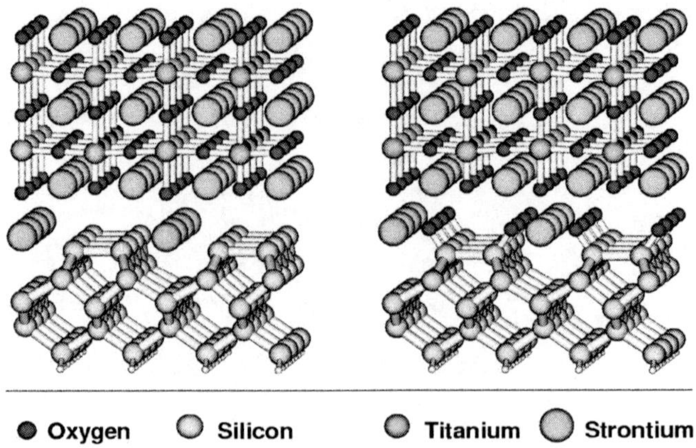

Fig. 7.12. The two relevant interface structures between silicon and $SrTiO_3$. *Left panel*: interface structure A with an Sr-passivated Si surface; *right panel*: interface structure B with oxidized dangling bonds

into the gate from the channel via the gate-oxide. For device applications, the conduction and valence band offsets have to be in the range of 1 eV or larger. However, for the interface just introduced (left panel of Fig. 7.12) we calculate a conduction band offset of only 0.2 eV, which is in line with measurements done on related interfaces [28].

Depending on growth and annealing conditions, the oxygen content of the interface can change. Oxygen can be introduced during growth, or it can diffuse in from the oxide. Therefore it is important to investigate the stability of the interface against oxidation [4]. We find that oxygen first attacks the filled dangling bonds of the silicon dimer atoms. In this way exactly one monolayer of oxygen can be selectively introduced at the interface. The resulting interface structure is shown in the right panel of Fig. 7.12 and labeled interface B. We confirmed that this interface can be formed without growing an interfacial SiO_2 layer by oxidizing the substrate. The phase diagram for oxygen at the interface is shown in Fig. 7.13. At very low chemical potentials, corresponding to a low oxygen partial pressure, interface A is stable. As the chemical potential is increased, dimer bonds get oxidized (above–0.60 eV). Within the region of interface B (−0.24 to 0.05 eV) all dimers are oxidized. When leaving the stability range of interface B, dimer bonds are transformed into Si–O–Si bridges, a phase which we label "dimer-oxidized interface B."

Interestingly we find that the oxygen monolayer at the interface increases the conduction band offset by 1.1 eV. The resulting injection barrier of 1.3 eV is in line with technological requirements. The origin of this effect is visualized in Fig. 7.14. The additional interface dipole formed by the Si^{+} and O^{2-} ions in interface B is responsible for a shift of 1.1 eV in the electrostatic potential within the oxide relative to the silicon substrate. This shift directly translates into a shift of the oxide band structure and thus increases the conduction band offset by the same amount.

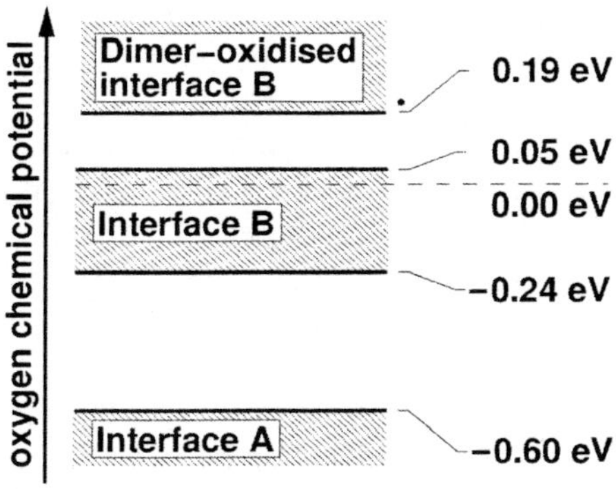

Fig. 7.13. Phase diagram for interface oxidation. *Shaded areas* indicate the stability regions of the defect-free interfaces A and B and the dimer-oxidized variant of interface B. The blank regions separating them correspond to disordered structures with an oxygen content that increases with increasing chemical potential. The external parameter is the oxygen chemical potential. The zero of the chemical potential corresponds to the coexistence of bulk Si and SiO_2 (a-quartz) in thermal equilibrium

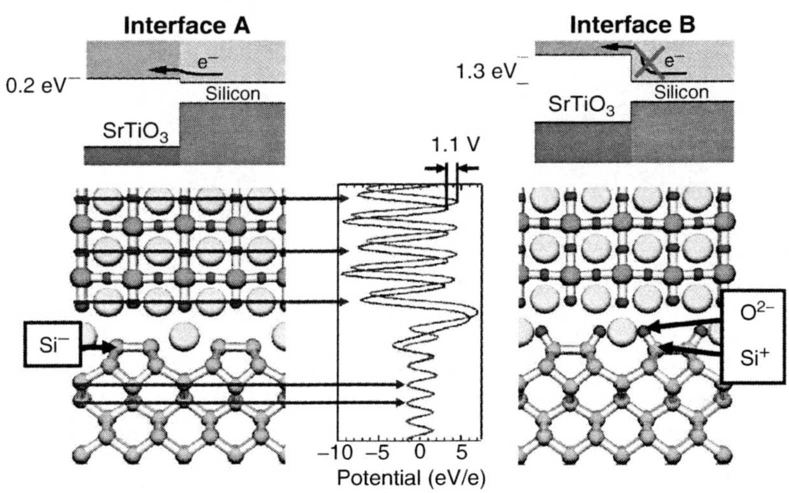

Fig. 7.14. Cause and effect of the interface oxidation. The *top row* shows a schematic drawing of the evolution of the band edges near the interface. The conduction band offset for interface A is only 0.2 eV. The *bottom row* shows interfaces A and B as well as the average electrostatic potential in planes parallel to the interface. The *blue curve* corresponds to interface A, the *red curve* to interface B. The *horizontal arrows* are a guide to the eye and mark the positions of the atomic planes in the potential curve. The relative shift in potential of 1.1 eV is observable and directly translated into a correspondingly increased conduction band offset of interface B (*top right*)

To the best of our knowledge, $SrTiO_3$ is the only high-k oxide, for which a controlled growth of an atomically well-defined interface with silicon has been demonstrated experimentally and theoretically. This system has nearly been discarded from the list of potential high-k oxides because of the small electron injection barrier. Our finding that the band offset may be adjusted by selective oxidation of the interface bears the hope that $SrTiO_3$ can still be made useful for device applications.

7.7 Conclusions

We have summarized the detailed picture of metal adsorption from groups II to IVa on silicon(001) that has emerged from our calculations. We have rationalized the sequence of phases seen during metal adsorption in a simple, intuitive and generalized picture. These simulations have led to a structure model for the interface of $SrTiO_3$ and Si(001), which differs from earlier proposals and is compatible with published experimental results. We furthermore investigated the chemical changes of the interface upon oxidation. We have shown that the electron injection barrier can be dramatically increased by controlled oxidation of the interface, in order to match the technological requirements. It is, however, crucial to first form the $SrTiO_3$/Si interface before the oxidation of the interface is done. Otherwise a too high oxygen pressure in the early stage of the growth process would unavoidable lead to SiO_2 formation that must be prevented.

Acknowledgments

We thank A. Dimoulas, J. Fompeyrine, J.-P. Loquet for useful discussions. Clemens Först acknowledges the support from S. Yip and J. Li. This work has been funded by the European Commission in the project Integration of Very High-K Dielectrics with CMOS Technology ("INVEST"), the European project ET4US "Epitaxial Technologies for Ultimate Scaling" and by the AURORA project of the Austrian Science Fond. Parts of the calculations have been performed on the Computers of the "Norddeutscher Verbund für Hoch- und Höchstleistungsrechnen (HLRN)." This work has benefited from the collaborations within the ESF Program on "Electronic Structure Calculations for Elucidating the Complex Atomistic Behavior of Solids and Surfaces."

References

1. 2003 Edition of the ITRS, http://public.itrs.net/
2. R.A. McKee, F.J. Walker and M.F. Chisholm, Phys. Rev. Lett. **81**, 3014 (1998)
3. C.R. Ashman, C.J. Först, K. Schwarz and P.E. Blöchl, Phys. Rev. B **69**, 75309 (2004)

4. C.J. Först, C.R. Ashman, K. Schwarz and P.E. Blöchl, Nature 427, **53** (2004)
5. C.J. Först, K. Schwarz and P.E. Blöchl, Phys. Rev. Lett **95**, 137602 (2005)
6. C.J. Först, P.E. Blöchl and K. Schwarz, Comp. Mater. Sci **27**, 70 (2003)
7. C.A. Ashman, C.J. Först, K. Schwarz and P.E. Blöchl, Phys. Rev. B **70**, 155330 (2004)
8. X. Zhang, A.A. Demkov, H. Li, X. Hu and Y. Wei, Phys. Rev. B **68**, 125323 (2003)
9. J. Robertson and P.W. Peacock, Mat. Res. Soc. Symp. Proc. 747, **99** (2002)
10. P.W. Peacock and J. Robertson, Appl. Phys. Lett. **83**, 5497 (2003)
11. P. Hohenberg and W. Kohn, Phys. Rev. **136**, B864 (1964)
12. W. Kohn and L.J. Sham, Phys. Rev. **140**, A1133 (1965)
13. P.E. Blöchl, Phys. Rev. B **50**, 17953 (1994)
14. P.E. Blöchl, C.J. Först and J. Schimpl, Bull. Mater. Sci. **26**, 33 (2003)
15. R. Car and M. Parrinello, Phys. Rev. Lett. **55**, 2471 (1985)
16. K.J. Hubbard and D.G. Schlom, J. Mater. Res. **11**, 2757 (1996)
17. D.G. Schlom and J.H. Haeni, MRS Bulletin **27**, 198 (2002)
18. Y.M. Sun et al., Appl. Surf. Sci. **161**, 115 (2000)
19. E. Zintl, Angew. Chem. **52**, 1 (1939)
20. W. Klemm, Proc. Chem. Soc. London, 329 (1958)
21. E. Bussmann, Z. Anorg. Allg. Chem. **313**, 90 (1961)
22. X. Yao et al., Phys. Rev. B**59**, 5115 (1999)
23. R.Z. Bakhtizin, J. Kishimoto, T. Hashizume and T. Sakurai, Appl. Surf. Sci. **94/95**, 478 (1996)
24. R.Z. Bakhtizin, J. Kishimoto, T. Hashizume and T. Sakurai, J. Vac. Sci. Technol. B **14**, 1000 (1996)
25. K. Ojima, M. Yoshimura and K. Ueda, Phys. Rev. B **65**, 075408 (2002)
26. R.A. McKee, F.J. Walker and M.F. Chisholm, Science **293**, 468 (2001)
27. Y. Liang, S. Gan. and M. Engelhard, Appl. Phys. Lett. **79**, 3591 (2001)
28. S.A. Chambers, Y. Liang, Z. Yu, R. Droopad and J. Ramdani, J. Vac. Sci. Technol. A **19**, 934 (2001)

8

Physical, Chemical, and Electrical Characterization of High-κ Dielectrics on Ge and GaAs

S. Spiga, C. Wiemer, G. Scarel, G. Seguini, M. Fanciulli,
A. Zenkevich, and Yu. Lebedinskii

Summary. The physical, chemical, and electrical properties of various high-κ oxides deposited using atomic layer deposition (ALD) on Ge and GaAs are presented. The choice of precursor combination for ALD, as well as the semiconductor surface preparation before film deposition, play a significant role in tailoring the properties of the high-κ oxide stacks. The results obtained for HfO_2, Lu_2O_3, and Al_2O_3 high-κ oxides are discussed and compared, when possible, with those reported in the literature for films grown using various deposition techniques. A review of the available data on the band offset of high-κ oxides deposited on Ge and GaAs is also presented.

8.1 Introduction

The introduction of advanced materials in future ultrascaled devices is a challenging task. In particular, a large worldwide effort is currently devoted to the investigation of high-dielectric constant materials (high-κ) to replace the ultrathin SiO_2 gate oxide [1]. Stringent requirements related to electrical and structural properties, interface properties and thermal stability must be fulfilled [1]. One among the most attractive approaches for high-performance devices is to substitute the silicon channel with high-mobility semiconductors, such as strained-Si, Si–Ge, Ge, and GaAs. However, for these substrates, the optimization of the gate oxide is still in the early stages. The combination of high-κ oxides with high-mobility substrates provides new opportunities for tailoring semiconductor devices and for interface engineering. Indeed, Ge- and GaAs-based metal oxide semiconductor (MOS) transistors, exhibiting good electrical properties with high-κ dielectrics as gate materials, have already been fabricated [2, 3].

High-κ dielectrics on Ge. Among various high-κ materials, HfO_2, and hafnium-based dielectrics are acknowledged to be very promising for ultrascaled Si-based devices. HfO_2 is presently the most studied gate dielectric material for Ge-based devices [3–21]; HfO_xN_y [22], ZrO_2 [23–25], Al_2O_3 [21,26,27], and, recently, rare earth-based oxides [28–31] are also investigated.

Amorphous, polycrystalline, or crystalline films deposited on Ge substrates using various deposition techniques, such as atomic layer deposition (ALD) [4–12, 23], metal organic chemical vapor deposition (MOCVD) [3, 9, 13–15], rapid thermal CVD [16, 17], molecular beam epitaxy (MBE) [18, 29], sputtering [19, 20], and ultraviolet ozone oxidation [21, 24, 25] are currently investigated. The key point for the implementation of high-κ dielectrics on Ge-based devices is to identify a proper Ge surface passivation method to promote low density of interfacial traps (D_{it}) and high channel mobility. Various strategies of Ge surface preparation before the high-κ dielectric layer deposition are therefore currently under investigation. The most promising ones are: growth of a thin GeO_xN_y layer on Ge [6,7,12,14], annealing of Ge in SiH_4 [3, 15, 17], deposition of a thin AlN_x layer [9, 11], and plasma treatment of Ge in PH_3 [11, 32]. An other issue for the integration of high-κ stacks on Ge is the diffusion of Ge inside the film itself [13, 16], either during the oxide deposition or during postdeposition annealing necessary for dopant activation during transistor fabrication. The presence of Ge in the high-κ dielectric film might be detrimental for electrical properties causing an increase of the leakage current [16].

Among the deposition methods of high-κ dielectric layers, ALD is acknowledged to be among the most well-performing ones on Si. On Ge, HfO_2 films deposited using this technique exhibit good electrical properties when a proper Ge surface passivation is obtained, e.g., upon annealing in NH_3 before the high-κ dielectric layer deposition [6, 7, 12]. The possibility to use various precursor combinations for the ALD growth of high-κ dielectrics on Ge is an additional advantage to improve film and interface properties, as already demonstrated for HfO_2 films deposited on Ge and Si [4, 5, 32].

High-κ dielectrics on GaAs. Despite the significant effort over the last 30 years, it is still necessary to identify, for GaAs, proper insulator, thermodynamically stable on this semiconductor, and promoting low D_{it} [33, 34]. Thermal, anodic or plasma oxidation of the GaAs surface produce highly defective interfaces [35, 36]. For this reason, various strategies combining proper surface preparation and deposition of epitaxial or amorphous oxides have been investigated. Ga_2O_3 and Gd_2O_3 or mixed $Ga_2O_3(Gd_2O_3)$ are among the most investigated oxides [34–46]. In particular, $(Ga_2O_3)_{1-x}(Gd_2O_3)_x$ dielectric films, deposited on GaAs by electron-beam evaporation from a single-crystal $Ga_5Gd_3O_{12}$ source, are shown to effectively passivate the GaAs surface [35, 44, 45]. Pure gallium oxide is not a good insulator due to the high leakage [35], even if it plays a crucial role in passivating the GaAs surface [37,38, 44]. It is speculated that the high leakage current of Ga_2O_3 might be due to Ga suboxides [35]. In fact, since Ga has three different oxidation states (i.e., $3^+, 2^+, 1^+$), it is difficult to obtain a pure Ga_2O_3 film. The addition of Gd to a Ga_2O_3 layer enables to reduce the film leakage current, because an electropositive element, such as Gd, can stabilize Ga in the 3^+ oxidation state [35]. Pure Gd_2O_3 films were also grown epitaxially on the GaAs(001) surface [41]. These films exhibit low leakage current and a dielectric constant of 10. The epitaxy of oxides on GaAs was also demonstrated for MgO [47], $SrTiO_3$ [48,49], TiO_2 [50], and NiO [51] oxides, but their electrical properties have not been extensively characterized. Finally, another approach to obtain low defective interfaces between GaAs and high-κ dielectric layers is to grow a thin Si interface control layer by MBE on GaAs before the deposition of any dielectric layer, such as SiO_2, Al_2O_3, and Si_3N_4 [52–55].

Recently, the structural and electrical properties of amorphous or polycrystalline HfO_2 [5,56], Al_2O_3 [2,56–58], Lu_2O_3 films [28,31], grown by ALD on GaAs have been reported. Ye et al. [2] fabricated a metal-oxide-semiconductor field-effect transistor (MOSFET) with thin ALD grown Al_2O_3 gate dielectric. The obtained results indicate that ALD is a promising technique, which can easily find application in device fabrication, to deposit high-κ dielectric layers on GaAs or, possibly, on other III–V semiconductors such as GaN [5].

This chapter reviews the physical, chemical, and electrical properties of high-κ dielectrics deposited on Ge and GaAs using ALD. The goal is to understand composition and structure of the dielectric stacks, and the interfacial layer properties. A comparison with the properties of dielectric stacks deposited on Ge and GaAs using techniques other than ALD is also proposed.

In the first part of this chapter, structural and chemical properties of HfO_2 and Lu_2O_3 films on GaAs and Ge are presented. The role of the precursor combination employed in the ALD growth is discussed. In the second part of this chapter, the electrical properties of various oxides grown by ALD, i.e., HfO_2, Al_2O_3, and Lu_2O_3 are examined. Finally, a review is presented of the results available on band alignment of high-κ dielectric oxides on Ge and GaAs, measured by X-ray photoelectron spectroscopy (XPS) and internal photoemission spectroscopy (IPE).

8.2 Experimental Methodology: ALD Deposition and Characterization Techniques

ALD is a very promising deposition technique, not just because it does not necessarily need ultra high vacuum (UHV), but also because it is already mature for microelectronic device production. Indeed, using adequate precursors and growth parameters, ALD generates conformal, stoichiometric, and smooth films. These good film characteristics are a consequence of the alternate anion and cation precursors injection in the reaction chamber (ALD cycle) [59]. In high base pressure reactors (~1 mbar), those preferred for industrial applications, precursors are transported toward the substrate in an inert carrier gas (usually N_2), and the reaction by-products and residual precursor species are purged away also in a N_2 flux after each precursor injection.

The Ge(001) substrates (Umicore), on which the films described in this chapter were deposited, are n-type (Sb doped, 0.073–0.150 Ω cm). To remove the native oxide, the Ge substrates were dipped for 30 s in a diluted HF solution (1 : 50 = HF : H_2O) at room temperature. Removal of native oxide on n-type GaAs(001) substrates (Wafer Technology Ltd., Si doped and with resistivity in the $2.3 - 5.2 \times 10^{-3}\ \Omega\,\mathrm{cm}$ range) was instead achieved through immersion in a $HCl : H_2O = 1 : 3$ solution for 2 min at room temperature. This procedure leads to an As rich surface with a minority concentration of Ga_2O_3 [60]. Some samples discussed in this chapter were deposited also on Si with a thin chemical oxide layer on top. This layer was obtained with a RCA cleaning ($HCl : H_2O_2 : H_2O = 1 : 1 : 5$ ratio, 10 min at 85°C) followed by a 30 s long dip in a diluted HF solution ($1 : 50 = HF : H_2O$) at room temperature. The chemical oxide generates during a further dip in the RCA cleaning step. A 30 s long rinse in deionized water followed each cleaning step for all semiconductor substrates.

Al_2O_3 films were deposited at 295°C from threemethylaluminum – $Al(CH_3)_3$ or TMA – and H_2O at a growth rate of about 0.09 nm/cycle. HfO_2 films were deposited at 375°C alternating pulses of $HfCl_4$ and H_2O [4,61], of $HfCl_4$ and O_3 [4], or of $HfCl_4$ and the alkoxide-type of compound $Hf(O^tBu)_2(mmp)_2$ [5]. $HfCl_4$ and H_2O on one hand, and $HfCl_4$ and O_3 on the other, generate HfO_2 layers at growth rates of about 0.08 nm/cycle, and 0.10 nm/cycle, respectively. $Hf(O^tBu)_2(mmp)_2$ acts both as metal and oxygen source because the organometallic ligand behaves as a basis inducing nucleophilic substitution of the oxygen with the halogen element in halogen-containing compounds [62]. The combination of $Hf(O^tBu)_2(mmp)_2$ with $HfCl_4$ therefore promotes a high growth rate (0.16 nm/cycle) for HfO_2 films. Lu_2O_3 films were deposited alternating the newly synthesized biscyclopentadienyl complex $[(\eta^5 - C_5H_4SiMe_3)_2LuCl]_2$ with H_2O at 360°C [63,64]. The growth rate was measured to be approximately 0.1 nm/cycle.

Advanced characterization techniques, such as X-ray reflectivity (XRR), X-ray diffraction (XRD), XPS, and low energy ion scattering (LEIS) were used to investigate the structural and chemical properties of the high-κ dielectric stacks, and of their interfaces with Ge and GaAs. Electrical properties were determined using capacitance–voltage (CV), current–voltage (IV), and conductance–voltage measurements on MOS capacitors. The high-κ/semiconductor band alignment was determined using IPE.

Crystallography and stack structure of high-κ dielectrics are usually investigated using direct local imaging techniques such as transmission electron microscopy (TEM), or nondestructive, model-dependent optical (ellipsometry) or X-ray based methods [1]. In contrast to direct imaging techniques, the X-ray based techniques provide structural information-averaged over a large sample volume [65]. Combined XRR and XRD have been demonstrated to be a very powerful tool for structural characterization of high-κ oxides on Si [66]. In particular, within some constraints, XRR measures electron density (ρ), thickness, and roughness of single or multilayered stacks [65,67,68]. XRR is sensitive to the electron density difference of two adjacent layers of different materials. The higher the electronic density difference between two different materials, the easier and the more reliable it is to investigate their respective properties. Therefore, since the electronic density difference ($\Delta\rho$) between

Ge and GaAs and their corresponding oxides is higher than the one between Si and SiO_2 $[\Delta\rho(\text{Ge–GeO}_2) = 0.37\,\text{e}^-\,\text{Å}^{-3}, \Delta\rho(\text{Ga–Ga}_2\text{O}_3) = 0.22\,\text{e}^-\,\text{Å}^{-3}, \Delta\rho(\text{Si–SiO}_2) = 0.06\,\text{e}^-\,\text{Å}^{-3}]$, the investigation of interfacial layer properties on Ge and GaAs is more effective than on Si. XRD and XRR experiments are performed with an X-ray source composed by a sealed copper tube, a parabolic multilayer monochromator able to select a parallel beam of Cu–K_α radiation and a system of crossed slits defining a beam of appropriate size. The sample is mounted on a four circle goniometer. XRR spectra are collected on a scintillator detector. The data are simulated using a model based on the matrix formalism corrected by the Croce–Nevot factor [69]. Grazing incidence XRD spectra are collected on a position sensitive detector (Inel CPS120) and simulated by Rietveld refinement using a freeware software [70]. The four circle goniometer, together with the position sensitive detector allows to perform high-precision and complete Bragg–Brentano, phi scans and reciprocal space maps of symmetric and asymmetric reflections in a relatively short time.

The chemical bonding at the interface formed between ultrathin high-κ dielectric films (HfO_2, Lu_2O_3), and semiconductor substrates (Ge, GaAs) was characterized by XPS using a XSAM-800 (Kratos) spectrometer (Mg K_α source). Auger Ge $L_3M_{45}M_{45}$, Ga L_2 M_{45} M_{45} Auger, and As $3p$ lines were used when Hf or Lu lines overlapped with the usually acquired Ge $3d$, Ga $3d$, and As $3d$ XPS lines. LEIS (He^+ ion beam at $E_0 = 1\,\text{keV}$, and angle between incident beam and detector $\theta = 125°$) was utilized to investigate the surface elemental composition of the oxide layers. To check the thermal stability of Lu_2O_3 films on Ge, one of the samples was annealed up to $T = 500°\text{C}$ in situ in vacuum and in O_2. The experimental details are described elsewhere [71]. It is worth noticing that to exclude detrimental effects of ion etching on the film/substrate interface, ultrathin ($\leq 5\,\text{nm}$) oxide films suitable for direct probing of the interface with XPS were chosen for the analysis.

MOS capacitors were fabricated by Al thermal evaporation through a shadow mask. In–Ga in the eutectic composition is used as back ohmic contact for n-Ge substrate and Au/Ge/Ni/Au for n-GaAs. CV and IV measurements are performed using, respectively, an HP4284A LCR meter and HP4140B picoamperometer.

The barrier energies at the high-κ oxide/semiconductor interface can be experimentally determined using various methods: IPE [72], XPS [73], and ballistic electron emission microscopy (BEEM) [74]. XPS measures directly only the valence band offset (VBO) of the oxide/semiconductor interface. Conduction band offset (CBO) is estimated indirectly if the oxide and semiconductor band gaps are known. On the other hand, IPE measures CBO directly. During IPE measurements, charge carriers are optically excited in the emitter (e.g., the semiconductor) by means of monochromatic radiation in the visible and near ultraviolet energy range. When the incident photon energy is higher than the barrier energy at the oxide/semiconductor interface, charges are excited from the valence band of the semiconductor to the conduction band of the insulator. The photocurrent that flows through the oxide is then measured as a function of photon energy. Quantum yield (Y) spectra are obtained from the ratio of photocurrent and incident photon flux [72, 75]. The barrier energies at the oxide/semiconductor interface are extracted from Y elevated to the (1/3) power. The IPE measurements were acquired at room temperature and in air. The light radiation was supplied by a 150-W Xe arc lamp. A source-meter femptoamperometer supplied the polarization voltage and detected the photocurrent across the MOS. The MOS devices were fabricated with a semitransparent (15 nm thick) aluminum layer.

8.3 Structural and Chemical Properties

8.3.1 HfO_2 Films Deposited by ALD on Ge and GaAs Using Various Precursor Combinations

The properties of HfO_2 films grown by ALD using $HfCl_4$ and H_2O on Ge [4–12], and GaAs [56] are extensively reported. In this paragraph we discuss the structural properties of films grown at 375°C from $HfCl_4$ and O_3, and from $Hf(O^tBu)_2(mmp)_2$ and $HfCl_4$, and we compare them with those of films grown from $HfCl_4$ and H_2O.

HfO_2 films grown at 375°C from $HfCl_4$ and H_2O on Si, Ge, and GaAs crystallize in a mixture of orthorhombic and monoclinic phases [4, 5, 76]. Films grown on an Si surface covered with chemical SiO_2, exhibit no preferential orientation of the crystallites, and the percentage of each detected phase depends on the HfO_2 film thickness. Figure 8.1 shows the grazing incidence XRD

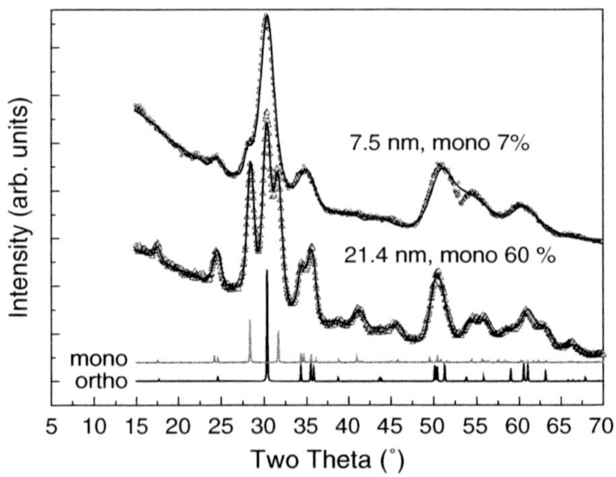

Fig. 8.1. Grazing incidence XRD (*dots*) and Rietveld refinement (*continuous lines*) of 7.5 nm (upper spectrum) and 21.4 nm (lower spectrum) thick HfO_2 film grown on chemical SiO_2. The powder spectra of monoclinic [80] and orthorhombic [81] HfO_2 are also added for comparison

patterns, and the corresponding Rietveld refinement of 7.5 and 21.4 nm thick HfO_2 films grown on Si with a chemical oxide layer on top. Since the films do not exhibit preferential orientation, quantitative Rietveld refinement can be applied. The obtained relative percentages of the two phases in the analyzed film volume are: 7% monoclinic versus 93% orthorhombic and 64% monoclinic versus 36% orthorhombic for the thinner and the thicker samples, respectively. These data demonstrate that the phase composition of a HfO_2 layer grown from $HfCl_4$ and H_2O on a substrate with an amorphous oxide on top varies with film thickness. Therefore, in order to understand the effect of the precursor combination on the structural properties of the HfO_2 films, samples of similar thickness have to be compared. Moreover, some authors report that the crystallization of ALD grown HfO_2 takes place in a mixture of monoclinic and cubic or tetragonal phases [77, 78]. We want to point out, however, that only high resolution grazing incidence X-ray diffraction, or a careful indexation of a selected area diffraction pattern of plan view TEM images [79] can discriminate among cubic, tetragonal and orthorhombic phases of few nanometer thick HfO_2 films.

The properties of HfO_2 films grown using ALD on Ge(001) and GaAs(001) alternating pulses of $Hf(O^tBu)_2(mmp)_2$ and $HfCl_4$ at 375°C were recently investigated [5]. These films have a remarkably smooth surface with roughness weakly dependent on film thickness. Their electronic density values are in the 2.25–2.40 $e^-/Å^3$ range, do not depend on film thickness, and are comparable, within the experimental error of $\pm 0.05\, e^-/Å^3$, with the value (2.42 $e^-/Å^3$) measured for a 12 nm thick HfO_2 film, grown using $HfCl_4$ and H_2O also at 375°C. Grazing incidence XRD spectra reveal similar crystallization for films deposited on Si, Ge, and GaAs. No signs of preferential orientations are found. Films in the 10–20 nm thickness range are weakly crystallized with grains in the monoclinic phase [80]. However, the presence of a minor component of a metastable orthorhombic phase [81] of HfO_2 cannot be ruled out. Actually, films grown with $Hf(O^tBu)_2(mmp)_2$ and $HfCl_4$ have lower grain size and higher microstrain than those deposited using $HfCl_4$ and H_2O at the same growth temperature.

Films grown with $HfCl_4$ and O_3 at 375°C on Ge and GaAs are less crystallized than films of similar thickness grown with $HfCl_4$ and H_2O at the same temperature [4]. Figure 8.2a shows the Bragg–Brentano analysis of a 15 nm thick HfO_2 films grown on GaAs(001) alternating $HfCl_4$ with either H_2O or O_3. The films are polycrystalline in a mixture of monoclinic and orthorhombic phases of HfO_2. The percentage of the orthorhombic component is higher in the films deposited using O_3 rather than H_2O as oxygen source. Using O_3, no signs of preferential orientation are found on both Ge(001) [4] and GaAs(001) (Fig. 8.2a), whereas a preferential orientation develops when H_2O is used as oxygen source (see next paragraph). Figure 8.2b shows the electronic density profiles at the semiconductor/high-κ oxide interface as obtained from XRR data fitting of HfO_2 films of similar thickness grown on Ge and on GaAs using $HfCl_4$ and O_3. In both cases, an interfacial layer (IL)

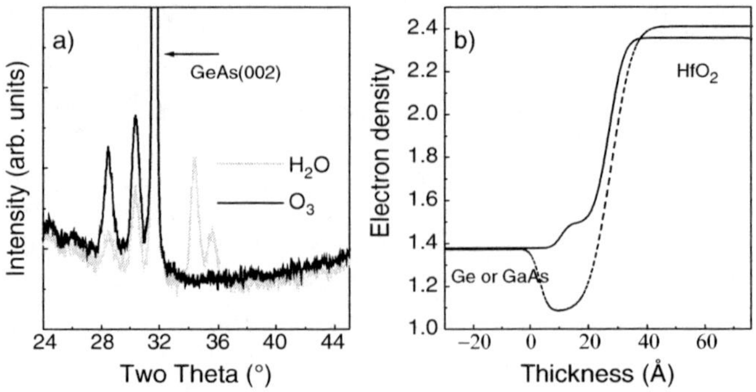

Fig. 8.2. (**a**) Bragg–Brentano analysis of a 15 nm thick HfO_2 film grown on GaAs from $HfCl_4$ and H_2O (*gray*) and $HfCl_4$ and O_3 (*black*). (**b**) Electronic density profile obtained from XRR in the interface region between HfO_2 grown alternating $HfCl_4$ and O_3 on Ge (*dotted line*) and on GaAs (*continuous line*)

develops. On Ge, the IL is ~2 nm thick and its electronic density is compatible with the one of amorphous GeO_2^4. On GaAs, its thickness is between 1 and 2 nm, and its electronic density is close to the one measured for amorphous $Ga_2O_3(1.59\,e^-/\text{Å}^3)$ [82]. although traces of As oxides cannot be ruled out. As in the case of films deposited on Si(001), this amorphous IL between the HfO_2 films and the Ge(001) or GaAs(001) substrates prevents any direct interaction between the HfO_2 crystallites and the semiconductor surface, thus inhibiting epitaxial growth.

XPS characterization indicates a strong dependence upon the precursor combination of the chemical state of the elements in an ultrathin HfO_2 films, and of its IL on top of Ge or GaAs substrates. Using $HfCl_4$ and H_2O, no GeO_2 is observed, i.e., no IL is formed (Fig. 8.3a, case 1. The Ge $L_3M_{45}M_{45}$ Auger line is sensitive to the Ge chemical state). In addition, LEIS results

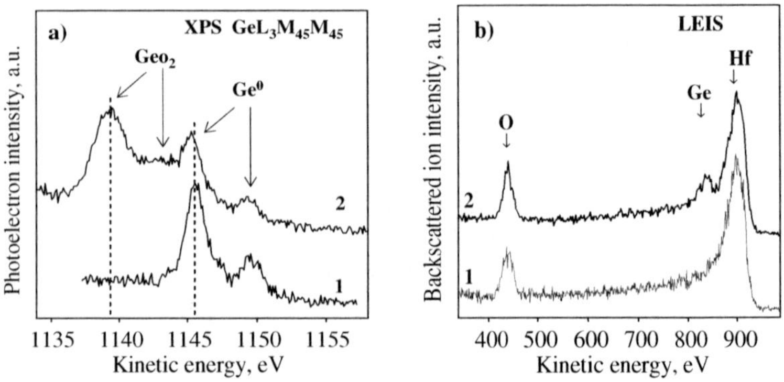

Fig. 8.3. Characterization of HfO_2 layers grown by ALD on Ge(001) with (**a**) XPS and (**b**) LEIS, for different oxygen precursors: (1) H_2O and (2) O_3

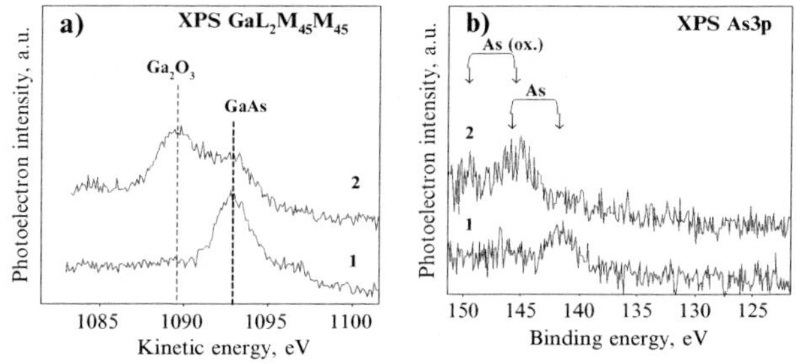

Fig. 8.4. Characterization of HfO_2 layers grown by ALD on GaAs with XPS for different oxygen precursors: (1) H_2O and (2) O_3: (**a**) GaL_2 M_{45} M_{45} Auger line; (**b**) As $3p$ lines

show that, in this case, Ge is missing at the film surface, therefore a 2–3 nm thick HfO_2 film is continuous (Fig. 8.3b, case 1). On the other hand, in HfO_2 films deposited on Ge using $HfCl_4$ and O_3, XPS detects GeO_2 (Fig. 8.3a, case 2). The latter can be located either at the IL (which would be ∼2 nm thick, as determined by XRR measurements) and/or at the film surface. Indeed, LEIS spectra show Ge at the film surface (see the small Ge peak in Fig. 8.3b, case 2), which could be due to Ge diffused through the HfO_2 layer [16], segregated at the film surface, and oxidized. Alternatively, the Ge signal detected in LEIS spectra could come from the substrate, thus revealing a noncontinuous HfO_2 film. Further analyses are necessary to conclude unambiguously between the two possibilities. It is worth noticing that films grown on Ge in the 3.5–10 nm thickness range exhibit good MOS behavior, nevertheless ultrathin films (<3 nm), as those measured by LEIS, were not electrically tested [4].

For HfO_2 films grown on GaAs using ALD and $HfCl_4$ as Hf source, the effect of the chosen oxygen source on the IL chemical nature is quite similar to the one revealed in the case of films on Ge. Indeed, neither Ga nor As oxides were detected at the HfO_2/GaAs interface for films deposited using H_2O as oxygen source, while an IL is clearly detected between HfO_2 and GaAs when O_3 was used to supply oxygen (Fig. 8.4). For films deposited in the latter case, XPS signals of the IL, collected from X-rays that traveled through an ultrathin HfO_2 layer, clearly show oxidized Ga and As. Unfortunately, the low relative sensitivity toward the As $3p$ line does not allow to quantify which oxide dominates at the IL. LEIS data (not shown) indicate that HfO_2 layers deposited on GaAs using $HfCl_4$ and either H_2O or O_3 as oxygen source, are continuous.

8.3.2 Local Epitaxy of HfO_2 Films Grown by ALD on Ge(001) and GaAs(001)

Despite the large lattice parameter mismatch, several authors observed local epitaxy of HfO_2 films on Ge(001) substrates [4, 6, 8, 83]. Gusev et al. [6] and Delabie et al. [8] reported on the epitaxial growth of HfO_2 deposited at 300°C

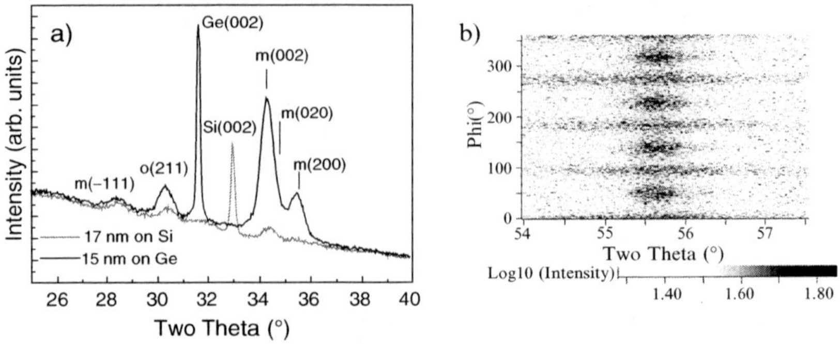

Fig. 8.5. (**a**) Bragg–Brentano analysis of a 15 nm thick HfO_2 film on Ge(001) (*black line*) and of a 17 nm thick HfO_2 film on Si (*gray line*). (**b**) Phi scan around the 013 reflection of monoclinic HfO_2 obtained on the 15 nm thick film grown on Ge(001)

with $HfCl_4$ and H_2O. Spiga et al. [4] reported on the local epitaxial growth when HfO_2 films are deposited at 375°C with the same precursors combination. Local epitaxial growth was also observed by Van Elshocht et al. [83] for HfO_2 films grown by MOCVD from tetrakis((diethyl)amido)hafnium and O_2 at 485°C. All these authors reported epitaxial growth on Ge(001) surfaces prepared by some HF-based cleaning procedures prior to HfO_2 layer deposition [4,6,8,83].

The Si lattice parameter (5.43 Å) is more favorable than the one of Ge (5.64 Å) to promote epitaxial growth of both HfO_2 and ZrO_2 layers [23]. Nevertheless, neither epitaxial growth of ZrO_2 nor of HfO_2 films on Si is reported. The successful occurrence of local epitaxial growth of HfO_2 films on Ge might instead be related to the dissolution of the interfacial Ge oxide layer eventually developed during film growth [23]. Ge oxides are unstable at temperatures higher than 400°C, moreover they are soluble in warm H_2O [84]. Therefore, we might assume that any GeO_x layer that might have formed during one ALD step, e.g., the first injection of H_2O, is then dissociated and removed during the following H_2O injection, leaving an IL-free interface. The lattice parameter of the substrate affects then the orientation of the HfO_2 grains. Moreover, it must be pointed out that no epitaxial growth takes place when O_3 is used as oxygen source.

Films in the 6–21 nm thickness range grown using $HfCl_4$ and H_2O on Ge(001) at 375°C exhibit the majority of the monoclinic {001} planes parallel to the Ge(001) surface [4]. In Fig. 8.5a, the Bragg–Brentano analysis of a 15 nm thick HfO_2 film on Ge(001) is compared with the one of a 17 nm thick HfO_2 film grown on Si(001). The intensities of the 002, 020, and 200 monoclinic reflections are higher for the film grown on Ge than for the one grown on Si. The epitaxial relationships are Ge(001)//mHfO_2{001} and Ge[111]//m$HfO_2\langle\bar{1}11\rangle$. The in-plane mismatch between Ge(001) and HfO_2(001), HfO_2(010) or HfO_2(100) is below 10%, which can actually lead

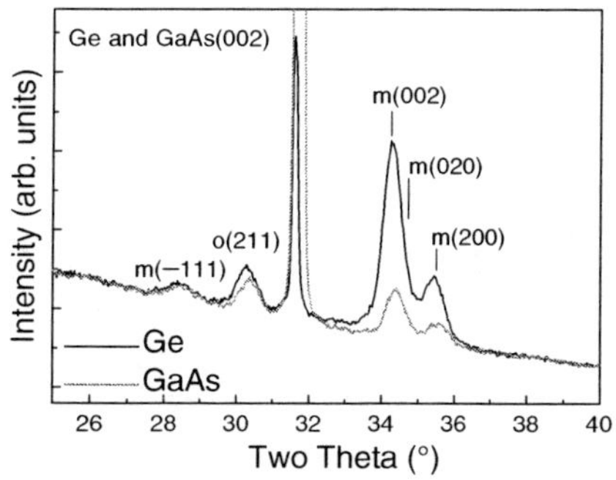

Fig. 8.6. Bragg–Brentano analysis of a 15 nm thick HfO_2 film on Ge (*black line*) and on GaAs (*gray line*)

to the observed epitaxial relationships. Moreover, due to the little difference between the a and b lattice parameters of the monoclinic phase ($a - b = 0.06$ Å [80]), the HfO_2(001)//Ge(001) orientation is more favored. In particular, to understand the in-plane orientation of the monoclinic (001) planes, a complete phi scan around the 013 asymmetric reflection was performed (Fig. 8.5b). The elongated spots at phi $= 0°, 90°, 180°$, and $270°$ are due to the Ge{311} reflections, whereas the four sharp spots at $45°, 135°, 225°$, and $315°$ are due to the monoclinic (013) and (103) HfO_2 planes. This finding suggests that there are two preferential in-plane orientations. Since the relative intensities of the four sharp spots are equal for all phi angles, we conclude that the growth of HfO_2 monoclinic (001) crystals oriented parallel to the Ge(001) plane takes place with the same probability in the two "cube on cube" in-plane orientations: Ge[100]//HfO_2(100) and Ge[100]//HfO_2(010).

HfO_2 films deposited by ALD using $HfCl_4$ and H_2O at 300°C on HF-treated GaAs(001) surfaces were studied by Frank et al. [56] A 4 nm thick HfO_2 film on GaAs is shown to exhibit monoclinic HfO_2 crystallites embedded in an amorphous matrix. Moreover, the authors report the formation of a 0.8 nm thick IL. Local epitaxy is observed in HfO_2 films grown at 375°C on GaAs(001), cleaned in a HCl-based solution, and using the same precursor combination as the above mentioned one (Fig. 8.6). The HCl-based cleaning procedure is reported to be very effective in removing the native oxide [60]. GaAs surfaces treated using this procedure indeed are shown to be As-rich, and with only a minor Ga_2O_3 component [60]. Moreover, at 375°C, almost all the mixed Ga and As oxides that might have formed during the ALD cycles could have desorbed [85]. Indeed, fittings of the XRR spectra are equally good using models either with or without an ~1 nm thick IL between the HfO_2 film and the GaAs substrate. Moreover, XPS measurements do not reveal any IL between the HfO_2 layer and the GaAs substrate, as previously discussed. Figure 8.6 compares the Bragg–Brentano spectra of 15 nm thick HfO_2 grown on Ge and on GaAs at 375°C. Since the GaAs lattice parameter (5.65 Å) is only 0.07% higher than the one of Ge, the same epitaxial relationships

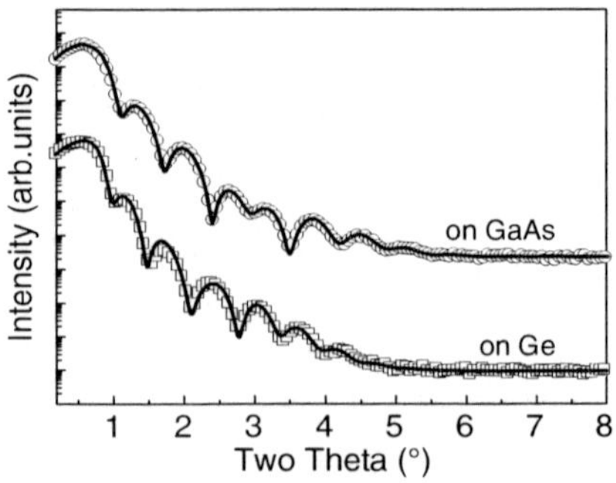

Fig. 8.7. XRR spectra and corresponding fittings (*lines*) of Lu_2O_3 films: 13.6 nm thick film on Ge (*squares*) and 10.7 nm thick film on GaAs (*circles*)

observed for HfO_2 films on Ge hold for those grown on GaAs. However, due either to some Ga oxides left on the GaAs surface and/or to the different oxidation states of Hf(4) on one hand, and Ga(3) and As(3) on the other, the monoclinic {001} out of plane orientations of the HfO_2 films are less developed on GaAs than on Ge. In particular, the monoclinic (001) reflection is much less intense for HfO_2 films grown on GaAs than on Ge, further suggesting that the effect of the substrate lattice parameter is screened by some effect related to the chemistry at the interface.

8.3.3 Lu_2O_3 Films Deposited by ALD on Ge and GaAs

The successful deposition of Lu_2O_3 layers on Si using ALD was recently obtained combining $[(\eta^5 - C_5H_4SiMe_3)_2LuCl]_2$ [63] and H_2O at 360°C [64]. Lu_2O_3 films deposited on GaAs and Ge clean surfaces exhibit the same structural properties than those grown on chemical SiO_2 [64]. They are nanocrystalline, with grain size always lower than film thickness. The crystallites exhibit the cubic bixbyite structure in the *Ia*-3 space group. Figure 8.7 shows the XRR spectra of Lu_2O_3 films grown on Ge and GaAs. XRR data fitting indicates either the existence of a rough Lu_2O_3/semiconductor interface, or of a low-density IL, whose thickness is below 1 nm.

The XPS and LEIS spectra of ultrathin Lu_2O_3 films deposited on Ge(001) using ALD are shown in Fig. 8.8. No Ge peak is detected in the LEIS spectrum (Fig. 8.8c), indicating that a ~3 nm thick Lu_2O_3 layer is continuous (film thickness was determined using XRR data). Cl, and some Si, both possibly released from the Lu precursor, are however detected on the surface of the as grown film. The Ge $L_3M_{45}M_{45}$ Auger and the O 1*s* XPS lines for the as grown and the annealed (in vacuum and O_2) Lu_2O_3/Ge(001) samples are shown in Fig. 8.8a,b, respectively. The XPS spectra reveal that: (a) as for Lu oxide films on Si [64], the as grown ultrathin Lu "oxide" on Ge is mostly a Lu hydroxide (see the O(II) peak in the O 1*s* line, Fig. 8.8b); (b) there is no Ge oxide-based IL between the film and the Ge substrate (compare the Ge $L_3M_{45}M_{45}$

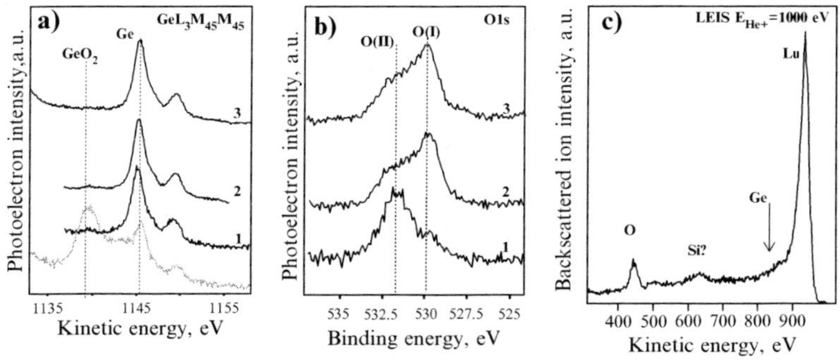

Fig. 8.8. XPS ((**a**) Ge $L_3M_{45}M_{45}$ Auger line, (**b**) O1s line) and LEIS (**c**) of Lu_2O_3 films grown by ALD on Ge(100): (1) as grown film, (2) film upon vacuum annealing up to $T = 500°C$, and (3) upon annealing in O_2 ($P_{O_2} = 10^{-6}$ Torr, $T = 500°C$). Ge $L_3M_{45}M_{45}$ line for oxidized Ge is also shown in gray color for comparison

line for the Lu_2O_3/Ge sample, with the one for oxidized Ge). This result is consistent with the XRR data; (c) the IL does not change upon annealing up to 500°C both in vacuum and in $O_2(P_{O_2} = 10^{-6}$ Torr). Moreover, upon annealing $Lu(OH)_3$ is mostly converted into Lu_2O_3, as the O(I) peak in the O 1*s* spectrum clearly indicates. In contrast to the case of films deposited on Si, transforming into Lu silicate upon annealing [71], a continuous ultrathin pure Lu_2O_3 layer can be formed on Ge upon moderate annealing, with no sign of interfacial reactions. Therefore, because of the absence of the IL also upon annealing, the Lu_2O_3/Ge system looks more promising than the Lu_2O_3/Si one to achieve a significant equivalent oxide thickness (EOT) reduction in future devices.

For Lu_2O_3 films deposited on GaAs using ALD, XPS (Fig. 8.9a,b) and LEIS (Fig. 8.9c) analyses indicate that no IL is present and that the ultrathin layers (2–3 nm thick) are continuous.

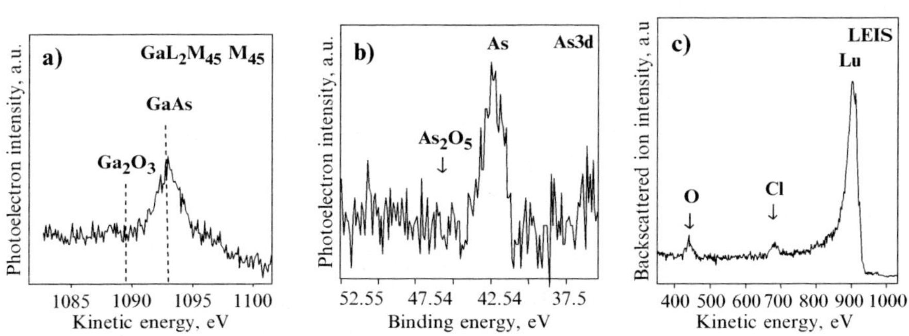

Fig. 8.9. XPS (**a**,**b**) and LEIS (**c**) spectra taken from as grown Lu_2O_3 films on GaAs. (**a**) GaL_2 M_{45} M_{45} Auger line; (**b**) As 3*d* line

8.4 Electrical Properties of High-κ Dielectrics on Ge and GaAs

8.4.1 Electrical Properties of High-κ Dielectrics Deposited on Ge: HfO_2, Al_2O_3, and Lu_2O_3

The electrical properties of HfO_2 films grown by ALD at 300–375°C on Ge from the combination of $HfCl_4$ and H_2O are reported by several authors [4, 6–12]. Ge surface preparation before growth is shown to strongly influence the measured leakage current, the EOT, the hysteresis and the D_{it} [6, 7, 9, 12]. The growth on clean Ge surface usually leads to poorly passivated interfaces [4, 6, 12]. On the other hand, HfO_2 films deposited by either ALD or other techniques on nitrided Ge surfaces exhibit a better interface quality. The nitridation process leads to a reduction in D_{it}, hysteresis and leakage current with respect to the direct deposition of films on HF-treated Ge surface or on native Ge oxide [6, 7]. Nevertheless, the measured D_{it} is still in the $10^{12}\,eV^{-1}\,cm^{-2}$ range or higher [18, 21], only Lu et al. [16] report a D_{it} as low as $8\times10^{10}\,eV^{-1}\,cm^{-2}$ for HfO_2/GeON/Ge stacks. Moreover, the nitridation process might also introduce additional trap levels and cause a significant flat band voltage shift [6, 21]. Recently, other surface preparation methods have been investigated, such as plasma-PH_3 treatment [11] or the deposition of 1 nm thick AlN_x layer at the HfO_2/Ge interface [9]. These surface treatments are shown to perform better than thermal nitridation to improve interface quality, reduce the leakage, suppress Ge oxides formation during HfO_2 deposition, and prevent Ge out-diffusion from the substrate into the HfO_2 films. The latter phenomenon is detrimental, because it might increase the leakage current in the oxide layer [9,83]. Finally, Bai et al. [17] reported a D_{it} of $7\times10^{10}\,eV^{-1}\,cm^{-2}$ for HfO_2 deposited by rapid thermal CVD at 400°C on a Ge surface passivated using a thin Si layer.

For ALD grown HfO_2 films, the use of precursors schemes alternative to $HfCl_4$ and H_2O is shown to improve the structural and electrical properties of the films themselves, even on HF-treated Ge surfaces [4, 5].

Figure 8.10a shows multifrequency CV characteristics of a 9.6 nm thick HfO_2 film grown at 375°C on a HF-last Ge surface. The CV characteristics exhibit significant frequency dispersion both in accumulation and in inversion. Sweeping the gate voltage from inversion to accumulation and back, a large hysteresis (>400 mV) is detected. The same behavior was reported for ZrO_2 films grown on HF-last Ge at 300°C by ALD [23], and was attributed to a highly defective interface.

HfO_2 films grown on HF-last Ge surfaces alternating pulses of $Hf(O^tBu)_2$ $(mmp)_2$ and $HfCl_4$ are mostly amorphous, with a low surface roughness (0.2 nm rms by atomic force microscopy) and with an extremely thin IL (few Å) between the oxide and Ge (see also previous paragraph) [5]. Unfortunately, the CV characteristics exhibit exactly the same behavior of HfO_2 films deposited from $HfCl_4$ and H_2O. On the other hand, films deposited on

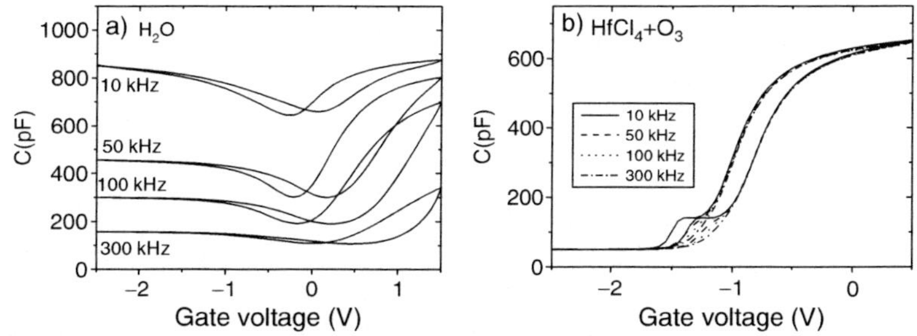

Fig. 8.10. CV curves acquired at different frequencies for HfO_2 films grown on Ge by ALD using the $HfCl_4+H_2O$ (**a**) or $HfCl_4+O_3$ (**b**) precursors combinations. MOS gate area: $7.8\times10^{-4}\,cm^2$

HF-last treated Ge using O_3 and $HfCl_4$ exhibit a better MOS behavior [4]. Figure 8.10b shows the multifrequency CV characteristics of Al/HfO_2/n-Ge MOS capacitors incorporating a 10 nm thick HfO_2 film deposited using O_3 as oxygen source. The leakage currents measured for HfO_2 films of various thicknesses are plotted in Fig. 8.11a. CV curves do not show any frequency dispersion in the inversion and accumulation regions in the 10–300 kHz frequency range. The D_{it} estimated using the Hill–Coleman method [86] is in the $0.5-1\times10^{12}\,cm^{-2}\,eV^{-1}$ range. A clockwise hysteresis (around 150 mV for a voltage sweep in the -2.5 to 0.5 V range), as well as a negative flat band voltage shift (positive fixed charges $\geq6\times10^{12}\,cm^{-2}$) are measured. This behavior is most likely due to the defective GeO_2 IL, which develops during HfO_2 deposition (from TEM [4] and XPS – see previous paragraph). It is worth noticing that the CV curves of Au/HfO_2/n-Ge MOS (not shown) and

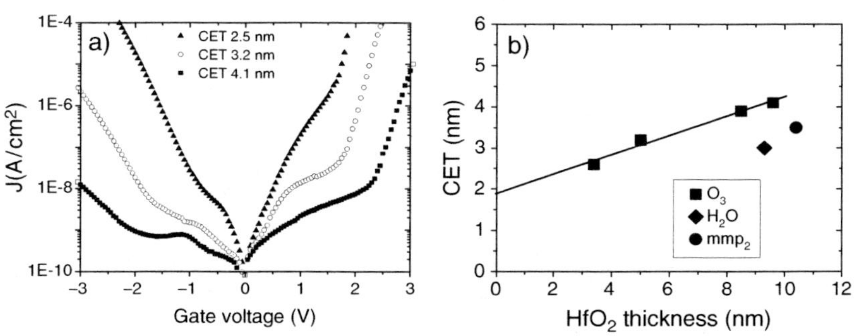

Fig. 8.11. (**a**) Leakage current measured for HfO_2 films of various thicknesses grown using O_3 as oxygen source (**b**) CET as a function of HfO_2 physical thickness for films grown using O_3. The CET (at 10 kHz) of a 9–10 nm thick HfO_2 films grown using H_2O or $Hf(O^tBu)_2(mmp)_2$ are also shown. For similar physical thicknesses, films grown using O_3 exhibit a slightly higher CET than those grown using H_2O or $Hf(O^tBu)_2(mmp)_2$, as expected due to the thicker IL

Al/HfO_2/n-Ge capacitors exhibit similar behavior and, within the experimental error, there is no influence of the metal gate on the measured accumulation capacitance (C_{ox}) and on the amount of hysteresis. Comparable hysteresis and flat band voltage shifts are reported also for as deposited HfO_2 films on nitrided Ge surface [6] and for GeO_xN_y films on Ge [87]. Capacitance equivalent oxide thickness (CET) values are extracted from C_{ox} at 10 kHz (without taking into account quantum mechanical corrections) for 3–10 nm thick films grown using O_3. The HfO_2 dielectric constant is determined to be 17 ± 1 from the slope of the linear fit of the CET values versus HfO_2 physical thickness (Fig. 8.11b). The intercept with the y-axis gives a CET of 1.9 nm for the IL. Taking into account the IL physical thickness, its dielectric constant turns out to be around 4.5, close to the one reported for GeO_2 (5–6) [88]. The leakage current (Fig. 8.11a) for the thinner films grown using O_3 and with a CET of 2.5 nm is comparable with the one reported for HfO_2 films grown by ALD at 300°C on GeO_xN_y or nitrided Ge surfaces [6, 12].

It is worth noticing that the measured dielectric constant for HfO_2 films deposited on Ge by various techniques ranges between 17 and 25 [4, 18, 20]. Comparable values are reported also for films deposited on Si [1, 20, 61, 89]. Moreover, various authors [18,20] have shown that the HfO_2/Ge stacks exhibit lower EOT values than oxide/Si ones, due to a better control (thickness, dielectric constant) of the high-κ/Ge interfacial layer. Indeed, EOT values lower than 1 nm have been demonstrated for oxides on Ge [3, 18]. On the other hand, the electrical properties of nonhafnium based oxides deposited on Ge have not been deeply investigated yet.

The multifrequency CV characteristics of Al_2O_3 (15 nm thick) and Lu_2O_3 films (14 nm thick) deposited by ALD are shown, respectively, in Fig. 8.12a,b. The films were deposited on HF-treated Ge surfaces, using H_2O as oxygen precursor. Compared to HfO_2 films grown using H_2O (Fig. 8.10a), Al_2O_3 and Lu_2O_3 films on Ge exhibit a lower frequency dispersion, and there is no variation of the accumulation capacitance between 10 and 500 kHz. This fact is

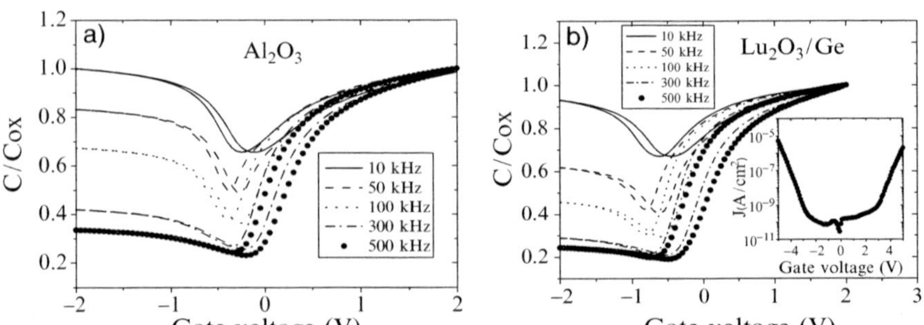

Fig. 8.12. (**a**) CV characteristics of Al/Al_2O_3/n-Ge capacitor acquired at different frequencies. The Al_2O_3 film is 15 nm thick. (**b**) CV and IV (inset) characteristics of Al/Lu_2O_3/n-Ge capacitor. The Lu_2O_3 film is 14 nm thick

an indication of a better interface quality for these oxides on Ge compared to the HfO_2/Ge system. The frequency dispersion in the inversion capacitance might be attributed to the high intrinsic carrier concentration in the Ge semiconductor [90] and/or to the enhanced generation of minority carrier due to interface traps [18].

The dielectric constants of Al_2O_3 ($\sim$9) [21] and Lu_2O_3 (12) [64,91] are significantly lower than the one of HfO_2, limiting the scalability of devices based on the former two oxides. Nevertheless, dielectrics alternative to HfO_2, such as Al_2O_3 and rare earth oxides deserve further investigation since they might be a better choice than HfO_2 for Ge-based devices in order to improve the interface quality [21,29]. Indeed, Chen et al. [21] reported that the D_{it} for Al_2O_3 films deposited on nitrided Ge ($4\times10^{11}\,\mathrm{eV}^{-1}\,\mathrm{cm}^{-2}$) is one order of magnitude lower than the one measured for HfO_2 films on Ge (6–$8\times10^{12}\,\mathrm{eV}^{-1}\,\mathrm{cm}^{-2}$). Moreover, recently, Dimoulas [29] reported that CeO_2, Gd_2O_3, Dy_2O_3 layers deposited by MBE directly on a clean Ge surface exhibit good CV characteristics and lower D_{it} than HfO_2/GeON/Ge stacks.

8.4.2 Electrical Properties of High-κ Dielectrics Deposited on GaAs

As discussed in the introduction, the deposition of mixed $Ga_2O_3(Gd_2O_3)$ dielectric on GaAs [35, 44, 45] or the use of a thin Si interlayer [52–55] have given the best results in terms of low D_{it} at the oxide/GaAs interface. On the other hand, ALD grown high-κ dielectrics have recently proven to be promising for the fabrication of GaAs-based devices [57]. Al_2O_3 films on GaAs exhibit a D_{it} aroud $5\times10^{11}\,\mathrm{cm}^{-2}\,\mathrm{eV}^{-1}$, low leakage and a high breakdown electric-field ($10\,\mathrm{MV\,cm}^{-1}$) [57,58]. Moreover, few studies have been published on HfO_2 films deposited by ALD on GaAs [5, 56], as opposed to the case of HfO_2 films on Ge.

Figure 8.13 shows the CV(a) and IV(b) characteristics of HfO_2 film of various thicknesses deposited on GaAs by ALD combining $HfCl_4$ and H_2O. The

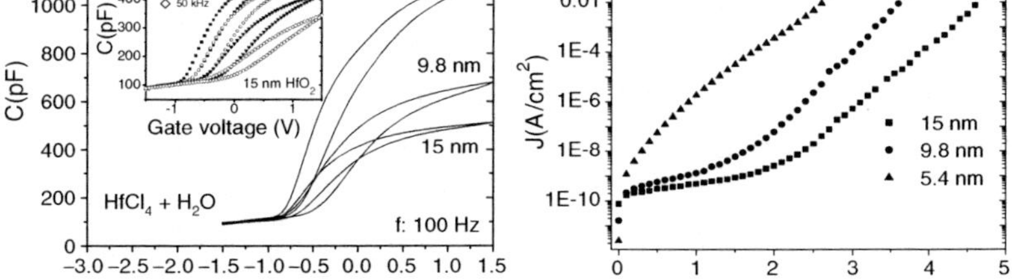

Fig. 8.13. (**a**) CV characteristics at 100 Hz of HfO_2 films of various thickness deposited on GaAs from $HfCl_4$ and H_2O. In the inset, multifrequency CV curves. (**b**) Leakage currents of HfO_2 films of various thickness. MOS gate area: $7.8\times10^{-4}\,\mathrm{cm}^2$

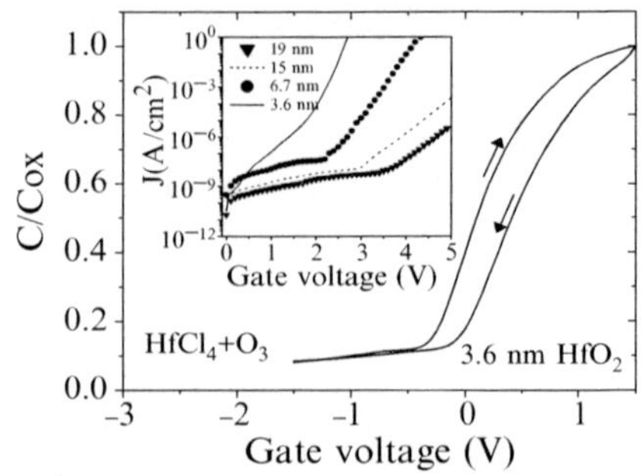

Fig. 8.14. CV characteristics acquired at 100 Hz for a Al/3.6 nm HfO_2/n-GaAs capacitor. HfO_2 films are deposited from $HfCl_4$ and O_3. In the inset the current density–voltage (JV) curves of HfO_2 films of various thicknesses. MOS gate area: $7.8\times10^{-4}\,cm^2$

CV clearly exhibit accumulation, depletion and inversion regions, indicating the unpinning of the GaAs Fermi level [42]. Nevertheless, the CV curves are stretched-out, and exhibit a significant dispersion with frequency of the accumulation capacitance (inset of Fig. 8.13a). CV characteristics of HfO_2 films deposited using O_3 (Fig. 8.14) as oxygen source exhibit a similar shape and frequency dispersion (not shown) than CV characteristics acquired on films grown using H_2O. Therefore, the use of ozone does not contribute to improve oxide/semiconductor interface quality for HfO_2 films deposited on GaAs, as opposed to the case of HfO_2 films on Ge. From CV characteristics acquired at 100 Hz on films of various thicknesses, the dielectric constant extracted for HfO_2 films on GaAs is around 15 ± 2. Within experimental errors, this value is compatible with the value measured for films deposited on Ge. The leakage current density measured for HfO_2 films with different thickness are reported in Fig. 8.13b ($HfCl_4$ and H_2O precursor combination) and on the inset of Fig. 8.14 ($HfCl_4$ and O_3). The HfO_2 films are good insulators showing low leakage at zero bias ($\sim10^{-10}\,A\,cm^{-2}$). For the thinner film grown using O_3 as oxygen source (physical thickness: 3.7 nm, estimated CET: 2.5 ± 0.2 nm), the leakage current is $\sim1\times10^{-7}\,A\,cm^{-2}$ at +1V accumulation (Fig. 8.14, inset).

8.5 Band Offset of High-κ Dielectrics Deposited on Ge and GaAs

One among the most important requirements imposed to high-κ oxides candidates to substitute SiO_2 in ultrascaled devices is that they must have CBO and VBO larger than at least 1.0 eV, to avoid high leakage current [1]. The band alignment between various high-κ oxides and Si have been extensively investigated using XPS and IPE techniques [28, 92–96]. The experimental CBO values reported for Al_2O_3 [92], ZrO_2 and HfO_2 [92–94], Lu_2O_3 [28, 75], and

Table 8.1. Experimental conduction band offset (CBO), valence band offset (VBO), and band gap values measured for various high-κ oxides deposited on Ge and GaAs

		CBO(eV)	VBO (eV)	Gap (eV)	
HfO_2/Ge	ALD	2.0	–	5.6	IPE [4,5]
HfO_2/Ge	MOCVD	2.0	3.0	5.6	IPE [97]
HfO_2/Ge	MBE	2.2	–	5.6	IPE [98]
Al_2O_3/Ge	ALD	2.1	–	–	IPE
Lu_2O_3/Ge	ALD	2.1	–	5.8	IPE [28]
Lu_2O_3/Ge	ALD	–	2.9	5.8	XPS [28,31]
ZrO_2/Ge	PLD	1.79	3.36	5.82	XPS [99]
$LaAlO_3$/Ge	MBE	2.2	2.9	5.7	IPE [98]
Lu_2O_3/GaAs	ALD	2.1	3.0	5.8	IPE [28]
Ga_2O_3/GaAs	MBE	0.8	2.6	4.8	IPE [100]
Gd_2O_3/GaAs	MBE	1.5	2.9	5.8	IPE [100]
$Ga_2O_3(Gd_2O_3)$/GaAs	MBE	1.4	2.6	5.4	XPS/IV [101]
$SrTiO_3$/GaAs	MBE	0.6	2.5	3.3	XPS [102]

The error on the CBO and VBO determined by IPE is ±0.1 eV

ternary compounds ($LaAlO_3$, $LaScO_3$, $GdScO_3$, $DyScO_3$, $La_2Hf_2O_7$) [95, 96] on Si are almost identical, ranging from 2.0 to 2.1 eV. Despite the increasing interest to integrate high-κ oxides in Ge- and GaAs-based devices, the band alignment has been measured only for a restricted number of oxides. The available results are summarized in Table 8.1.

In the following, we will present the experimental IPE determination of CBO for the Al_2O_3/Ge and the HfO_2/Ge systems. Figure 8.15a shows the cube root ($Y^{1/3}$) of IPE quantum yield versus photon energy measured for an Al/Al_2O_3/n-Ge capacitor biased at positive voltages. The spectral threshold around 2.75 eV is related to the electron transition from the Ge valence band to the Al_2O_3 conduction band. Figure 8.15b shows the IPE thresholds as a function of the square root of the average applied field (Schottky plot).

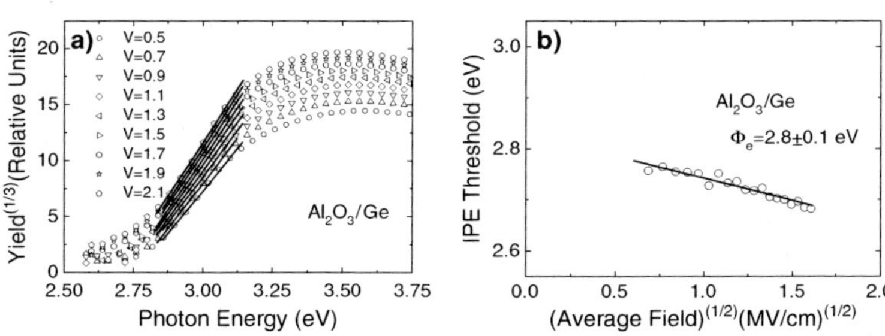

Fig. 8.15. (**a**) IPE spectra of an Al/Al_2O_3/Ge stack for various positive applied voltages. (**b**) Schottky plot of the spectral thresholds for IPE from the Ge valence band into the Al_2O_3 conduction band. *The lines* represent the linear fittings

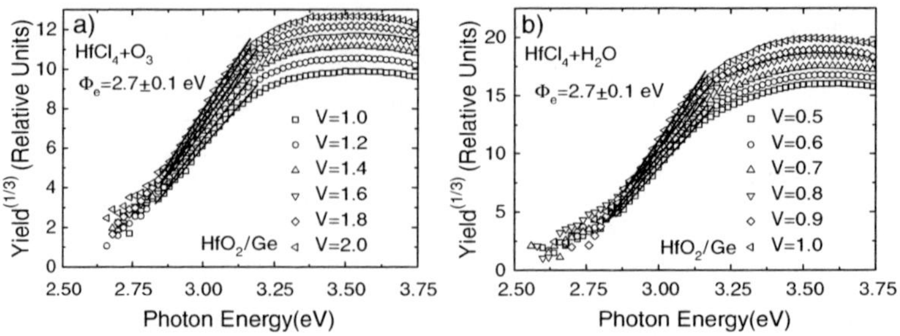

Fig. 8.16. IPE spectra of an Al/HfO_2/Ge stack for various positive applied voltages. (**a**) O_3 as oxygen precursor of HfO_2 (**b**) H_2O as oxygen precursor of HfO_2. *The lines* represent the linear fittings

The linear fitting of the data allows to extrapolate the barrier energy at zero applied field (Φ_e), which is equal to 2.8 ± 0.1 eV. By subtracting the Ge band gap to Φ_e, the CBO value is determined to be 2.1 ± 0.1 eV.

The $Y^{1/3}$ versus the photon energy measured for HfO_2 films grown on Ge by ALD using O_3 or H_2O are reported, respectively, in Fig. 8.16a,b [4]. The extracted CBO values are 2.0 ± 0.1 eV in both cases. Moreover, the same CBO value is extracted for films grown at 375°C using ALD alternating pulses of $Hf(O^tBu)_2(mmp)_2$ and $HfCl_4$ [5]. It was already pointed out in this chapter that HfO_2 films deposited by ALD using various precursor combinations exhibit different structural and chemical properties. In particular, fixing $HfCl_4$ as Hf precursor, when O_3 is used as oxygen source, the films develop a 2 nm thick GeO_2 IL between the oxide and Ge, whereas, using H_2O, HfO_2 films develop local epitaxy on Ge. Therefore, the obtained IPE results for the ALD grown HfO_2/Ge systems indicate that the CBO at the HfO_2/Ge interface is independent from the oxide structure (amorphous, polycrystalline) and oxide/semiconductor interface (presence or absence of GeO_2). Within experimental errors, the same CBO value is reported also for films grown using other deposition techniques. Seguini et al. [98] measured a CBO of 2.2 ± 0.1 eV for HfO_2 films deposited by MBE on nitridated Ge surfaces (Table 8.1). Afanas'ev et al. [97] reported a CBO of 2.0 ± 0.1 eV and VBO of 3.0 ± 0.1 eV for HfO_2 films deposited by MOCVD on Ge. Moreover, Afanas'ev measured a reduction of $\sim$1 eV of the VBO due to the growth of an GeO_2 IL upon postdeposition thermal treatments [97]. The CBO values measured for Al_2O_3, HfO_2, Lu_2O_3, $LaAlO_3$ oxides on Ge (see Table 8.1) are identical, within the experimental errors, to those measured for the same oxides on Si [75, 92–95].

The reported experimental CBO values at the high-κ dielectrics/GaAs interfaces range from 0.6 eV for $SrTiO_3$ to 2.1 eV for Lu_2O_3 (Table 8.1). The CBO value obtained for the Lu_2O_3/GaAs interface by IPE is equal, within the experimental errors, to the one reported for the Lu_2O_3/Si and the Lu_2O_3/Ge systems [28, 75]. The data available in the literature are not sufficient to compare CBO values on GaAs, Ge and Si for other high-κ dielectrics. Finally, it is worth noticing that the slope of Lu_2O_3/GaAs Schottky plot is higher than those observed at the interface with Si or Ge [28]. This effect can be related to interfacial charge, as already reported for the Gd_2O_3/GaAs interface [100].

8.6 Conclusions

We discussed and compared with data available in the literature the chemical, physical, and electrical properties of HfO_2, Lu_2O_3, and Al_2O_3 deposited using ALD on Ge and GaAs. Effects of substrate surface preparation before film deposition are reported. The choice of precursor combinations and growth parameters for ALD are also shown to significantly affect the structural and electrical properties of high-κ oxides on semiconductors.

HfO_2 layers (3–10 nm thick) deposited on Ge and GaAs from $Hf(O^tBu)_2(mmp)_2$ and $HfCl_4$ precursor combination are mostly amorphous. In the thicker films (10 nm), small and stressed crystallites are present in an amorphous matrix. HfO_2 films grown using $HfCl_4$ and either O_3 or H_2O as oxygen sources are polycrystalline with different percentages of orthorhombic and monoclinic phases. For films of similar thickness, the percentage of the orthorhombic phase is higher in films grown using O_3 than in those deposited using H_2O. The use of ozone as oxygen source promotes the growth of ~2 nm thick IL at the HfO_2/Ge and HfO_2/GaAs interfaces. On the other hand, for films grown using H_2O as oxygen source, XPS and XRR data evidence a sharp HfO_2/semiconductor interface, without any IL. Moreover, HfO_2 films exhibit local epitaxial growth on both Ge(001) and GaAs(001). The majority of the monoclinic {001} planes are parallel to Ge(001) and GaAs(001). For the out of plane HfO_2(001)//Ge(001) orientation, the in-plane epitaxial relationships correspond to a "cube on cube" growth with the lattice parameter a (or b) of monoclinic HfO_2 parallel either to Ge[100] or to Ge[010] with the same probability. Due to the different chemistry at the interface, either related to cleaning effectiveness and/or to the different oxidation state of Ge, Ga, and As on one hand, anf Hf on the other, epitaxial growth is more pronounced on Ge than on GaAs. Lu_2O_3 films are deposited on Ge and GaAs using the bis-cyclopentadienyl complex $[(\eta^5\text{-}C_5H_4SiMe_3)_2LuCl]_2$ with H_2O at 360°C. As grown films are nanocrystalline with grains in the bixbyte structure of Lu_2O_3. The Lu_2O_3/Ge and Lu_2O_3/GaAs interfaces are sharp, with no IL layer (XPS and XRR data). Thin as grown (<3 nm) films are mostly hydroxide ($Lu(OH)_3$). Postdeposition annealing at 500°C in vacuum or O_2 ($P_{O_2} = 10^{-6}$ Torr) is effective in transforming a large amount of $Lu(OH)_3$ into Lu_2O_3, without generating any interfacial layer.

HfO_2 films deposited on Ge using O_3 as oxygen source exhibit better MOS behavior and lower D_{it} than those deposited using H_2O or $Hf(O^tBu)_2(mmp)_2$ combined with $HfCl_4$. Al_2O_3 and Lu_2O_3 films grown directly on Ge using H_2O as oxygen source exhibit a better MOS behavior than HfO_2 films on Ge. These results, together with those reported in the literature for Al_2O_3 and rare earth oxides deposited on Ge, indicate that dielectrics alternative to HfO_2 might be a better choice for Ge-based devices and deserve further investigations. HfO_2 films deposited on GaAs using either O_3 or H_2O as oxygen precursor are good insulator characterized by low leakage, but their interface properties must be improved. Ga_2O_3, Gd_2O_3, and mixed $Ga_2O_3(Gd_2O_3)$ oxides deposited on GaAs by MBE have been among the most investigated oxides for GaAs-devices. Nevertheless, recent results on the structural and electrical properties of ALD grown Al_2O_3 and HfO_2 films on GaAs are also very promising.

Finally, a review on band offsets of various high-κ dielectrics on Ge and GaAs is presented. For the HfO_2/Ge system fabricated using ALD, MOCVD, and MBE, the CBO values ($\sim$2.0 eV) depend on neither oxide structure (amorphous, polycrystalline), nor on the presence or absence of a GeO_2 IL. The CBO values (measured by IPE) for HfO_2, Lu_2O_3, Al_2O_3, and $LaAlO_3$ are identical, within the experimental errors, for films deposited on Si and Ge. The measured CBO at the Lu_2O_3/GaAs interface is also equal to the one measured at the Lu_2O_3/Si and Lu_2O_3/Ge interfaces.

The structural and electrical properties, as well as band offsets of high-κ dielectric films deposited on Ge and GaAs by ALD are promising for the application in high-performance devices. Nevertheless, the control of high-κ dielectric/Ge (GaAs) interface must be further addressed and optimized.

Acknowledgments

The authors would like to thank Prof. I.L. Fedushkin of the G. A. Razuvaev Institute of Organometallic Chemistry in Nizhny Novgorod, Russia, for supplying the $[(\eta^5 - C_5H_4SiMe_3)_2LuCl]_2$ precursor. Mr. M. Alia (MDM) is acknowledged for help in sample processing. The Italian Ministry of Foreign Affairs, through Joint Projects between Italy and Russia, and the National Institute for the Physics of Matter (INFM), through the PAIS-REOHK project, are acknowledged for funding.

References

1. G.D. Wilk, R.M. Wallace, and J.M. Anthony. High-κ gate dielectrics: Current status and materials properties considerations. J. Appl. Phys. **89**, 5243 (2001).
2. P.D. Ye, G.D. Wilk, B. Yang, J. Kwo, S.N.G. Chu, S. Nakahara, H.-J.L. Gossmann, J.P. Mannaerts, M. Hong, K.K. Ng, and J. Bude. GaAs metal-oxide-semiconductor field-effect transistor with nanometer-thin dielectric grown by atomic layer deposition. Appl. Phys. Lett. **83**, 180 (2003).
3. N. Wu, Q. Zhang, C. Zhu, D.S.H. Chan, A. Du, N. Balasubramanian, M.F. Li, A. Chin, J.K.O. Sin, and D.-L. Kwong. A TaN-HfO_2-Ge pMOSFET with novel SiH_4/surface passivation. IEEE Electron Device Lett. **25**, 631 (2004).
4. S. Spiga, C. Wiemer, G. Tallarida, G. Scarel, S. Ferrari, G. Seguini, and M. Fanciulli. Effects of the oxygen precursor on the electrical and structural

properties of HfO_2 films grown by atomic layer deposition on Ge. Appl. Phys. Lett. **87**, 112904 (2005).

5. M. Fanciulli, S. Spiga, G. Scarel, G. Tallarida, C. Wiemer, and G. Seguini. Structural and Electrical Properties of HfO_2 films grown by atomic layer deposition on Si, Ge, GaAs and GaN. Mater. Res. Soc. Symp. Proc. Vol. **786**, E6.14 (2004).
6. E.P. Gusev, H. Shang, M. Copel, M. Gribelyuk, C. D'Emic, P. Kozlowski, and T. Zabel. Microstructure and thermal stability of HfO_2 gate dielectric deposited on Ge(100). Appl. Phys. Lett. **85**, 2334 (2004).
7. H. Kim, P.C. McIntyre, Chi. O. Chui, K.C. Saraswat, and M.-H. Cho. Interfacial characteristics of HfO_2 grown on nitrided Ge(100) substrates by atomic-layer deposition. Appl. Phys. Lett. **85**, 2902 (2004).
8. A. Delabie, R.L. Puurunen, B. Brijs, M. Caymax, T. Conard, B. Onsia, O. Richard, W. Vandervorst, C. Zhao, M.M. Heyns, and M. Meuris. Atomic layer deposition of hafnium oxide on germanium substrates. J. Appl. Phys. **97**, 064104 (2005).
9. F. Gao, S.J. Lee, J.S. Pan, L.J. Tang, and D.-L. Kwong. Surface passivation using ultrathin AlN_x film for Ge-metal-oxide-semiconductor devices with hafnium oxide gate dielectric. Appl. Phys. Lett. **86**, 113501 (2005).
10. B. De Jaeger, R. Bonzom, F. Leys, O. Richard, J. Van Steenbergen, G. Winderickx, E. Van Moorhem, G. Raskin, F. Letertre, T. Billon, M. Meuris, and M. Heyns. Optimisation of a thin epitaxial Si layer as Ge passivation layer to demonstrate deep sub-micron n- and p-FETs on Ge-On-Insulator substrates. Microelectron. Eng. **80**, 26 (2005).
11. S.J. Whang, S.J. Lee, F. Gao, N. Wu, C.X. Zhu, J.S. Pan, L. J. Tang, and D.-L. Kwong. Germanium p- & n-MOSFETs fabricated with novel surface passivation (plasma-PH_3 and thin AlN) and TaN/HfO_2 Gate stack. *IEEE IEDM 2004 Technical Digest* 2004, p. 307 (2004).
12. C.O. Chui, H. Kim, P.C. McIntyre, and K.C. Saraswat. Atomic layer deposition of high-κ dielectric for germanium MOS applications – substrate. IEEE Electron Device Lett. **25**, 274 (2004).
13. S. Van Elshocht, B. Brijs, M. Caymax, T. Conrad, S. De Gendt, S. Kubicek, M. Meuris, B. Onsia, O. Richard, I. Teerlinck, J. Van Steenbergen, C. Zhao, and M. Heyns. Physical characterization of HfO_2 deposited on Ge substrates by MOCVD. Mater. Res. Soc. Symp. Proc. Vol. **809**, B5.4.1/D5.4.1 (2004).
14. N. Wu, Q. Zhang, C. Zhu, C.C. Yeo, S.J. Whang, D.S.H. Chan, M.F. Li, B.J. Cho, A. Chin, D.-L. Kwong, A.Y. Du, C.H. Tung, and N. Balasubramanian. Effect of surface NH_3 anneal on the physical and electrical properties of HfO_2 films on Ge substrate. Appl. Phys. Lett. **84**, 3741 (2004).
15. N. Wu, Q. Zhang, C. Zhu, D.S.H. Chan, M.F. Li, N. Balasubramanian, A. Chin, and D.-L. Kwong. Alternative surface passivation on germanium for metal-oxide-semiconductor applications with high-κ gate dielectric. Appl. Phys. Lett. **85**, 4127 (2004).
16. N. Lu, W. Bai, A. Ramirez, C. Mouli, A. Ritenour, M.L. Lee, D. Antoniadis, and D.L. Kwong. Ge diffusion in Ge metal oxide semiconductor with chemical vapor deposition HfO_2 dielectric. Appl. Phys. Lett. **87**, 051922 (2005).
17. W.P. Bai, N. Lu, and D.-L. Kwong. Si interlayer passivation on germanium MOS capacitors with high-κ dielectric and metal gate. IEEE Electron Device Lett. **26**, 378 (2005).

18. A. Dimoulas, G. Mavrou, G. Vellianitis, E.K. Evangelou, N. Boukos, M. Houssa, and M. Caymax. HfO_2 high-κ gate dielectrics on Ge(100) by atomic oxygen beam deposition. Appl. Phys. Lett. **86**, 032908 (2005).
19. K. Kita, K. Kyuno, and A. Toriumi. Growth mechanism difference of sputtered HfO_2 on Ge and on Si. Appl. Phys. Lett. **85**, 52 (2004).
20. K. Kita, M. Sasagawa, M. Toyama, K. Kyuno, and A. Toriumi. Retarded growth of sputtered HfO_2 films on Germanium. Mater. Res. Soc. Symp. Proc. Vol. **809**, B5.5.1/D5.5.1 (2004).
21. J.J.-H. Chen, N.A. Bojarczuk, Jr., H. Shang, M. Copel, J.B. Hannon, J. Karasinski, E. Preisler, S.K. Banerjee, and S. Guha. Ultrathin Al_2O_3 and HfO_2 gate dielectrics on surface-nitrided Ge. IEEE Tans. Electron Device **51**, 1441 (2004).
22. C.-C. Cheng, C.-H. Chien, C.-W. Chen, S.-Lu Hsu, M.-Y. Yang, C.-C. Huang, F.-L. Yang, and C.-Y. Chang. Impact of post-deposition-annealing on the electrical characteristics of HfO_xN_y gate dielectric on Ge substrate. Microelectron. Eng. **80**, 30 (2005).
23. H. Kim, C.O. Chui, K.C. Saraswat, and P.C. McIntyre. Local epitaxial growth of ZrO_2 on Ge(100) substrates by atomic layer epitaxy. Appl. Phys. Lett. **83**, 2647 (2003).
24. C.O. Chui, S. Ramanathan, B.B. Triplett, P.C. McIntyre, and K. C. Saraswat. Germanium MOS capacitors incorporating ultrathin high-κ gate dielectric. IEEE Electron Device Lett. **23**, 473 (2002).
25. D. Chi, C.O. Chui, K.C. Saraswat, B.B. Triplett, and P.C. McIntryre. Zirconia grown by ultraviolet ozone oxidation on germanium (100) substrates. J. Appl. Phys. **96**, 813 (2004).
26. D.S. Yu, C.H. Huang, A. Chin, C. Zhu, M.F. Li, B.J. Cho, and D.-L. Kwong. Al_2O_3/-Ge-on-insulator n- and p-MOSFETs with fully NiSi and NiGe dual gates. IEEE Electron Device Lett. **25**, 138 (2004).
27. Z. Benamara, S. Tizi, M. Chellali, and B. Gruzza. Experiments in MIS structure based on germanium and improvements of the interfacial properties. Mater. Chem. Phys. **62**, 273 (2000).
28. G. Seguini, M. Perego, and M. Fanciulli. Experimental determination of the band offset of rare earth oxides on different semiconductors, in "*Rare Earth Oxide Thin Films: Growth, Characterization, and Applications*", Eds. M. Fanciulli and G. Scarel, Springer, Berlin Heidelberg New York (2006).
29. A. Dimoulas. Electrically active interface and bulk semiconductor defects in high-κ/germanium structures, in *Defects in Advanced High-κ Dielectric Nano-Electronic Semiconductor Devices*, Ed. E. Gusev, Springer Berlin Heidelberg New York (2006)
30. A. Dimoulas. Rare earth oxides grown by molecular beam epitaxy for ultimate scaling, in *Rare Earth Oxide Thin Films: Growth, Characterization, and Applications*, Eds. M. Fanciulli and G. Scarel, Springerm Berlin Heidelberg New York (2006).
31. M. Perego, G. Seguini, G. Scarel, and M. Fanciulli. X-ray photoelectron spectroscopy study of energy-band alignments of Lu_2O_3 on Ge. Surf. Interface Anal. **38**, 494 (2005).
32. H.B. Park, M. Cho, J. Park, S.W. Lee, and C.S. Hwang, J.-P Kim, J.-H. Lee, N.-H. Lee, H.-K. Kang, J.-C. Lee, and S.-J. Oh. Comparison of HfO_2 films grown by atomic layer deposition using $HfCl_4$ and H_2O or O_3 as the oxidant. J. Appl. Phys. **94**, 3641 (2003).

33. J. Kwo, M. Hong, B. Busch, D.A. Muller,Y.J. Chabal, A.R. Kortan, J.P. Mannaerts, B. Yang, P. Ye, H. Gossmann, A.M. Sergent, K.K. Ng, J. Bude, W.H. Schlte, E. Garfunkel, and T. Gustafsson. Advances in high-κ gate dielectrics for Si and III–V semiconductors. J. Cryst. Growth **251**, 645 (2003).
34. J.R. Kwo and M. Hong. High-κ gate dielectrics for Si and compound semiconductors by molecular beam epitaxy. Mater. Res. Soc. Symp. Proc. Vol. **745**, N8.1.1/T6.1.1 (2003).
35. J. Kwo, D.W. Murphy, M. Hong, R.L. Opila, J.P. Mannaerts, A. M. Sergent, and R.L. Masaitis. Passivation of GaAs using $(Ga_2O_3)_{1-x}(Gd_2O_3)_x$, $0 < x < 1.0$ films. Appl. Phys. Lett. **75**, 1116 (1999).
36. M. Hong. Low interface state density oxide-GaAs structures fabricated by in situ molecular beam epitaxy. J. Vac. Sci. Technol. B **14**, 2297 (1996).
37. A. Callegari, P.D. Hoh, D.A. Buchanan, and D. Lacey. Unpinned gallium oxide/GaAs interface by hydrogen and nitrogen surface plasma treatment. Appl. Phys. Lett. **54**, 332 (1989).
38. M. Passlack, M. Hong, and J.P. Mannaerts. Quasistatic and high frequency capacitance–voltage characterization of Ga_2O_3–GaAs structures fabricated by in-situ molecular beam epitaxy. Appl. Phys. Lett. **68**, 1099 (1996).
39. M. Passlack, M. Hong, J.P. Mannaerts, R.L. Opila, S.N.G. Chu, N. Moriya, F. Ren, and J.R. Kwo. Low D_{it}, thermodynamically stable Ga_2O_3–GaAs interfaces: Fabrication, characterization, and modelling. IEEE Trans. Electron Devices **44**, 214 (1997).
40. Z. Yu, C.D. Overgaard, R. Droopad, M. Passlack, and J.K. Abrokwah. Growth and physical properties of Ga_2O_3 thin films on GaAs(001) substrate by molecular-beam epitaxy. Appl. Phys. Lett. **82**, 2978 (2003).
41. M. Hong, J. Kwo, A.R. Kortan, J.P. Mannaerts, and A.M. Sergent. Epitaxial cubic gadolinium oxide as a dielectric for gallium arsenide passivation. Science **283**, 1897 (1999).
42. J. Kwo, D.W. Murphy, M. Hong, J.P. Mannaerts, R.L. Opila, R. L. Masaitis, and A.M. Sergent. Passivation of GaAs using gallium–gadolinium oxides. J. Vac. Sci. Technol. B **17**, 1294 (1999).
43. M. Hong, Z.H. Lu, J. Kwo, A.R. Kortan, J.P. Mannaerts, J.J. Krajewski, K.C. Hsieh, L.J. Chou, and K.Y. Cheng. Initial growth of $Ga_2O_3(Gd_2O_3)$ on GaAs: Key to the attainment of a low interfacial density of states. Appl. Phys. Lett. **76**, 312 (2000).
44. M. Passlack, N. Medendorp, R. Gregory, and D. Braddock. Role of Ga_2O_3 template thickness and gadolinium mole fraction in $Gd_xGa_{0.4-x}O_{0.6}/Ga_2O_3$ gate dielectric stacks on GaAs. Appl. Phys. Lett. **83**, 5262 (2003).
45. M. Passlack. Development methodology for high-κ gate dielectrics on III–V semiconductors: $Gd_xGa_{0.4-x}O_{0.6}/Ga_2O_3$ dielectric stacks on GaAs. J. Vac. Sci. Technol. B **23**, 1773 (2005).
46. R. Droopad, M. Passlack, N. England, K. Rajagopalan, J. Abrokwah, and A. Kummel. Gate dielectrics on compound semiconductors. Microelectonic Eng. **80**, 138 (2005).
47. E.J. Tarsa, X.H. Wu, J.P. Ibbetson, J.S. Speck, and J.J. Zinck. Growth of epitaxial MgO films on Sb-passivated (001)GaAs: Properties of the MgO/GaAs interface. Appl. Phys. Lett. **66**, 3588 (1995).
48. Y. Liang, J. Kulik, T.C. Eschrich, R. Droopad, Z. Yu, and P. Maniar. Heteroepitaxy of perovskite oxides on GaAs(001) by molecular beam epitaxy. Appl. Phys. Lett. **85**, 1217 (2004).

49. R.F. Klie, Y. Zhu, E.I. Altman, and Y. Liang. Atomic structure of epitaxial $SrTiO_3$–GaAs(001) heterojunctions. Appl. Phys. Lett. **87**, 143106 (2005).
50. X. Liu, X.Y. Chen, J. Yin, Z.G. Liu, J.M. Liu, X.B. Yin, G.X. Chen, and M. Wang. Epitaxial growth of TiO_2 thin films by pulsed laser deposition on GaAs(100) substrates. J. Vac. Sci. Technol. A **19**, 391 (2001).
51. K. Nishita, A. Koma, and K. Saiki. Growth of NiO films on various GaAs faces via electron bombardment evaporation. J. Vac. Sci. Technol. A **19**, 2282 (2001).
52. Z. Chen, D.-G. Park, F. Stengal, S. Noor Mohammad, and H. Morkoç. Metal-insulator-semiconductor structures on *p*-type GaAs with low interface state density. Appl. Phys. Lett. **69**, 230 (1996).
53. S. Anantathanasarn and H. Hasegawa. Pinning-free GaAs MIS structures with Si interface control layers formed on (4×6) reconstructed (001) surface. Appl. Surf. Sci. **216**, 275 (2003).
54. S. Tiwari, S.L. Wright, and J. Batey. Unpinned GaAs MOS capacitors and transistors. IEEE Electron Device Lett. **9**, 488 (1998).
55. L. Wu. Tu, E.F. Schubert, M. Hong, and G.J. Zydzik. In-vacuum cleaving and coating of semiconductor laser facets using thin silicon and a dielectric. J. Appl. Phys. **80**, 6448 (96).
56. M.M. Frank, G.D. Wilk, D. Starodub, T. Gustafsson, E. Garfunkel, Y.J. Chabal, J. Grazul, and D.A. Muller. HfO_2 and Al_2O_3 gate dielectrics on GaAs grown by atomic layer deposition. Appl. Phys. Lett. **86**, 152904 (2005).
57. H.C. Lin, P.D. Ye, and G.D. Wilk. Leakage current and breakdown electric-field studies on ultrathin atomic-layer-deposited Al_2O_3 on GaAs. Appl. Phys. Lett. **87**, 182904 (2005).
58. P.D. Ye, G.D. Wilk, J. Kwo, B. Yang, H.-J.L. Gossmann, M. Frei, S.N.G. Chu, J.P. Mannaerts, M. Sergent, M. Hong, K.K. Ng, and J. Bude. GaAs MOSFET with oxide gate dielectric grown by atomic layer deposition. IEEE. Electron Device Lett. **24**, 209 (2003).
59. M. Ritala and M. Leskelä. Atomic layer epitaxy - a valuable tool for nanotechnology? Nanotechnology **10**, 19 (1999).
60. Z. Liu, Y. Sun, F. Machuca, P. Pianetta, W.E. Spider, and R.F. W. Pease. Optimization and characterization of III–V surface cleaning. J. Vac. Sci. Technol. B **21**, 1953 (2003).
61. G. Scarel, S. Spiga, C. Wiemer, G. Tallarida, S. Ferrari, and M. Fanciulli. Trends of structural and electrical properties in atomic layer deposited HfO_2 films. Mater. Sci. Eng. B **109**, 11 (2004).
62. P.A. Williams, J.L. Roberts, A.C. Jones, P.R. Chalker, N.L. Tobin, J.F. Bickley, H.O. Davies, L.M. Smith, and T.J. Leedham. Novel mononuclear alkoxide precursors for the MOCVD of ZrO_2 and HfO_2 thin films. Chem. Vap. Dep. **8**, 163 (2002).
63. H. Schumann, I.L. Fedushkin, M. Hummert, G. Scarel, E. Bonera, and M. Fanciulli. Crystal and molecular structure of $[(\eta^5-C_5H_4SiMe_3)_2LuCl]_2$ suitable precursor for Lu_2O_3 films. Z. Naturforsch. **59b**, 1035 (2004).
64. G. Scarel, E. Bonera, C. Wiemer, G. Tallarida, S. Spiga, M. Fanciulli, I.L. Fedushkin, H. Schumann, Y. Lebedinskii, and A. Zenkevich. Atomic-layer deposition of Lu_2O_3. Appl. Phys. Lett. **85**, 630 (2004).
65. P. Zaumseil. X-ray reflectivity characterization of thin-film and multilayer structures, in *Materials for Information Technology*, Eds. E. Zschech, C.M. Whelan, and T. Mikolajick, Springer, Berlin Heidelberg New York (2005).

66. C. Wiemer, S. Ferrari, M. Fanciulli, G. Pavia, and L. Lutterotti. Combining grazing incidence X-ray diffraction and X-ray reflectivity for the evaluation of the structural evolution of HfO_2 thin films with annealing. Thin Solid Films **450**, 134 (2004).
67. P.S. Lysaght, B. Foran, G. Bersuker, P.J. Chen, R. Murto, and H. R. Huff. Physicochemical properties of HfO_2 in response to rapid thermal anneal. Appl. Phys. Lett. **82**, 1266 (2003).
68. J.M. Howard, V. Craciun, C. Essary, and R.K. Singh. Interfacial layer formation during high-temperature annealing of ZrO_2 thin films on Si. Appl. Phys. Lett. **81**, 3431 (2002).
69. A. Gibaud. Specular Reflectivity from smooth and rough surfaces, in *X-ray and Neutron Reflectivity: Principles and Applications*, Eds. J. Daillant and A. Gibaud, Springer, Berlin Heidelberg New York (1999).
70. http://www.ing.unitn.it/∼maud/
71. Yu. Lebedinskii, A. Zenkevich, G. Scarel, and M. Fanciulli. Film and interface layer composition of rare earth (Lu, Yb) oxides deposited by ALD, in *Rare earth oxide thin films: growth, characterization, and applications*, Eds. M. Fanciulli and G. Scarel, Springer, Berlin Heidelberg New York (2006).
72. V.K. Adamchuk and V.V. Afanas'ev. Internal photoemission spectroscopy of semiconductor-insulator interfaces. Prog. Surf. Sci. **41**, 111 (1992).
73. S.A. Chambers, T. Droubay, T.C. Kaspar, and M. Gutowski. Experimental determination of valence band maxima for $SrTiO_3$, TiO_2, and SrO and the associated valence band offsets with Si(001). J. Vac. Sci. Technol. B **22**, 2205 (2004).
74. R. Ludeke, M.T. Cuberes, and E. Cartier. Local transport and trapping issues in Al_2O_3 gate oxide structures. Appl. Phys. Lett. **76**, 2886 (2000).
75. G. Seguini, E. Bonera, S. Spiga, G. Scarel, and M. Fanciulli. Energy-band diagram of metal/Lu_2O_3/silicon structures. Appl. Phys. Lett. **85**, 5316 (2004).
76. J. Aarik, A. Aidla, A.-A. Kiisler, T. Uustare, and V. Sammelselg. Influence of substrate temperature on atomic layer growth and properties of HfO_2 thin films. Thin Solid Films **340**, 110 (1999)
77. J. Aarik, A. Aidla, H. Mandar, Teet Uustare, K. Kukli, and M. Schuisky. Influence of substrate temperature on atomic layer growth and properties of HfO_2 thin films. Appl. Surf. Sci. **173**, 15 (2001).
78. M.-H. Cho, Y.S. Roh, C.N. Whang, K. Jeong, S.W. Nahm, D.-H. Ko, J.H. Lee, N.I. Lee, and K. Fujihara. Thermal stability and structural characteristics of HfO_2 films on Si (100) grown by atomic-layer deposition. Appl. Phys. Lett. **81**, 472 (2002).
79. H. Kim, P.C. McIntyre, and K.C. Saraswat. Effects of crystallization on the electrical properties of ultrathin HfO_2 dielectrics grown by atomic layer deposition. Appl. Phys. Lett. **82**, 106 (2002).
80. Inorganic Crystal Structure Database, file 27313 Fachinformationzentrum Karlsruhe, 2005.
81. Inorganic Crystal Structure Database, file 79913 Fachinformationzentrum Karlsruhe, 2005.
82. C. Wiemer, unpublished. (*Results obtained in the framework of the FP6 IST project ET4US "Epitaxial Technologies for Ultimate Scaling"*).
83. S. Van Elshocht, B. Brijs, M. Caymax, T. Conard, T. Chiarella, S. De Gendt, B. De Jaeger, S. Kubicek, M. Meuris, B. Onsia, O. Richard, I. Teerlinck, J.

Van Steenbergen, C. Zhao, and M. Heyns. Deposition of HfO_2 on germanium and the impact of surface pretreatments. Appl. Phys. Lett. **85**, 3824 (2004).
84. K. Prabhakaran and T. Ogino. Oxidation of Ge(100) amd Ge(111) surfaces: and UPS and XPS sudy. Surf. Sci. **325**, 263 (1995) and references therein.
85. T. Van Buuren, M.K. Weilmeier, I. Athwal, K.M. Colbow, J.A. Mackenzie, T. Tiedje, P.C. Wong, and K.A.R. Mitchell. Oxide thickness effect and surface roughening in the desorption of the oxide from GaAs. Appl. Phys. Lett. **59**, 464 (1991).
86. W.A. Hill and C.C. Coleman. A single-frequency approximation for interface state density determination. Solid State Electron **23**, 987 (1980).
87. Chi. O. Chui, F. Ito, and K.C. Saraswat. Scalability and electrical properties of germanium oxynitride MOS dielectrics. IEEE Electron Device Lett. **25**, 613 (2004).
88. S. Zhaoqi and L. Chunrong. Preparation and C–V characteristic of the Ge-MOS structure using plasma-anodized film as gate insulator. J. Phys. D **24**, 1624 (1991).
89. Y.-S. Lin, R. Puthenkovilakam, and J.P. Chang. Dielectric property and thermal stability of HfO_2 on silicon. Appl. Phys. Lett. **81**, 2041 (2002).
90. A. Dimoulas, G. Vellianitis, G. Mavrou, E.K. Evangelou, K. Argyropoulos, and M. Houssa. Short minority carrier response time in HfO_2/Ge metal-insulator-semiconductor capacitors. Microelectr. Eng. **80**, 34 (2005)
91. E. Bonera, G. Scarel, M. Fanciulli, P. Delugas, and V. Fiorentini. Dielectric properties of high-κ oxides: theory and experiment for Lu_2O_3. Phys. Rev. Lett. 94, 027602 (2005).
92. V.V. Afanas'ev, M. Houssa, A. Stesmans, and M.M. Heyns. Electron energy barriers between (100)Si and ultrathin stacks of SiO_2, Al_2O_3, and ZrO_2 insulators. Appl. Phys. Lett. **78**, 3073 (2001).
93. V.V. Afanas'ev, M. Houssa, A. Stesmans, and M.M. Heyns. Band alignments in metal-oxide-silicon structures with atomic-layer deposited Al_2O_3 and ZrO_2. J. Appl. Phys. **91**, 3079 (2002).
94. V.V. Afanas'ev, A. Stesmans, F. Chen, X. Shi, and S.A. Campbell. Internal photoemission of electrons and holes from (100)Si into HfO_2. Appl. Phys. Lett. **81**, 1053 (2002).
95. V.V. Afanas'ev, A. Stesmans, C. Zhao, M. Caymax, T. Heeg, J. Schubert, Y. Jia, D.G. Scholm, and G. Lucovsky. Band alignment between (100)Si and complex rare earth/transition metal oxides. Appl. Phys. Lett. **85**, 5917 (2004).
96. G. Seguini, S. Spiga, E. Bonera, M. Fanciulli, A. Reyes Huamantinco, C.J. Först, C.R. Ashman, P.E. Blöchl, A. Dimoulas, and G. Mavrou. Band alignment at the $La_2Hf_2O_7$/(001)Si interface. Appl. Phys. Lett. **88**, 202903 (2006).
97. V.V. Afanas'ev and A. Stesmans. Energy band alignment at the (100)Ge/HfO_2 interface. Appl. Phys. Lett. **84**, 2319 (2004).
98. G. Seguini, unpublished. (*Results obtained in the framework of the FP6 IST project ET4US "Epitaxial Technologies for Ultimate Scaling"*).
99. S.J. Wang, A.C.H. Huan, Y.L. Foo, j. W. Chai, J.S. Pan, Q. Li, Y.F. Dong, Y.P. Feng, and C.K. Ong. Energy-band alignments at ZrO_2/Si, SiGe, and Ge interfaces. Appl. Phys. Lett. **85**, 4418 (2004).

100. V.V. Afanas'ev, A. Stesmans, M. Passlack, and N. Medendorp. Band offsets at the interfaces of GaAs(100) with $Gd_xGa_{0.4-x}O_{0.6}$ insulators. Appl. Phys. Lett. **85**, 597 (2004).
101. T.S. Lay, M. Hong, J. Kwo, J.P. Mannaerts, W.H. Hung, and D. J. Huang. Energy-band parameters at the GaAs- and GaN – $Ga_2O_3(Gd_2O_3)$ interfaces. Solid State Electron. **45**, 1679 (2001).
102. Y. Liangh, J. Curless, and D. McCready. Band alignment at epitaxial $SrTiO_3$–GaAs(001) heterojunction. Appl. Phys. Lett. **86**, 082905 (2005).

9

Point Defects in Stacks of High-κ Metal Oxides on Ge: Contrast with the Si Case

A. Stesmans and V.V. Afanas'ev

Summary. The results are overviewed of an ESR analysis in combination with an electrical capacitance–voltage and conductance–voltage study of point defects and traps in (100)Ge/GeO_xN_y/HfO_2 and (100)Ge/GeO_2 structures. Comparative study suggests drastic differences in the interface defect properties of the (100)Ge/GeO_xN_y/HfO_2 and (100)Ge/GeO_2 interfaces from the seemingly isomorphic interfaces of (100)Si with the HfO_2 and SiO_2. ESR fails to detect dangling bond centers associated with Ge crystal surface atoms – only paramagnetic defects in the near-interfacial Ge oxide or Ge (oxy)nitride layers are observed which show no correlation with the major portion of electrically active traps; their atomic nature remains unknown. In contrast with the amphoteric traps related to the dangling bonds (P_b-type centers) commonly observed at the silicon/insulator interfaces, the major component of the Ge/insulator interface trap spectrum comes from slow acceptor states which show no immediate correlation with the observed paramagnetic centers. The influence of thermal passivation in H_2 is addressed as well.

9.1 Introduction

The Si-based complementary metal-oxide-semiconductor (CMOS) integrated circuit (IC) technology accounts for the immense development of the microelectronic industry leading into the information age. But industrial activity is not static, and as almost any human activity, the restless strive to progress appears as a basic constant of motion. Progress in the past and current world of IC technology means improvement in performance of the elemental CMOS field effect transistor (FET), of which continuous miniaturization has emerged as the main route forward. Recent history over about 4 decades has witnessed an exponential decrease in minimum feature size in a transistor with time, resulting in a doubling of the number of active elements on an IC in about every 2–3 years (denoted as one element of Moore's law [1]). This trend of continuous scaling has been categorically outlined in the International Technology Roadmap for Semiconductors (ITRS) [2] of the Semiconductor Industry Association (SIA), projecting as one prerequisite to its realization, the further reduction of the SiO_2 gate dielectric toward the sub-1.1–1.5 nm range for the next sub-100-nm nodes to be introduced in nearing years. The superb Si-dioxide, an insulator of unprecedented quality native to Si, possesses an almost perfect set of electrical and manufacturability properties that has enable device scaling for over about 3 decades: it has had a major contribution to the success story of the Si-based IC technology. Yet, such extreme thinning of crucial SiO_2 layers cannot continue forever and will run into major problems. Fundamental limits are encountered such as excessive gate leakage currents due to direct tunneling of electrons through the gate insulator, raising circuit power dissipation to intolerable levels [3,4]. Other issues include dopant penetration, electron mobility degradation, and, of prime technological importance also, device reliability [3–6].

As known, tunneling currents decrease exponentially with distance. Hence, the obvious route chosen to counter the fundamental tunneling issue is to replace SiO_2 by a physically thicker layer of an insulator of higher dielectric constant κ. In doing so, tunneling currents may be drastically limited while not giving in on the specific gate capacitance $C = \varepsilon_0 \kappa A/t$, where ε_0 is the permittivity of vacuum, κ the dielectric constant, A the area, and t the insulator thickness. Thus, in summary, it is the continuous device scaling in IC technology based on the MOSFET that provides the motivation for the currently immense worldwide efforts in studying dielectrics with a static dielectric constant higher than that of SiO_2 ($\kappa = 3.9$) – the so-called "high-κ" materials. The issue has been overviewed in several excellent articles and books [4–9].

Numerous high-κ metal oxides have been or are investigated as potential alternative gate dielectrics, including $Al_2O_3, Y_2O_3, ZrO_2, SrTiO_2, TiO_2, Ta_2O_5, HfO_2$, and La_2O_3, and many of their composites such as, e.g., silicates, nitrided silicates, etc. Several have already been disfavored, among others, because of stability reasons; see articles in this book and Refs. [8, 9] for more details. For the first generation of newly introduced alternative high-κ gate dielectrics, the Hf-oxide base dielectrics appear a finest choice, explaining the large research effort in these materials.

Currently, to keep scaling on track, Si-oxynitride serves as the "transitory" gate dielectric in replacement of the classic SiO_2 gate dielectric toward the introduction of high-κ metal oxides. As mentioned, here it appears that the nitrided Hf-silicate ($HfSi_xO_yN_z$) has emerged as the current best compromise for several key reasons, as extensively outlined elsewhere [7], i.e., in terms of κ-value, thermodynamic stability, preservation of amorphous structure compatible with required thermal budget in device manufacturing, bandgap and valence and conduction band offsets with respect to Si, reliability, bulk defect density, and interface quality in contact with Si. It actually has already been introduced or is currently eagerly being incorporated as new gate dielectric, herewith being well on schedule with the SIA's projected introduction of high-κ metal oxides in future technology nodes by the year 2007. For the more distant future, $LaAlO_3$ is put forward as a much valuable candidate [2].

The decided replacement of conventional SiO_2 in Si CMOS technology by a new alternative gate insulator of higher dielectric constant κ than of $SiO_2(\kappa \sim 3.9)$ has resulted in strong progress in the development of high-κ gate dielectrics nonnative to Si. This newly acquired experience opens new ways to address the possible development of CMOS technology for other semiconductors, less fortunate in terms of high quality native insulators, with the view to enhance MOS device performance. One such semiconductor is Ge, for which it is known that the Ge MOS technology using (native) Ge oxide – a nonprotecting hygroscopic oxide – as gate insulator appears unfeasible because of poor thermodynamic and electrical properties [10] of $GeO_{2(x)}$ and its interfaces with Ge: GeO_2 species transform into GeO and desorb at temperatures >420°C, while GeO_2 in hexagonal phase is water soluble [11]. Hence, effective surface passivation of Ge has appeared a classic obstacle for CMOS device realization with Ge. So, if to be successful, alternative insulators will be required, for which the newly acquired knowledge on high-κ layers may come to the rescue. As a result, Ge (and its alloys with Si) has regained much interest.

With respect to their potentiality as suitable gate dielectrics for Ge MOS devices, several dielectrics have recently been addressed [12–14]. These include the high-κ metal oxides HfO_2, ZrO_2, and Al_2O_3, as well as Ge_3N_4 and Ge oxy-nitride with many aspects of the newly conceived Ge MOS structures studied: electrical properties [15–20] (charge trapping, interface traps, gate leakage current, carrier density), band offset and electronic structure. Without SiO_xN_y gate dielectric in Si MOS devices (at least, HfO_2-based dielectrics), the HfO_2 dielectric on Ge in particular enjoys most interest. Here, it is hoped that much doubt highly stimulated by its evidenced success as (near future) replacement for the current of the beneficial properties of the HfO_2-based dielectrics on Si can be transferred to the promising Ge substrate. Properties already addressed or under investigation include electronic structure and band offsets [21, 22], Ge diffusion into deposited HfO_2 layers [23], interface properties [18–20], and influence of postdeposition (PD) annealing [21]. Of particular importance for successful HfO_2 layer growth appears the predeposition (growth) Ge surface pretreatment and passivation [24–29], a main issue of research. Various deposition and growth methods are applied such as ALD, MOCVD, and VUV oxidation of e-beam deposited Hf layers. A more detailed and complete overview can be found in the various chapters of this book.

Germanium is a high-performance device material for various reasons: In particular, as compared to Si, Ge application in MOS transistors promises improved channel mobility (intrinsically 3–4 times higher in Ge bulk than in Si), while the narrower bandgap (0.67 eV versus 1.1 eV for Si at 300 K) enables a low-voltage operation and, in turn, a reduced power consumption [30, 31]. Also, Ge requires lower dopant activation temperatures. This fact of lower processing temperature required for the Ge-based devices ($T < 600$°C as compared to $900 < T < 1{,}000$°C for Si) also suggests attractive integration of high-κ dielectrics in the Ge MOS technology. Among the high-κ insulators,

HfO_2 was recently shown to offer several advantages when applied on (100)Ge including a thinner interlayer than in (100)Si/SiO_2 [12] and high barriers for electrons and holes in Ge [21]. However, the Ge/HfO_2 interface suffers from a high trap density [18], which requires identification of the corresponding imperfections and exploration of ways to eliminate them.

A basic requirement for device grade MOS entities concerns the tight control of point defects, at the origin of detrimental charge traps. Their formation should be either ad hoc prevented, or, when introduced, post hoc efficiently inactivated electrically, e.g., through binding to H. As demonstrated by electron spin resonance (ESR), in the case of conventional Si/SiO_2 it is known that the interfacial trivalent Si dangling bond (DB) defects constitute a dominant source of detrimental interface traps [32, 33], termed P_b-type centers in ESR jargon. They are inherently generated during thermal oxidation as a result of network-lattice mismatch. Specifically, these are P_b (identified as $Si_3 \equiv Si^{\bullet}$) in (111)Si/SiO_2 and P_{b0} in (100)Si/SiO_2, with naturally incorporated site densities [34] of $\sim 5 \times 10^{12}\,cm^{-2}$ and $1 \times 10^{12}\,cm^{-2}$, respectively, for conventional oxidation temperatures (~800–960°C). The identical P_b, P_{b0} traps must be efficiently passivated by H down to the $10^{10}\,cm^{-2}\,eV^{-1}$ level – the goal of the standard forming gas anneal in device processing.

Remarkably, these P_b-type centers will likely remain as the major interface threat in the Si/high-κ insulator issue, as for the currently intensely investigated metal oxides (e.g., HfO_2, Al_2O_3, ZrO_2), it has been found [35, 36] that the realized Si/dielectric interface is basically Si/SiO_2 like also. In fact, the insertion of a well-controlled, sub-nm thin SiO_2 interlayer is adopted as an acceptable route of progress. One may then wonder how this situation would translate for other semiconductors, e.g., Ge, where the possible integration of high-κ layers for MOSFET applications is currently widely investigated.

The scope of this work is to overview results on point defects and traps at the interfaces of bulk Ge with nm-thin layers of HfO_2 and $GeO_x(N_y)$ as obtained from ESR probing of spin-active point defects in combination with standard electrical analysis, i.e., capacitance–voltage (C–V) and conductance–voltage (G–V) measurements. With Si/insulator entities being the technological basis of reference, these results are discussed in comparison with well known properties of the interfaces of (100)Si with HfO_2 and SiO_2. As a main finding, it will be outlined that the interfaces of (100)Ge with HfO_2 and GeO_x differ drastically from the nominally structurally isomorphic (100)Si/insulator (HfO_2, SiO_2) interfaces both in terms of the interface trap properties and the observed paramagnetic defects. No measurable density of dangling bonds of the semiconductor surface atoms (analog of paramagnetic P_b-type centers in Si/oxide structures) could be traced in Ge/oxide structures, while the dominant contribution to the interface trap density (D_{it}) stems from diamagnetic acceptor centers in the insulator. The paramagnetic centers tentatively associated with dangling bonds of Ge atoms in the interlayer between Ge and HfO_2 are found to be resistant to passivation by hydrogen and, therefore, may account for the high D_{it} still observed in Ge MOS structures after annealing in H_2. Rather than hydrogen passivation, a low-temperature oxidation of Ge/HfO_2 structures is found to reduce D_{it} to a level of $\approx 1 \times 10^{12}\,cm^{-2}\,eV^{-1}$ suggesting the modification of the Ge/high-κ oxide interface to be an effective method for trap density reduction.

9.1.1 Previous ESR Results

For the sake of comparison, we briefly overview ESR works so far carried out on de novo conceived semiconductor/high-κ insulator structures. So far, these only concern Si with high-κ dielectrics; First results on high-κ layers in combination with Ge are the subject of the current work.

A limited number of ESR studies have so far reported on newly composed Si/high-κ insulator entities with ZrO_2 and/or Al_2O_3 layers [35–39]. Initial K-band work [35, 36] studied stacks of (100)Si with nm-thick ALCVD layers of ZrO_2 and Al_2O_3, and simultaneously addressed the role of SiO_x and Al_2O_3 interlayers. In the as grown state, the sole defects observed, in enhanced densities as compared to Si/SiO_2, were the P_b-type (trivalent Si) interface defects P_{b0} and P_{b1} – the archetypal defects for the (100)Si/SiO_2(SiO_xN_y) interfaces [40]. This finding was confirmed by a subsequent X-band ESR works [37] on ALCVD (100)Si/Al_2O_3 and an electrically detected magnetic resonance investigation on ALCVD (100)Si/Al_2O_3 and (100)ZrO_2 entities [38]. The latter ESR works also reported an additional signal, the D center, which is generally ascribed to unpaired Si bonds in disordered/amorphous Si and has likely originated from damage in the Si substrate [35, 38]. Five more works addressed the Si/HfO_2 entity. A first one [41] compared (100)Si/HfO_2 entities grown by three variants of CVD: also here, in the as-grown state, the P_b-type defects were reported as predominant defects. In agreement, a second X-band work [42] reported the observation of P_b defects at the interface of nominally (111)Si/HfO_2(14.5 nm) entities manufactured via ALCVD using the nitrato precursor $Hf(NO_3)_4$ (NCVD). A third-one, studying the effect of charge injection in NCVD (100)Si/HfO_2(42.7 nm) entities using a UV/corona ion charging technique, found evidence for ESR-active centers (O_2^-) in the HfO_2 layers after electron injection [43]. A recent work revealed the incorporation of N into the HfO_2 layer of NCVD (100)Si/HfO_2 (100 nm) entities through identification of embedded NO_2 radicals upon ^{60}Coγ-irradiation [44]. One more work used X-band ESR to study the influence of the presence of hydrogen peroxide on the etching of layers and powders ZrO_2 and HfO_2 in aqueous solution of HF, reporting the observation of the O_2^- superoxide radical on the ZrO_2 surface [45]. Finally, in a separate work [46], interlayer-related paramagnetic defects were studied in stacks of ultrathin layers of SiO_x, Al_2O_3, ZrO_2, and HfO_2 on (100)Si-subjected to VUV irradiation. The observation was reported of typical SiO_2-associated defects, i.e., the E' and EX centers, in all studied entities, and an additional 95-G-split doublet in (100)Si/SiO_x/ZrO_2 entities, tentatively interpreted as involving a H-split doublet related with an impurity in the ZrO_2 layer.

9.2 Experimental Methodology and Samples

9.2.1 ESR Spectroscopy

Generally, ESR is considered as the technique of choice when it comes to atomic identification of point defects. As such, one would hope to apply this technique as a standard when investigating semiconductor/insulator structures where point defects emerge as the origin of detrimental charge trapping, carrier recombination, and leakage currents. Yet, practice is different as the application of the technique faces some obstacles. One is that the defects envisioned need to be in a spin-active (paramagnetic) state (suitable charge state), which appears often not the case. Second, the sensitivity is limited: Current top performance ESR spectrometers may detect $\sim 1 \times 10^9$ centers (spin S = 1/2) of 1 G line width at low T within acceptable averaging time. Many signals appear much broadened, which strongly impairs their detection, thus generally rendering the conventional ESR technique less sensitive than typical state-of-the-art electrical observations, such as, e.g., C–V.

The ESR data presented in this work have been obtained by conventional CW derivative–absorption measurements at 4.3 K using a locally constructed K-band (~20.2 GHz) spectrometer, as described elsewhere [47]. Angular dependence of ESR parameters was investigated by rotating the applied magnetic field **B** in the $(0\bar{1}1)$ substrate plane over a range of 0–90° with respect to the [100] surface normal **n**. The applied microwave power P_{μ} was varied in the range 0.25–100 nW to enable avoiding saturation effects on detected signals and/or to boost successful ESR detection. The applied modulation amplitude B_m(~100 kHz) of the magnetic field was restricted to such levels that no signal distortion was observable. More details can be found in Ref. [34]. Defect (spin S = 1/2) densities were determined relative to the signal of a comounted Si:P intensity marker through comparison of the signal intensities (I) obtained by orthodox double numerical integration of the detected derivative–absorption $dP_{\mu}/d\mathbf{B}$ spectra. The attained absolute accuracy on spin densities determinations is estimated at ~20 %. The Si:P marker of $g(4.3\,\mathrm{K}) = 1.99869$ was also used for g calibration purposes. ESR samples were (100)Si slices of $2 \times 9\,\mathrm{mm}^2$ area with the 9-mm edge along a $[0\bar{1}1]$ direction. The backside of the sample slices was treated in a HF–H_2O mixture immediately before each ESR observation. Typically, an ESR sample was comprised of ~12–16 slices.

In order to enhance ESR detectivity, after an initial ESR analysis, some Ge/HfO_2 samples were additionally subjected to VUV irradiation in air at RT to photo-dissociate H from potentially H-inactivated point defects [48,49]. Indeed, there is an abundance of H during MOCVD growth, which may have resulted in ESR inactivation of occurring point defects through interaction with hydrogen – a phenomenon well known, e.g., for the Si dangling bond P_b-type defects in thermal Si/SiO_2 [36,50,51].

9.2.2 Electrical Analysis

The density and energy distribution of Ge/HfO_2 interface traps were determined from analysis of capacitance–voltage (C–V) and conductance–voltage (G–V) curves measured at 77 or 300 K in the frequency range 100 Hz–1 MHz using a HP4284A bridge [52].

9.2.3 Samples

The studied samples were prepared on (100)Ge wafers of both n- (Sb-doped) and p-type (Ga-doped) conductivity supplied by Umicore (Belgium). After wet chemical cleaning, the Ge surface was exposed to NH_3 at 600°C, known to result in the formation of Ge–N bonds [53], with the intention to minimize oxidation of Ge during subsequent chemical vapor deposition (CVD) of 10-nm thick HfO_2 films from the metallo-organic precursor tetrakis-diethylaminohafnium ($[(C_2H_5)_2N]_4Hf$) and O_2 at 485°C. A nitridation temperature of 600°C was selected because of yielding the best electrical results in the temperature range studied (500–700°C) [18].

To enable comparative measurements, HfO_2 layers were also deposited under identical conditions onto (100)Si substrates. Some of Ge/HfO_2 samples were subjected to a postdeposition anneal (PDA) in O_2 (99.9995%, 1.1 atm) at 650°C for 10 min or passivated in H_2 (1.1 atm) at 400°C for 30 min. To compare the properties of interfaces of (100)Si and (100)Ge with their native oxides, the analysis was extended to Si and Ge substrates thermally oxidized 280°C (i.e., low-T oxide) or oxidized at 300 K by exposure to UV-generated (10 eV photons; flux $\sim 10^{15}\,cm^{-2}s^{-1}$) ozone and atomic oxygen [54].

Electrical studies were carried out on MOS capacitors of 0.4 mm^2 area, fabricated by thermoresistive evaporation of Au electrodes.

9.3 Experimental Results

9.3.1 Electrical Analysis

Figure 9.1 compares observed 100-kHz C–V curves of n- and p-type Ge/HfO_2/Au capacitors measured at 300 and 77 K. In addition to the as-grown samples (open circle, closed circle) data, results are also shown for samples exposed to vacuum ultraviolet (VUV, $h\nu = 10$ eV) photons prior to metallization (open square, filled square) intended to photodissociate hydrogen from dangling bond defects potentially present at the interface [48, 49]. These are to be compared to the samples annealed in H_2 (open triangle, filled triangle) or subjected to the oxidizing PDA (*open diamond, filled diamond*). Among other features, one may notice a ledge in the room temperature C–V curves of the as-deposited and the VUV-depassivated samples in Fig. 9.1a indicative of a high density of interface traps. The marginal effect of H-photodissociation

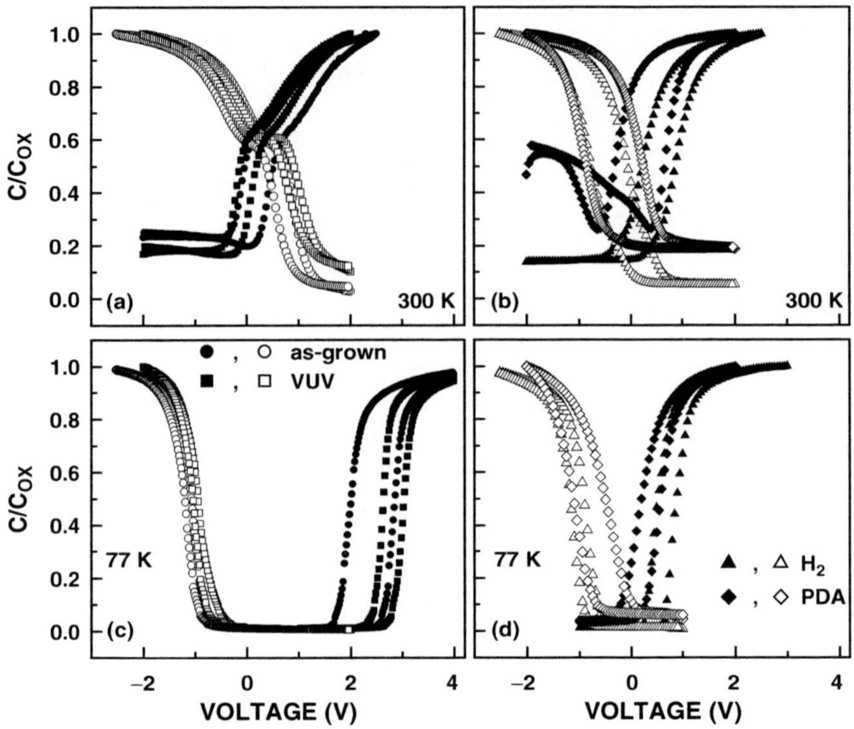

Fig. 9.1. Capacitance–voltage (100-kHz) curves measured at 300 K (**a**,**b**) and 77 K (**c**,**d**) on n- (filled symbols) and p-type (empty symbols) (100)Ge/HfO_2/Au capacitors. Data are shown for the as-grown (*opencircle, closedcircle*), VUV-depassivated (*open square, filled square*), and H_2-passivated (*open triangle, filled triangle*) samples, and samples subjected to the oxidizing PDA (*open diamond, filled diamond*)

treatment suggests a negligible interface trap passivation by H-containing byproducts of the HfO_2 deposition. The trap-related ledge disappears upon annealing in hydrogen (open triangle, filled triangle) or oxidizing PDA (open diamond, filled diamond) but, instead, now a large hysteresis appears in the C–V curves as exposed in Fig. 9.1b pointing to an enhanced density of slow traps. This hysteresis hampers the trap density determination from C–V curvesElectrical Analysis recorded at room temperature, prompting us to carry out low-temperature measurements. Upon cooling to 77 K the C–V curves show (Fig. 9.1c) a remarkable asymmetry with a large shift to positive voltages in the n-type as-grown and VUV-exposed samples indicating a high density of acceptor-type interface traps in the upper part of the Ge bandgap. From the difference in flat-band voltages of the n- and p-type capacitors at 77 K measured with the voltage swept from accumulation to depletion, one can infer the total density of states recharged at the surface when the Fermi level is shifted from its position close to the valence band (in p-type semiconductors) to near the conduction band (n-type samples) because the thermal emission of

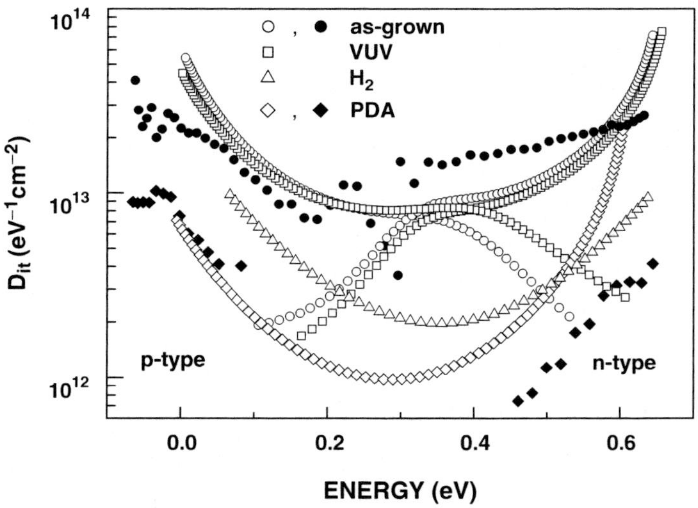

Fig. 9.2. Observed interface trap density as a function of energy in the Ge band gap as inferred from low- and high-frequency *C–V* (*open symbols*) and *C–V* (*filled symbols*) curves of p- and n-type (100)Ge/HfO_2/Au MOS capacitors, spanning the lower and upper half of the band gap, respectively. Data are given for the as-grown (*open circle, closed circle*), VUV-irradiated (*open square*), H_2-passivated (*open triangle*) samples, and structures subjected to PDA in O_2 (*open diamond, filled diamond*). The origin of the energy axis is taken at the top of the Ge valence band

carriers from most of the traps is negligible [33]. In this way, the total density of traps (integrated over the ~0.6 eV wide central portion of the Ge band gap) is inferred as $(2.5 \pm 0.5) \times 10^{13}\,\mathrm{cm}^{-2}$. Passivation in H_2 (open triangle, filled triangle) and oxidation (open diamond, filled diamond) are seen to significantly reduce the density of acceptor traps leading to symmetric *C–V* curves of n- and p-type capacitors, as illustrated in Fig. 9.1d. Noteworthy here is that the flatband voltages in the p-type MOS structures are barely affected by the trap-eliminating anneals which suggests a negligible density of the donor-type interface traps.

More details are provided by the (100)Ge/HfO_2 interface state density (D_{it}) distribution versus energy derived from the *C–V* (open symbols) and *C–V* curves (closed symbols), shown in Fig. 9.2 for the as-grown (open circle, closed circle), VUV-depassivated (open square), H_2-passivated (open triangle), and O_2-annealed (open diamond, filled diamond) samples. Both the as-grown and VUV-exposed samples *C–V* data yield a U-shaped D_{it} distribution and suggest a broad peak near the midgap (Fig. 9.2), which is consistent with the observed ledge in the room temperature *C–V* curves of these samples (cf. Fig. 9.1a). The *C–V* results are less revealing of the presence of a peak, possibly due to preferential detection of the fast interface states by this method. Indeed, the G/ω-versus-ω curves ($\omega = 2\pi f$ is the angular frequency of the probing signal) exhibit an increase at $f < 500\,\mathrm{Hz}$ (not shown) suggestive of a high slow trap density. Thus, in the as-deposited samples the dominant interface traps are acceptors with a large time constant for recharging.

The trap density is significantly reduced after hydrogen annealing (open triangle in Fig. 9.2) possibly hinting at the presence of some dangling bond centers. However, the absence of well-defined energy levels in the gap and the predominantly acceptor-type behavior of the interface states exclude the simple explanation for the observed D_{it} distribution as originating from an amphoteric defect similar to the P_{b0} center in (100)Si/SiO_2 [32, 33] or (100)Si/HfO_2 [41, 52]. More likely is that there are some defects located in the near-interfacial insulator layer. This hypothesis is indirectly supported by the results of the postdeposition oxidation in H-free conditions resulting in even lower D_{it} values than observed after hydrogen passivation, as also shown in Fig. 9.2 (open diamond, filled diamond). This indicates that the dominant traps are physically removed by oxidation and/or modification of the near-interfacial insulator region. The remaining D_{it} is U-shaped and reaches $\approx 1 \times 10^{12}\,\mathrm{cm^{-2}\,eV^{-1}}$ near the Ge midgap point.

9.3.2 ESR Measurements

As well established for both the Si/SiO_2 [55] and Si/HfO_2 [41,52] structures, a considerable density of interface traps is related to P_b-type centers ($Si_3 \equiv Si^{\bullet}$ defects located at the interfacial Si crystal plane, where the dot symbolizes an unpaired electron) exhibiting a characteristic two-peak D_{it} pattern. Examples of observed K-band ESR spectra for **B** along the [100] surface normal **n** are shown in Fig. 9.3 for three types of (100)Si/HfO_2 entities manufactured using different HfO_2 deposition methods. In short, these three types of (100)Si/HfO_2 entities (labeled A, B, C) were obtained by depositing 5–7 nm thick HfO_2 layers on n and p-type (100)Si substrates using three variants of the CVD method. Type A was prepared by AL-CVD at 300 °C using $HfCl_4$ and H_2O precursors, while type B was produced by metallo-organic chemical vapor deposition (MO-CVD) at 485°C from tetrakis-diethylaminohafnium and O_2; The modified IMEC predeposition Si surface cleaning was applied for both cases. The third type was prepared by CVD on HF-dipped last (100)Si surfaces at 350°C using the nitrato precursor $Hf(NO_3)_4$ (referred to as N-CVD), that is, nominally under H and C-free conditions [56]. The observed P_{b0} and P_{b1}-type signals correspond to $(2.3 \pm 0.4) \times 10^{12}\,\mathrm{cm^{-2}}$ and $(1.6 \pm 0.2) \times 10^{12}\,\mathrm{cm^{-2}}$ for the ALCVD sample (type A), while for the MOCVD sample (type B), these values are $(2.4 \pm 0.4) \times 10^{12}\,\mathrm{cm^{-2}}$ and $(1.2 \pm 0.3) \times 10^{12}\,\mathrm{cm^{-2}}$, respectively. The presence of P_b-type interface defects in Si/high-κ insulator structures has been confirmed in several other ESR works [35–39, 41, 46]. It refers to the presence of an $SiO_{2(x)}$-type interlayer.

In the wake of previous research and in search for the possible source of Ge/HfO_2 interface states, we resorted to ESR measurements for possible elucidation. On overview of the results is presented in Fig. 9.4, showing a set of representative K-band ESR spectra observed for **B**//**n** on various samples at

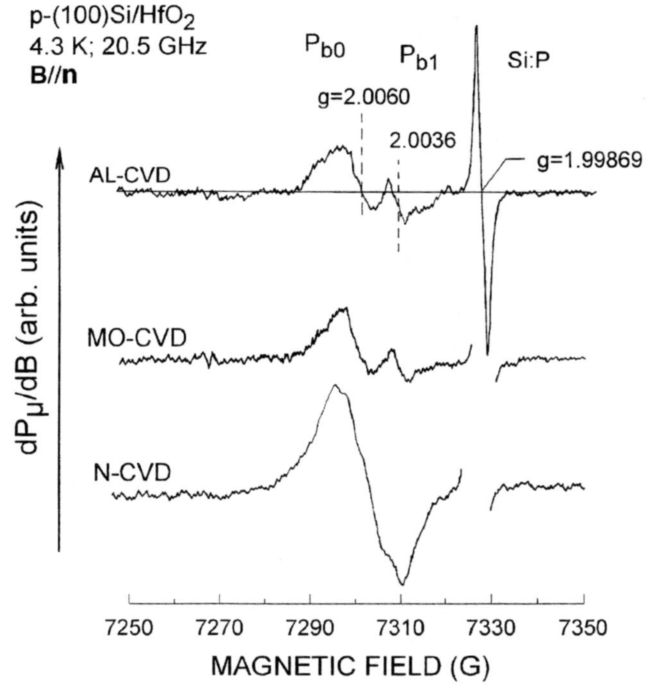

Fig. 9.3. Derivative–absorption K-band ESR spectra observed at 4.3 K on (100)Si/HfO_2 entities of types A, B, and C, subjected to room temperature VUV irradiation (hydrogen photo dissociation) prior to observation. The signals observed at $g = 2.0060$ and 2.0036 stem from P_{b0} and P_{b1} interface centers, respectively, while the signal at $g = 1.99869$ stems from a comounted marker sample. The applied modulation field amplitude was 0.4 G, and incident $P_\mu \sim 0.8$ nW. Spectrum A was observed on a sample prepared using ALCVD at 300 °C from $HfCl_4$ and H_2O precursors, while type B was produced by the MOCVD technique at 485 °C from the tetrakisdiethylaminohafnium and O_2 precursors Type C was prepared by CVD at 385 °C from $Hf(NO_3)_4$

4.3 K. Quite in contrast to all the previously analyzed (100)Si/oxide systems (Si/SiO_2, Si/SiON/HfO_2, Si/SiO_x/HfO_2, Si/SiN_x/HfO_2) universally exhibiting high densities of P_{b0} centers [36,41,52,57], the (100)Ge/HfO_2 entity, likely with an oxynitride interlayer as denoted in Fig. 9.4, shows only one broad line ($\Delta B_{pp} = 8 \pm 1$ G) at zero crossing g value $g_c = 2.0022 \pm 0.0001$. As no additional signal appears after VUV exposure (cf. Fig. 9.4), we conclude that no significant H-passivation has occurred during fabrication. The observed ESR signal is insensitive to the orientation of applied magnetic field with respect to the sample surface. Thus, there exists no registry between the ESR-active defects and the Ge substrate crystal lattice suggesting the defects to be located in an amorphous oxide network (GeO_xN_y or HfO_2 layers). Furthermore, we noticed that no such signal is observed in the Si/HfO_2 structures prepared using similar surface preparation and deposition procedures indicating that

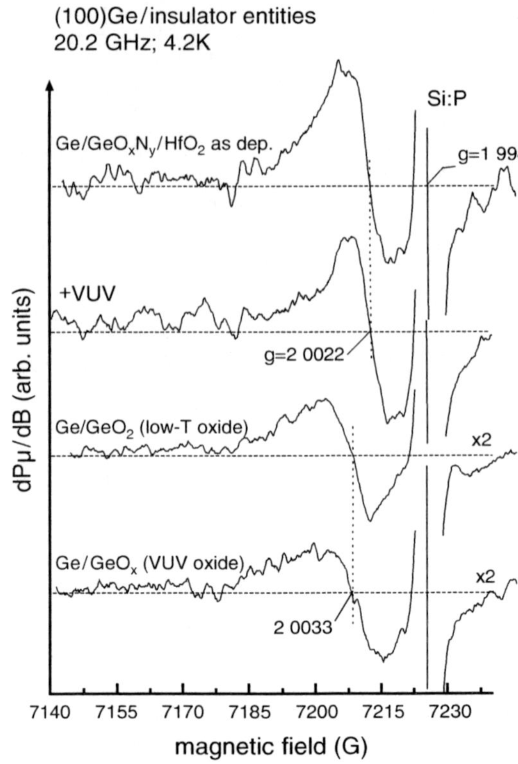

Fig. 9.4. Typical K-band ESR spectra observed at 4.2 K on (100)Ge samples with nm-thick dielectric overlayers of the indicated composition. The signal at $g = 1.99869$ stems from a comounted calibration Si : P marker sample. Observations were made with **B** closely along the [100] sample normal. The applied P_{μ} and modulation field amplitude were ∼10 nW and ∼0.3 G, respectively. Note that, due to unavoidable (slight) shifts in observational microwave frequency from sample to sample, the added magnetic field axis can of course absolutely be correct only for one spectrum; the other spectra have been shifted slightly to make the marker signals coinciding

the originating defects likely pertain to the germanium (oxy)nitride interlayer, GeO_xN_y. The areal density of the unpaired spins (S = 1/2) is found to be $\sim 5 \times 10^{12}\,cm^{-2}$ both in the as-deposited and VUV-exposed samples. One more interesting result concerns the postmanufacturing treatment in molecular H_2 at 400 °C. The treatment hardly affects the signal: A close density and the same g-value are observed after the treatment in hydrogen (spectra not shown) bearing out these defects to be resistant to passivation in H_2, quite in contrast the behavior of Si DB (P_b) type defects.

Finally, in contrast to passivation by H, the oxidizing PDA strongly affects the ESR spectra of Ge/HfO_2: It leads to an isotropic ESR signal at $g \approx 2.0033$ and $\Delta B_{pp} = 10 \pm 1$ G which is very similar to the signals observed after oxidation of a clean (100)Ge at 280 °C in O_2 or under VUV excitation, as illustrated in Fig. 9.4. No signs of anisotropic ESR signals expected for the P_b-type centers could be detected in the oxidized (100)Ge, which, again, deviates from the general (100)Si/oxide results [55,57]. The corresponding spin density in the oxidized Ge and Ge/HfO_2 after PDA appears to be about four times smaller than in the as-deposited Ge/HfO_2 structures. When comparing with the results obtained on the Ge/GeO_xN_y/HfO_2 structures, this may point to an impact of the presence of N in the interlayer on the defect density. At the same time, as no ^{14}N (nuclear spin I = 1; 99.63 % natural abundance) hyperfine (hf) structure is resolved. If deemed appropriate to compare with other known N-centered defects, such as the $Si_2 = N^\bullet$ center observed in Si_3N_4, which shows a well resolved ^{14}N splitting ($A_{//}$~35 G) [58], the finding would suggest the unpaired electron not to be mainly localized at one nitrogen site. But of course, albeit less conceivable, without actual knowledge of possible ^{14}N hf splitting, hf signatures could have remained unresolved just because of unfavorable line shape parameters (e.g., line broadening, g matrix). Probably then, as a first working model, the defects in both cases are likely to originate from some Ge-related imperfections in the Ge (oxy)nitride.

9.4 Discussion

With the ESR results available, we may compare the observed interface trap densities to the density of ESR-active dangling bond defects. In the as-deposited samples the inferred density of fast interface traps D_{it} is $\sim 1 \times 10^{13}\,cm^{-2}\,eV^{-1}$ in the probed 0.6-eV wide portion of the Ge bandgap yields a total trap density of $6 \times 10^{12}\,cm^{-2}$ which is close to the density of $5 \times 10^{12}\,cm^{-2}$ of centers observed by ESR at $g = 2.0022$. The stability of these trap densities upon VUV exposure and H-passivation is consistent with the marginal changes of the corresponding ESR signal. However, the much higher trap density derived from the difference in flat band voltage of n- and p-type MOS capacitors at 77 K, namely $N_{it} = (2.5 \pm 0.5) \times 10^{13}\,cm^{-2}$, suggests a dominance of slow traps not related to the dangling bond defects. Their acceptor behavior may be explained by trapping of an electron by an imperfection containing only paired electrons (saturated bonds) which, therefore, before hopping escape ESR detection. Potentially, the trapping of the extra electron would provide an unpaired spin in principle detectable by ESR, but now it may result in an absorption signal broadened beyond detection. Noteworthy here is the effect of annealing in hydrogen which significantly reduces the trap density (cf. Figs. 9.1 and 9.2) without affecting the ESR spectrum. Clearly then, the removed traps have no relationship to the observed paramagnetic centers. The nature of the hydrogen effect is unlikely to be related to a simple passivation because no reverse behavior (depassivation) is observed after subsequent photo- or thermodissociating treatments of the H_2-annealed samples. Apparently, the H_2 anneal modifies the insulator network leading to a lower trap density.

After the oxidizing PDA, the density of spins observed in Ge/HfO_2[(1 ± 0.2) × 10^{12} cm^{-2}] is numerically consistent with the density of fast interface states integrated over the Ge bandgap (cf. Fig. 9.2). Though this may appear coincidental, both the ESR signal intensity and the Ge/HfO_2 trap density are insensitive to the subsequent annealing in hydrogen, similarly to that in the as-deposited samples, suggesting that the insulator-related defects may give rise to a continuum of electrically detected interface traps.

Next, there is the unanticipated fact that apparently conventional ESR has so far failed in resolving DB type interface defects which could be attributed to Ge. Regardless of the origin and nature of the observed isotropic signals, it comes as a key overall observation is that only *isotropic* signals are observed. This implies that there is no evidence for an anisotropic Ge DB type center such as the trigonal Ge-centered defect [59] ($g_{//} = 1.9998, g_{\perp} = 2.026$) previously isolated in O-implanted SiGe alloys (10–40% Ge) and ascribed to a threefold coordinated central Ge atom back bonded to Si and Ge atoms-designated as the Ge P_b center of C_{3v} symmetry with $g_{//}$~2.005 and $g_{\perp}$~2.022 in other work on oxidized porous $Si_{0.8}Ge_{0.2}$ [60].

In a critical attitude, one may surmise that no ESR signal from Ge dangling bonds could be resolved plainly because of insufficient ESR sensitivity, e.g., brought about by excessive signal broadening, an apparent characteristic for Ge-related defects because of enhanced spin–orbit coupling as compared to Si. Indeed, as compared to Si, the line widths of Ge-surface related dangling bonds appear generally distinctly broader: X-band ESR observations reported a signal at $g = 2.023 \pm 0.003$ with $\Delta B_{pp} \sim 50$ G for the Ge dangling bond at the surface of c-Ge (crushed Ge powder) [61], while a signal of width ~39 G was observed at $g = 2.021$ in rf sputtered a-Ge [62]. This has been verified in the current work by K-band ESR on freshly crushed Ge powder; In agreement, a signal at $g = 2.021$ with $\Delta B_{pp} \sim 70$ G was readily detected. This is a broad signal indeed. But in the same breath, it should be added that as demonstrated previously for O-implanted SiGe (10–40% Ge) alloys [59], and as well known for the Si case [34], ESR signals that would pertain to interfacial Ge DB type defects are expected to be in registry with the substrate's crystallinity, i.e., depending on the orientation of **B** with respect to the Ge surface, various different drastically narrower signals are expected corresponding to the different branches of the g map pertaining to the various equivalent orientations of the defect in the Ge crystal. All in all, taken together with known K-band ESR spectroscopy sensitivity and extremized detectability, it is estimated that even under the "worst case" assumption of the line width to scale linearly with the microwave frequency, the detection limit for Ge-type defects in the current low temperature experiments is estimated to be of the order of $\sim 2 \times 10^{12}$ cm^2.

But, of course, here extreme care has to be exercised as the nonobservation of a signal means a negative result. Even within the measures of careful estimates about sensitivity and analysis of the kind of signals that might reasonably be expected, there may still intervene one or more hidden (overlooked) reasons conspiring to destructively obstruct ESR observation, even with all defects residing in the paramagnetic state. Quite disappointingly, the answer about the veracity of such premise cannot be given before any such Ge-related DB-type related defect, should it occur at all in any substantial amount, would have been detected in the investigated Ge/insulator structures. Within this consideration, it should also be remarked that the available literature on ERS observation of putative Ge-related DB centers is still limited. For one, apart perhaps what concerns the observations on Ge powder and a-Ge layers [61,62], the atomic nature of the defects ascribed to Ge-related DB type centers at the origin of the ESR signals observed from SiGe alloys is still uncertain.

Finally, we may wonder about the possible observation of paramagnetic defects residing in the oxide layers. Noteworthy is that, with respect to the presence of GeO_x layers, we also failed to detect known characteristic defects for glassy GeO_2. First of all, this includes the oxygen vacancy, termed the Ge E' center ($O_3 \equiv Ge^\bullet$; $g_{//} = 2.0016$; $g_\perp \sim 1.996$), previously identified [61] in glassy GeO_2 and Ge-doped silica, not even after VUV excitation. Neither could we observe after VUV irradiation any evidence for the presence of the Ge peroxyradical $\equiv$ Ge–O–O$^\bullet$($g_1 = 2.002; g_2 = 2.08; g_3 = 2.051$) [63]. Perhaps, the latter comes less as surprise because of the generally widely spread powder pattern distributed over an extended g range characteristic of oxygen associated hole centers in vitreous Si dioxide [63]. Though convincingly identified in bulk glassy SiO_2, such defects are less easily revealed in thermal Si/SiO_2 structures, with only a few known reports; It generally requires thick oxide layers (>100 nm) subjected to intense damage by energetic particles (ions, γ-rays) (See, e.g., [64]). As to the nonobservation of the Ge E' signal, this may also be due to reduced ESR sensitivity as a result of increased signal broadening as compared to the well known Si E' signal. In a different opinion it could also refer to an inherently less density of O vacancies in the currently studied thin Ge-oxide layers vis-à-vis bulk glassy germania.

9.5 Conclusions

As a main finding, the presented results of electrical and ESR analysis reveal important differences between the semiconductor/insulator interfaces of seemingly similar group IV semiconductor surfaces, (100)Si and (100)Ge. This concerns several related aspects. First, the dangling bonds at the semiconductor crystal surface universally observed at the interfaces of Si with different insulators are not present in any detectable density in the case of Ge. Recalling that the dangling bond formation in the oxidized Si is associated with accommodation of network mismatch between the substrate crystal and the oxide phase [47], the nondetection of such centers in Ge/insulator structures may suggest a fundamental structural difference between these interfaces (interlayers) for (100)Si and (100)Ge. Yet, as large densities of interface traps ($N_{it} \sim 10^{13}\,cm^{-2}$) are detected electrically in the Ge/HfO_2

system also, this means that the origin of the dominant interface trap for the two semiconductors would be basically different. Possibly, the observation of insulator-related defects in Ge/oxide and Ge/(oxy)nitride systems indicate that the energy of defect formation in the Ge-based interlayer is lower than the energy needed to create the stable dangling bond at the surface of Ge.

Second, the interface trap spectrum in Ge/oxide systems appears to be dominated by slow acceptor states with a broad energy distribution, which is also consistent with the major contribution coming from imperfections located in the insulating layer.

Finally, the absence of a measurable influence of hydrogen passivation on the ESR signal intensity would be consistent with the mentioned lack of observation of semiconductor dangling-bond defects. This may have fundamental technological consequences: The difference in interfacial point defect genealogy may reflect in the approach how to realize stable device grade interfaces in terms of interface traps. For one, when aiming to improve the Ge/HfO_2 interface properties, one may want to look for a solution addressing proper interlayer engineering rather than attempting to improve on the hydrogen passivation procedure.

Acknowledgments

The authors are thankful to Dr. M. Meuris, Dr. I. Teerlinck, and Dr. S. Van Elshocht (IMEC, Leuven, Belgium) for providing the HfO_2/Ge samples investigated in the present study, and to Dr. G. Raskin (Umicore, Belgium) for the supply of Ge substrate crystals.

The Flanders Fund for Scientific Research is acknowledged for financial support.

References

1. G.D. Hutcheson, Interface **14**, 17 (2005); H. Wong and H. Iwai, Physics World **18**, 40 (2005).
2. The *International Technology Roadmap for Semiconductors* edn 2003 (SIA, San Jose, CA); http://public.itrs.net/
3. S.H. Lo, D.A. Buchanan, Y. Taur, and W. Wang, IEEE Electron. Device Lett. **18**, 209 (1997).
4. G. Wilk, R.M. Wallace, and J.M. Anthony, J. Appl. Phys. **89**, 5243 (2001).
5. R.M. Wallace and G. Wilk, Crit. Rev. Solid State **28**, 231 (2003).
6. M.L. Green, E.P. Gusev, R. Degraeve, and E. Garfunkel, J. Appl. Phys. **90**, 2057 (2001).
7. J. Robertson, Eur. Phys. J. Appl. Phys. **28**, 265 (2004).
8. *High-κ gate dielectrics*, edited by M. Houssa (Institute of Physics Publishing, Bristol, 2004).

9. *High Dielectric Constant Materials: VLSI MOSFET Applications*, edited by H.R. Huff and D.C. Gilmer, (Springer Series in Advanced Microelectronics, 2004).
10. M.-A. Nicolet and W.-S. Liu, Microelectron. Eng. **28**, 185 (1995).
11. K. Prabhakaran and T. Ogino, Surf. Sci. **325**, 263 (1995).
12. K. Kita, K. Kyuno, and A. Toriumi, Appl. Phys. Lett. **85**, 52 (2004).
13. S.J. Wang, A.C.H. Huan, Y.L. Foo, J.W. Chai, J.S. Pan, Q. Li, Y.F. Dong, Y.P. Feng, and C.K. Ong, Appl. Phys. Lett. **85**, 4418 (2004).
14. T. Maeda, T. Yasuda, M. Nishizawa, N. Miyata, Y. Morita, and S. Takagi, Appl. Phys. Lett. **85**, 3181 (2004).
15. C.O. Chui, H. Kim, D. Chi, B.B. Triplett, P.C. McIntyre, and K.C. Saraswat, Techn. Dig. – Int. Electron Devices Meet. 2002, 437.
16. H. Shang, H. Okorn-Schmidt, K.K. Chan, M. Copel, J.A. Ott, P. M. Kozlowski, S.E. Steen, H.-S.P. Wong, E.C. Jones, and W.E. Haensch, Techn. Dig. – Int. Electron Devices Meet. 2002, 441.
17. W.P. Bai, N. Lu, J. Liu, A. Ramirez, D.L. Kwong, D. Wristers, A. Ritenour, L. Lee, and D. Antoniadis, Techn. Dig. VLSI Symp. 2003, 121.
18. M. Houssa, B. DeJaeger, A. Delabie, S. Van Elschocht, V.V. Afanas'ev, J.L. Autran, A. Stesmans, M. Meuris, and M.M. Heyns, J. Non-Cryst. Solids **251**, 1902 (2005).
19. A. Dimoulas, G. Vellianitis, G. Mavrou, E.K. Evangelou, and A. Sotiropoulos, Appl. Phys. Lett. **86**, 223507 (2005).
20. H. Kim, P.C. McIntyre, C.O. Chui, K.C. Saraswat, and M.-H. Cho, Appl. Phys. Lett. **85**, 2902 (2004).
21. V.V. Afanas'ev and A. Stesmans, Appl. Phys. Lett. **84**, 2319 (2004).
22. K.-III Seo, P.C. McIntyre, S. Sun, D.-I. Lee, P. Pianetta, and K.C. Saraswat, Appl. Phys. Lett. **87**, 042902 (2005).
23. N. Lu, W. Bai, A. Raminez, C. Mouli, A. Ritenour, M.L. Lee, D. Antoniadis, and D.L. Kwong, Appl. Phys. Lett. **87**, 051922 (2005).
24. S. Van Elshocht, B. Brijs, M. Caymax, T. Conard, T. Chiarella, S. De Gendt, B. De Jaeger, K. Kubicek, M. Meuris, B. Onsia, O. Richard, I. Teerlinck, J. Van Steenbergen, C. Zhao, and M. Heyns, Appl. Phys. Lett. **85**, 3824 (2004).
25. C.O. Chui, H. Kim, P.C. McIntyre, and K.C. Saraswat, IEEE Electron Device Lett. **25**, 274 (2004).
26. N. Wu, Q. Zhang, C. Zhu, D.S.H. Chan, M.F. Li, N. Balasubramanian, A. Chin, and D.-L. Kwong, Appl. Phys. Lett. **85**, 4127 (2004).
27. N. Wu, Q. Zhang, C. Zhu, C.C. Yeo, S.J. Whang, A. Chin, Dim-Lee Kwong, A.Y. Du, C.H. Tung, and N. Balasubramanian, Appl. Phys. Lett. **84**, 3741 (2004).
28. F. Gao, S.J. Lee, J.S. Pan, L.J. Tang, and D.-L. Kwong, Appl. Phys. Lett. **86**, 113501 (2005).
29. A. Delabie, R.L. Puurunen, B. Brijs, M. Caymax, T. Conard, B. Onsia, O. Richard, W. Vandervorst, C. Zhao, M.M. Heyns, M. Meuris, M.M. Vitanen, H.H. Brongersma, M. de Ridder, L.V. Goncharova, E. Garfunkel, T. Gustafsson, and W. Tsai, J. Appl. Phys. 97, 064104 (2005).
30. C.O. Chui, S. Ramanathan, B.B. Triplett, P. McIntyre, and K.C. Saraswat, IEEE Electron Dev. Lett. **23**, 473 (2002).
31. H. Shang, H. Okorn-Schmidt, J. Ott, P. Kozlowski, S. Steen, E.C. Jones, H.S. Wong, and W. Hanesch, IEEE Electron Dev. Lett. **24**, 242 (2003).
32. G.J. Gerardi, E.H. Pointdexter, P.J. Caplan, and N.M. Johnson, Appl. Phys. Lett. **49** 348 (1986).

33. A. Stesmans and V.V. Afanas'ev; Phys. Rev. B. **57**, 10030 (1998).
34. A. Stesmans and V.V. Afanas'ev, J. Appl. Phys. **83**, 2449 (1998).
35. A. Stesmans and V.V. Afanas'ev, J. Phys.: Condens. Matter **13**, L673 (2001).
36. A. Stesmans and V.V. Afanas'ev, Appl. Phys. Lett. **80**, 1957 (2002).
37. J.L. Cantin and H.J. von Bardeleben, J. Non-Cryst. Solids **303**, 175 (2002)
38. S. Baldovino, S. Nokrin, G. Scarel, M. Fanciulli, T. Graf, and M. S. Brandt, J. Non-Cryst. Solids **322**, 168 (2003).
39. B.J. Jones and R.C. Barklie, Microelectron. Eng. **80**, 74 (2005).
40. R. Helms and E.H. Poindexter, Rep. Prog. Phys. **83**, 2449 (1998).
41. A. Stesmans and V.V. Afanas'ev, Appl. Phys. Lett. **82**, 4074 (2003).
42. A.Y. Kang, P.M. Lenahan, J.F. Conley, Jr, and R. Solanski, Appl. Phys. Lett. **81**, 1128 (2002).
43. A.Y. Kang, P.M. Lenahan, and J.F. Conley, Jr, Appl. Phys. Lett. **83**, 3407 (2003).
44. A. Stesmans, V. Afanas'ev, F. Chen, and S.A. Campbell, Appl. Phys. Lett. **84**, 4574 (2004).
45. V. Lowalekar and S. Raghavan, J. Non-Cryst. Solids **351**, 1559 (2005)
46. A. Stesmans and V.V. Afanas'ev, Appl. Phys. Lett. **85**, 3792 (2004); J. Appl. Phys. **97**, 033510 (2005).
47. A. Stesmans, Phys. Rev. B **48**, 2418 (1993).
48. A. Pusel, U. Wetterauer, and P. Hess, Phys. Rev. Lett **81**, 645 (1998).
49. T. Vondrak and X.Y. Zhu, J. Phys. Chem. B **103**, 4892 (1999).
50. K.L. Brower, Phys. Reb. B 38, 9657 (1988).
51. A. Stesmans, Appl. Phys. Lett. **68**, 2076 (1996); **68**, 2723 (1996).
52. Y.G. Fedorenko, L. Truong, V.V. Afanas'ev, and A. Stesmans, Appl. Phys. Lett. **84**, 4771 (2004).
53. W. Ranke and J. Wasserfall, Surf. Sci. **303**, 45 (1994).
54. A. Stesmans and V.V. Afanas'ev, Appl. Phys. Lett. **77**, 1469 (2000).
55. E.H. Poindexter, Semicond. Sci. Technol. **4**, 961 (1989).
56. S.A. Campbell, T.Z. Ma, R. Smith, W.L. Gladfelter, and F. Chen, Microelectron. Eng. **59**, 361 (2001).
57. A. Stesmans and V.V. Afanas'ev, Appl. Phys. Lett. **77**, 1469 (2000); **82**, 2835 (2003).
58. W.L. Warren, F.C. Rong, E.H. Poindexter, G.J. Gerardi, and J. Kanicki, J. Aoppl. Phys. **70**, 346 (1991).
59. M.E. Zvanut, W.E. Carlos, M.E. Twigg, R. Stahlbush, and D.J. Godbey, J. Vac. Sci. Technol. B **10**, 2026 (1992).
60. S. Lebib, M. Schoisswohl, J.L. Cantin, and H.J. von Bardeleben, Thin Solid Films **294**, 242 (1997).
61. G.K. Walters and T.L. Estle, J. Appl. Phys. **32**, 1854 (1961).
62. M.H. Brodsky and R.S. Title, Phys. Rev. Lett. **23**, 581 (1969).
63. T.-E. Tsai, D.L. Griscom, E.J. Friebele, and J.W. Fleming, J. Appl. Phys. **62**, 2264 (1987).
64. A. Stesmans, J. Braet, J. Witters, and R.F. DeKeersmaecker, J. Appl. Phys. **55**, 1551 (1984).

10

High κ Gate Dielectrics for Compound Semiconductors

J. Kwo and M. Hong

Summary. The ability of controlling the growth and interfaces of ultra-thin dielectric films on compound semiconductors by ultrahigh vacuum physical vapor deposition has led to comprehensive studies of gate stacks employing high κ gate oxide $Ga_2O_3(Gd_2O_3)$ and rare earth oxide Gd_2O_3. These oxides as gate dielectrics on GaAs have been shown to possess a low interfacial density of states, thus solving a problem which has puzzled researches for almost four decades. The electrical, thermal, chemical, and structural properties of these novel oxides and their interfaces with GaAs are reviewed. Particularly the achievement of low interfacial density of states (D_{it}) and thermodynamic stability upon high temperature annealing is discussed. The interfacial oxide layers on GaAs were found to be a single crystal of pure Gd_2O_3. The ultra-thin Gd_2O_3 on GaAs has given a very low leakage current and low D_{it}, a first time achieved by a single crystal oxide. Various GaAs metal-oxide-semiconductor field-effect-transistors (MOSFETs) and their device performance are reviewed. The mechanism of Fermi-level unpinning in ALD-Al_2O_3 on InGaAs was studied and understood. The epitaxy and the interfaces of Gd_2O_3 on GaN were characterized, and show strong tendency to conform to the underlying substrate, thus providing insight into the fundamental mechanism for low interfacial state density and effective passivation. These gate stacks of abrupt interfaces and controlled microstructures were employed as a model system to elucidate critical issues of materials integration in the CMOS process.

10.1 Introduction

The Si technology is entering the age of nano-meters, with the gate length of 90 nm in production and devices of 50 nm or smaller in research and development. Up to now, the level of perfection in the well known Si–SiO_2 interface enables the design and large-scale applications of complementary metal-oxide-semiconductor (CMOS) transistors and integrated circuits. The rapid shrinkage of transistor feature size in Si CMOS scaling has forced the channel length to decrease to be around 15 nm by year 2010, and the SiO_2 gate oxide thickness is correspondingly reduced to be close to the quantum tunneling limit of

1.0 nm. Beyond this point leakage current due to tunneling, $\sim$1–10 A cm^{-2}, becomes the dominant leakage mechanism in device designs. There is another ultimate physical limit below which SiO_2 no longer maintains its bulk electronic structure [1], and this appears to be about 0.7 nm.

The current trend of Si CMOS scaling thus calls for replacing SiO_2 with high κ dielectrics in gate related applications [2]. Over the last five years of intense research on high κ gate dielectrics, a number of binary oxides and silicates in amorphous form have emerged [2–7], and shown impressive dielectric properties with an equivalent oxide thickness (EOT), defined as teq ($\kappa_{SiO2}/\kappa_{oxide}$), as thin as 1.0 nm. To achieve performance comparable to SiO_2, the new high κ dielectric materials must satisfy very stringent requirements for the fundamental properties such as dielectric constant, band gap, conduction band offset, leakage, mobility, and good thermodynamic stability in contact with Si up to 1,000°C. It is equally critical to address device processing and integration issues such as morphology, interfacial structure and reactions, gate and process compatibility, and reliability.

Among these challenges, the mobility degradation due to the high κ gate dielectrics is the most difficult issue. High Coulomb scattering rate from charge trapping may lead to poor channel mobility. In order to overcome the degraded channel mobility encountered in the high κ gate stacks on Si, channel materials with higher carrier mobility such as strained Si, Si–Ge alloys, and Ge are being studied, and the strained Si is now being used in the production.

It is known that electrons move much faster in GaAs (and other III–V compound semiconductors) than those in Si, and Ge, an important aspect for building high-speed devices. Furthermore, semi-insulating substrates, not available in Si and Ge, will reduce cross talks between high-speed signal lines in dense circuits. A mature compound semiconductor technology (particularly III–V MOS devices) with electron mobilities at least 10 times higher than that in Si and with dielectrics having κ several times higher that that of SiO_2 would certainly enable the electronic industry to continue pushing its new frontiers for a few more decades. Furthermore, bandgap engineering and direct bandgaps, not available in Si-based material systems, provide novel designs and make highly performed integrated optoelectronic circuits (combining MOS and photonic devices) a reality.

For microwave and digital applications, III–V MOSFET's promise the advantage of low power consumption and circuit simplicity, comparing with the present devices based on MESFET or HEMT technologies, which have encountered inevitable current leakage through the Schottky metal gates. In the area of high power devices, the high band-gaps in the compound semiconductors (e.g., 1.42 eV in GaAs and 3.2 eV in GaN, comparing with 1.1 eV in Si) have provided intrinsic advantages such as larger bulk breakdown fields over the present Si technology (1.1 eV in Si). Their intrinsic high bandgap and breakdown fields make them natural candidates for high power electronic devices operated at high temperatures. Particularly, GaN-based MOSFET or MOHEMT are the choice of devices for the very high power applications.

Intensive efforts in searching and identifying electrically and thermodynamically stable insulators on GaAs with a low interfacial density of states (D_{it}), one of the key challenges in the compound semiconductor devices over the past four decades [8,9], have found a solution in ultra high vacuum (UHV) deposition of $Ga_2O_3(Gd_2O_3)$[10] and pure Gd_2O_3[11] dielectric films on GaAs surfaces. With such novel gate dielectric, MOS diodes have shown inversion and accumulation with a low $D_{it}(<10^{11}\,cm^{-2}\,eV^{-1})$ [10, 12]. Subsequent employment of $Ga_2O_3(Gd_2O_3)$ as a gate dielectric along with an ion implantation process led to the demonstration of the first inversion-channel GaAs MOSFETs in both n- and p-configurations [13, 14], and many other novel devices [15–18]: For example, the oxide was also used for fabricating inversion-channel InGaAs MOSFETs on InP [15] and depletion-mode GaN MOSFET's [16]. The InGaAs devices showed excellent performances with a transconductance of $190\,mS\,mm^{-1}$ and an effective mobility of $470\,cm^2\,V^{-1}\,s^{-1}$, and the GaN device was operated at 400°C [16]. Depletion-mode GaAs MOSFETs of 0.8 μm gate-length have been shown to have a drain current of $450\,mA\,mm^{-1}$ and a transconductance of $130\,mS\,mm^{-1}$, and moreover, to have exhibited negligible drain current drift and hysteresis, the first achievement in this class of transistors and an important technological advance for manufacturing consideration [17]. A power GaAs MOSFET exhibited excellent performance and is a promising candidate for microwave applications [18]. *Is it possible to develop these new compound devices into a feasible and/or mature technology in the next few years?*

Unexpectedly, the interfacial layer in $Ga_2O_3(Gd_2O_3)$/GaAs was found to be a single crystal of pure Gd_2O_3 epitaxially grown on the substrate [11, 19]. Moreover, single crystal pure rare earth oxides (such as Gd_2O_3 and Y_2O_3) have been grown epitaxially on GaAs [11]. A detailed study on the interface of Gd_2O_3/GaAs has resulted in a new structure [20] of Gd_2O_3 with a new stacking sequence [21, 22], which is similar to that of the underlying GaAs. Unlike SiO_2 grown on Si, which is amorphous, the heterostructure of Gd_2O_3/GaAs is all single crystal. This then opens up a possibility of peering into the interfacial structure using experimental tools such as diffraction methods.

While numerous previous approaches using thermal, anodic and plasma oxidation of GaAs surfaces, have failed to produce an insulator-GaAs heterointerface with a low D_{it}, they have given valuable scientific and technological knowledge and information. A review on these research activities is given in [8, 9]. The effective passivation will find applications in many other electronic and photonic devices, as many of those devices now suffer performance instability and reliability problems without an adequate passivation.

Recently, an 1 μm gate-length depletion-mode n-channel GaAs MOSFET with an 8 nm thick Al_2O_3 gate oxide has shown a maximum transconductance of $120\,mS\,mm^{-1}$ and a drain current of $400\,mA\,mm^{-1}$ [23]. The Al_2O_3 gate oxide, ex situ deposited on GaAs, was grown using atomic layer deposition (ALD), a technique commonly used for depositing high κ gate dielectrics for

Si. The ALD-Al_2O_3 gate oxide was later applied to fabricate InGaAs/GaAs MOSFET. Annealing at 600–650°C in O_2 was found to be necessary for improving the device properties [23, 24].

In this paper we review the key features of these new dielectrics grown on GaAs and GaN surfaces in terms of their crystal-chemical, electronic, and transport properties. The mechanism of the Fermi-level unpinning will be discussed, with emphases being given to the materials characteristics of the interfacial structures. The experimental details of the published work are found in the cited references.

10.2 High κ Gate Dielectrics for GaAs and its Related Compounds: $Ga_2O_3(Gd_2O_3)$ Approach

$Ga_2O_3(Gd_2O_3)$ has been successfully extended from passivating GaAs to several compound semiconductors including AlGaAs, InGaAs, and InP. The formation of low leakage, insulating barrier on their (100) surfaces has been demonstrated. We have conducted studies of the MOS diodes consisting of $Ga_2O_3(Gd_2O_3)$ mostly on InGaAs including the electrical performance in order to achieve better control of this novel oxide–semiconductor interface, to further optimize the growth and processing parameters essential to optimize device process, and to further establish a viable III–V MOSFET technology [25]. Systematic postannealing studies of varying temperature and gas species were carried out to improve the electrical performance, such as substantial reduction of problematic features of frequency dispersion, voltage hysteresis, and D_{it} value.

Deposition of GaAs epilayers and oxide films as the gate dielectrics was carried out in a multichamber ultra high vacuum UHV/MBE system, and the details were given in [10, 11].

The materials requirements for selecting the high κ dielectric oxides are listed in the Table 10.1 for a series of binary/ternary dielectric oxide candidates of increasing dielectric constant κ. Strong oxygen affinity property is essential to prevent direct oxidation of the As-based compound semiconductors that will form arsenic oxides, and cause Fermi surface pinning. Good thermal stability in contact with GaAs up to 850°C is also necessary in order to maintain an abrupt oxide/semiconductor interface. The hydroxide formation is common for many transition metal or rare earth oxides upon air exposure, and that often affects the stability, and reduces the reliability of the oxide for device processing.

An atomically abrupt interface between the gate dielectric/GaAs interface was achieved by our MBE growth, as clearly shown from the medium energy ion scattering. Since the back scattered ion intensity from the Ga element in the $Ga_2O_3(Gd_2O_3)$ oxide tends to interfere with those from the underlying GaAs, we have studied instead another dielectric Al_2O_3 film 8.5 nm thick on GaAs grown by similar manner to simplify the spectrum analysis.

Table 10.1. Basic characteristics of binary oxide dielectrics for GaAs

Dielectrics	SiO_2	Al_2O_3	Ga_2O_3 (Gd_2O_3)	Gd_2O_3	Sc_2O_3	HfO_2
Dielectric constant	3.9	8.0	15	14	14	20
Band gap (eV)	9.0	8.8	5.4	5.4	6.5	5.7
Breakdown field ($MV\,cm^{-1}$)	10	5–8	3–5	3.5	5	5
Hydroxide formation	Slight	Some	High	High	High	Some
Recrystallization temperature	>1200 C	>1000 C	>950C	>950 C	?	∼900 C
Stability in contact with GaAs at high T	?	?	Good	Good	?	?

The chemical depth profile of each species obtained from the best fit to the measured MEIS spectra indicated an abrupt Al_2O_3/GaAs interface with the presence of a slightly As rich, amorphous GaAs interfacial layer about 0.35 nm thick on the Ga-terminated GaAs surface.

Electrical characterizations of $Ga_2O_3(Gd_2O_3)$ gate oxide grown on the (100) AlGaAs, InGaAs, and InP surfaces have demonstrated the formation of low leakage, insulating barrier. Representative J–E, and C–V curves of an MOS diode made of Au/$Ga_2O_3(Gd_2O_3)$/($Ga_{0.85}In_{0.15}$) As after 750°CN_2 annealing for 3 min are shown in Fig. 10.1a, b, respectively. The oxide thickness is 13 nm with a κ of 13, and the breakdown field of 2.8 $MV\,cm^{-1}$ in the forward direction. The C–V curves displayed a frequency dispersion of 10% with the frequency increasing from 1 to 100 kHz, and a voltage hysteresis of ∼0.2 V when sweeping bias from depletion to accumulation, and in the reverse direction.

Systematic postannealing studies were carried out by varying annealing parameters including temperature (from 450°C to 700°C), and annealing gas species including He, O_2, N_2, and forming gas in a quartz tube furnace. The study was aimed to improve the dielectric performance, and reduce the problematic features such as frequency dispersion and voltage hysteresis observed in the C–V data. For instance, He gas anneal anneals at 450–650°C was performed to make the oxide dense and to heal the growth-induced defects. We observed notable frequency dispersion reduction from 25% (1 kHZ to 1 MHz) prior to anneal to 12% after the 600°C He anneals due to removal of the slow-moving charge carriers.

The D_{it} value was deduced from the C–V and G–V traces by including a series resistance effect. In general D_{it} showed an order of magnitude of reduction when increasing the frequency from 1 kHz to 1 MHz. The dependence of D_{it} (100 kHz and 1 MHz) on the annealing temperature and the annealing gas species are plotted in Figs. 10.2 and 10.3, respectively. Excellent D_{it} value approaching low $10^{10}\,cm^{-2}\,V^{-1}$ was achieved through a 600°C He anneal for

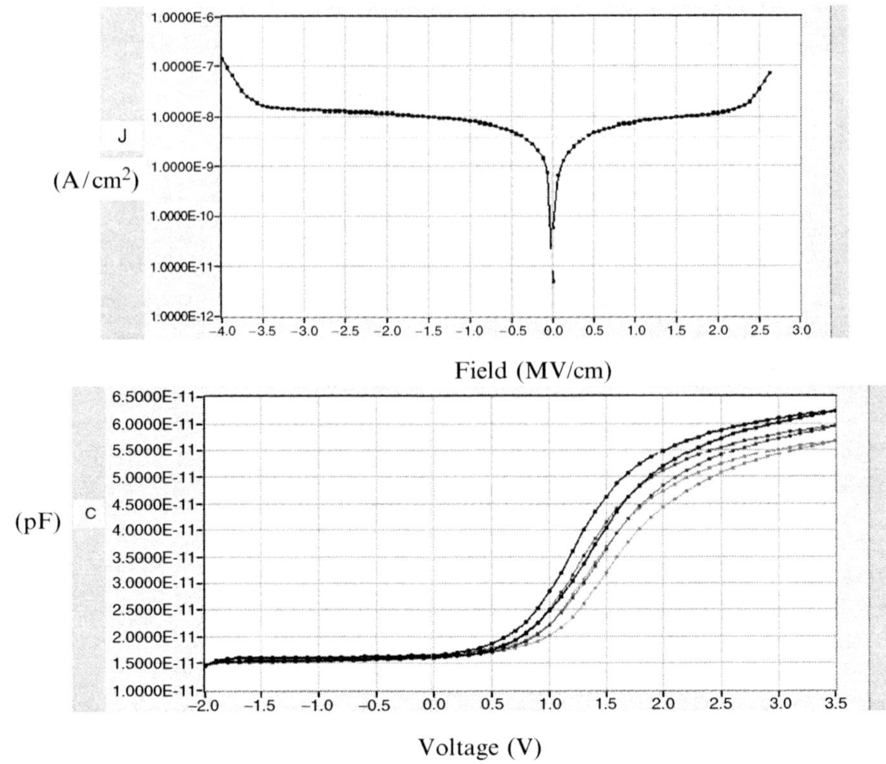

Fig. 10.1. (**a**) Leakage current density J (A cm^{-2}) vs. electrical field E (MV cm^{-1}) for an MOS diode made of Au/Ga_2O_3(Gd_2O_3)/($Ga_{085}In_{015}$). As after 750°CN_2 annealing. (**b**) Capacitance (in pF) vs. voltage (V) for an MOS diode made of Au/Ga_2O_3(Gd_2O_3)/($Ga_{085}In_{015}$). As after a 750°CN_2 anneal with frequency increasing from *top* (1 kHz), *mid* (10 kHz), to *bottom* trace (100 KHz)

15 min. Annealing temperature exceeding 650°C may promote chemical reactions at the interface, and cause D_{it} to rise.

In addition, Fig. 10.3 data suggest that annealing in a reducing gas stream resulted in a preferred decrease of D_{it} as opposed to annealing in an oxidizing gas. However, the forming gas (15% H_2 in a N_2 gas mixture) anneal at 450°C may have overly reduced the oxides, and adversely increased D_{it} to 10^{13} cm^{-2} V^{-1}. From Fig. 10.4, we observed that the forming gas anneal at 375°C has effectively reduced the voltage hysteresis δV to 0.1–0.2 V. Hence a postannealing process of combining a He anneal at 600°C with a forming gas anneal at 375°C are important to achieve best electrical performance for Ga_2O_3(Gd_2O_3), and should be incorporated in the III–V MOSFET fabrication routine.

The frequency dispersion common for Ga_2O_3(Gd_2O_3) and Gd_2O_3 high κ oxide are recently accounted for by employing an improved two frequency method, where the equivalent circuit model for the high κ MOS capacitor

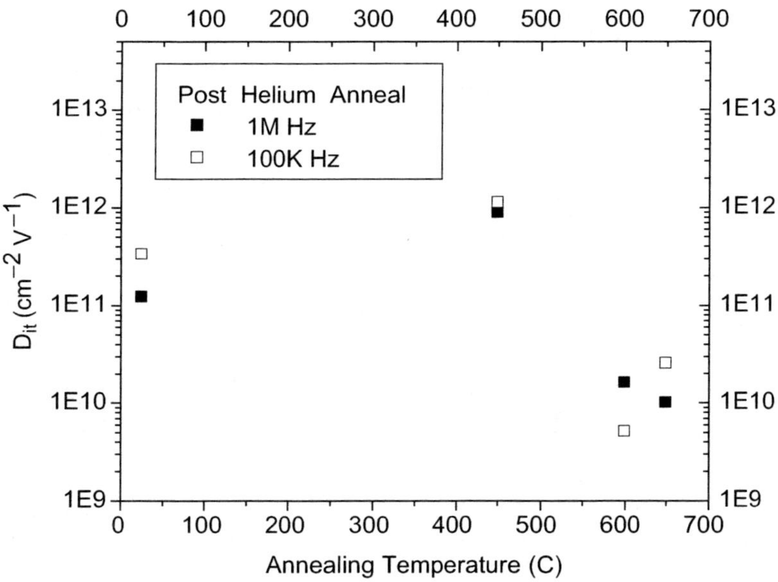

Fig. 10.2. The dependence of $D_{\rm it}$ on the annealing temperature of the He gas as deduced from the C–V data at 100 kHz and 1 MHz

includes four parameters of intrinsic capacitance, loss tangent, parasitic series inductance, and series resistance [25]. The result of the calculated capacitances based on this model indicated the calibrated capacitances at each pair of frequency coincide with other very well, and the variation of capacitance is

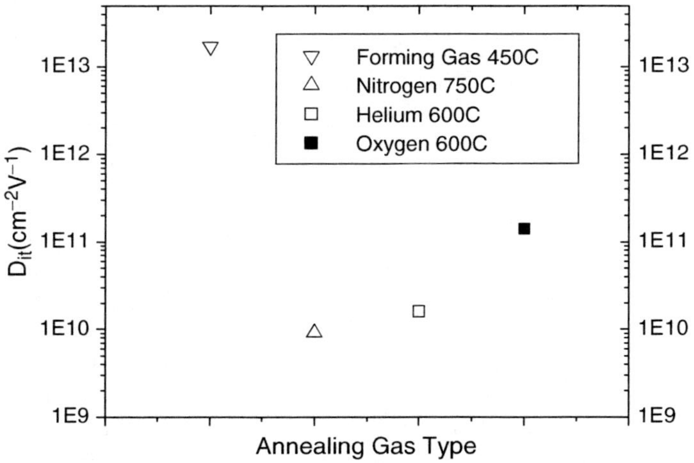

Fig. 10.3. The dependence of the interfacial state density $D_{\rm it}$ on the annealing gas species from oxygen to forming gas

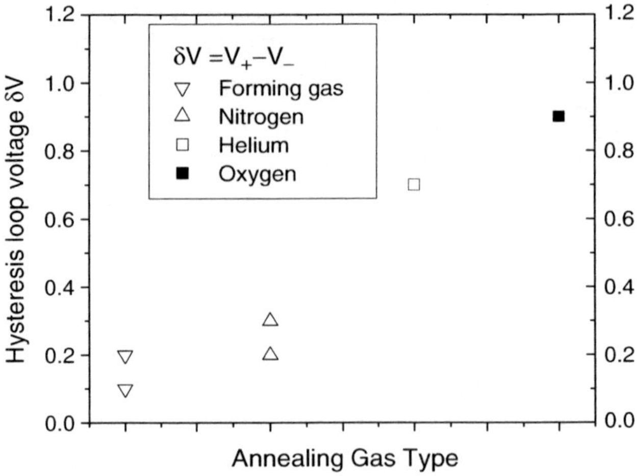

Fig. 10.4. The systematic reduction of the hysteresis loop voltage δV on the annealing gas species varying from oxygen to forming gas

reduced to be less than 2% [26]. The calculated D_{it} value from this analysis varies in the range of $4\text{–}9 \times 10^{10}\,\mathrm{cm^{-2}\,V^{-1}}$.

Fundamental studies of the MOS diodes consisting of $Ga_2O_3(Gd_2O_3)$ mostly on GaAs, InGaAs, and AlGaAs have been conducted, and the dielectric performance and the oxide reliability were significantly improved after systematic postannealing studies. We have demonstrated that these results are important to MOSFET device fabrication, and to further establish a viable GaAs MOSFET technology in near future.

10.3 Thermodynamic Stability of $Ga_2O_3(Gd_2O_3)$/GaAs Interface at High Temperatures [26]

For achieving high device performance such as mobility in GaAs MOSFET using $Ga_2O_3(Gd_2O_3)$ as the gate dielectric, the interfacial roughness in $Ga_2O_3(Gd_2O_3)$/GaAs has to be controlled and minimized to a few Å, as was witnessed in the case of the perfected SiO_2–Si interface.

Previously, the oxide–GaAs interface was found to be roughened in a high temperature (>750°C) annealing for fabricating the inversion-channel GaAs MOSFETs [13,14], in which the annealing was inevitably needed to activate the ion implantation for ohmic contacts at source and drain regions. Efforts were, therefore, then taken to circumvent the difficulties by implanting and activating dopant ions before the oxide growth. For preserving the GaAs surface and preventing evaporation of As from the surface during the high temperature activation annealing, several approaches were employed [13,14]: One was to grow AlGaAs, SiO_2, and other insulators as cap layers on GaAs with the

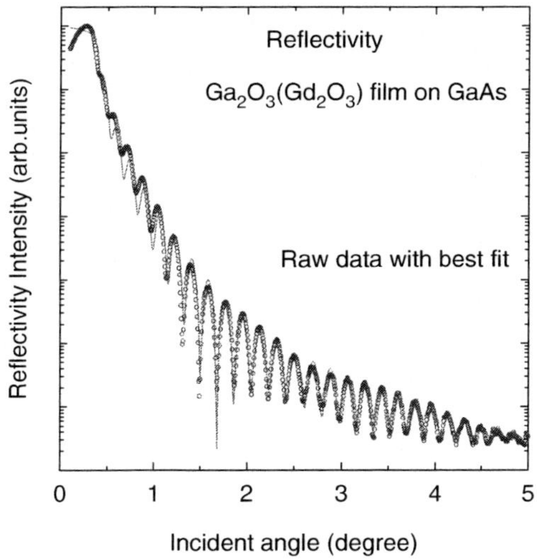

Fig. 10.5. Low angle X-ray reflectivity of $Ga_2O_3(Gd_2O_3)$ on GaAs annealing in UHV, with experimental data (*dots*), and a theoretical fit (*line*)

etching of those cap layers after the activation. Another way was to activate the implantation with the implanted GaAs wafers annealed at the high temperature under AsH_3 flux in a gas source molecular beam epitaxy (GSMBE) chamber. Using both approaches, the annealed GaAs surface was still rough, lost the atomic ordering, and was not recovered with annealing to ∼600°C under an arsenic overpressure in an MBE chamber, as evidenced from the observation of almost no RHEED (reflection high-energy electron diffraction) patterns, or very faint spotty ones. Note that a good GaAs surface should have streaky 2×4 reconstructed RHEED patterns with annealing under arsenic environment. The rough GaAs surface perhaps was caused by interaction between the cap layers and GaAs during the high temperature annealing. Another drawback of ion-implanting and activating prior to the oxide growth is that the devices can not be reduced to a small scale because no self-aligned process was allowed with the approach.

According to free energy consideration, $Ga_2O_3(Gd_2O_3)$ should be thermodynamically stable with GaAs at temperatures of ∼750°C or above. However, it was found out later that when the samples are exposed to air, they absorb water and form hydro-oxides [27]. During the annealing process, the hydro-oxides, not the pure $Ga_2O_3(Gd_2O_3)$, react with GaAs, resulting in rough interfaces.

In this work, the thermodynamic stability of $Ga_2O_3(Gd_2O_3)$ (not the hydro-oxides) with GaAs with UHV annealing has been studied using structural and morphological probing tools of X-ray reflectivity (Fig. 10.5), atomic force microscopy (AFM) (Fig. 10.6), and cross-sectional high-resolution transmission electron microscopy (HRTEM) (Fig. 10.7). The results have revealed

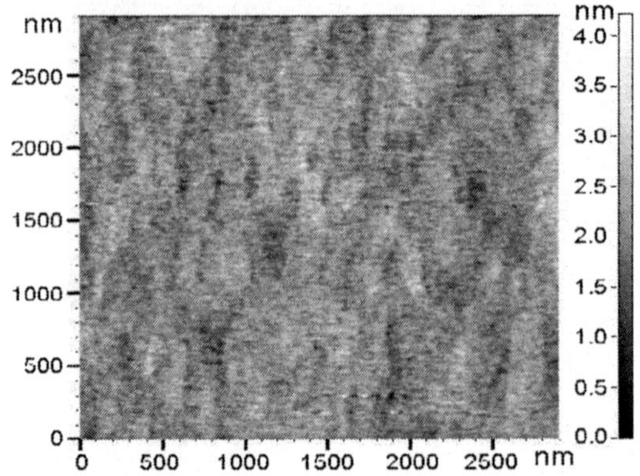

Fig. 10.6. AFM image of $Ga_2O_3(Gd_2O_3)$ grown on GaAs annealing in UHV system

that the interface between $Ga_2O_3(Gd_2O_3)$ and GaAs remains intact with the annealing temperatures up to 780°C and the interfacial roughness is less than 0.2 nm, a value close to that of SiO_2–Si interface. Moreover, $Ga_2O_3(Gd_2O_3)$ remains amorphous with the high temperature annealing, an important aspect for high κ gate dielectrics. I–V (current–voltage) (Fig. 10.8) and C–V (capacitance–voltage) (Fig. 10.9) measurements showed that the leakage currents (10^{-8} to 10^{-9} A cm^{-2}) through the oxide, high dielectric constants of 15, and the interfacial density of states (D_{it}) between gate dielectrics and GaAs (Fig. 10.10) remain low with the samples annealed at high temperatures.

The attainment of a smooth interface between $Ga_2O_3(Gd_2O_3)$ and GaAs, even after high temperature annealing for activating implanted dopant, is a must to ensure the low D_{it} and to maintain a high carrier mobility in the channel of the MOSFET. Our results have provided a critical step for implementing an inversion-channel GaAs MOSFET technology.

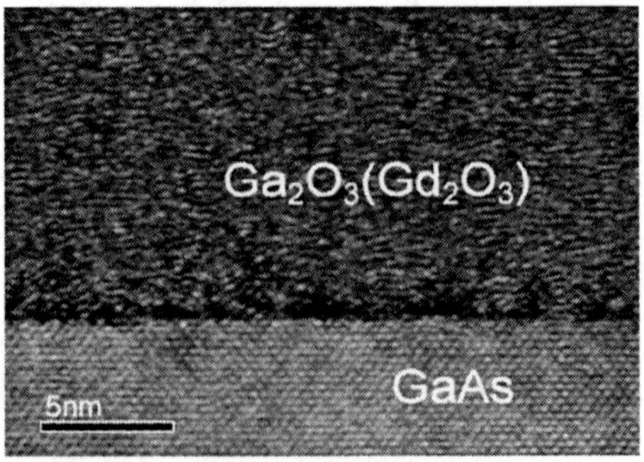

Fig. 10.7. High resolution cross sectional TEM picture of $Ga_2O_3(Gd_2O_3)$ on GaAs annealing in UHV system

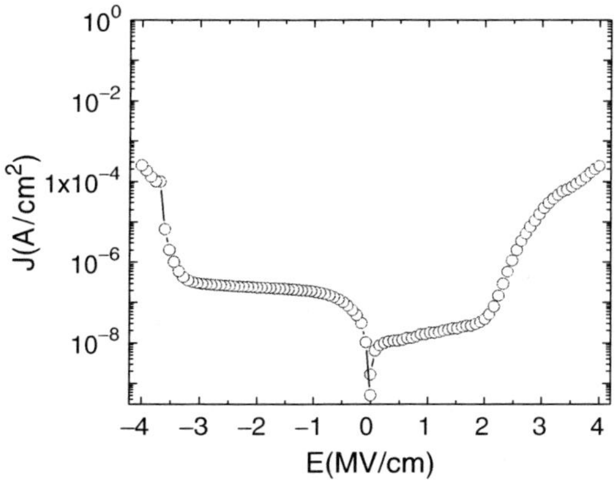

Fig. 10.8. Leakage current density J ($A\,cm^{-2}$) vs. E ($MV\,cm^{-1}$) for $Ga_2O_3(Gd_2O_3)$ films annealed in UHV system

10.4 Single Crystal Gd_2O_3 on GaAs and Interfaces

Pure Gd_2O_3 film is a single crystal normally grown in the body centered cubic structure isomorphic to α–Mn2O_3, and this is easily detected during the growth by a major change of symmetry in RHEED [11]. Figure 10.11a shows (2 × 4) reconstructed RHEED patterns along [011] and [01$\bar{1}$] directions of GaAs(100) surface. Deposition of a Gd_2O_3 film 25 Å thick resulted in streaky patterns of two fold symmetry shown in Fig. 10.11(b). The typical oxide growth rate is about 10 Å min^{-1}. Further RHEED and X-ray diffraction analysis indicated that the Gd_2O_3 film is (110) oriented and grown in single domain. The in-plane epitaxial relationship between (100) GaAs substrate and (110) Gd_2O_3 film is [001]Gd_2O_3 [011]GaAs and [$\bar{1}$10]Gd_2O_3 [01 $\bar{1}$]GaAs. Gd_2O_3 is an ionic crystal with a large lattice constant of 10.81 Å, and GaAs

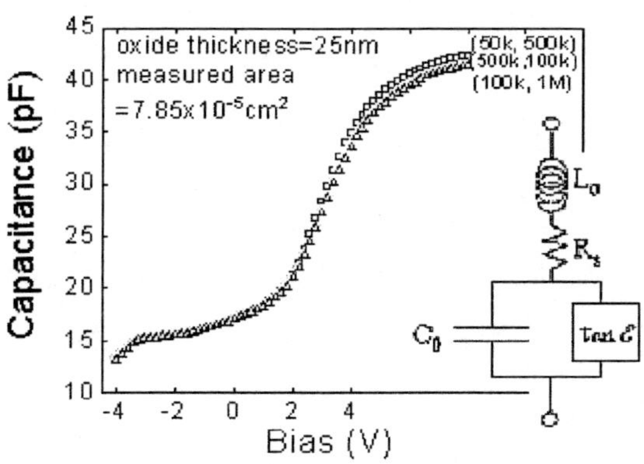

Fig. 10.9. *C–V* curves after two-frequency corrections

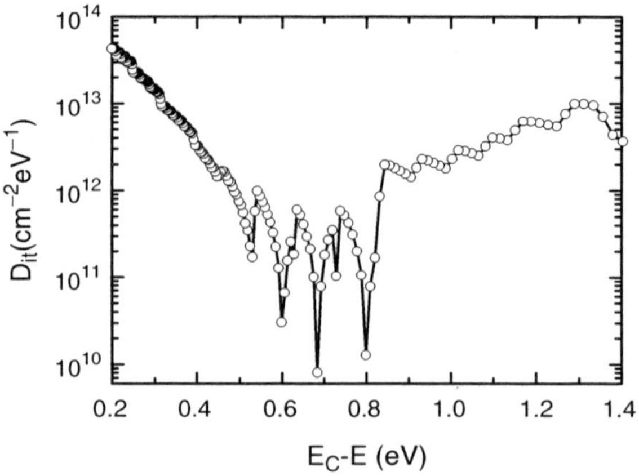

Fig. 10.10. Distribution of $D_{\rm it}$ vs. energy for $Ga_2O_3(Gd_2O_3)$/GaAs interface

is covalent bonding with a lattice constant of 5.65 Å. The in-plane epitaxy of [100] of Gd_2O_3 being in parallel with [011] of GaAs suggests a super-cell lattice match, i.e. the spacing of three Gd_2O_3 [100] lattices (32.4 Å) matched to that of four GaAs [011] lattices (32 Å).

Interestingly when the film growth rate is lowered to ~5 Å min^{-1}, this transformation does not take place, and the film remains in the four fold symmetry, the same as that of the substrate [20–22]. Figure 10.11c shows

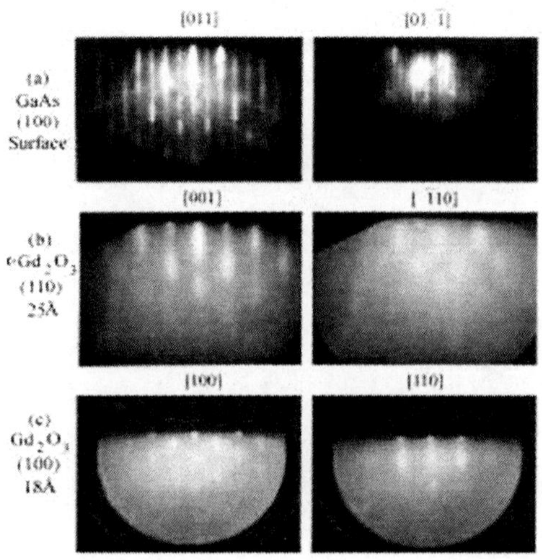

Fig. 10.11. (**a**) (100) GaAs surface along [011] and [01 $\bar{1}$] axes, (**b**) (110) cubic α-Gd_2O_3 film 25 Å thick along [001] and [$\bar{1}$10] axes, and (**c**) (100) fluorite-related Gd_2O_3 film 18 Å thick along the two highly symmetrical directions of [010] and [011]

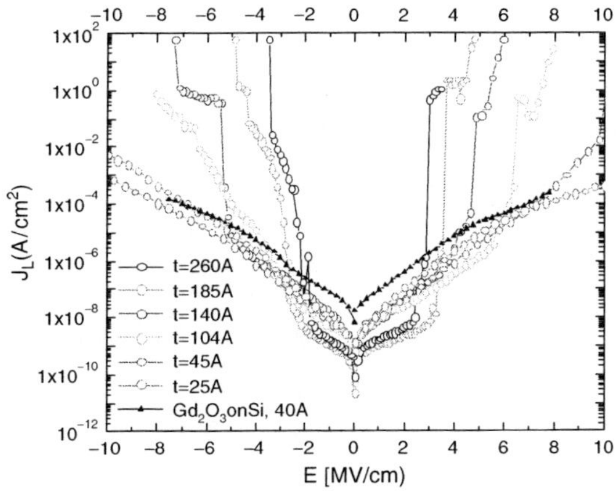

Fig. 10.12. Leakage current density J_L vs. electrical field E for Gd_2O_3 films with decreasing thickness

the RHEED patterns of such a film 18 Å thick as the sample is rotated to match the in-plane [100] and [110] directions. X-ray analysis showed that the structure is consistent with a fully strained epitaxial film of Gd_2O_3 with a tetragonal unit cell of a = 5.65 Å and c = 5.37 Å. The unit cell is best described as a deformed fluorite-related structure under in-plane tensile strain due to the substrate epitaxy. This unit cell with 1/4 oxygen atom sites vacant is distorted to the extent that most of the atomic planes are not very well defined.

The Gd_2O_3 dielectric films are highly electrically insulating, showing very low leakage current densities of $\sim 10^{-9}$ to 10^{-10} A cm^{-2} at zero bias. This may have to do with that fact that Gd is electropositive +3 and has a strong affinity to oxygen. We measured the dependence of the leakage current density (J_L) on the applied field (E) for a set of Gd_2O_3 samples with the oxide thickness (t) systematically reduced from 260 to 25 Å (Fig. 10.12). The positive bias means that the metal electrode is positive with respective to GaAs. As t is decreased from 260 to 45 Å, the respective breakdown field E_{br} increases systematically from 3 to 10 MV cm^{-1}, yet J_L increases merely by one order of magnitude. The maintenance of low electrical leakage even for films as thin as 25 Å suggests that a high degree of structural integrity is sustained through epitaxy. Note that the leakage current density for the 25 Å film is in mid 10^{-4} A cm^{-2} at 10 MV cm^{-1}, and 10^{-4} A cm^{-2} at 7 MV cm^{-1}. Considering the dielectric constant of ∼14, the EOT of the 25 Å film is ∼7 Å. In comparison, the leakage current density is 1 A cm^{-2} at 7 MV cm^{-1} for an SiO_2 film 15 Å thick. Our results suggest that a thin and highly perfected single crystal oxide film gives better dielectric characteristics over its amorphous counterpart, in terms of leakage currents and breakdown fields.

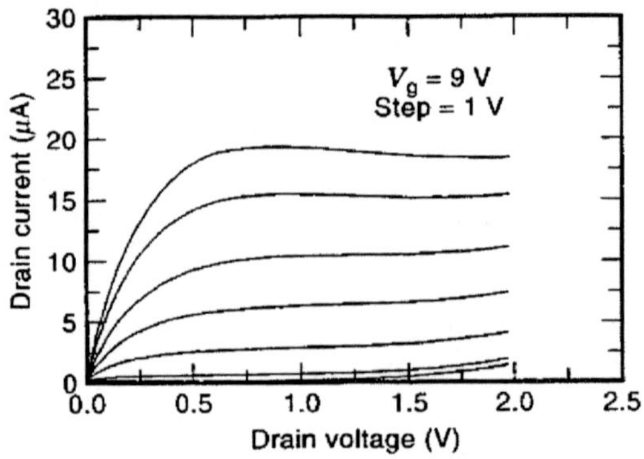

Fig. 10.13. Drain *I–V* of inversion-channel enhancement-mode GaAs MOSFET's with a gate dimension of $4 \times 50\,\mu m^2$

10.5 GaAs MOSFETs

10.5.1 Enhancement-Mode with Inversion

Since the carriers in this type of devices are confined to the top of GaAs, which is within 50–100 Å near the oxide–GaAs interface, the interfacial roughness becomes a very important issue. Concerning the interaction between $Ga_2O_3(Gd_2O_3)$ and GaAs, which may occur during the high-temperature anneal (>750 and >700°C for implants of Si and Be, respectively) for activating ion implants in the well, source and drain of the MOSFET, one approach in the device processing was taken by ion implantation on GaAs wafers, followed by activation annealing, and then gate-oxide deposition [13]. During the high temperature anneal, however, the GaAs surface became rough and not atomically ordered as observed using RHEED, despite extra efforts to protect the surface. Even the surface of the device wafers was rough before the oxide deposition, p- and n-channel GaAs MOSFETs unexpectedly showed inversion. For the n-MOSFET, the gate voltage varies from 9 to 0 V in steps of −1 V and the V_I is around 2 V. The saturation drain current is proportional to $(V_g – V_I)^2$, typical characteristics of an enhancement-mode device with inversion. Figure 10.13 illustrates the drain I-V curves of such devices with a dimension of $4 \times 50\,\mu m^2$. The p-MOSFET was operated as a three-terminal device and the gate voltage varies from −9 to 0 V in steps of 1 V (not shown). The threshold voltage, V_I, is around −0.5 V and the inversion channel is clearly demonstrated. The magnitude of the maximum drain currents for both devices was low, in the range of 20–25 µA.

Since the free GaAs surface could not be maintained smooth during the high temperature annealing, another processing approach was employed, in which gate dielectrics were deposited on GaAs prior to the ion-implant, and activation-annealing [14]. It was hoped that due to the thermodynamic stability between $Ga_2O_3(Gd_2O_3)$ and GaAs, the interface would remain smooth

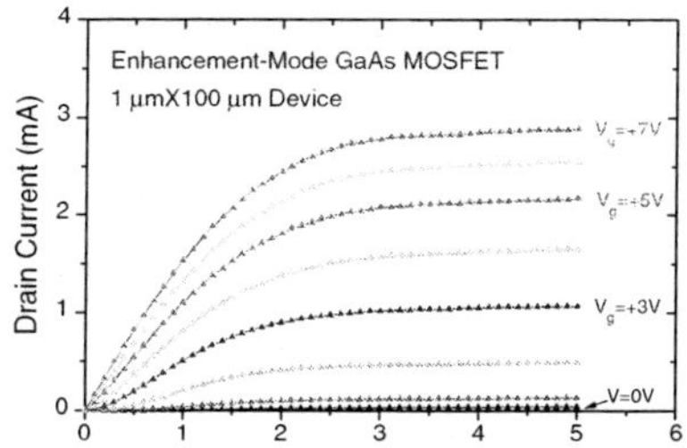

Fig. 10.14. Drain I–V characteristics of inversion-mode n-channel GaAs MOSFET using the second processing.

during the activation anneal at the high temperatures. With such a device processing, the drain current of an inversion n-channel GaAs MOSFET has now been increased to 3 mA for devices with 1 µm gate length. Figure 10.14 shows typical drain current versus drain voltage characteristics of such a device, with the gate voltage varying from 0 to 7 V in steps of 1 V. The maximum drain current density and the extrinsic transconductance are $30\,\mathrm{mA\,mm^{-1}}$ and $4\,\mathrm{mS\,mm^{-1}}$, respectively. The increase in drain currents of the present inversion-channel GaAs MOSFETs is more than 100 times over those of the previous devices. CMOS circuits have also been demonstrated [28]. Efforts are now being taken to further increase the drain currents to the range of 20–30 mA for 1 µm gate length devices. This is very likely judging from what has been achieved in the inversion-channel InGaAs MOSFET's on InP substrates [15]. Moreover, what we have recently accomplished in achieving an atomically smooth $Ga_2O_3(Gd_2O_3)$/GaAs interface during high temperature annealing as discussed in Sect. 10.3 in this paper will make such devices a reality.

10.5.2 Depletion-Mode MOSFET and Power Devices

GaAs MOSFETs feature a large logic swing, which gives a greater flexibility for digital IC designs. In contrast, GaAs metal-semiconductor FETs (MESFETs) and high electron mobility transistors (HEMTs) exhibit small forward gate voltages limited by the Schottky barrier heights. However, the drain current drift and hysteresis in the past has hindered the deployment of GaAs MOSFETs due to inadequate insulating films with significant bulk trapped charges and a high D_{it} on GaAs [29,30]. Now with $Ga_2O_3(Gd_2O_3)$ as the gate dielectric on GaAs, in which a D_{it} in the mid $10^{10}\ \mathrm{cm^{-2}\ eV^{-1}}$ was attained, the depletion-mode MOSFET may show device performance not attainable previously. In this section, depletion-mode GaAs MOSFET's of 0.8 – µm gate-length, for the first time, exhibit negligible drain current drift and hysteresis.

Both drain and gate I–V characteristics of a $0.8\,\mu\mathrm{m} \times 60\,\mu\mathrm{m}$ device are shown in Fig. 10.15, with the breakdown voltage of 24 V (corresponding to a breakdown field of $6.3\,\mathrm{MV\,cm^{-1}}$). The output characteristics measured with a

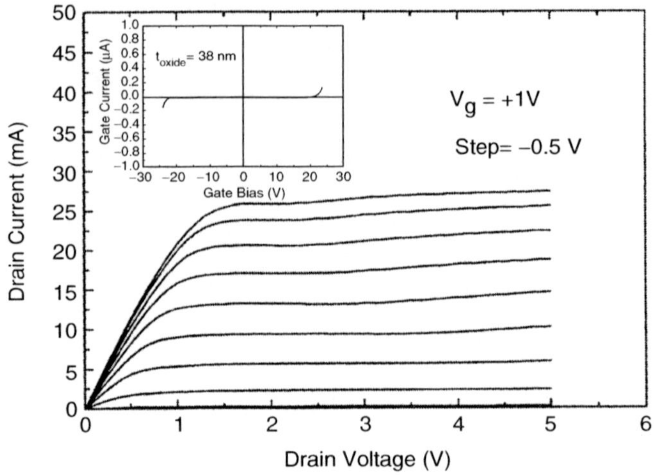

Fig. 10.15. Drain *I–V* characteristics of a $0.8 \times 60\,\mu m^2$ depletion mode GaAs MOSFET. *Inset* shows the gate *I–V* characteristics of such device

curve tracer show no *I–V* hysteresis and drain–current drift (not shown). The drain *I–V* characteristics are not sensitive to light, either. These observations indicate insignificant bulk oxide trapped charges as well as a low D_{it}. The device shows a clean pinch-off at a threshold voltage of -3.5 V with the off-state drain–source breakdown voltage of 12.5 V. Same threshold voltage was measured in a quasi-dc condition. No discrepancy between *I–V* curves measured under quasi-dc condition and measured by 120 Hz curve tracer was found. The very high on-resistance r_{on} was caused by high source resistances, which are due to the low doping concentration (4×10^{17} cm^{-3}) in the channel layer. The calculated effective channel mobility is 1,100 $cm^2\,V^{-1}\,s^{-1}$. The maximum drain current density (I_{max}) and the peak extrinsic transconductance (g_m) are 450 $mA\,mm^{-1}$ and 130 $mS\,mm^{-1}$, respectively. The drain current density as a function of gate bias in both forward and reverse sweep directions shows negligible hysteresis, again indicating low mobile charge density and no charge injection. The short-circuit current-gain cut-off frequency (f_T) and the maximum oscillation frequency (f_{max}) were measured by biasing the devices at $V_{ds} = 2$ V and $V_{gs} = -1.5$ V. The $f_T = 17$ GHz and the $f_{max} = 60$ GHz were determined by extrapolating the short-circuit current-gain (H_{21}) and the maximum stable gain (MSG) curves, respectively, using -20 dB per decade slopes, as shown in Fig. 10.16. The long-term drain current drifting behavior of the MOSFET was tested with the devices biasing at an extreme stress condition of $V_{ds} = 4$ V and $V_{gs} = +1$ V. No detectable short-term current drift was observed in a period shorter than 1 s after the device was turned on. The long-term drain current drift is less than 1.5% during operation for a period of over 150 h.

The GaAs depletion-mode MOSFTE's with $Ga_2O_3(Gd_2O_3)$ as gate dielectrics show excellent device performance, including long lifetime and for

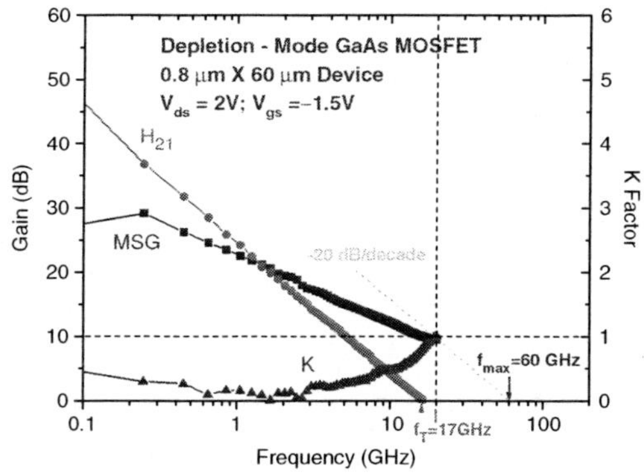

Fig. 10.16. Microwave performance of the device in Fig. 10.15 measured at a drain voltage of 2 V and a gate voltage of −15 V

the first time negligible hysteresis and drain current drift in the I-V characteristics. The flat transconductance profile reveals the advantage of MOSFET's for linearity consideration. These results present a significant advance towards the manufacture of commercially useful devices.

A power GaAs MOSFET (1 μm × 2.4 mm), measured from a curve tracer operating at 120 Hz, shows clean pinch-off drain current–voltage characteristics at a threshold voltage of −5 V with no significant hysteresis [18]. The maximum drain current density I_{max} and peak extrinsic transconductance g_m are 550 mA mm^{-1} and 125 mS mm^{-1}, respectively. When measured at 850 MHz under 3 V operation tuned for maximum output power, a peak power-added efficiency (PAE) of 45% was obtained. Under the same condition, a linear gain G_L = 20 dB and a saturated output power P_{sat} = 23 dB m were measured. When the drain bias increased to 5 V, the maximum PAE, G_L, and P_{sat} are 56%, 20 dB, and 26.5 dB m (power density=186 mW mm^{-1}), respectively. The low doping concentration (2×10^{17} cm^{-3}) in the channel layer gives high contact resistance, which in turn results in the high on-resistance r_{on} (5.1 Ω · mm). It is expected that the PAE can be further improved by reducing the source and drain contact resistances as described earlier. A short-circuit current gain cutoff frequency f_T of 15 GHz and a maximum oscillation frequency f_{max} of 50 GHz were obtained from on-wafer S-parameter measurements of a 1 μm × 100 μm process control module. Larger devices (gate periphery up to 2 cm) were also fabricated. However, the load impedances of those devices are too small (< 10 Ω), which are beyond the tuning range of our load-pull system for optimum matching. Further improvement on PAE is expected by reducing the contact resistance and optimizing the layer structure.

Recently, new depletion-mode GaAs and $In_{015}Ga_{085}As$/GaAs MOSFETs (with $Ga_2O_3(Gd_2O_3)$ as a gate dielectric and a dimension of 1.6 μm × 100 μm) were successfully fabricated. Well-behaved I–V characteristics were measured for these devices. The strong accumulation current of 335 mA mm^{-1} at gate bias of 4 V and 510 mA mm^{-1} at 2 V of the depletion-mode GaAs

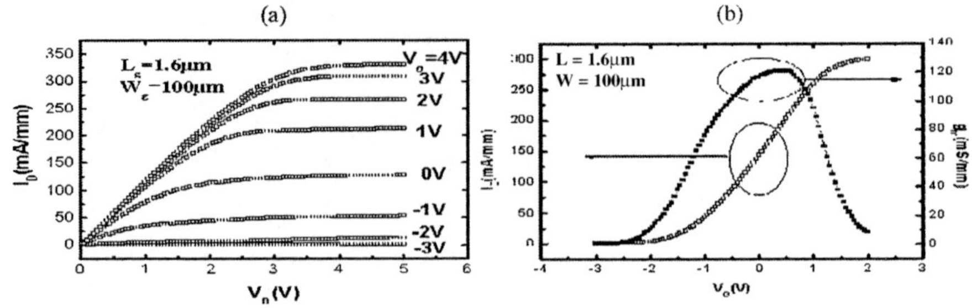

Fig. 10.17. Drain current characteristics (**a**) and transconductance (**b**) of GaAs MOSFET

(Fig. 10.17(a)) and $In_{0.15}Ga_{0.85}As/GaAs$ MOSFETs (not shown), respectively, strongly indicates a high-quality of the interface between $Ga_2O_3(Gd_2O_3)$ and the n-channel. It is significant that the gate of the GaAs MOSFET has sustained a gate bias up to 4 V. The transconductance of the GaAs and $In_{0.15}Ga_{0.85}As/GaAs$ MOSFET reaches 130 (Fig. 10.17b) and $170\,\mathrm{mS\,mm^{-1}}$, respectively. Again, no noticeable drain current hysteresis and drift was observed in both forward and reverse gate–voltage sweep for these devices. This indicates that no significant bulk oxide charge is present and the density of interfacial traps is low.

10.6 High κ Gate Dielectrics for GaAs and its Related Compounds: ALD Al_2O_3 Approach and its Mechanism of Unpinning the Fermi Level [31]

The recent development of high-quality ALD-grown high-κ gate dielectrics on Si justifies some attempts to integrate the ALD grown oxides on III–V substrates. Depletion mode MOSFETs with $D_{\mathrm{it}} < 10^{12}\,\mathrm{cm^{-2}\,eV^{-1}}$ were recently fabricated by ALD Al_2O_3 onto native oxide covered GaAs [23, 24]. A 600–650°C anneal in O_2 minimized current–voltage hysteresis and frequency dispersion, and maximized gate stack capacitance. Transmission electron microscopy (TEM) pointed to a remarkably sharp $Al_2O_3/GaAs$ interface, prompting the speculation that some native oxide may be removed during the ALD process. ALD-grown Al_2O_3 also results in good gate stack properties on InGaAs [24] and AlGaN/GaN [24] and may be used to coat compound semiconductor nano-wires conformally. By contrast, very few studies have been published regarding the mechanism of unpinning the Fermi level in the ALD-Al_2O_3 grown on III–V compound semiconductors.

Here, we characterize the structure and composition of Al_2O_3/InGaAs to help develop an understanding of the impact of material and processing conditions on the quality of ALD-grown high-k/III–V stacks [31]. Al_2O_3 was grown on oxide-covered MBE grown InGaAs /GaAs. For high-k dielectric growth, we followed a procedure that has yielded high-quality Al_2O_3 on Si.

Fig. 10.18. High-resolution cross sectional TEM picture of Al_2O_3 on $In_{0.15}Ga_{0.85}$ As after nitrogen annealing at 500°C

Depositions were performed using alternating exposures of the common ALD precursors $Al(CH_3)_3 + H_2O$ in a N_2 carrier gas at 300°C, using a Taiwan-made ALD reactor. Films were characterized by TEM, X-ray photoelectron spectroscopy (XPS), current–voltage (I–V), and capacitance–voltage (C–V) measurement.

Figure 10.18 shows a high-resolution cross-section TEM picture of Al_2O_3/$In_{02}Ga_{08}$As/GaAs structure which was after 500°C nitrogen annealing. The oxide thickness measured by the TEM is 8 nm and a sharp transition from InGaAs to Al_2O_3 was observed.

High-resolution X-ray photoelectron spectroscopy (XPS) using synchrotron radiation was performed to determine the interfacial chemistry. The XPS data were taken at the U5 undulator beamline of National Synchrotron Radiation Research Center in Hsinchu, Taiwan. The U5 beamline operates over the photon energies from 60 to 1,500 eV using a 6 m spherical grating monochromator with four interchangeable gratings. The energy resolving power at 400 eV is better than 10,000 with slit openings set at 10 μm. Photoelectrons were detected at a take-off angle of 53° with respect to the sample surface by a PHI 279.4-mm diameter hemispherical electron analyzer. The pass energy of electron analyzer was fixed at 5.85 eV, and overall energy resolution was better than 0.15 eV. Before measurements, the surface of samples was cleaned in situ by Ar^+ sputtering at 500–1,000 eV primary energy.

Two samples, Al_2O_3/$In_{0.2}Ga_{0.8}$As/GaAs and InGaAs/GaAs, were studied with the latter as a reference (Fig. 10.19). The As 3d spectra for the as-grown Al_2O_3/$In_{0.2}Ga_{08}$As/GaAs showed a small, but very profound peak of As_2O_3. The peak quickly disappeared with a slight Ar^+ sputtering, indicating a very small amount of arsenic oxides on top of the as-grown sample. There is no detection of any arsenic oxides during the continuous sputtering. After the removal of Al_2O_3 with sputtering, the peak belonging to InGaAs was revealed.

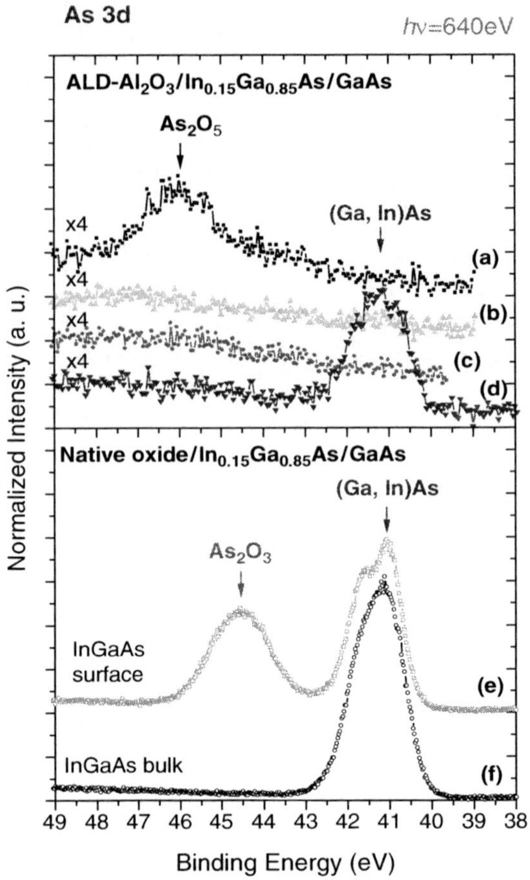

Fig. 10.19. As 3d core level spectra recorded from two samples, $Al_2O_3/In_{0.15}Ga_{0.85}As/GaAs$ (*top*) and native oxide/$In_{0.15}Ga_{085}As/GaAs$ (*bottom*): (**a**) at the surface of Al_2O_3; (**b**) immediately below the Al_2O_3 surface; (**c**) in the bulk of Al_2O_3; (d) at the interface of $Al_2O_3/In_{0.15}Ga_{085}As$; (**e**) at the surface of air-exposed $In_{015}Ga_{085}As$; and (**f**) in the bulk of $In_{0.15}Ga_{0.85}As$

In comparison, the native oxides on the reference sample are As_2O_3, different than the arsenic oxide on the ALD grown Al_2O_5 on InGaAs. The XPS studies on the above two samples clearly showed that (1) there is a native arsenic oxide on the MBE grown InGaAs after being exposed to air, which is As_2O_3; (2) during the ALD process, As_2O_3 diffuses (via reduction and re-oxidation) through Al_2O_3 to the top and becomes As_2O_5; and (3) there is no residue of arsenic oxides in the oxide or at the oxide/InGaAs interface. The removal of arsenic oxides from the oxide/InGaAs heterostructures ensures the Fermi level unpinning, which was observed in the C–V measurements as described in the following.

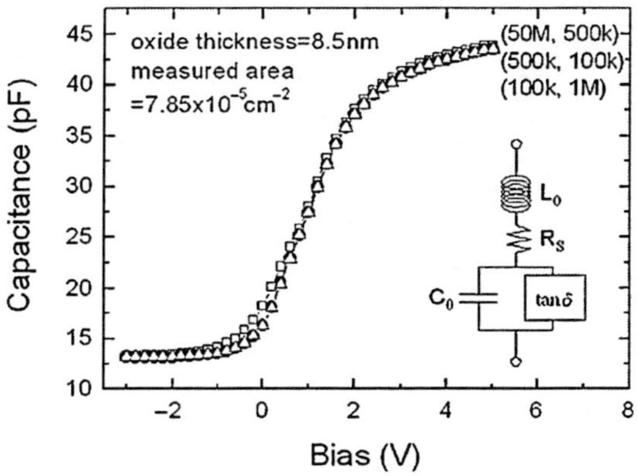

Fig. 10.20. *C–V* curves of an MOS diode made of Au/Al_2O_3 (8.5 nm)/$In_{015}Ga_{085}$ As after two-frequency corrections in different thermal processes

The MOS diode structure was fabricated by evaporating Au dots 0.1 mm in diameter. *I–V* and *C–V* characteristics were measured using Agilent 4156C and 4284, respectively. The dielectric constant of 8.5 nm Al_2O_3 is calculated to be about 8.4 using the CV measurements (Fig. 10.20). Current density–voltage (*J–V*) curves show a leakage current density of 8.5 nm Al_2O_3 on $In_{0.15}Ga_{0.85}$ As. Both as- and postannealing samples reveal leakage current about 10^{-8} to 10^{-9} A cm^{-2} at a bias of 1 V (Fig. 10.21).

The D_{it} was calculated to be around 10^{12} cm^{-2} eV^{-1} at the midgap using the Terman method as shown in Fig. 10.22. In comparison, the D_{it} from $Ga_2O_3(Gd_2O_3)$ deposited on GaAs in UHV is one order of magnitude lower in the range of ~10^{11} cm^{-2} eV^{-1} [26]. The higher D_{it} in the ALD-Al_2O_3 on GaAs or InGaAs is probably caused by the existence of the native oxides of

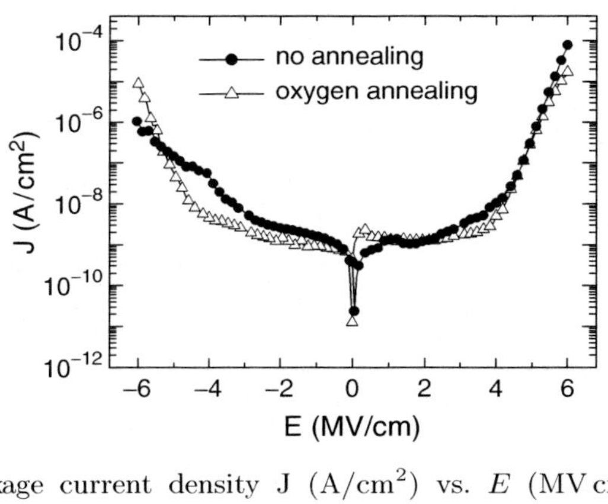

Fig. 10.21. Leakage current density J (A/cm^2) vs. *E* (MV cm^{-1}) for Al_2O_3/ $In_{015}Ga_{085}$As heterostructure in different thermal processes

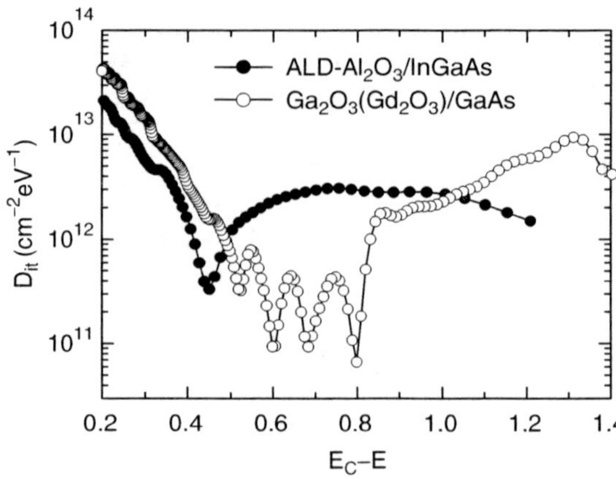

Fig. 10.22. D_{it}'s determined using the Terman Method for ALD- Al_2O_3/InGaAs and for UHV deposited $Ga_2O_3(Gd_2O_3)$ on GaAs

In_2O_3 and Ga_2O_3 at the interface, while there are no such native oxides at the UHV prepared $Ga_2O_3(Gd_2O_3)$ on GaAs or InGaAs.

10.7 GaN Passivation

We now turn to a brief discussion of the epitaxial growth of the rare earth oxide films on GaN for surface pssivation [32]. RHEED patterns in Fig. 10.23 upper panel are the UHV annealed GaN surface of sixfold symmetry along the ⟨100⟩ and ⟨110⟩ azimuthal direction with a 30° separation. During the initial growth of $Gd_2O_3$18 $\mathring{A}$ thick on GaN, intense and streaky RHEED patterns of sixfold symmetry were observed, indicating a smooth two-dimensional growth. The two major in-plane directions of the oxide, separated by 30° in Fig. 10.23 lower panel are aligned with those of GaN. This is in sharp contrast to the two- and fourfold symmetry observed in the growth of Gd_2O_3 on GaAs.

X-ray diffraction studies show that Gd_2O_3 grows epitaxially on GaN in a hexagonal phase (Fig. 10.24), rather than the common cubic phase observed in case of rare earth oxides on GaAs or Si [32]. Despite a large lattice mismatch between the hexagonal rare earth oxides and the wurtzite GaN, the epitaxy occurs in the very initial stage of the film growth, and there is no in-plane rotation between the rare earth oxides and GaN. The bulk values of in- and out-of-plane lattice constants for hexagonal GaN shown in Fig. 10.25 are 3.189 and 5.185 Å, respectively. Those for bulk hexagonal Gd_2O_3 are 3.86 and 6.16 Å, respectively (Fig. 10.26) [33]. The in-plane lattice constant of the Gd_2O_3 films 18 Å thick is close to the bulk value, indicating that the Gd_2O_3 film was relaxed and not constrained to the GaN. Gd_2O_3 and Y_2O_3 appear to wet the GaN surface very well, despite the large lattice

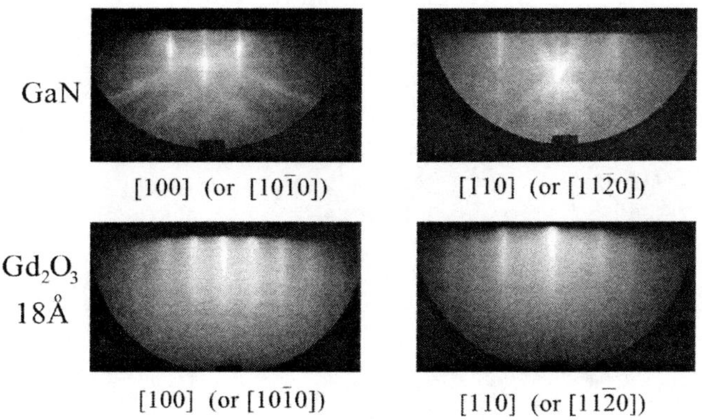

Fig. 10.23. (*upper*) RHEED patterns of GaN surface and (*lower*) RHEED of Gd_2O_3 film 18 Å thick deposited on GaN

mismatch. These thin epitaxial oxide films are fully relaxed and are of excellent structural quality, indicating that misfit dislocations are trapped near the interface.

We have obtained a low D_{it} of $\sim 10^{11}\,\mathrm{cm^{-2}\,eV^{-1}}$ from the capacitance–voltage data for the rare earth oxide/GaN interface [34]. The attainment of a low interfacial density of states is critical to the passivation of the GaN surface to reduce the surface state density, and to improve the lifetime of the MESFET and HEMT devices for electronic and optoelectronic applications. Furthermore, we have successfully carried out the overgrowth of single crystalline GaN film on Gd_2O_3/GaN after ex situ transferring to another MBE system [35]. A polycrystalline growth of preferred orientation was observed during the initial overgrowth of GaN on the rare earth oxides. The GaN film surface then becomes smoother and better ordered in the hcp wurtzite structure as the film grows thicker. Again, there was no inplane rotation between GaN and the oxide underneath.

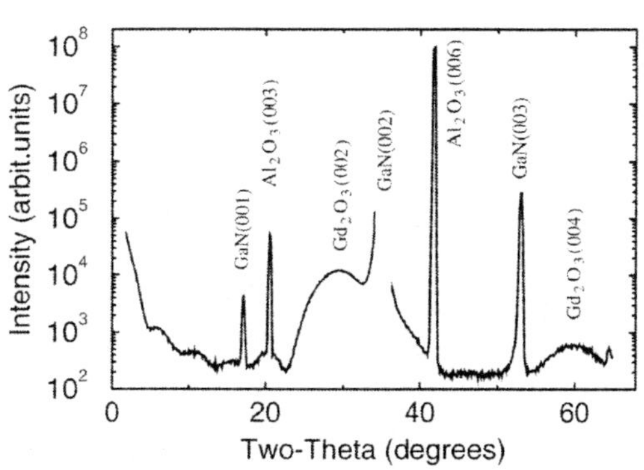

Fig. 10.24. X-ray diffraction normal scan on the Gd_2O_3 film as discussed in Fig. 10.23

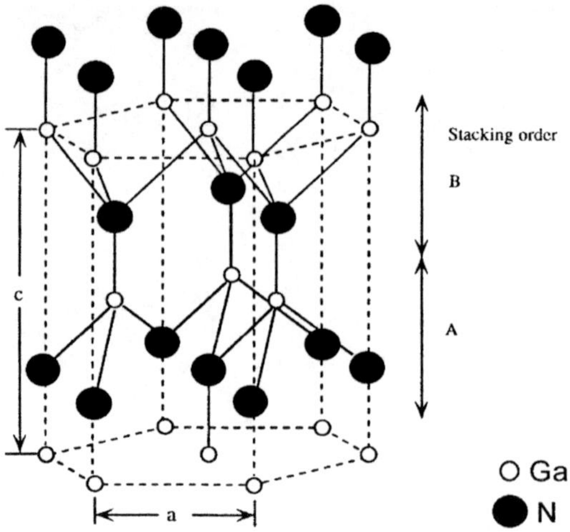

Fig. 10.25. Structure and stacking sequence of the wurtzite GaN hcp phase

In the area of GaN surface passivation, research efforts of depositing or growing insulators (and methods to prepare them) on GaN to give a low D_{it} at the insulator/GaN interface have shown that appropriate insulators and proper steps for cleaning GaN surfaces are critical to achieve a low D_{it}. GaN surfaces are not as robust and inert as generally thought when exposed to atmosphere such as room air. A summary of the efforts was given in a review article [36]. Cleaning procedures for preparing GaN surfaces for deposition of insulators are also included in the reference. Among the insulating materials being studied for GaN passivation, AlN, rare earth oxide Gd_2O_3, and Sc_2O_3

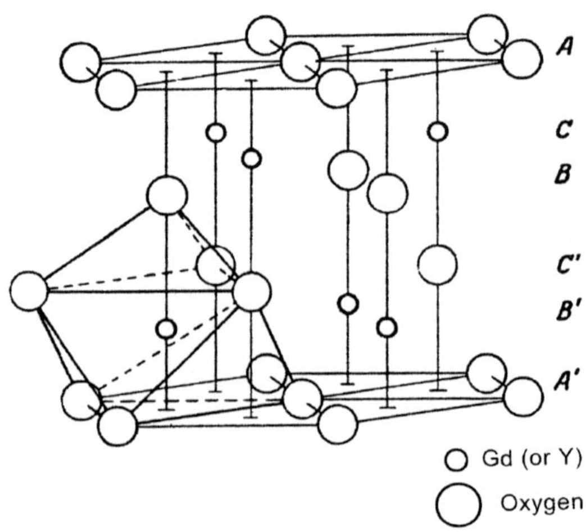

Fig. 10.26. Structure and stacking sequence of the hcp rare earth Gd_2O_3 sesquioxide

are single crystals. Previously, Sc_2O_3 was found to effectively passivate GaN [37] and to grow single crystal on GaN and sapphire (Al_2O_3), but with the in-plane growth in two degenerate orientations [38].

However, we have also achieved an excellent epitaxial growth of single-domain, single-crystal Sc_2O_3 films on Si(111), an unexpected result [39]. The structural perfection of the Sc_2O_3 films 3 and 18 nm thick is evidenced from very bright streaky and reconstructed RHEED patterns, very narrow rocking curves in the high-resolution x-ray diffraction, and a flat smooth film both at the interface and in the film interior observed using X-ray reflectivity and HR-TEM. Decent electrical characteristics of the oxide film were measured with a low leakage current and a high breakdown of $>5\,\mathrm{MV\,cm^{-1}}$.

10.8 Conclusion

We have given a review on some of the important recent work on high κ gate dielectrics on compound semiconductors. $Ga_2O_3(Gd_2O_3)$, the novel oxide, which was electron-beam evaporated from a gallium–gadollium–garnet target in UHV, has unpinned the GaAs Fermi level for the first time. Systematic heat treatments under various gases were studied to achieve low leakage currents and low D_{it}'s. Thermodynamic stability of the $Ga_2O_3(Gd_2O_3)$/GaAs heterostructures and the interfaces were achieved with high temperature annealing, which is needed for fabricating inversion-channel MOSFET's. More importantly, $Ga_2O_3(Gd_2O_3)$ remains amorphous and the interface remains intact with atomic smoothness and sharpness. Single crystal Gd_2O_3 was found to grow epitaxially on GaAs, for the first time, single crystal oxide epitaxially grown on single crystal semiconductor with excellent electrical properties and device performance. We have studied the mechanism of Fermi-level unpinning in ALD-Al_2O_3 *ex-situ* deposited on GaAs (InGaAs). Note that ALD is commonly used to deposited high-k gate dielectrics such as Al_2O_3 and HfO_2 on Si. We have also reviewed the work of inversion-channel, depletion-mode, and power GaAs MOSFETs using $Ga_2O_3(Gd_2O_3)$ as the gate dielectric. Finally, we have discussed the work of GaN passivation using single crystal rare earth of Gd_2O_3, which will play an important role in high temperature and high power electronic device application.

Acknowledgments

The authors thank Department of Nature Sciences at National Science Council, Taiwan, Republic of China for supporting our work on high-k gate dielectrics. We also thank the discussion with our able student Y.C. Chang.

References

1. D.A. Muller, T. Sorsch, S. Moccio, F.H. Baumann, and G. Timp, *Nature*, **399**, 758, (1999).
2. For a complete review, Materials Research Bulletin, March 2002 issue, on "*Alternative Gate Dielectric for Microelectronics*", Ed. by R.M. Wallace and G.D. Wilk.
3. J. Kwo, M. Hong, A.R. Kortan, K.L. Queeney, Y.J. Chabal, J.P. Mannaerts, T. Boone, J.J. Krajewski, A.M. Sergent, and J.M. Rosamilia, Appl. Phys. Lett. **77**, 130, (2000).
4. J. Kwo, M. Hong, A.R. Kortan, K.L. Queeney, Y.J. Chabal, R.L. Opila, Jr., D.A. Muller, et al., J. Appl. Phys. **89**, 3920, (2001).
5. B.W. Busch, J. Kwo, M. Hong, J.P. Mannaerts, and B.J. Sapjeta, W.H. Schulte, E. Garfunkel, and T. Gustafsson, Appl. Phys. Lett. **79**, 2447, (2001).
6. J. Kwo, M. Hong, B. Busch, D.A. Muller, Y.J. Chabal, A.R. Kortan, J.P. Mannaerts, B. Yang, P. Ye, H. Gossmann, A.M. Sergent, K.K. Ng, J. Bude, W.H. Schulte, E. Garfunkel, and T. Gustafsson, J. Crystal Growth, **251**, 645, (2003).
7. "High κ Gate Dielectrics For Si and Compound Semiconductors By MBE", J. Kwo and M. Hong, MRS Proceeding, Vol. **745**, (2003).
8. "Semiconductor-insulator interfaces", M. Hong, C.T. Liu, H. Reese, and J. Kwo in "Encyclopedia of Electrical and Electronics Engineering", Vol. **19**, pp. 87–100, Ed. J.G. Webster, Wiley, New York, 1999, and references therein.
9. "Phys. & Chem. of III–V Compound Semiconductor Interfaces", Ed. by C.W. Wilmsen, Plenum, New York, 1985.
10. M. Hong, M. Passlack, J.P. Mannaerts, J. Kwo, S.N.G. Chu, N. Moriya, S.Y. Hou, and V.J. Fratello, J. Vac. Sci. Technol. **B 14**(3), May/Jun, 2297, (1996).
11. M. Hong, J. Kwo, A.R. Kortan, J.P. Mannaerts, and A.M. Sergent, Science, **283**, pp. 1897–1900, (1999).
12. M. Passlack, M. Hong, and J.P. Mannaerts, Appl. Phys. Lett. 68(8), 1099, (1996).
13. F. Ren, M. Hong, W.S. Hobson, J.M. Kuo, J.R. Lothian, J.P. Mannaerts, J. Kwo, Y.K. Chen, and A.Y. Cho, IEEE Int. Electron Dev. Mtg (IEDM) Technical Digest, p.943, (1996), and also in Solid State Electron., **41** (11), 1751, (1997).
14. Y.C. Wang, M. Hong, J.M. Kuo, J.P. Mannaerts, J. Kwo, H.S. Tsai, J.J. Krajewski, J.S. Weiner, Y.K. Chen, and A.Y. Cho, Mater. Res. Soc. Symp. Proc. **573**, 219, (1999).
15. F. Ren, J.M. Kuo, M. Hong, W.S. Hobson, J.R. Lothian, J. Lin, W.S. Tseng, J.P. Mannaerts, J. Kwo, S.N.G. Chu, Y.K. Chen, and A.Y. Cho, IEEE Electron Device Lett., **19**(8), 309, (1998).
16. F. Ren, M. Hong, S.N.G. Chu, M.A. Marcus, M.J. Schurman, A. Baca S.J. Pearton, and C.R. Abernathy, Appl. Phys. Lett., 73, 3893–3895, (1998).
17. Y.C. Wang, M. Hong, J.M. Kuo, J.P. Mannaerts, J. Kwo, H.S. Tsai, J.J. Krajewski, Y.K. Chen, and A.Y. Cho, IEEE Int. Electron Devices Mtg (IEDM) Technical Digest, 67, (1998), and also in Electron Dev. Lett., **20**, 457, (1999).
18. Y.C. Wang, M. Hong, J.M. Kuo, J.P. Mannaerts, H.S. Tsai, J. Kwo, J.J. Krajewski, Y.K. Chen, and A.Y. Cho, Electron. Lett., **35**(8), 667, April 15th (1999).
19. M. Hong, Z.H. Lu, J. Kwo, A.R. Kortan, J.P. Mannaerts, J.J. Krajewski, K.C. Hsieh, L.J. Chou, and K.Y. Cheng, Appl. Phys. Lett. 76(3), 312, (2000).

20. A.R. Kortan, M. Hong, J. Kwo, J.P. Mannaerts, J.J. Krajewski, N. Kopylov, C. Steiner, B. Bolliger, and M. Erbudak, J. Vac. Sci. Technol. B 19(4), 1434, (2001).
21. M. Sowwan, Y. Yacoby, J. Pitney, R. MacHarrie, M. Hong, J. Cross, D.A. Walko, R. Clarke, R. Pindak, and E.A. Stern, Phys. Rev. B 66, 205311, (2002).
22. Y. Yacoby, M. Sowwan, E. Stern, J. Cross, D. Brewe, R. Pindak, J. Pitney, E. Dufresne, and R. Clarke, Nat. Mater. **1**, 99, (2002).
23. P.D. Ye, G.D. Wilk, B. Yang, J. Kwo, S.N.G. Chu, S. Nakahara, H.-J.L. Gossmann, J.P. Mannaerts, M. Hong, K.K. Ng, and J. Bude, Appl. Phys. Lett. **83**, 180, (2003).
24. M.M. Frank, G.D. Wilk, D. Starodub, T. Gustafsson, E. Garfunkel, Y.J. Chabal, J. Grazul, and D.A. Muller, Appl. Phys. Lett. **86**, 152904, (2005), and references therein.
25. J. Kwo, M. Hong, J.P. Mannaerts, Y.D. Wu, K.Y. Lee, B. Yang, and T. Gustafsson, in Integration of Advanced Micro- and Nanoelectronic Devices—Critical Issues and Solutions, Ed. by J. Morais, D. Kumar, M. Houssa, R.K. Singh, D. Landheer, R. Ramesh, R.M. Wallace, S. Guha, and H. Koinuma (Mater. Res. Soc. Symp. Proc. 811, Warrendale, PA, 2004), E1.12. (MRS Spring Meeting 2004.).
26. Y.L. Huang, P. Chang, Z.K. Yang, Y.J. Lee, H.Y. Lee, H.J. Liu J. Kwo, J.P. Mannaerts, and M. Hong, Appl. Phys. Lett. **86**, 191905, 2005, and references therein.
27. M. Hong, Z.H. Lu, J. Kwo, A.R. Kortan, J.P. Mannaerts, J.J. Krajewski, K.C. Hsieh, L.J. Chou, and K.Y. Cheng, Appl. Phys. Lett. **76** (3), 312, (2000).
28. M. Hong, J.N. Baillargeon, J. Kwo, J.P. Mannaerts, and A.Y. Cho, Proceedings of 2000 IEEE International Symposium on Compound Semiconductors, 345, 00TH 9498.
29. T. Mimura and M. Fukuta, IEEE Trans. Electron Dev., **27**(6), 1147, (1980).
30. D.S.L. Mui, Z. Wang, and H. Morkoç, Thin Solid Films, **231**, 107, (1993).
31. M.L. Huang, Y.C. Chang, C.H. Chang, Y.J. Lee, P. Chang, J. Kwo, T.B. Wu, and M. Hong, Appl. Phys. Lett. **87**, 252104, 2005.
32. M. Hong, A.R. Kortan, J. Kwo, J.P. Mannaerts, C.M. Lee, and J. I. Chyi, in IEEE International Symposium on Compound Semiconductors, IEEE Publication Series 00TH8498, 495, (2000).
33. L. Pauling, Z. Kristallogr, **69**, 415, (1928).
34. T.S. Lay, W.D. Liu, M. Hong, J. Kwo, and J.P. Mannaerts, Electronics Letters, **37**, 595, (2001).
35. M. Hong, A.R. Kortan, H.M. Ng, J. Kwo, S.N.G. Chu, J.P. Mannaerts, A.Y. Cho, C.M. Lee, J.I. Chyi, and K.A. Anselm, 20th NA MBE October 1–3, 2001, J. Vac. Sci.Technol. **B 20**, 1274, (2002).
36. M. Hong, H.M. Ng, J. Kwo, A.R. Kortan, J.N. Baillargeon, S.N. G. Chu, J.P. Mannaerts, A.Y. Cho, F. Ren, C.R. Abernathy, S. J. Pearton, and J.I. Chyi, Invited talk at 197th Electrochemical Society Meeting, Toronto, Ontario, Canada, May 14-19, 2000. The paper is on p. 103 of "Compound semiconductor power transistors II and state-of-the-art program on compound semiconductors (SOTAPOCS XXXII), Ed. by R. Kopf, A.G. Baca, and S.N.G. Chu ECS Proc. volume 2000–1.
37. B. Luo, J.W. Johnson, J. Kim, R.M. Mehandru, F. Ren, B.P. Gila, A.H. Onstine, C.R. Abernathy, and S.J. Pearton, A.G. Baca, R.D. Briggs, R.J. Shul, and C. Monier, J. Han, App. Phys. Lett. **80**, 1661 (2002).

38. A.R. Kortan, M. Hong, J. Kwo, P. Chang, C.P. Chen, J.P. Mannaerts, and S.H. Liou, in *Integration of Advanced Micro- and Nanoelectronic Devices—Critical Issues and Solutions*, Ed. by J. Morais, D. Kumar, M. Houssa, R.K. Singh, D. Landheer, R. Ramesh, R.M. Wallace, S. Guha, and H. Koinuma (Mater. Res. Soc. Symp. Proc. **811**, Warrendale, PA , 2004), p. E1.2. (MRS Spring Meeting 2004.).
39. C.P. Chen, M. Hong, J. Kwo, H.M. Cheng, Y.L. Huang, S.Y. Lin, J. Chi, H.Y. Lee, Y.F. Hsieh, and J.P. Mannaerts, J. Cryst. Growth **278,** 638–642, (2005).

11

Interface Properties of High-k Dielectrics on Germanium

A. Toriumi, K. Kita, M. Toyama, and H. Nomura

Summary. Ge CMOS with high-k dielectric films is very attractive as one of the post-Si device candidates with low-power as well as high-performance. The most important issue for Ge CMOS technology is how to achieve a superior-quality interface between the high-k dielectric film on Ge. This chapter first describes the effects that GeOx volatility during the high-k film growth process have on Ge. Another interesting feature of the high-k/Ge interface is the effect that Ge surface orientation has on interface properties. Finally, HfO_2 and Y_2O_3 on Ge are compared and discussed in terms of the interface quality of high-k on Ge.

11.1 Introduction

Although Ge CMOS devices have historically been left behind the recent research has brought them to the forefront again due by Si field effect devices to the trend of using deposited high-k films instead of thermally grown SiO_2 to achieve further CMOS scaling. This is very attractive because of both Ge's intrinsically higher carrier mobilities in part than those of Si, which can provide a larger driving current, and its smaller band gap to enable operation at a lower voltage. In addition, high-k dielectrics are needed irrespective of the channel materials to reduce the gate leakage current for lower-power operation. However, the challenges actually posed by high-k/Ge CMOSs are still poorly understood.

The interface layer in a high-k/Ge system is associated with GeOx growth, the same as for the high-k/Si interface in an O_2 ambient, although it has been reported that GeOx on Ge is thermally unstable and worse in electrical characteristics compared with SiOx grown on Si. We have recently found that MIS capacitors with no interface layer can be achieved in a high-k/Ge system [1]. There are, however, concerns that electrical characteristics strongly depend on the interfacial properties as well as the high-k materials employed. Therefore, we urgently need to select high-k materials that form a good interface with Ge.

Careful development of (100) Si substrates has been necessary to achieve industrial application of Si microelectronics. This should also be the case for Ge devices. Here, we report distinct differences in the surface chemistry of (111) and (100) Ge surfaces in terms of the oxidation rate, surface roughening effect, and MIS capacitor characteristics. Based on the results, we discuss the advantages of the (111) Ge surface. It is also important to investigate which high-k material is suitable for Ge devices. We report that Y_2O_3 forms an excellent interface with Ge in contrast with HfO_2. We do not intend to analyse the experimental results in great detail in this chapter but would like to show new findings on Ge interface-related characteristics.

11.2 Experimental

Three kinds of experimental results are presented in this chapter. HfO_2 films were processed on Ge and Si wafers simultaneously for comparison, except for the wafer cleaning steps. The Ge wafers were degreased with methanol, immersed in HCl solution to remove the native oxides, and re-oxidized in H_2O_2 solution, which was followed by dipping in diluted HF solution and rinsing in deionized water. Almost no Ge oxides were left on the Ge surface after these cleaning steps, which was confirmed by X-ray photoelectron spectroscopy (XPS) measurements. The Si wafers were cleaned with H_2SO_4/H_2O_2 and HCl/H_2O_2 solutions, and immersed in diluted HF solution.

In the first experiment, HfO_2 films (2–15 nm-thick) were deposited in a two-step sputtering process for the Hf metal target. A 1.5-nm-thick Hf metal layer was deposited in Ar, followed by reactively sputtering Hf metal in O_2/Ar. The ultra-thin metallic Hf layer was expected to be oxidized during the reactive sputtering to restrict the growth of the interface layer. For comparison, HfO_2 films without pre-depositing metallic Hf layers were also prepared. The films on both substrates were simultaneously annealed in $O_2(0.1\%) + N_2$ at 400 and 500°C. The interface layer and HfO_2 thicknesses were determined by both cross-sectional transmission electron microscopy (TEM) and the combination of glazing incidence X-ray reflectivity (GIXR) with spectroscopic ellipsometry (SE) measurements [2].

In the second experiment, the differences in the oxidation rates between (100) and (111) Ge wafers were investigated. Moreover, HfO_2 films were deposited on (100) and (111) Ge wafers by reactive sputtering and investigated in terms of thermal stability and interface quality, because there has been no a priori consensus on an appropriate surface orientation for device applications, though there have been debates on the optimum surface orientation from the device performance viewpoint [3, 4].

In the third experiment, we prepared HfO_2/Ge and Y_2O_3/Ge MIS capacitors to study suitable high-k films on the (100) Ge substrate. We deposited 5 nm-thick Y_2O_3 or HfO_2 films by rf-sputtering on p- (100) Ge wafers, and then annealed them in N_2 or in O_2. Au was deposited by vacuum evaporation to form the MIS capacitor gate electrodes.

11.3 Results and Discussion

11.3.1 Effect of Hf Metal Pre-deposition Prior to HfO_2 Deposition [5]

Both the HfO_2 and interface-layer thicknesses for samples processed simultaneously on Ge and Si were evaluated by a combination of GIXR and SE. Figure 11.1 plots the difference in thickness of as-deposited HfO_2 on Si from that on Ge ($\Delta T_{\mathrm{HfO2}} = T_{\mathrm{HfO2(on\ Si)}} - T_{\mathrm{HfO2(on\ Ge)}}$) as a function of the reactive sputtering time after the Hf metal pre-deposition. This difference can also be observed in cross-sectional TEM micrographs in Fig. 11.2. Note that the HfO_2 thickness does not include the interface layer but denotes only the upper high-k layer of the dielectric film. We can see from Fig. 11.1 that the ΔT_{HfO2} values seem nearly constant (0.5–0.8 nm) during the reactive sputtering process. Therefore, we surmised that the difference in the thickness of ΔT_{HfO2} originated from the difference in the HfO_2 film-growth mechanism in the very early stages on these substrates, where the role of the ultra-thin Hf metal layer should be taken into consideration to better understand the difference in the HfO_2 film thickness. It is not likely that Ge diffused into the HfO_2 and densified and shrank the bulk HfO_2 film. We suspect a volatile Ge compound might be associated with this effect. In fact, we noted that even

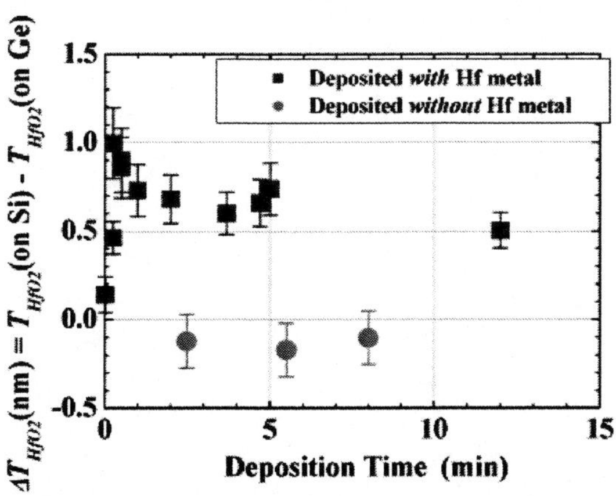

Fig. 11.1. Difference in thickness of as-deposited HfO_2 on Si from that on Ge as function of reactive sputtering time after Hf metal pre-deposition. Here, HfO_2 thickness does not include interface layer. Results with and without Hf metal pre-deposition are shown. HfO_2 thickness for Ge with Hf metal is always thinner than that for Si

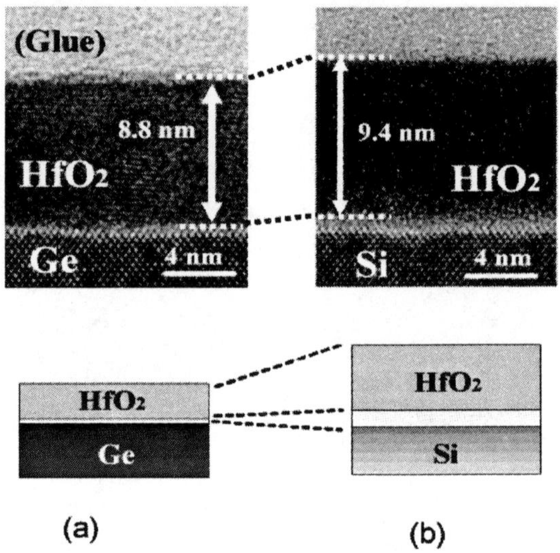

Fig. 11.2. Cross-sectional TEM micrographs of both (**a**) HfO_2/Ge and (**b**) HfO_2/Si deposited by reactive sputtering in Ar/O_2 after Hf metal pre-deposition. Schematics are also shown. Note that the upper HfO_2 thicknesses are different despite HfO_2 being deposited simultaneously in this experiment

without annealing, HfO_2 films on Ge were also thinner than those on Si, which suggests that Hf–Ge–O volatilization occurs not during thermal-annealing but during the film-deposition processes. Although the mixing process of Hf with Ge and/or O may explain the substantial interface layer thickness reduction, the change in upper high-k layer thickness is hard to be understood. So, it is likely that the volatilization of Ge oxides may be enhanced by the presence of the Hf metal acting as a catalyst, since GeO_x volatilization will not occur under moderate pressure at room temperature. Chui et al. reported that they had observed no Ge oxides at the ZrO_2/Ge interface after the UV-ozone low-temperature oxidation of a metallic Zr layer on Ge [6], and we noted from our present results that the thickness of upper HfO_2 film was also affected by the growth conditions during the reactive sputtering deposition process. Furthermore, the difference of HfO_2 film growth on both Ge and Si substrates in the initial stages of the reactive sputtering process can be seen in Fig. 11.3. Film growth on the Ge substrate clearly reveals a retardation at the beginning of reactive sputtering, contrary to monotonic film growth on the Si substrate. The retarded film growth on Ge is consistent with our reaction model, if the desorption of Hf–Ge–O compounds is assumed to occur only before the ultra-thin metallic Hf layer is fully oxidized. These results strongly suggest that Hf–Ge–O volatilization may be involved in the very early stages of film growth during the reactive sputtering process of HfO_2, and that the Ge interface needs to be treated more carefully than the Si one.

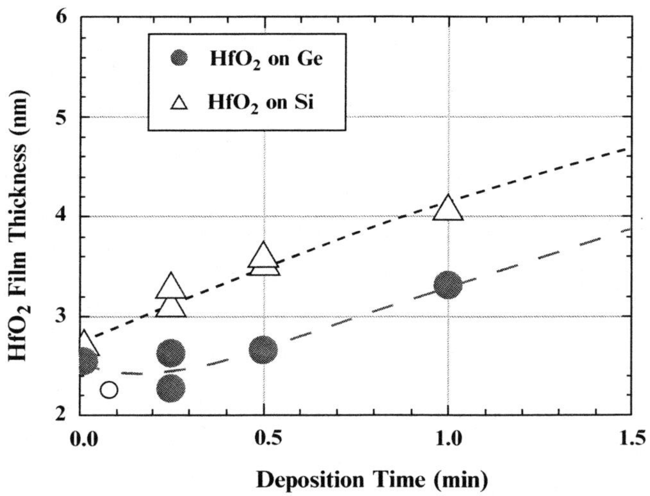

Fig. 11.3. HfO_2 film thickness on both Ge and Si substrates as a function of reactive sputtering time after Hf metal pre-deposition. Obvious film growth retardation at the initial stage is observed on Ge

11.3.2 Effects of Ge Surface Orientation

The surface orientation definitely affects the transport properties of the Ge MIS inversion layer, similar to the case for Si MOSFETs. The fact that the interface quality depends on the surface orientation employed should be considered more carefully, since there may be no interface layer in a high-k/Ge system. Here, we have focused on the differences in the surface and interface properties between (100) and (111) Ge. Figure 11.4 plots the difference in the thickness of the oxide on (100) and (111) Ge wafers as a function of annealing temperature in O_2 [7]. Ge oxidation occurs at around 450°C, and (111) Ge has a lower oxidation rate than (100) Ge. This means that the (111) Ge wafer is more robust than the (100) Ge against oxidation in O_2.

The difference in surface roughness between (100) and (111) Ge wafers was also evaluated [8]. Table 11.1 compares the roughness RMS values measured by AFM of GeOx grown at 600°C for an as-cleaned, GeOx top, and Ge surface after GeOx was removed. We can clearly see that (100) Ge is rougher than (111) Ge, although the initial values are the same. This is quite important from the viewpoint of selecting the Ge wafer orientation for device applications, since the surface roughness should degrade the carrier mobility and/or device reliability.

Figure 11.5a,b shows C–V characteristics of Au/HfO_2/Ge MIS capacitors on (100) and (111) Ge, in which HfO_2 was prepared by reactive sputtering. No differences in C–V characteristics can be observed after annealing at 400°C, while after annealing at 650°C there is significant stretch-out

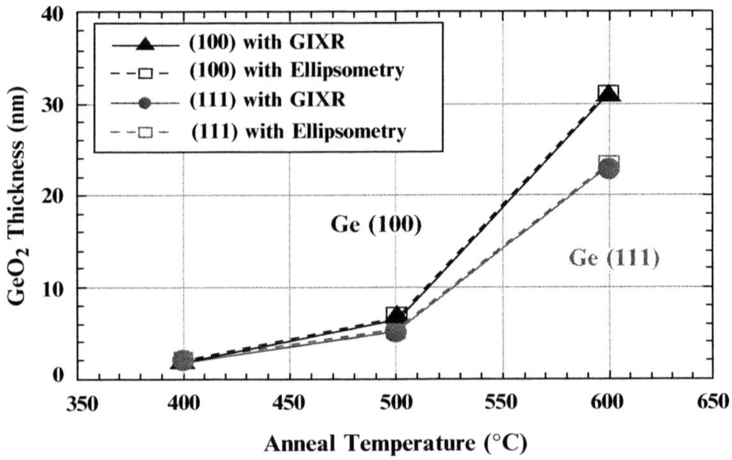

Fig. 11.4. GeOx thickness on (100) and (111) Ge wafers as a function of oxidation temperature. Thickness was determined independently by GIXR and SE

in *C–V* curves for HfO_2/(100) Ge both in 1-kHz and 1-MHz measurements. Although *C–V* hysteresis was observed in the results, they are not shown here for clear comparison between the two different substrate orientations. In addition, one should note that the saturated capacitance on (111) Ge is higher that that on (100) Ge. This indicates that the interface on (100) Ge is degraded more severely (imperfectly oxidized) compared to that on (111) Ge in terms of electrical properties as well as surface morphology. Furthermore, the stretch-out becomes worse at higher annealing temperatures. The microscopic mechanism responsible for the surface orientation dependence of degradation is now under investigation.

11.3.3 Y_2O_3 and HfO_2 on (100) Ge [9,10]

The interface quality is expected to depend on the high-k materials employed. We compared two kinds of high-k materials on (100) Ge from the viewpoint of MIS capacitor characteristics. Figure 11.6a plots both the thicknesses of the upper high-k and interface layers as a function of the annealing temperature in N_2. In both Y_2O_3 and HfO_2 cases, the high-k layer thicknesses are almost constant, while no interface layers are identified above annealing at 400°C. The

Table 11.1. Roughness RMS values measured by AFM for as-cleaned, GeOx top, and Ge surface after removing GeOx

Ge surface	As-cleaned (nm)	Oxide top (nm)	Ge top (nm)
(100)	0.22 ± 0.01	0.71 ± 0.06	0.28 ± 0.03
(111)	0.22 ± 0.02	0.58 ± 0.03	0.15 ± 0.01

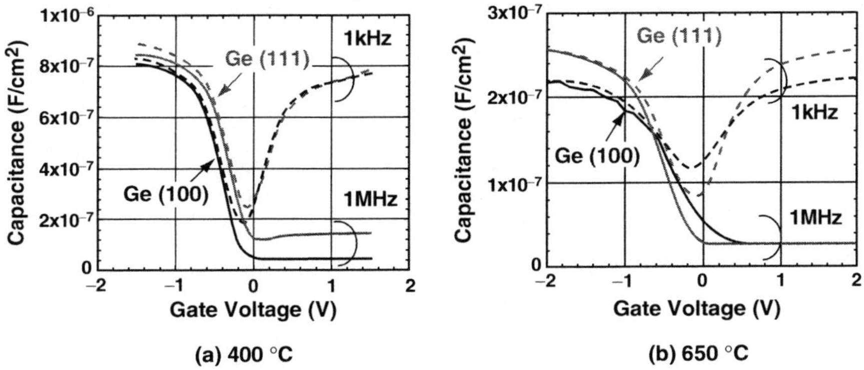

Fig. 11.5. Comparison of high frequency *C–V* characteristics of Au/HfO_2/Ge MIS capacitors on (100) and on (111) Ge wafers, annealed at (**a**) 400°C, and (**b**) 650°C, in O_2 for 30 s

TEM micrograph in Fig. 11.6b shows direct evidence of an abrupt interface at the Y_2O_3/Ge boundary annealed at 600°C in N_2. Thus, we conclude that the interface layers in Y_2O_3/Ge and HfO_2/Ge stacks disappeared due to the reaction between high-*k* films and GeOx. This is one advantage of sputter-deposited high-*k*/Ge systems in terms of EOT scaling.

We next discuss the MIS capacitor characteristics of Au/Y_2O_3/Ge and Au/HfO_2/Ge capacitors. Figure 11.7 plots the high frequency *C–V* characteristics of Y_2O_3/Ge and HfO_2/Ge MIS capacitors annealed in N_2 at 600°C at 1 MHz at 300 K. The *C–V* characteristics of the Y_2O_3/Ge MIS capacitors seem much better than those of the HfO_2/Ge system. This indicates that Y_2O_3/Ge

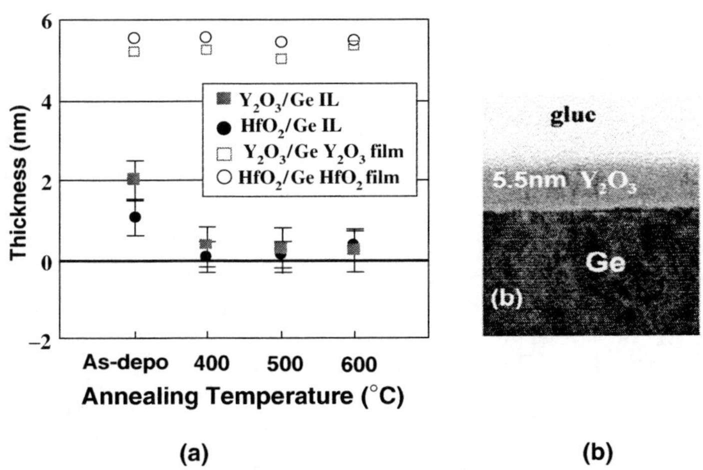

Fig. 11.6. (**a**) Thicknesses of upper high-*k* and interface layers as function of annealing temperature in N_2. (**b**) TEM micrograph of abrupt interface at Y_2O_3/Ge annealed at 600°C in N_2 showing direct evidence of no interface layer

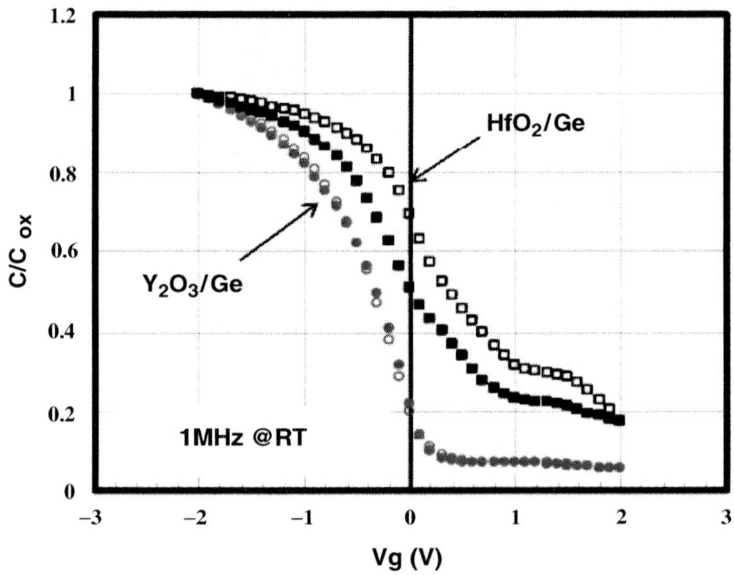

Fig. 11.7. Normalized 1 MHz *C–V* characteristics at room temperature of Y_2O_3/Ge and HfO_2/Ge MIS capacitors annealed in N_2 at 600°C. Large hysteresis is observed in the HfO_2/Ge case

is more robust than HfO_2/Ge up to a higher process temperature (600°C). It is surprising that the interface properties of Y_2O_3/Ge are not significantly degraded despite the lack of an interface layer in Y_2O_3/Ge.

To investigate what effects thermal processes had on the Ge substrate, Zerbst analysis was performed [11], because chemical reaction at the interface might degrade the minority carrier life time in the substrate. This is a classical method of monitoring the minority carrier generation time ($\tau_{g,eff}$) in the substrate by measuring the change in transient capacitance from accumulation to the deep depletion state in MOS capacitors. Here it should be taken into consideration that the minority carrier generation at room temperature can follow a small amplitude surface potential modulation even in high-frequency measurements in the case of Ge because of its small energy band gap [12]. So, we measured the transient capacitance characteristics at low temperatures in order to achieve a deep depletion layer in the Ge bulk. Typical results for transient capacitance at 120 K are plotted in Fig. 11.8. The effective minority carrier generation time ($\tau_{g,eff}$) can be calculated from the slope of the "Zerbst plot", as shown in Fig. 11.9. The $\tau_{g,eff}$ of the Y_2O_3/Ge case calculated from the slope in Zerbst plot is similar to that for HfO_2/Ge, while the initial part of transient capacitance is different between them. The results indicate that the HfO_2/Ge interface is deteriorated more severely than the Y_2O_3/Ge one, while the bulk Ge quality is almost the same for both cases.

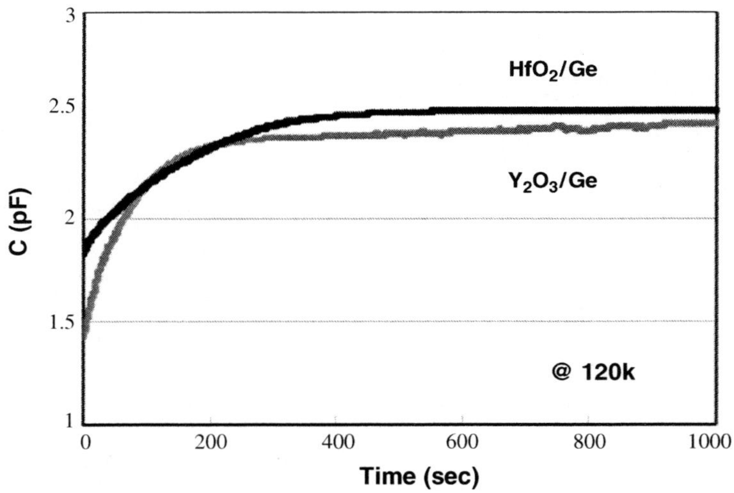

Fig. 11.8. Transient capacitance characteristics when gate bias was changed from accumulation to inversion at 120 K both for Y_2O_3/Ge and HfO_2/Ge MIS capacitors

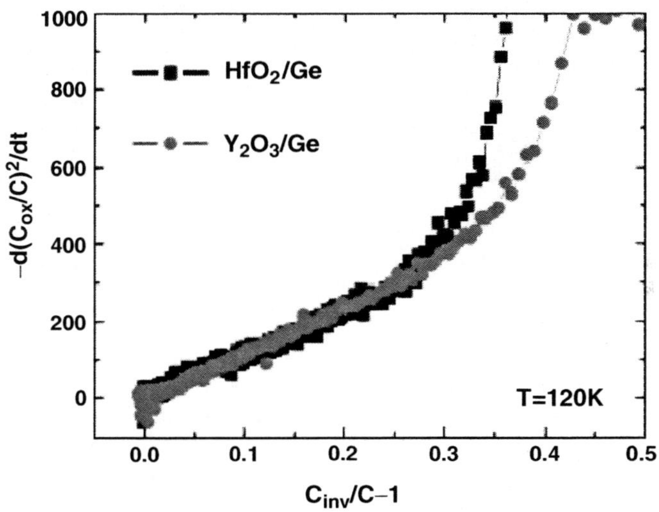

Fig. 11.9. Low temperature Zerbst plots for transient capacitance results in Fig. 11.8. It is noticed that a difference of the initial transient behavior is observed

We are now investigating possible reasons why Y_2O_3/Ge MIS capacitors exhibited excellent C–V characteristics at room temperature even after annealing at 600°C in contrast to HfO_2/Ge case. The SIMS results in Fig. 11.10 indicate that Ge atoms diffused into Y_2O_3, in addition to there being no apparent interface layer at the Y_2O_3/Ge. This fact suggests that Ge diffusion into Y_2O_3 may relax interface structural disorder, while that into HfO_2 should

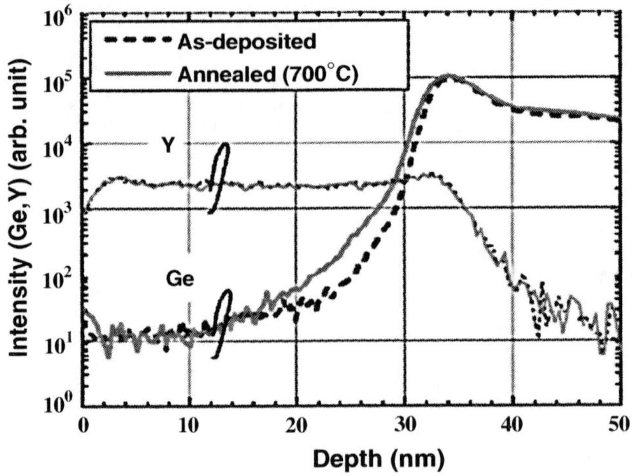

Fig. 11.10. SIMS profiles of Y_2O_3/Ge samples as-deposited and annealed at 700°C

degrade electronic properties. Moreover, the XPS analysis suggests that the interface quality difference may be due to the different reaction process with GeOx between Y_2O_3 and HfO_2, though further study is obviously needed. Thus, it can be stated that the high-k material's inherent properties definitely affect MIS interface quality and appropriate selection of the high-k material is key to developing Ge-CMOS technology.

11.4 Conclusion

We found that not only the interface layer but the HfO_2 film itself was thinner on Ge than on Si during an identical reactive sputtering process with Hf metal pre-deposition. We found that the metallic Hf layer, deposited before HfO_2 reactive sputtering, played a crucial role in the different thicknesses of both interface layers and HfO_2 on Ge, and that controlling reactions among Hf, Ge, and O, more generally that among metal, Ge, and O, is important to achieve practical high-k/Ge MOS systems. Several advantages of Ge(111) were demonstrated. In addition to the possible advantages of low field mobility in n-MOSFETs, Ge(111) has superior properties from the viewpoints of device and process stability combined with high-k dielectrics. Furthermore, we investigated the characteristics of Ge/high-k MIS capacitors with Y_2O_3 and HfO_2. Y_2O_3/Ge MIS capacitors were found to have excellent properties compared to HfO_2/Ge after annealing at 600°C in N_2 ambient, despite the apparent lack of an interface layer. We thus conclude that Y_2O_3 is better than HfO_2 as a high-k material on Ge in terms of its thermal robustness during the device fabrication process.

Acknowledgments

This work was partly supported by a Grant-in-Aid for Scientific Research from the Ministry of Education, Culture, Sports, Science and Technology, Japan.

References

1. K. Kita, M. Sasagawa, K. Tomida, M. Toyama, K. Kyuno, and A. Toriumi, Ext. Abst. Int. Workshop of Gate Insulators (IWGI'03), p. 186 (2003, Tokyo)
2. H. Shimizu, K. Kita, K. Kyuno, and A. Toriumi, Jpn. J. Appl. Phys. Pt.1, **44**, 6131 (2005)
3. S. Takagi, Dig. Sym. VLSI Technology, p. 115 (2003, Kyoto)
4. T. Low, M.. Li, W.J. Fan, S.T. Ng, Y.-C. Yeo, C. Zhu, A. Chin, L. Chan, and D.L. Kwong, Tech. Dig. IEDM'04, p. 151 (2004)
5. K. Kita, K. Kyuno, and A. Toriumi, Appl. Phys. Lett. **85**, 52 (2004)
6. C.O. Chui, S. Ramanathan, B.B. Triplett, P.C. McIntyre, and K.C. Saraswat, IEEE Electron Dev. Lett. **23,** 473 (2002)
7. M. Toyama, K. Kita, K. Kyuno, and A. Toriumi, Ext. Abst. Solid State Dev. and Mat. (SSDM'04), p. 226 (2004, Tokyo)
8. M. Toyama, Master Thesis, The University of Tokyo (2006)
9. H. Nomura, K. Kita, and A. Toriumi, Ext. Abst. Solid State Dev. and Mat. (SSDM'05), p. 858 (2005, Kobe)
10. H. Nomura, K. Kita, T. Nishimura and A. Toriumi, Ext. Abst. Solid State Dev. and Mat. (SSDM'06), p. 406 (2006, Yokohama)
11. D.K. Schroder, "*Semiconductor Material and Device Characterization*," (2nd Ed.) (John Wiley & Sons) (1998)
12. H. Nomura, T. Nishimura, K. Kita and A. Toriumi, to be presented at IWDTF 2006 (Kawasaki)

12

A Theoretical View on the Dielectric Properties of Crystalline and Amorphous High-κ Materials and Films

V. Fiorentini, P. Delugas, and A. Filippetti

Summary. We review recent and current theoretical work on several aspects of the dielectric response and dynamical properties of oxides used as dielectric layers in microelectronics. A personal choice of studies are singled out, on crystalline, amorphous, and alloy phases as well as thin films, mostly with focus on rare-earth and transition-metal "high-κ" compounds, as well as a selection of important work on silica.

12.1 Introduction

The so-called "high-κ" oxides, i.e., bearing large values of the static dielectric permittivity, have become the focus of intense research for their use as insulating layers in transistors, capacitors, and memories. This is because they are expected to replace silica, in the next few years, as the gate insulator in nanometric technology nodes of integrated silicon-based circuits. The high-κ is intended to help increase the series capacitance of the conducting channel stack upon size downscaling [1, 2], by reducing the effective oxide thickness $d\,\kappa_{\text{silica}}/\kappa_{\text{oxide}}$ without reducing the physical layer thickness so as to cause tunneling leakage through the layer. Static dielectric constants in the range of about 20 are needed for the current scaling-down.

Of course, this paradigm shift is faced with integration, manufacturing, and basic problems. Here we discuss those posed by the theoretical understanding of the dielectric properties of these materials. The logical sequence is as follows. Theoretical studies start invariably, and for good reasons, from bulk crystalline phases. On the other hand, while in principle crystal layers would be desirable, they actually tend not to be single-crystal, but polycrystalline, which generally entails poor electrical properties related to grain boundary conduction. Therefore, the analysis of amorphous phases is generally a natural, although daunting, follow-up to crystal studies. A slightly different but equally mind-numbing problem is alloying, which is interesting as a way to tweak the properties of a material by mixing in another compounds: One component may dictate, say, the structure, the other the vibrational properties. But of course the study of alloys is no less demanding than that of amorphous systems. Finally, though not as thin as silica films in current technology nodes, high-κ oxide layers in transistors are hardly thicker than a couple of nanometers or so. Do the desirable dielectric properties of the bulk carry over to, or do they change appreciably in such a film? A partial answer has been provided by studies addressing yet another technology problem, i.e., the properties of the ultrathin layers of silicon suboxide that are often, and usually unintentionally, interposed between the Si substrate and the oxide layer, be it silica or a high-κ stack.

High-κ oxides are produced only with selected transition-metal and rare-earth cations. Here, with no pretence of completeness and with somewhat arbitrary choices where needed, we cover a selection of works on Hf and Zr dioxides, on rare-earth sesquioxides and aluminates, and on silicates. Some of these are dealt with in both the crystal and amorphous (or disordered). We will also mention work on silicon dioxide which, though of course not a high-κ, is indeed "the mother of all oxides" in the electronics context; besides its persistent relevance as a generally unavoidable interlayer in actual gate stacks, it serves as well-established playground for new concepts. We will touch upon work on silica in amorphous and ultrathin-film variants. Some of the materials in focus (e.g., rare-earth compounds) are not commonly studied with respect to their dielectric and dynamical properties, so that a review of work in this area is premature. However the general aspects of dielectric response and the trends that can be extracted from first-principles calculations are qualitatively illuminating and, occasionally, quantitative, and establish relations between structure and expected dielectric properties.

12.1.1 Linear Response Theory and Dielectric Properties

At the GHz frequencies relevant to present Si-based devices, the dielectric permittivity ϵ (i.e., κ: we will be using ϵ henceforth to avoid notation problems) results from both electronic and ionic contributions. For both these mechanisms, a GHz-frequency field in a transistor structure is essentially static and uniform. The electronic contribution is the partial derivative of the polarization with respect to electric field, and in high-κ materials is generally a factor 2–5 smaller than the ionic one, and typically of order 4–5 in magnitude. One therefore generally focuses mostly on ionic screening, which is due to polarization by ionic displacements. The appearance of a polarization is possible in the absence of symmetry (e.g., in amorphous materials) or when symmetry allows for polar collective displacement. Under the action of a static electric field, ions relax preserving symmetry (if any) to minimize their potential energy. These collective displacements having lattice periodicity are linear combinations of vectors related to the eigenmodes of the dynamical matrix at the Brillouin zone center (Γ).

In the harmonic approximation, each normal mode j is an independent harmonic oscillator interacting with the electric field $\mathcal{E}$ via a potential energy $-\mathbf{M}_j \cdot \mathcal{E}$, where the vector $\mathbf{M}_j$ is the electric polarization per unit cell produced by a unit ionic displacement. $\mathbf{M}$ is proportional to the partial derivative of macroscopic polarization with respect to a normal coordinate Q_j (at zero electric field):

$$M_{j\alpha} = \Omega_0 \left[\frac{\partial \mathcal{P}_\alpha}{\partial Q_j}\right]_{\mathcal{E}=0} . \tag{12.1}$$

Similarly, the partial derivatives

$$Z^*_{\kappa,\alpha\beta} = \Omega_0 \cdot \frac{\partial \mathcal{P}_\alpha}{\partial \tau_{\kappa\beta}} \tag{12.2}$$

of the macroscopic polarization component α with respect to component $\tau_{\kappa\beta}$ of the position vector of atom κ in the primitive cell are usually named the Born or dynamical effective charges. They are in general tensor quantities (a 3×3 matrix for each atom). The displacement of the κth ion, with mass m_κ, in the jth vibrational mode is the mass-weighted normalized phonon eigenvector

$$U_j(\kappa,\alpha) = \frac{1}{\sqrt{m_\kappa}} \cdot e_j(\kappa,\alpha). \tag{12.3}$$

In terms of the Born charges the mode polarization is then

$$M_{j\alpha} = \sum_{\kappa\beta} Z^*_{\kappa,\alpha\beta} U_j(\kappa,\beta). \tag{12.4}$$

The dielectric permittivity is by definition

$$\epsilon_{\alpha\beta} = \delta_{\alpha\beta} + 4\pi \frac{\mathrm{d}\mathcal{P}_\alpha}{\mathrm{d}\mathcal{E}_\beta}. \tag{12.5}$$

If no spontaneous (i.e., zero-field) polarization is present, under vanishing strain conditions (see [3] and [4] for another formulation and [5] for the vanishing stress case), the polarization $\mathcal{P}_\alpha$ may be expanded to linear order in the macroscopic electric field as

$$\mathcal{P}_\alpha = \sum_\beta \mathcal{E}_\beta \left(\frac{\partial \mathcal{P}_\alpha}{\partial \mathcal{E}_\beta} + \frac{1}{\Omega_0} \sum_j \frac{M_{j\beta} \cdot M_{j\alpha}}{\omega_j^2} \right), \tag{12.6}$$

and hence

$$\epsilon_{\alpha\beta} = \epsilon^\infty_{\alpha\beta} + \frac{4\pi}{\Omega_0} \sum_j \frac{M_{j\beta} \cdot M_{j\alpha}}{\omega_j^2}. \tag{12.7}$$

$\epsilon^\infty_{\alpha\beta}$ is the permittivity originating from electronic polarizability alone with clamped ions. This is the only permittivity measured at frequencies well above vibrational ones, i.e., when the response of ionic motion to electric field is completely suppressed. The coupling of each polar mode with electric fields is proportional to the squared modulus of the mode dipole moment. This coupling is usually quantified by the oscillator strengths, defined as

$$S_{j,\alpha\beta} = \sqrt{M_{j\alpha} M_{j\beta} / \Omega_0}. \tag{12.8}$$

Thus $4\pi(\mathrm{S}_j/\omega)^2$ yields directly the individual contribution of mode j to the ionic dielectric tensor, and is known as mode dielectric intensity. Of course, this intensity, and hence the ionic component of the dielectric constant, can become large when M is large or ω is small, or both.

Therefore, large ϵ's result from large effective charges and soft vibrations (as well as from the efficient alignment of dipole contributions from each ion). Effective charges exceeding significantly the nominal ionic charge of the relevant atom in a given compound are usually named anomalous; dynamical charges up to four times the nominal ionic charge have been reported for ferroelectrics. Due to their effects on the dielectric constant, anomalies are also important in high-κ's. Recently it has been pointed out [6] that density-functional level calculations may overestimate dynamical charge anomalies, and that self-interaction corrections [7] may be a cure for the problem. Too large a dynamical charge would of course lead to overestimated ionic permittivities. In the high-κ context, this is supported by calculations on $LaAlO_3$ [8].

We note in passing that the phonons involved in optical transitions correspond to the eigenmodes and frequencies of the dynamical matrix in the limit $\mathbf{q} \to 0$. Approaching Γ, the dynamical matrix is given by its value at Γ plus a nonanalytical term [9] dependent on the $\hat{\mathbf{q}}$ direction whence the $\mathrm{q} \to 0$ limit is reached. Since this additive term does not act on modes whose dipoles are orthogonal to $\hat{\mathbf{q}}$, the oscillators strengths of zone center modes may be obtained by infrared absorption and reflectivity measurements revealing only transverse optical (TO) modes. Longitudinal optical (LO) modes can also be obtained by including nonanaliticity, and compared with those measured at grazing incidence or on films grown on metallic substrates [10].

All the quantities discussed above are related to second-order derivatives of the total energy with respect to ionic positions and electric field. In the DFT Kohn–Sham approach, the second-order derivatives can be calculated by the linear-response technique, which enables one to calculate the first-order derivatives of the wavefunctions, and hence of the density, with respect to external parameters [11]. A similar formulation is based on the $2n+1$ theorem [12], which produces all mixed derivatives up to third order from the first-order perturbed density [9,13]. The calculations mentioned in this paper are mostly performed using variants and evolutions of pseudopotential plane wave codes for Kohn–Sham DFT total-energy calculations [14], such as abinit [15] and Espresso [16] or custom variants thereof for the linear response, and the same plus VASP [17] for total-energy calculations. All calculations employ either LDA or GGA functionals, with some prevalence of the latter for exotic cations.

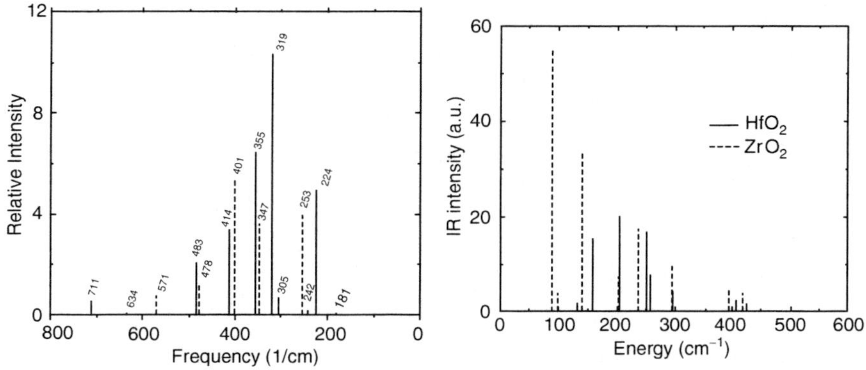

Fig. 12.1. *Left*: Dielectric intensity in crystalline monoclinic ZrO_2 (*solid*: *A* modes; *dashed*: *B* modes). Figure reproduced with permission from [3], copyright 2002 American Physical Society. *Right*: Dielectric intensity in Si-epitaxial, approximately tetragonal, crystalline ZrO_2 and HfO_2. Compare with Fig. 12.1 (frequency axis inverted). Note the enhancement at low frequencies due to new (*backfolded*) low-frequency IR modes. Figure reproduced with permission from [19], copyright 2002 American Physical Society

12.2 A Crystal Selection: Dioxides, Sesquioxides, Aluminates

12.2.1 Multiphase and Epitaxial Transition-Metal Dioxides

The first and foremost candidates as gate oxides are hafnia and zirconia, the dioxides of the transition metals Zr and Hf. They exist in two phases, the stable monoclinic and the metastable tetragonal, both deriving from distortions of the fluorite (CaF_2) structure; the latter is only explicitly adopted by Ce oxide. The monoclinic variant, with its cation coordination of 7, strikes a balance for these "undecided" cations between the eightfold coordination of fluorite and the sixfold of X_2O_3 bixbyite or hexagonal, while keeping the stoichiometry of a dioxide The structures and experimental data thereon are discussed in detail in [3, 4] and [18]. Here we mention the salient points of the investigations. Fluorite-structure hafnia and zirconia have very large dielectric constants [3, 4, 18, 19], in the order of 35 including the electronic part of about 5. Average values are also large ($\sim$40) in the tetragonal phase, resulting however from a highly anisotropic ϵ tensors, the zz (i.e., the axial component) being only around 15. The average value drops to about 20 in the monoclinic phase, mostly because of reduced effective charge anomalies; this values compares reasonably with typical experimental values [3, 4, 18, 19].

It is interesting to compare the dielectric intensity for free-standing monoclinic zirconia [4] in Fig. 12.1, left panel, with that of a hypothetical Si-epitaxial zirconia, which turns out to be tetragonal, reported in the same figure, right panel. Clearly, new dielectric intensity lines appear in the latter at low frequency (amplified by the $1/\omega^2$ dependence). These originate from the unstable zone-border Raman modes of fluorite–zirconia which backfold at zone center in the larger simulation cell of [19] and acquire IR character due to symmetry lowering implied by the epitaxial relation. This foreshadows the analogous effects produced by low-energy distortions and by IR-Raman mixing in crystalline and amorphous aluminates, and the appearance of low-energy dielectric intensity in amorphous zirconia (both discussed below).

12.2.2 Sesquioxides: Lutetia, Lanthana, and the Hex–Bix Difference

Sesquioxides have the formula X_2O_3, with X a trivalent metal. We discuss La and Lu sesquioxides, pointing out the close analogy of the latter with most of the lanthanide (Ln-) sesquioxides. The latter adopt either the cubic bixbyite (C-phase) and hexagonal (A-phase) structures [20]. Indeed, the hex phase is only observed for La, while it is metastable and with modest abundance for mixed-valent Pr and Nd. Ce sesquioxide can only be stabilized in strongly reducing atmosphere, so we can consider Ce as tetravalent, i.e., dioxide-forming. Calculations on structural preferences are very scarce; recent first-principles work [21] found an energy ordering of the structures consistent with that just outlined. Our own earlier calculations [22] show that for La_2O_3 the hex phase is favored by 0.2 eV per formula unit compared to the cubic; for Lu, the opposite happens, with a difference of 0.25 eV. This indicates again the increasing cubic-hex difference across the Ln-series in favor of the cubic phase, as expected experimentally.

Building on the bixbyite vs. hex structural theme we analyzed the differences in dielectric response of the two phases and deduced a general trend for all Ln-sesquioxides on a purely structural basis. Bixbyite lutetia has a diagonal and isotropic static dielectric tensor of 11.98 (which hardly qualifies as "high" in the present context), the electronic part being 4.2, consistently with the expected predominance of the ionic component in high-κ's. For hexagonal lutetia, the dielectric tensor is anisotropic, with two identical in-plane components of 19.8 and an axial component of 17.2. The "inverse average" that would be measured via series capacitance of a stack, i.e.,

$$3/\overline{\epsilon}_{\mathrm{s}} = 1/\epsilon_{\mathrm{s}}^{11} + 1/\epsilon_{\mathrm{s}}^{22} + 1/\epsilon_{\mathrm{s}}^{33} \tag{12.9}$$

is over 18. The electronic tensor is essentially the same as that of bixbyite (hence a factor 3 smaller than the ionic).

Interestingly, the ionic component is nearly twice as large that of bixbyite. Analyzing the Born effective charge tensors, we find an average of 3.6 for Lu and −2.4 for oxygen in bixbyite, and 3.65, −2.43 in the hex phase. Clearly, these charges are equally anomalous (or rather, nonanomalous); the different ionic constant must therefore be of vibrational origin. A related issue is, how

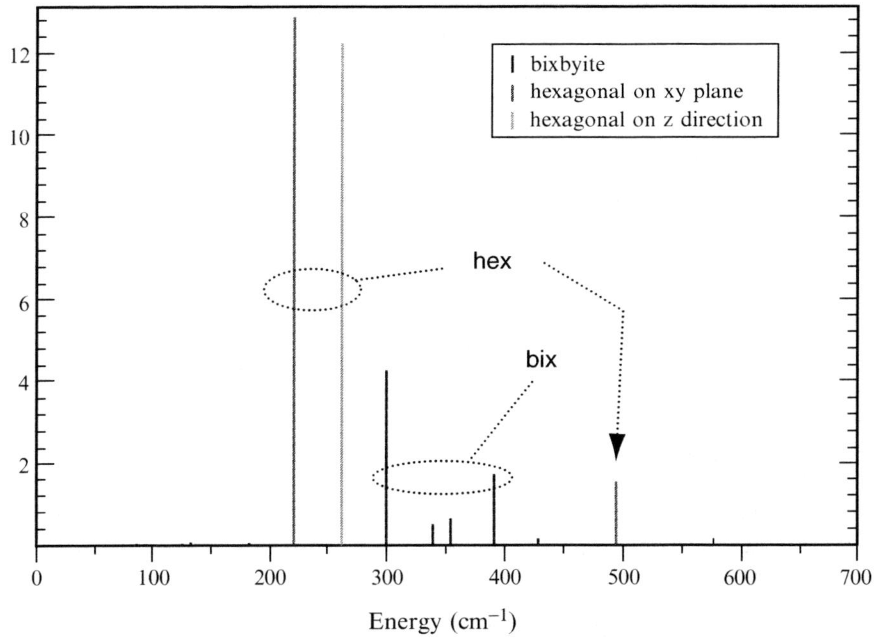

Fig. 12.2. Dielectric mode intensity in bixbyite (lower, above 300 cm^{-1}) and hex Lu_2O_3

general is this behavior for sesquioxides at large. To answer this we analyze the vibrational modes. Group theory shows that the sum (12.4) may be non-vanishing only for modes belonging to certain representations, while it must vanish for others, so that one can discriminate between polar and nonpolar modes simply on the basis of symmetry arguments. In the hex structure a pair of E_u modes contribute isotropically to the in-plane component and two A_{2u} contribute to the axial component of the ionic tensor. The planar modes are at 221 cm^{-1}, contributing 90% of the in-plane component, and 494 cm^{-1}. The axial modes are at 261 cm^{-1}, accounting for over 95% of the axial component, and 497 cm^{-1}. In the bixbyite structure the main contributions are from modes at 300–400 cm^{-1}, building up 93% of the total. Bixbyite is the experimental structure, and these data compared extremely well [10] with IR absorption on films [23]. The experimental value of $\epsilon_0 = 12 \pm 1$ obtained both optically and electrically in [10] is also in excellent agreement with the predicted one.

The spectra of the two structures are compared explicitly in Fig. 12.2, displaying the dielectric intensities vs. energy for the two structures. Clearly, the near doubling of the ionic component is due to the softer modes supported by the hex structure, despite the equally nonanomalous charges. Is this conclusion exportable to other sesquioxides with different cations? If so, the vibrational modes should be independent of the cation. For the hex phase, by inspection, one sees that indeed the relevant eigenmodes involve practically only oxygen

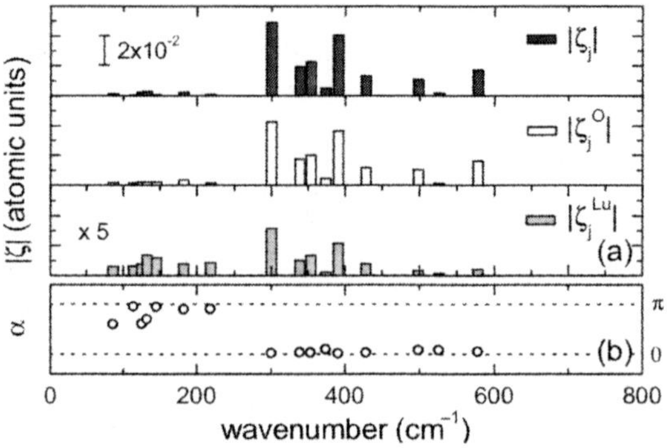

Fig. 12.3. Upper panel: moduli of the dipole vectors building up the ionic contribution; From *top* to *bottom*: total, O, and Lu. *Lower panel*: angle between the O and Lu dipole vectors; small angles indicate efficient alignment of the dipoles. The quantities named ζ in this figure reproduced from [10] (copyright 2005 by the American Physical Society) are the sum 12.4 evaluated over cations, anions, or all atoms

motions, with the cation just keeping the center of mass still. For bixbyite, the eigenvectors are hard to visualize. We therefore display in Fig. 12.3 the mode electric polarization vectors defined in (12.4). From top to bottom in the upper panel, the modulus of total, O, Lu vectors (note that the sum in (12.4) is also over atoms). Below the angle between the O and Lu dipoles is shown: if the angle is small the mode is IR-efficient. Ostensibly O dipoles are dominant. Lu dipoles are an order of magnitude smaller, and whenever they are comparable in weight (at low energies) they are inefficient (large α) or have small amplitude. Therefore, one concludes that ionic IR screening in bixbyite is also oxygen-driven. The reduced ionic constant in a bixbyite vs. hex sesquioxide is due to a purely structural effect, as the different modes are due to the energy response of the structure (i.e., the vibrational modes). The independence on the cation will hold as long as the cation is not too light; the similarity of the IR spectra of yttria and lutetia in the O-related region indicates this to be a good approximation for any transition-metal or lanthanide cation. Another evidence is that the Raman spectrum of yttria and lutetia are again almost identical in the 300–400 cm^{-1} range, and only differ in some peaks at low frequency. We could identify the former with O-related displacements, and the latter with cation-related ones. The calculated Raman modes compares very favorably with the experimental spectrum [24].

A calculation for La_2O_3, which actually has the hex structure, confirms this conclusion. Once the somewhat larger cation Born charge of 4.2 (-2.8 for O) is taken account of, the IR spectrum is indeed lutetia-like, with an in-plane doublet at 200 cm^{-1} and an axial singlet at 220 cm^{-1}. The high-frequency component, 5.2, is not substantially larger for lutetia. As a consequence of the softer modes and larger charges, the final value of ϵ_0 is 22.5, not as larger than in hex lutetia as suggested by charge anomaly, due to the reduced dipole density. Similar results were obtained recently in [25]. Despite quantitative deviations, our inference is confirmed that hex-phase sesquioxides have a larger dielectric constant than their bixbyite counterpart. The details may depend on cation polarizability, but the main discriminant is structure. Indeed, this conclusion is unfortunate, as most Ln-sesquioxides are bixbyites: Dy and Yb oxides have dielectric constants of ∼11, in line with the expectations just outlined. As we mentioned earlier DFT may overestimate effective charge anomalies, but, while reducing slightly the differences related thereto, this would not affect the core of the argument.

12.2.3 Rare-Earth and Transition-Metal Aluminates

A quite different class of materials is that of Ln-aluminates. Useful informations can be extracted from a limited set of explicit studies; we concentrate here on lanthanum aluminate, $LaAlO_3$. The central point is that due to the smaller size of Al vs. all Ln, and its preference for sixfold coordination with O (despite the larger bond enthalpies of Ln–O bonds) the primary bonding in $LnAlO_3$ is Al–O, and Al coordination is octahedral. The large Ln ions fit well in the cubic-symmetry network interstices, and a perovskite result: most $LnAlO_3$'s are indeed distorted variants of perovskites with Al at the octahedral B-site (for the same reasons, even $LnScO_3$ seem to obey this rule to a large extent). While distortions affect the details of the dielectric response, the general picture does not quite change.

From the discussion below, a few general predictions on all Ln-aluminates will become apparent:

(a) In the crystal, the dominant IR modes have low frequency and correspond to vibrations of the Ln cation at the cubic A-site; thanks to the charge anomaly of the cation, this results in ionic constants of around 20 and total static values of 25; Al–O motions are high frequency and there is no Al charge anomaly (this should apply to Sc too in scandates); low-frequency (40 and 130 cm^{-1}) Raman modes are associated to rotational distortions in $LaAlO_3$: similar modes may appear in other Ln-aluminates in association with similar low-energy distortions; in some cases (such as yttrium aluminate, see below) they may indirectly cause a lowering of the IR energies.

(b) The simulated amorphous phase (see below) conserves Al–O short range order; Ln charge anomalies are reduced by about 10–15%; due to disorder the Ln IR vibrations in the cubic cage and the formerly Raman, but now weakly IR-active modes couple, and a large dielectric intensity at low (∼100 cm^{-1}) frequency appears, related to Ln vibrations. The resulting ionic component of the dielectric tensor is comparable to that of the crystal.

$LaAlO_3$ is observed [26] to be a perovskite with a small rotational rhombohedral distortion. Calculations [8] find no ferroelectric distortions. The main

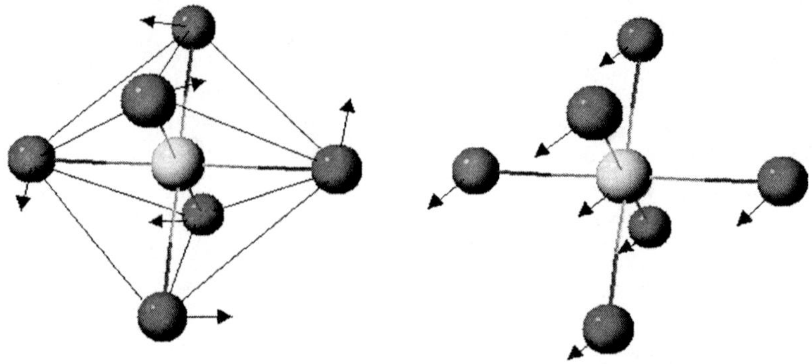

Fig. 12.4. The displacement patterns of the main Raman (A_{1g}, *left*) and IR (A_{2u}, *right*) modes (the latter is in a reference system where La atoms are at rest). Only one Al–O octahedron is shown for clarity

IR mode is at 167 cm^{-1} (80% of the intensity) in undistorted perovskite. A soft F_{2u} mode at the R point causes an instability of the cubic phase toward a rotational distortion producing a rhombohedral phase, whereby octahedra are rotated by ∼6° away from their ideal orientation, with a small energy gain (10 meV per formula).

Distortion is found to transform the soft mode into stable low-energy Raman modes E_g (33 cm^{-1}) and A_{1g} (129 cm^{-1}), closely agreeing with experiment [27], and associated rotations of the O octahedra around Al. The IR singlet A_{2u} at 168 cm^{-1} and doublet E_u at 179 cm^{-1} deriving from the IR triplet of the perovskite, provide 82% of the total dielectric intensity. This causes the rather high ionic dielectric constant of $LaAlO_3$, along with the charge anomaly on the La cation. The distortion-induced Raman doublet repels upward the IR doublet by approximately 10 cm^{-1}, giving rise to a uniaxial dielectric anisotropy.

In Fig. 12.4 we display the displacement patterns for the low-energy Raman A_{1g} and IR A_{2u} modes. The Raman mode is a rotation of the O–Al octahedron directly related to the rhombohedral rotational distortion. The dominant IR mode, displayed here in the reference system of La being immobile, is essentially a vibration of La atoms in the cubic cages against the Al–O octahedra backbone.

Neglecting for simplicity their axial anisotropy, the GGA-calculated average effective charges are $Z^*_{La} = 4.37$, $Z^*_{Al} = 2.98$, and $Z^*_{O} = -2.45$. The Al effective charges are strictly nonanomalous. The La charges are appreciably anomalous, and therefore such are the O ones. Unlike other perovskites (e.g., $BaTiO_3$), where the anomaly is due to the primary directional bonding between O and the octahedrally coordinated cation, here the same role is played by the hybridization and charge transfer between the Ln-cation (here La) and O, second-neighbors at about 2.7 Å. This is expected as the states involved are prevailingly La d and O p. The dominance of La anomaly and of La modes in the dielectric response are consistent with the high dieletric constant of 25 for $LaGaO_3$ [28], which has a closely similar structure. Given the similarity

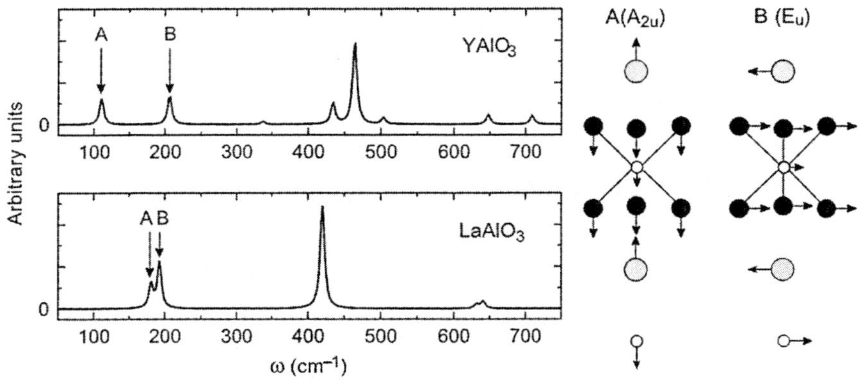

Fig. 12.5. IR spectrum of distorted-perovskite yttrium and lanthanum aluminates, and displacement patterns of the IR modes. Note the down shift of mode A in $YAlO_3$. Figure reproduced from [29], copyright 2005 American Physical Society

of Al and Ga in this context, this suggests that also Ln-gallates will share to some extent the properties of Ln-aluminates at issue here.

The average electronic dielectric tensor is again small and largely structure independent, $\epsilon_\infty = 4.77$. The static value is large, 25.5 in-plane and 28.4 axially. The ionic component is dominant, and determines the accuracy of the final static value. The inverse orientational average (12.9) is $\overline{\epsilon}_s = 26.4$ using the GGA-calculated values, 11% larger than experiment. We thus considered the effect of the pseudoself-interaction-correction [7] on effective charges, and the resulting changes in dielectric constant. As in [6], we find reduced average charges as compared to GGA: $Z^{*(\mathrm{SIC})}_{\mathrm{La}} = 4.06$, $Z^{*(\mathrm{SIC})}_{\mathrm{Al}} = 2.87$, $Z^{*(\mathrm{SIC})}_{\mathrm{O}} = -2.31$. As expected, the nonanomalous Al charge hardly changes, while the La charges decrease sizably (~8%), due to the increased localization, hence reduced polarizability, of hybridized La-d/O-p states. The smaller pseudo-SIC charges cause a 15% drop of the dielectric constant, entirely due to the ionic part. The orientational average is now $\overline{\kappa}_s = 23.3$, within 2% of the measured 23.7. This remarkable agreement suggests that DFT calculations supplemented by SIC in the case of strongly anomalous Born charges are predictive of the dielectric constant of high-κ oxides.

A recent study [29] of dielectric screening in $La_xY_{1-x}AlO_3$ alloys has brought out an interesting feature related to the topic of the present section: yttrium aluminate hypothetically crystallizing in the distorted perovskite structure of lanthanum aluminate would have a maximum axial value (along the (111) axis) of the static ϵ of 60! This is due to the much lower frequency of the axial A_{2u} singlet IR vibration in distorted-perovskite $YAlO_3$, about $100\,\mathrm{cm}^{-1}$ compared to that of $LaAlO_3$, $170\,\mathrm{cm}^{-1}$; the two spectra are compared in Fig. 12.5. A reason for this huge downshift may be that the repulsive interaction of the IR A_{2u} mode with the Raman A_{1g} mode pushes the former *downward*, rather than upward as in $LaAlO_3$. This is quite plausible as the Raman downfolded mode is at about the same energy as the IR mode of the undistorted perovskite, and that the partner IR doublet stays at about the same position as in $LaAlO_3$. Alloying with La [29] may stabilize the yttrium-containing aluminate in the "useful" structure, as discussed in Sect. 12.3.4.

12.3 Amorphous and Alloyed Systems: Silica, Aluminates, Silicates

Amorphous systems are interesting in this context because crystal layers are usually polycrystalline and tend to leak electrically due to grain boundary conduction, a problem not present in disordered layers. Alloys are relevant both because they occur unintentionally at or near the oxide–silicon interface (typically, silicates), and because alloying proportions can be used (at least in principles) to tune the properties of the mixture. Also, it appears that the dielectric properties of some high-κ amorphous systems present intriguing surprises, and are therefore of basic interest.

12.3.1 A Pioneering Study of Silica

The first-principles simulation of amorphous systems, and especially of subtle aspects such as dielectric and polarization properties, is a formidable task. If one had to mention a single key paper in this area, this would probably be Pasquarello and Car's [30] study of effective charges and IR spectra in amorphous silica. Besides its practical interest, and the agreement it reached with experimental IR data, this paper has provided two important messages. Firstly, it proved the feasibility of a realistic ab initio simulation of complex properties of complex materials; secondly, it proved that such simulations are *necessary* to achieve a detailed understanding of polar response (among others).

We recall here two main points. In Fig. 12.6a the effective charges of oxygens are reported vs. the Si–O–Si angle centered on those O atoms, and in panel b of the same figure, the Si average charge is compared with the charge neutralizing those in panel a. These data quantify the structurally induced changes in polarizabilities due to the local distortions of the tetrahedral network (panel a), and show that each tetrahedral unit in the network is locally neutral in the dynamical sense.

As a second important point, Fig. 12.7 reports the IR spectrum calculated with different approximations to the effective charge tensors: The message of this result is that while the magnitude of the effective charge only uniformly changes the absolute intensity of the various peaks, the *relative intensities* are correct only when the anisotropies of the charges are accounted for. This is of course essential for amorphous systems, but it is qualitatively relevant also for crystals with complex basis and vibrational patterns such as bixbyite lutetia, where the correct intensities can only be obtained [10] with the full anisotropic charge tensors.

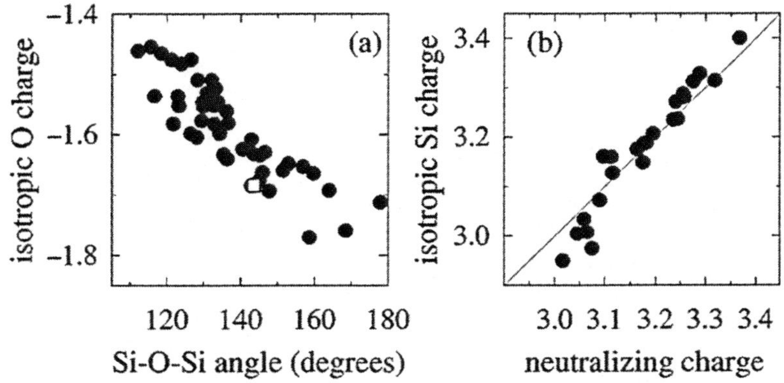

Fig. 12.6. (a) Isotropic component of the effective charge of O atoms in amorphous SiO_2 vs. the Si–O–Si angle centered on those specific atoms (*left*); (b) neutralizing charge vs. isotropic Si charge. Figure reproduced from [30], copyright 1997 American Physical Society

12.3.2 Amorphous Zirconia

Amorphous zirconia has been studied in detail by Zhao et al. [31], following a study of the crystal phases [4]. The samples were obtained by a melt-and-quench procedure using ab initio dynamics, and exhibit a typical 20% density decrease and a reduction in average coordination compared to the monoclinic crystal phase (cations sevenfold, oxygen threefold, and fourfold), visible in Fig. 12.8.

Figure 12.9 reports instead the phonon density of states (top), and the dielectric intensity (bottom), i.e., the DOS weighted by the ratio of scalar

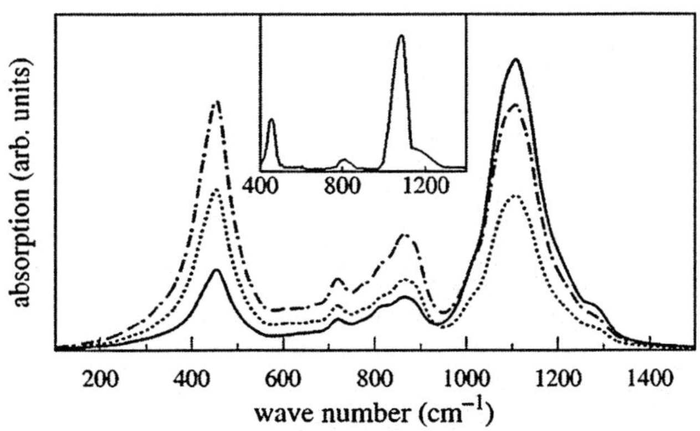

Fig. 12.7. IR spectrum of amorphous silica calculated using different approximations to effective charges (full: *solid*; isotropic: *dashed*; nominal point charges: *dash-dotted*). Figure reproduced from [30], copyright 1997 American Physical Society

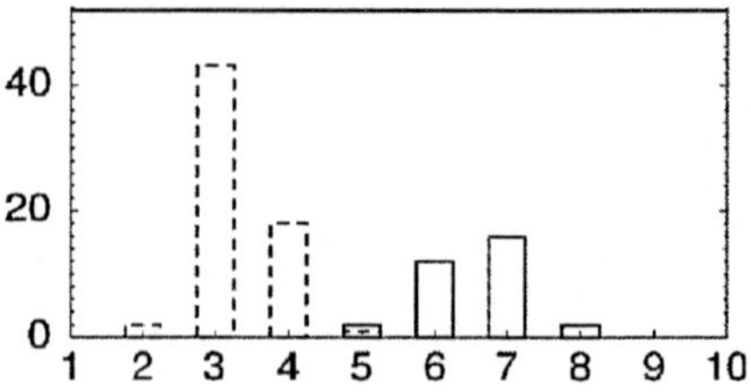

Fig. 12.8. Number of atoms vs. coordination for amorphous zirconia at 0 K (from data in [31]). *Dashed* profiles: O atoms; solid profiles: Zr atoms

effective mode charges to the square of the mode frequency. The scalar mode charge is essentially the projection of the dynamical charge tensors onto the mode eigenvectors (see (6)–(9) of [4]). The dielectric activity of specific atom classes (say, oxygen, threefold coordinated, and such) in the structure was investigated, but there appears to be no obvious dominance on the part of any single one of them.

The striking feature in Fig. 12.9 is that the dielectric activity is peaked in a quite lower frequency range than in the crystal. Since the effective charges are

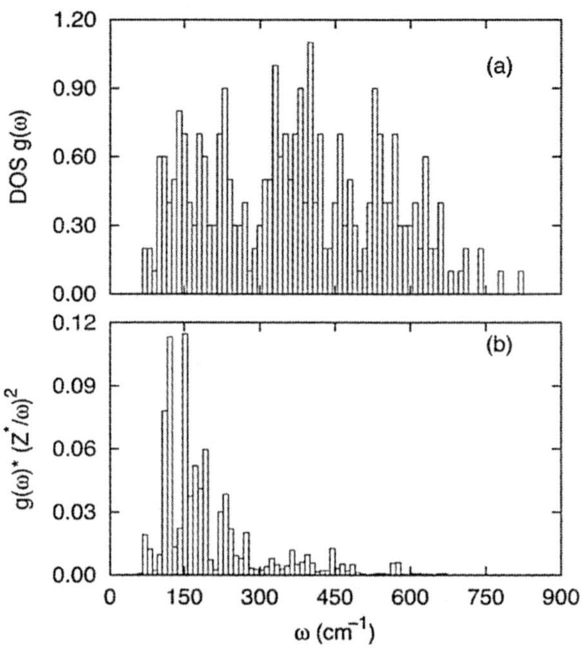

Fig. 12.9. Dielectric intensity in amorphous zirconia. Note the considerable weight at low energy compared to the IR spectrum in Fig. 12.1. This spectrum has also an intriguing similarity with the same spectrum of epitaxial zirconia on Si of Fig. 12.1, right panel. Figure reproduced from [31], copyright 2005 American Physical Society

found to be sizably anomalous, with values of about 5 for the cation (compared to 5.2–5.4 of the crystal phases), the total average dielectric constant is about 22, i.e., comparable to that of the monoclinic crystal (as mentioned earlier, in the tetragonal and cubic phases, the values are much larger). Experiments report values of 16–18, which can be considered a satisfactory agreement given the poor control of stoichiometry and microscopic structure in the observed layers. Amorphous zirconia, and by extension amorphous hafnia too (given the usual similarity between the two materials) seems therefore promising in the high-κ context.

12.3.3 Conservation of Permittivity in Amorphous Lanthanide Aluminates?

Epitaxially grown amorphous $LaAlO_3$ has been reported to have dielectric constants in the same range (20–24) as the crystalline phase [32], which is qualitatively surprising since the large effective charges coresponsible for the high-κ stem from a delicate ionicity–covalency balance that one expects to be disrupted in amorphous structures. It has been suggested that XRD measurements may, due to limited resolution, be observing as amorphous a system which is actually nanocrystalline. However, in view of the results discussed above for zirconia [31], it is also plausible that the charge reduction may be compensated by a downshift of the IR modes; we thus analyzed [33] the dielectric properties of model amorphous lanthanum aluminate constructed by ab initio melt-and-quench to verify the plausibility of this idea.

The main indication of radial distribution functions and bonding angle distributions (Fig. 12.10) on the structure of amorphous $LaAlO_3$ is that short-range Al–O order is preserved (with angles centered around 90° as in the crystal), but that La-octahedra symmetry relations are removed. Considering the O–La–O angle distribution, we see that the orientation of Al–O octahedra with respect to La is very disordered. We expect therefore that the main IR modes (La against Al–O octahedra backbone) will shift and mix with octahedra rotation modes – besides of course a reduction of the anomalous O–La dynamical charge transfer: The cation charges that were anomalous (4.35) in the crystal, are indeed reduced sizably (to 3.88). The electronic permittivity component is the same as in the crystal, and therefore the static value is once again determined by ionic screening.

The dielectric intensities of ionic origin shown in Fig. 12.11, left, along with their frequency integral giving the value of the ionic constant accumulated as a function of frequency, suggest that the hypothesis made earlier is correct. Allowing for effective charge reduction by SIC corrections, we end up with an estimate of the ionic part of around 20, and hence $\epsilon_0 = 24$–25, about the same as in the crystal. The dielectric intensity is not only broadened around the crystal IR modes due to disorder, but is also large at lower frequencies down to 50–100 cm^{-1}: this softening is related to disorder in the structure around La ions. The 50–100 cm^{-1} peaks may be due to disorder-induced mixing of

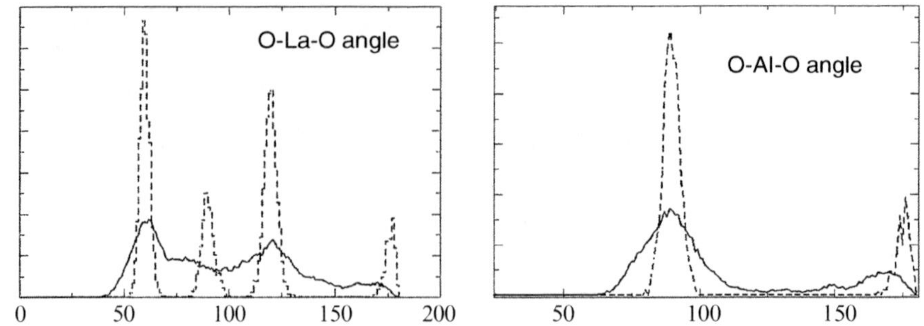

Fig. 12.10. Average O–La–O (*left*) and O–Al–O (*right*) angle distributions in amorphous (*solid*) and crystalline (*dashed*) $LaAlO_3$ collected in an ab initio dynamics run at 500 K. Similar results are obtained in both small and medium-sized samples

IR La translations (crystal frequency 160 cm^{-1}) and Raman Al–O octahedra rotations (40 and 130 cm^{-1}), and their large dielectric intensity is due to the inverse quadratic energy dependence.

Indeed, the mode displacement amplitudes vs. energy in Fig. 12.11, right, are large for La as in the crystal IR modes all the way from 50 to 200 cm^{-1}. In addition, they decrease for first low energy modes, with O amplitudes growing concurrently. While a quantitative evaluation of the weight of these low-frequency modes would require better statistics, this result would match the idea of a mixing of low energy, formerly Raman modes with IR La vibrations, and therefore supports the idea that the ionic (and therefore the total) ϵ, may

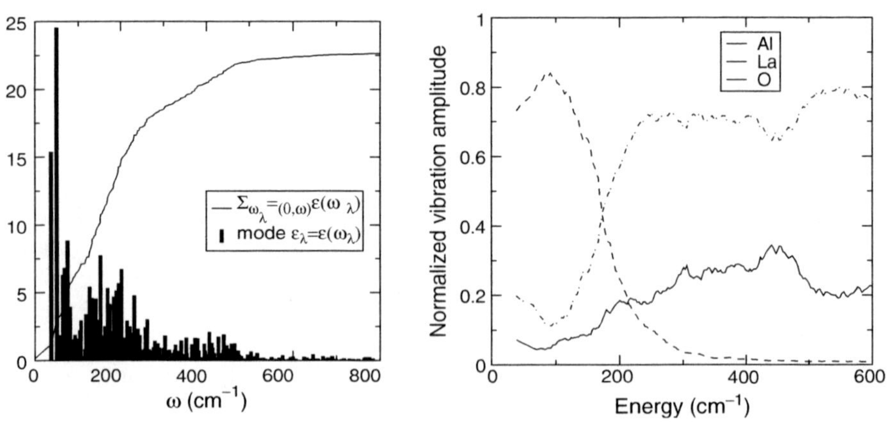

Fig. 12.11. *Left*: mode dielectric intensities and cumulant ionic dielectric constant in amorphous $LaAlO_3$. *Right*: Normalized displacements of the three species in the IR vibrational modes vs. frequency. Modes below 200 cm^{-1} are dominated by La displacements, hence related to the dominant crystal IR modes. Displacements increasing for O, decreasing for La, and constant for Al at low frequency are consistent with a mixing crystal IR and Raman modes

well be as large in the amorphous phase as in the crystal phase. This result echoes that on zirconia in Sect. 12.3.2 [31], although with entirely different details. As the central ingredient is the structure of $LaAlO_3$ as a octahedral backbone with rare-earth ion in the interstices, it is natural to suppose that Ln-aluminates with similar structure will behave similarly.

Recent unpublished work in our group on crystalline and amorphous $DyScO_3$ has confirmed the general picture obtained for $LaAlO_3$, and specifically that the dielectric constant is on average conserved upon amorphization. Further, investigations on the amorphous phase of the sesquioxides discussed in Sect. 12.2.2 have shown that the dielectric constant is *enhanced* by amorphization (about 20–22 vs. 15–16), due again to the enhancement of IR intensity of low-frequency cation-related modes. These results are overly interesting, as they apply to a large class of compounds, and because amorphous layers are preferable to polycrystals from the electrical viewpoint.

12.3.4 Dielectric Enhancement in Aluminate Alloys

We now come back to the study [29] on $La_xY_{1-x}AlO_3$ alloys mentioned earlier. With all its fantastic potential ϵ, distorted perovskite $YAlO_3$ is unstable. The solid solution with lanthanum aluminate may be envisaged such that the latter dictates the structure and $YAlO_3$ may provide an enhanced dielectric screening. The behavior of the static ϵ as a function of composition is shown in Fig. 12.12. The reported stability range of the solid solution is in the $x \sim 0.4$–0.5 composition range; the maximum axial component would then be somewhat higher than for lanthanum aluminate, and hence the enhancement might be exploited for 111-oriented single crystals. The average ϵ which would be observed in polycrystalline material, remains at the the $LaAlO_3$ level. This work [29] has nevertheless the merit of pointing out another way to exploit the peculiar and to some extent unexpected properties of this class of materials.

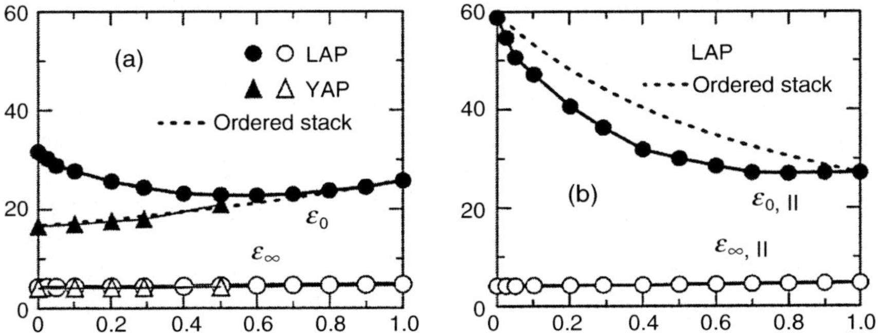

Fig. 12.12. Electronic and static dielectric constant of $La_xY_{1-x}AlO_3$ vs. x in various structures. *Left*: average values; *right*: maximum component). Figure adapted from [29], copyright 2005 American Physical Society

12.3.5 Models vs. Ab Initio Predictions in Transition-Metal Silicates

A frequent occurrence at electronic-grade oxide–silicon interfaces is the formation of silicates (an exception being of course silicon oxide itself). Zirconium and hafnium silicates have been studied in a standard fashion in their crystalline form (as discussed in the review by Rignanese [18]). For as regards disordered or amorphous silicates, the story is obviously much more complex. Clearly (besides structure generation which is a formidable problem even for amorphous systems with well-defined stoichiometry) a major drawback of ab initio approaches in this context is the essential impossibility of accumulating sufficient configurational statistics. Recently, a series of ab initio calculations and a structural and response model built thereupon [18, 34] have partially circumvented this problem, shedding light on the relation of local structure and coordination with dielectric behavior in disordered Zr silicates.

Nine crystal structures containing different amounts and local configurations of Zr–O and Si–O structural units, for a total of five different units (SiO_n, $n = 4{,}6$; ZrO_m, $m = 4{,}6{,}8$), are studied from first principles. From the nine calculated ϵ_∞'s and the Clausius–Mossotti relation, the effective polarizabilities of the five structural units are determined, and (assuming as reasonable their locality and additivity) they are plugged into a model expression to calculate the ϵ_∞ of the nine structures, and of an amorphous phase separately simulated (and not entering the determination of the polarizabilities). The agreement of the directly calculated and model values is excellent, and confirms the correctness of the local/additive hypothesis.

For the phonon, i.e., ionic component, no single local and additive quantity such as the electronic polarizability can be determined. Rignanese et al. [34] expresses the difference $\epsilon_0 - \epsilon_\infty$ as $\Delta\epsilon \propto \overline{Z}^2/\overline{C}$, with $\overline{Z}$ and $\overline{C}^{-1}$ the "characteristic" (i.e., average) dynamical charges and force constants, which can be calculated with the ab initio ingredients already handy. Although additivity is not guaranteed to hold, similarly to the electronic part the nine calculated $\overline{Z}$ and $\overline{C}^{-1}$ are decomposed in a sum of five (as many as the structural units) stoichiometry-weighted components. Analyzing these parameters the major role of ZrO_6 units emerges as the main feature (similarly to amorphous zirconia [31], as mentioned earlier). These quantities are then used to calculate again the $\Delta\epsilon$. The agreement with ab initio data here is less striking that for the electronic part, but still overall very good, as shown by Fig. 12.13a (left). The locality of $\overline{C}$ is related to local dynamical charge neutrality (see also [30]) and the quality of one affects that of the other: In Fig. 12.13b, right, the anionic and cationic charges for each structural unit are compared; it can be seen that those very few structures that give poor results for $\Delta\epsilon$ are indeed those failing to satisfy local neutrality satisfactorily.

Finally, using the model data accumulated, the dielectric constant is calculated for the alloy $(ZrO_2)_x(SiO_2)_{1-x}$ up to $x = 0.5$. The predicted ϵ_∞ is in excellent agreement with experiment. Applying the scheme to the ionic

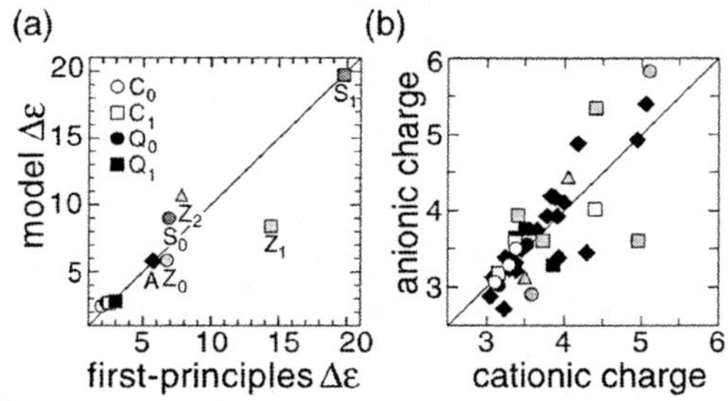

Fig. 12.13. *Left*: model vs. ab initio $\Delta\epsilon$ for the various structures studied. *Right*: assessment of local dynamical charge neutrality for the same structures. Figure reproduced from [34], copyright 2002 American Physical Society

component requires the determination of the local cation coordination, which has been found experimentally to rise from about 4 to about 8 with concentration. In Fig. 12.14, ϵ_0 is drawn as a band of values reflecting the uncertainties in the coordination to be plugged into the model: the upper limit is obtained with only ZrO_6 present, the lower limit with only ZrO_4. The agreement is overall good given the complexity of the system and the difficulties in extracting reliable thicknesses, and hence dielectric constants, of silicates layers in dielectric stacks.

12.4 Local Microscopic Screening in Ultrathin Films

Although they may be useful and illuminating, bulk calculations often do not agree very well with theory regarding epsilon and for this reason they are questioned especially considering that real structures are very thin films, not bulk. Addressing the question, do ultrathin films behave bulk-like in term of their dielectric properties, is quite difficult. Besides the obvious considerations

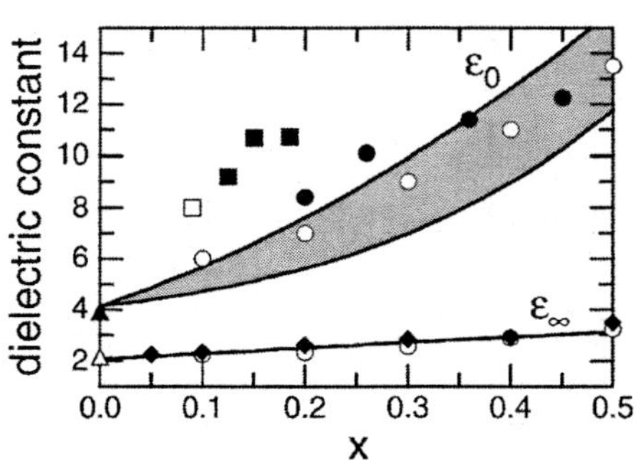

Fig. 12.14. Electronic and static dielectric constants for $Zr_xSi_{1-x}O$ alloys predicted by a model of dielectric response based on ab initio calculations for crystal structural units [34]. Figure reproduced from [34], copyright 2002 American Physical Society

that the selection or validation of a candidate dielectric must start from its bulk properties and that films used in microelectronics need not necessarily be extremely thin (e.g., dielectric stacks in FLASH memories), one may take the attitude that they do, on the basis of arguments related to the dominance of ionic screening, which depends on effective charges and phonon frequencies. First, modifications and energy shifts of infrared-active phonon modes due to "confinement" in a specific layer of a dielectric stack are expected to be modest (may be ∼10%). Second, as we discuss next, the microscopic polarization (and hence the dynamical charges) at Si/oxide interfaces is found to change over length scales comparable to bond lengths, so that oxide layers significantly exceeding that thickness (which they should exceed to qualify as films at all) are effectively bulk-like in dielectric terms. It turns out, in fact, that the real discriminating factor is the local chemical composition or grading.

The study by Giustino et al. [35], the only one so far dealing with these issues at the microscopic level, approaches the dielectric behavior of the "mother of all interfaces," i.e., Si/SiO_2, but it is nevertheless relevant to the high-k business. The interface model contains, interposed between Si and SiO_2 a thin suboxide interlayer in which Si atoms are not fully oxidized (i.e., less than fourfold ionized). The calculated dielectric constant of the full overlayer (suboxide plus oxide) is found to be larger than that of silica, and to decrease asymptotically with layer thickness to the silica value (at about 10–12 Å from the nominal interface). This behavior is then rationalized by a classical electrostatic model with three dielectric layers (Si, suboxide, silica) which reproduces accurately the ab initio results: The idea is that the change in dielectric screening between Si and silica, which is relatively abrupt, occurs essentially all within the suboxide region, in which screening is stronger (more Si-like) than in silica.

The reason why this model works at all is shown in Fig. 12.15, left panel. The high frequency and ionic polarization (obtained from the induced charge density in a finite field using a technique developed by the authors [36]) change according to the progressive oxidation of Si atoms in the dielectric transition layer (shaded area): It is the thickness and the chemical (i.e., oxidation, in this case) grading of this layer that determines the rate of variation of the polarization. For this specific interface, the polarization changes to that of silica over just about 2.5 Å, and its value in the interlayer is well described by the value used by the classical model (of course some care is needed in defining properly the thicknesses of the layers involved).

Finally, the change in polarization can be further rationalized by defining the effective polarizabilities of appropriately chosen polarizable structural units. This polarizability is defined via the modern theory of polarization [37] in terms of Wannier functions; as these are associated to bonds or nonbonding orbitals, it is straightforward to attach a dipole moment to each structural unit. Again one finds that polarizability changes drastically with the ionization state of the Si atom in the specific unit, and therefore is determined by the oxidation degree, and hence chemical grading in the suboxide.

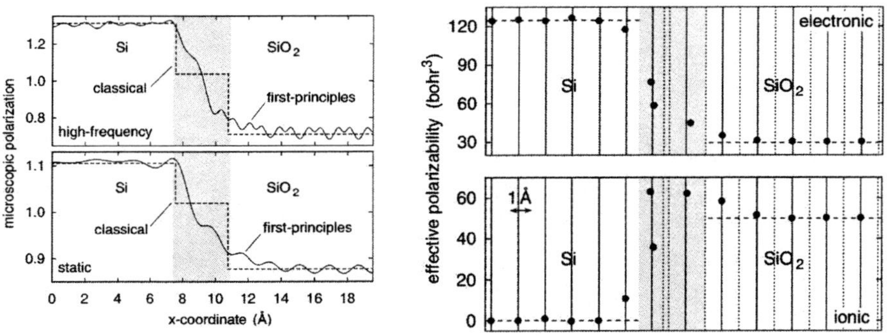

Fig. 12.15. *Left*: microscopic polarization in the vicinity of an Si SiO_2 junction calculated ab initio and with a classical electrostatic model. *Right*: effective electronic (*top panel*) and ionic (*bottom panel*) polarizability in the vicinity of an Si SiO_2 junction calculated ab initio using a Wannier function decomposition. Figure reproduced from [35], copyright 2003 American Physical Society

As can be appreciated from Fig. 12.15, right panel, both the electronic and the ionic polarization change strongly in the transition layer. The former decreases drastically (a factor 4) from Si to silica, while the former rises from zero in Si to a large value in silica (about a half of the electronic polarization in Si). The authors also show that interface-induced gap states in Si (analogous to those coming into play at metal–semiconductor interfaces) contribute just 10% of the polarization in the interface region, and are therefore secondary. Of course, the main practical message of this work is that the effective dielectric constant of a thin layer depends on chemical composition and grading (which indeed could be regarded as obvious in retrospective!), and that thin silicon oxide interlayers in high-κ stacks, due to their likely suboxide nature, may well have dielectric constants larger than usually assumed. This would of course affect experimental estimates of dielectric constants by capacitance methods.

12.5 Conclusions

We have summarized some recent work on a few classes of oxides in focus as dielectrics for microelectronics. The main message of the present work is that rather general conclusions about wide classes of materials (e.g., Ln-sesquioxides and Ln-aluminates, transition-metal silicates) can be drawn from a limited, although demanding, set of calculations. While valuable work already exists in this area, more effort will be needed in the direction of characterizing microscopically the transition layers in dielectric stacks on Si to obtain predictive-level results useful in technology. Actual layers are now in the 10-nm thickness regime, so that a direct description thereof may be – for once – not too far from the system size affordable in simulation. While work has progressed since the completion of this manuscript (late 2005), only a minor addendum has been included (end of Sect. 12.3.3).

Calculations by the present authors were done on the SLACS-HPC cluster `ichnusa` at CASPUR Rome, and at CINECA Bologna. A. Filippetti thank the Ministry of Research for a "Cervelli per la ricerca" grant. Fiorentini thanks Dimoulas for the invitation to write this review. This work was supported in part by MIUR through PON-CyberSar project and by Fondazione BdS.

References

1. Alternative Gate Dielectrics for Microelectronics, MRS Bulletin Vol. 27, edited by G. Wilk and R.M. Wallace (Material Research Society, Warrendale, 2002)
2. G.D. Wilk, R.M. Wallace, and J.M. Anthony, High-κ gate dielectrics: Current status and materials properties considerations, J. Appl. Phys. **89**, 5243 (2001)
3. X. Zhao and D. Vanderbilt, Phonons and lattice dielectric properties of zirconia, Phys. Rev. B **65**, 075105 (2002)
4. X. Zhao and D. Vanderbilt, First-principles study of structural, vibrational, and lattice dielectric properties of hafnium oxide, Phys. Rev. B **65**, 233106 (2002)
5. F. Bernardini, V. Fiorentini, and D. Vanderbilt, Polarization-based calculation of the dielectric tensor of polar crystals, Phys. Rev. Lett. **79**, 3958 (1997)
6. A. Filippetti and N.A. Spaldin, Strong-correlation effects in Born effective charges, Phys. Rev. B **68**, 045111 (2003)
7. A. Filippetti and N.A. Spaldin, Self-interaction-corrected pseudopotential scheme for magnetic and strongly-correlated systems, Phys. Rev. B **67**, 125109 (2003)
8. P. Delugas, V. Fiorentini, and A. Filippetti, Dielectric properties and long-wavelength optical modes of the high-κ oxide $LaAlO_3$, Phys. Rev. B **71**, 134302 (2005)
9. X. Gonze and C. Lee, Dynamical matrices, Born effective charges, dielectric permittivity tensors, and interatomic force constants from density-functional perturbation theory, Phys. Rev. B **55**, 10355 (1997)
10. E. Bonera, G. Scarel, M. Fanciulli, P. Delugas, and V. Fiorentini, Dielectric properties of high-κ oxides: theory and experiment for Lu_2O_3, Phys. Rev. Lett. **94**, 027602 (2005)
11. S. Baroni, P. Giannozzi, and A. Testa, Green's-function approach to linear response in solids, Phys. Rev. Lett. **58**, 1861, (1987); see also S. Baroni, S. de Gironcoli, A. Dal Corso, and P. Giannozzi, Phonons and related crystal properties from density-functional perturbation theory, Rev. Mod. Phys. **73**, 516 (2001)
12. X. Gonze, Adiabatic density-functional perturbation theory, Phys. Rev. A **52**, 1096 (1995)
13. X. Gonze, First-principles responses of solids to atomic displacements and homogeneous electric fields: Implementation of a conjugate-gradient algorithm, Phys. Rev. B **55**, 10337 (1997)
14. W. Pickett, Pseudopotential methods in condensed matter applications, Comput. Phys. Rep. **9**, 115 (1989)

15. The abinit code a common project of Universitè Catholique de Louvain, Corning Incorporated, and other contibutors. Available from: http://www.abinit.org; X. Gonze, J.-M. Beuken, R. Caracas, F. Detraux, M. Fuchs, G.-M. Rignanese, L. Sindic, M. Verstaete, G. Zerah, F. Jollet, M. Torrent, A. Roy, M. Mikami, Ph Ghosez, J. -Y. Raty, and D.C. Allan, First-principles computation of material properties: the ABINIT software project, Comp. Mater. Sci. **25**, 478, (2002)
16. S. Baroni, A. Dal Corso, S. de Gironcoli, P. Giannozzi, C. Cavazzoni, G. Ballabio, S. Scandolo, G. Chiarotti, P. Focher, A. Pasquarello, K. Laasonen, A. Trave, R. Car, N. Marzari, and A. Kokalj, http://www.pwscf.org/
17. G. Kresse and J. Furthmüller, Efficient iterative schemes for ab initio total-energy calculations using a plane-wave basis set, Phys. Rev. B **54**, 11169 (1996); http://cms.mpi.univie.ac.at/vasp/
18. G.-M. Rignanese, Dielectric properties of crystalline and amorphous transition metal oxides and silicates as potential high-κ candidates: the contribution of density-functional theory, J. Phys.: Condens. Matter **17** 357, (2005)
19. V. Fiorentini and G. Gulleri, Theoretical evaluation of zirconia and hafnia as gate oxides for Si microelectronics, Phys. Rev. Lett. **89**, 266101 (2002)
20. D. Bloor and J.R. Dean, Spectroscopy of rare earth oxide systems. I. Far infrared spectra of the rare earth sesquioxides, cerium dioxide and nonstoichiometric praseodymium and terbium oxides, J. Phys. C: Solid State Phys. **5**, 1237 (1972)
21. N. Hirosaki, S. Ogata, and C. Kocer, Ab initio calculation of the crystal structure of the lanthanide Ln_2O_3 sesquioxides, J. Alloys and Comp. **35**, 31–34 (2003); C. Kocer, private communication
22. L. Marsella and V. Fiorentini, Structure and stability of transition-metal and rare-earth oxides, Phys. Rev. B **69**, 172103 (2004)
23. G. Scarel, E. Bonera, C. Wiemer, G. Tallarida, S. Spiga, M. Fanciulli, I.L. Fedushkin, H. Schumann, Yu. Lebedinskii, and A. Zenkevich, Atomic-layer deposition of Lu_2O_3, Appl. Phys. Lett. **85**, 630 (2004)
24. L. Laversenne, Y. Guyot, C. Goutaudier, M.T. Cohen-Adad, and G. Boulon, Optimization of spectroscopic properties of Yb^{3+}-doped refractory sesquioxides: cubic $Y2O_3$, Lu_2O_3, and monoclinic Gd_2O_3, Opt. Mater. **16**, 475 (2001)
25. R. Vali and S.M. Hosseini, First-principles study of structural, dynamical, and dielectric properties of A-La_2O_3, Comp. Mat. Sci. **31**, 125 (2004)
26. D. du Boulay, *Structure, Vibration and Electron Density in Neodymium-Iron-Boride and some Rare-Earth Perovskite Oxides*, PhD Thesis, University of Western Australia (1999)
27. J.F. Scott, Raman study of trigonal-cubic phase transitions in rare-earth aluminates Phys. Rev. **183**, 823 (1969)
28. R. L. Sandstrom, E. A. Giess, W.J. Gallagher, A. Segmller, E.I. Cooper, M.F. Chisholm, A. Gupta, S. Shinde, and R.B. Laibowitz, Lanthanum gallate substrates for epitaxial high-temperature superconducting thin films, Appl. Phys. Lett. **53**, 1874 (1988)
29. S. Shevlin, A. Curioni, and W. Andreoni, Ab initio design of high-κ dielectrics: $La_xY_{1-x}AlO_3$, Phys. Rev. Lett. **94**, 146401 (2005)
30. A. Pasquarello and R. Car, Dynamical charge tensor and infrared spectrum of amorphous SiO_2, Phys. Rev. Lett. **79**, 1766–1769 (1997)

31. X. Zhao, D. Ceresoli, and D. Vanderbilt, Structural, electronic, and dielectric properties of amorphous ZrO_2 from ab initio molecular dynamics, Phys. Rev. B **71**, 0085107 (2005)
32. B.E. Park and H. Ishiwara, Formation of LaAlO3 films on Si(100) substrates using molecular beam deposition, Appl. Phys. Lett. **82**, 1197 (2003); L.F. Edge, D.G. Schlom, S.A. Chambers, E. Cicerrella, J.L. Freeouf, B. Holländer, and J. Schubert, Meas urement of the band offsets between amorphous LaAlO3 and silicon, Appl. Phys. Lett. **84**, 726 (2004)
33. P. Delugas and V. Fiorentini, unpublished
34. G.-M. Rignanese, F. Detraux, X. Gonze, A. Bongiorno, and A. Pasquarello, Phys. Rev. Lett **89**, 177601 (2002)
35. F. Giustino, P. Umari, and A. Pasquarello, Dielectric discontinuity at interfaces in the atomic-scale limit: permittivity of ultrathin oxide films on silicon, Phys. Rev. Lett **91** 267601 (2003)
36. P. Umari and A. Pasquarello, Ab initio molecular dynamics in a finite homogeneous electric field, Phys. Rev. Lett. **89**, 157602 (2002)
37. R.D. King-Smith and D. Vanderbilt, Phys. Rev. B **47**, 1651 (1993); R. Resta, Ferroelectrics **136**, 51 (1992); R. Resta, Rev. Mod. Phys. **66**, 899 (1994)

13

Germanium Nanodevices and Technology

C.O. Chui and K.C. Saraswat

Summary. It is believed that below the 65 nm node although conventional bulk CMOS can be scaled, it will be without appreciable performance gains. To continue the scaling of Si CMOS in the sub-65 nm regime, innovative device structures and new materials have to be created in order to continue the historic progress in information processing and transmission. One such promising channel material is Ge due to its higher source injection velocity. However, the lack of a sufficiently stable gate dielectric and prior knowledge of doping Ge challenged the demonstration of a MOSFET device. In this chapter, we review various advanced Ge MOS device technologies on nanoscale gate stack, shallow junction, and low thermal budget integration process, which together have enabled functional metal-gated Ge MOSFETs with high-κ dielectric for the first time.

13.1 Introduction

For over three decades, there has been a quadrupling of transistor density and a doubling of electrical performance every 2–3 years. Silicon (Si) transistor technology, in particular CMOS has played a pivotal role in this. It is believed that continued scaling will take the industry down to the 35-nm technology node, at the limit of the "long-term" range of the International Technology Roadmap for Semiconductors (ITRS) [1]. However, it is also well accepted that this long-term range of the 65-nm to 35-nm nodes remains solidly in the "no-known solution" category. The difficulty in scaling the conventional MOSFET makes it prudent to search for alternative device structures. This will require new structural, material and fabrication technology solutions that are generally compatible with current and forecasted installed Si manufacturing.

13.2 Challenges to Scaling Conventional CMOS

It is believed that continued scaling of conventional bulk CMOS will take the industry down to the 65-nm technology node. It is also well accepted that

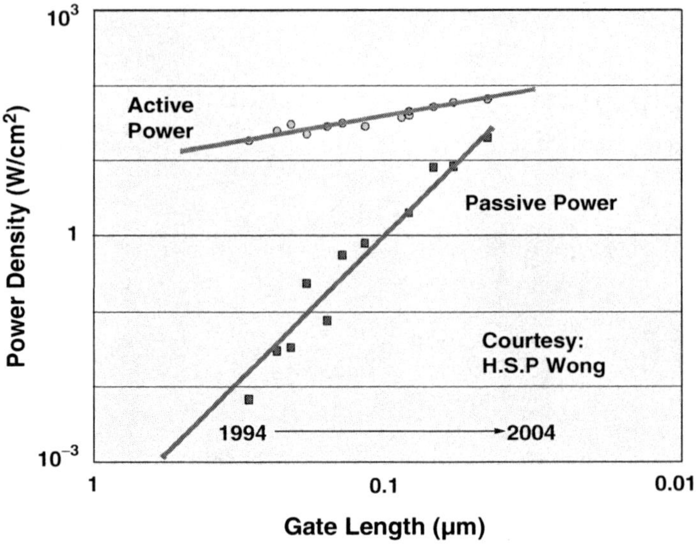

Fig. 13.1. Active and standby power density trends plotted from industry data. The extrapolations indicate a cross-over below 20 nm gate length. As devices scale towards that point, it is questionable if the traditional approaches and reasons for scaling will still be valid (Courtesy H.S.P. Wong, Standford Univ.)

below the 65 nm node although the conventional bulk CMOS can be scaled, it will be without appreciable performance gains. New materials and/or structures will certainly be needed to supplement or even replace the conventional bulk MOSFET in future technology generations [1].

The enhanced speed and complexity of IC chips has been accompanied by an increase in power dissipation [2]. Figure 13.1 depicts the evolution of power density as the gate length is scaled. The active power arises due to the dissipative switching of charge between the transistor gates and supply/ground terminals during logic operations. The subthreshold power, also known as static or standby power, is dissipated even in the absence of any switching operation. It arises due to the fact that the MOS transistor is not a perfect switch – there is some leakage current that flows through it in the off-state.

While the active power density has steadily increased with gate length scaling, the static power density has grown at a much faster rate [2]. The latter was a relatively insignificant component of power just a few generations back, but it is now comparable in magnitude to the active power. Management and suppression of static power is one of the major challenges to continued gate length reduction for higher switching speed. Traditional MOSFET scaling has begun to face impediments of both a fundamental as well as practical nature. It is now widely accepted that novel (i.e., non-classical) transistors will be needed to prolong device scaling with commensurate improvements in performance. The double-gate (DG) FET in conjunction with a high mobility channel and high-κ gate dielectric is a promising device structure that can potentially replace conventional transistors in future technology generations.

13.3 Why High Mobility Channel?

The saturation of bulk Si MOSFET drive current (I_{Dsat}) upon dimension shrinkage is limiting the prospect of future scaling. To understand this saturation phenomenon, numerous theoretical and experimental analyses were carried out [3–5]. First of all, the I_{Dsat} (and transconductance) in very short-channel MOSFETs is believed to be limited by carrier injection from the source into the channel [3]. In order words, the source injection velocity (v_{src}) saturates during scaling and that its limit is set by the thermal injection velocity (v_{inj}) [4]. Also, the carrier density at the top of the source to channel barrier is fixed by MOS electrostatics and the scattering in a short region near the beginning of the channel limits the I_{Dsat}. In deeply scaled MOSFETs, v_{src} was experimentally shown to be at most 40% of v_{inj} [5].

The lower effective mass and lower valley degeneracy of high mobility materials like germanium (Ge) [6] could alleviate the problem by providing a higher v_{inj}, which translates into higher drive current and smaller gate delay:

$$I_{\mathrm{DS}} = W \times Q_{\mathrm{inv}} \times v_{\mathrm{inj}} \quad , \tag{13.1}$$

$$\frac{C_{\mathrm{LOAD}} V_{\mathrm{DD}}}{I_{\mathrm{DS}}} = \frac{L_{\mathrm{gate}} \times V_{\mathrm{DD}}}{(V_{\mathrm{DD}} - V_{\mathrm{T}}) \times v_{\mathrm{inj}}} \quad , \tag{13.2}$$

where W is the channel width, Q_{inv} is the inversion charge, C_{LOAD} is the load capacitance, V_{DD} is the supply voltage, L_{gate} is the gate length, and V_{T} is the threshold voltage.

13.4 Which High Mobility Channel Material?

Table 13.1 shows properties of several high mobility materials which are possible candidates for the channel of the nanoscale MOSFETs. Due to their small Γ-valley electron mass, III–V materials like GaAs, InAs and InSb are being investigated as high mobility channel materials for high performance NMOS [8]. Under ballistic conditions, the main advantage of a semiconductor with a small transport mass is its high injection velocity. However, these materials also have a very low density of states in the Γ-valley, which tends to greatly reduce the inversion charge and hence reduce drive current. Further, the very high mobility III–V materials like InAs and InSb, have a much smaller direct band gap which gives rise to high band to band tunneling (BTBT) leakage. Materials like InAs and InSb have a high dielectric constant and hence are more prone to short-channel effects. All the III–V materials have a severe problem in surface passivation and hence fabrication of a MOSFET is problematic. High performance HEMT and MODFETs have been demonstrated

Table 13.1. Properties of high mobility semiconductors [7]

Material property	Si	Ge	GaAs	InAs	InP	InSb
Electron mobility ($cm^2\,V^{-1}\,s^{-1}$)	1,400	3,900	8,500	40,000	5,400	77,000
Hole mobility ($cm^2\,V^{-1}\,s^{-1}$)	450	1,900	400	500	200	850
Bandgap (eV)	1.12	0.66	1.42	0.35	1.34	0.17
Lattice constant (Å)	5.431	5.658	5.653	6.058	5.869	6.749
Dielectric constant	11.7	16.2	12.9	15.2	12.5	16.8

in these materials but suffer from high gate leakage and hence are not too useful for conventional logic applications.

Belonging to the same group in the Periodic Table as Si, Ge offers several attractive physical properties over Si. In Ge, the lower electron transverse and light hole (and heavy) effective masses are primarily responsible, respectively, for the higher electron and hole drift mobility. This property is the most advantageous over Si for deeply scaled MOSFET applications as previously discussed regardless of the higher Si saturation velocity. The more symmetric electron and hole mobility in Ge would not only reduce the real estate of p-MOSFETs, but would permit more CMOS logic gates as well. Moreover, its smaller bandgap is more compliant with the supply voltage scaling as specified in ITRS [1]; at the same time, this also broadens the optical absorption spectrum to cover telecommunication wavelengths (1.3 and 1.55 μm) allowing optoelectronic integration [9] to enhance CMOS functionality. Furthermore, its lower melting point reflects a possibility to fabricate Ge MOSFETs with much lower thermal budget processes while relaxing some stringent thermal stability requirements in integrating novel materials like metal gate electrode and high-κ dielectric into advanced transistors.

A combination of high mobility channel like Ge to enhance the transport, a double gate structure (Fig. 13.2) to address the problem of poor electrostatics in the channel region, and high-κ gate dielectric to minimize the gate leakage appears to be a very promising solution. Full band Monte Carlo

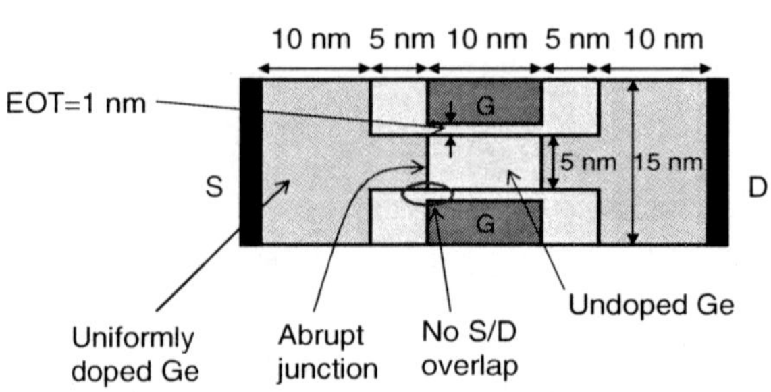

Fig. 13.2. The double-gate MOSFET structure used for Monte Carlo device simulation in both Ge and Si [10]

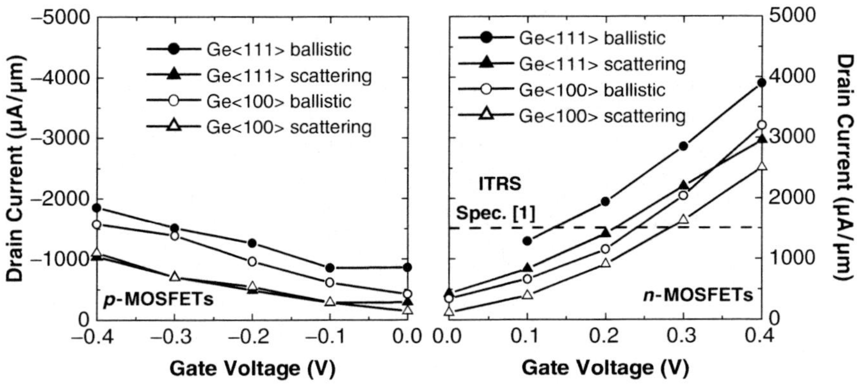

Fig. 13.3. Full band Monte Carlo (DAMOCLES^{TM}) simulation [10] of the transfer characteristics at $V_{DS} = V_{DD}$ on the double-gate Ge MOSFET (shown in Fig. 13.2)

simulations were performed (Fig. 13.3) on such a device structure, which exhibited a higher ballistic transport limit (I_{Dsat}) and ballistic ratio (Table 13.2) on the Ge channel vs. the Si counterpart.

Nonetheless, surface passivation for gate dielectric and field isolation, and insufficient understanding of dopant incorporation are the two classic problems that obstruct CMOS device realization in Ge. Furthermore, for Ge to become mainstream, heterogeneous integration of crystalline Ge layers on Si must be achieved. In this review, we present various advanced Ge MOSFET technology on heteroepitaxial Ge growth on Si, nanoscale high-κ gate dielectrics as well as shallow source/drain junctions. In addition, we have demonstrated functional metal-gated Ge MOSFETs using either a conventional or a simple self-aligned gate-last process.

13.5 Heteroepitaxial Ge Growth on Si

One of the most difficult and continuing research challenges in the semiconductor industry is the ability to grow high quality films using lattice mismatched materials, called *heteroepitaxy*. For Ge to become a viable candidate to augment Si for CMOS device and optoelectronic applications, it is essential to develop new methods for heteroepitaxial growth of Ge on Si. This is not straightforward because of the large lattice mismatch (4%)

Table 13.2. Ballistic ratios on different channel [10]

	Si⟨ 100 ⟩	Ge⟨100⟩	Ge⟨111⟩
n-MOS	0.68	0.78	0.76
p-MOS	0.48	0.70	0.56

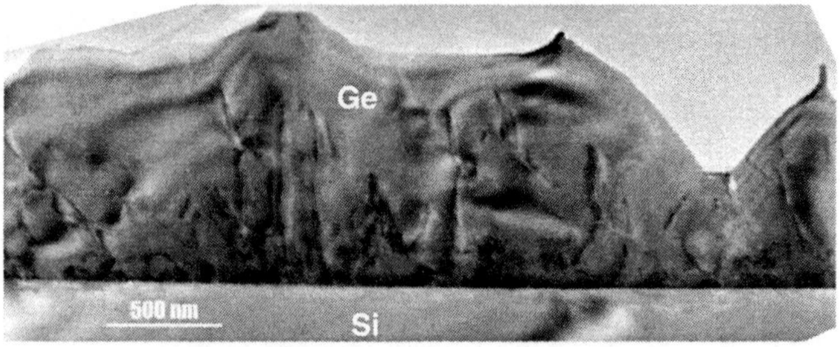

Fig. 13.4. Cross-sectional TEM image of a 1.6 μm Ge layer grown directly on silicon. Ge "islands" and misfit dislocations dominate the growth

between Ge and Si, which limits the quality of the heteroepitaxial growth. Below a critical growth thickness, the lattice-mismatch between Ge and Si causes the grown film to succumb to the lattice structure of the underlying Si substrate and hence strain the layer. However, above the critical thickness, it is energetically favorable for the layer to create dislocations to relieve this strain. Therefore, above the critical thickness, the layer will have many misfit dislocations making it unusable for any practical applications. Second, the growth of Ge on Si results in island morphology, or the so-called "Stranski–Krastanow" (S–K) growth. Such growth is associated with large surface roughness, causing difficulties in process integration, such as bonding for Ge-on-insulator (GOI). This can lead to degradation in device properties. An example of direct growth of Ge on Si is shown in Fig. 13.4. Misfit dislocations are formed at the substrate/film interface and typically terminate at the film surface as threading dislocations, thus degrading device performance.

Historically, heteroepitaxy of germanium on silicon has been a challenge. Many novel ideas and techniques have been introduced to grow high quality Ge layers. These layers have resulted in threading dislocation densities in the range of a $1\times10^7\,\text{cm}^{-2}$ to $1\times10^9\,\text{cm}^{-2}$. Bean et al. at Bell Labs studied pseudomorphic growth of Ge_xSi_{1-x} on Si using molecular beam epitaxy (MBE) [11] and developed comprehensive information on critical thickness of the grown film as a function of Ge content and growth temperature. Eaglesham et al. studied low temperature heteroepitaxial MBE growth of Ge on Si (100) [12]. They showed that growth is planar for all temperatures below 300°C and can be used near 200°C to suppress island formation. Currie et al. demonstrated a method of controlling threading dislocation densities in Ge on Si involving graded SiGe layers and chemical–mechanical polishing (CMP) [13]. This method allowed them to grow a relaxed graded buffer to 100% Ge without the increase in threading dislocation density normally observed in thick graded structures. Luan et al. demonstrated high-quality Ge epitaxial layers on Si with low threading-dislocation densities achieved by two-step ultrahigh vacuum chemical-vapor-deposition (UHVCVD) process followed by cyclical thermal annealing [14].

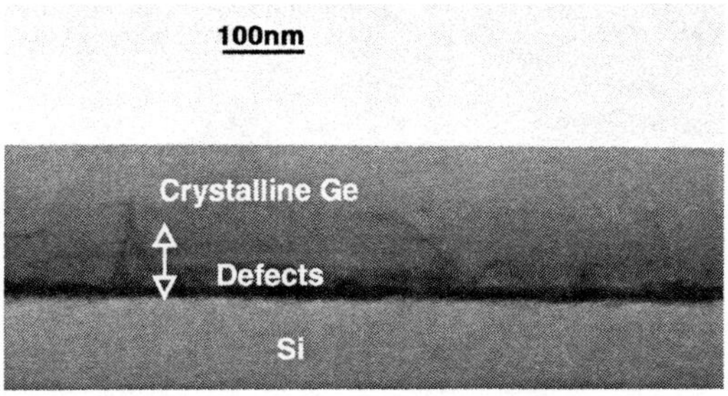

Fig. 13.5. Cross-sectional TEM image of a heteroepitaxial-Ge layer on Si grown by the MHAH method [16]

Nayfeh et al. developed a novel technique to achieve high quality heteroepitaxial Ge layers on Si [15, 16]. The technique involves CVD growth of Ge on Si, followed by in situ hydrogen (H_2) annealing with subsequent growth and anneal steps and hence the name Multiple Hydrogen Annealing for Heteroepitaxy (MHAH). A thin layer (~200 nm) of Ge is grown by CVD at 400°C, followed by a H_2 bake at 825°C for 1 h. After that a second layer of Ge is grown at 400°C for 15 min followed by a H_2 bake at 825°C for 1 h yielding a 400 nm layer. Following the first Ge growth and hydrogen anneal, the Ge surface roughness from "islanding" is reduced by 90%. As the layer is annealed at a high temperature in H_2, H is continuously adsorbed and desorbed on the surface. Ge and H atoms are highly mobile and the Ge–H bond is unstable. Due to the high temperature, Ge diffuses from top of the valley to the crest due to the curvature of the surface and the system's least energy state. As a result, the Ge surface flattens to a smooth layer. Hydrogen drives the Ge diffusion by disallowing any surface oxide from forming on the surface. Use of hydrogen annealing to reduce the surface roughness was first demonstrated by Sato and Yonehara [17] where they were able to reduce the surface roughness of epitaxial Si grown on anodized Si by hydrogen annealing. At the same time during the H_2 annealing, threading dislocations glide away from the surface and can be annihilated or they can terminate at the Ge/Si interface thus reducing the dislocation density at the surface. An additional CVD Ge layer is grown on the low roughness Ge layer followed by the final hydrogen annealing to further reduce the density of dislocations. From cross-sectional transmission electron microscopy (TEM), misfit dislocations are confined to the Ge/Si interface or bend parallel to this interface, rather than threading to the surface as expected in this 4% lattice mismatched heteroepitaxial system (Fig. 13.5). A threading dislocations density as low as $1 \times 10^7\,\mathrm{cm}^{-2}$ was achieved by using the MHAH method. The resulting Ge layer is a low-defect metamorphic single crystal. Atomic force microscopy (AFM) indicates the final layer surface roughness is reduced to device quality, while X-ray diffraction (XRD) confirms the Ge layer is fully relaxed and single crystal. This heteroepitaxial growth technique, MHAH, can be used to fabricate Ge based MOS devices, GOI substrates, or for the eventual integration of GaAs/Ge/Si for optoelectronics.

13.6 Nanoscale Gate Stacks on Germanium

The native insulators on the Ge surface are of poor quality that either desorbs at low temperature or dissolves in water. Over the last four decades, a variety of grown and deposited approaches have been suggested for Ge MOS dielectrics. Thermally grown GeO_2 and GeO_xN_y, and deposited SiO_2, Si_3N_4, GeO_2, Ge_3N_4, Al_2O_3, and AlP_xO_y were experimentally attempted. Nevertheless, none of them seems promising in the nanoscale regime.

13.6.1 Grown Germanium Oxynitride Dielectrics

The electrical properties and scalability of GeO_xN_y have been investigated for MOS applications [18, 19]. Tungsten-gated GeO_xN_y MOS capacitors were fabricated using rapid thermal oxidation followed by nitridation in NH_3 (RTN) [19]. Typical C–V characteristics are shown in Fig. 13.6. Well-behaved MOS capacitors were obtained with minimal dielectric/substrate interfacial charge trapping and frequency dispersion. The kinks that showed up near inversion in lower frequency scans suggest the presence of slow interface states. Additionally, this GeO_xN_y dielectric was shown to be scalable down to an EOT of 1.9 nm without suffering from gate leakage-induced C–V distortion.

13.6.2 Deposited High-Permittivity Dielectrics

In order to scale the EOT to below 1.0 nm, deposited high-κ dielectrics should be considered as inspired by the experience on Si. Favored by the thermodynamically unstable nature of Ge oxides, an interlayer-free high-κ dielectric stack on Ge could plausibly be achieved. Merging both the thermodynamic stability and material accessibility criteria, ZrO_2 and HfO_2 were employed in our study. From the dielectric leakage perspective, a large conduction offset of 1.63 and 1.65 eV at the ZrO_2/Ge and HfO_2/Ge interfaces,

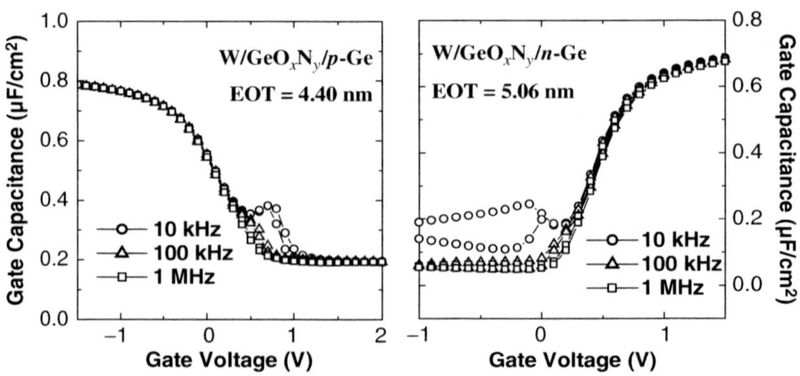

Fig. 13.6. Typical W/GeO_xN_y/Ge C–V characteristics. GeO_xN_y was grown by RTO for 15 s and RTN for 5 min at 600°C

respectively, can be predicted [20] using

$$\Delta E_C = \chi_{Ge} - \chi_{MO_x} + (S - 1)\left(\Phi_{CNL,Ge} - \Phi_{CNL,MO_x}\right), \tag{13.3}$$

where χ_{Ge} and χ_{MO_x} are the electron affinities of Ge and MO_x, respectively, S is an empirically fitted slope accounting for dielectric screening, and both $\Phi_{CNL,Ge}$ and Φ_{CNL,MO_x} are the charge neutrality levels of Ge and MO_x respectively.

Compared to other deposition techniques, atomic layer deposition (ALD) is particularly attractive as a method for preparing ultrathin high-κ layers with excellent electrical characteristics and near-perfect film conformality. The typical ALD process was performed at 300°C using alternating surface-saturating reactions of H_2O and metal tetrachloride precursors. Pt-gated MOS capacitors were fabricated with ~3 nm of HfO_2 on Ge substrate with various surface pre-parations [21]:

(1) Cyclic rinsing between HF and DI water (CHF Ge)
(2) Rinsing in DI water (DIW Ge)
(3) RTN of thermal GeO_2 (Thick GeO_xN_y)
(4) RTN of CHF Ge (Thin GeO_xN_y).

The gate leakage current density, EOT and normalized hysteresis were extracted from these capacitors as illustrated in Fig. 13.7. The leakages were similarly low for different splits with the thick GeO_xN_y method giving the lowest leakage. RTN of CHF Ge was found to produce the best electrical results with the lowest EOT and minimal C–V hysteresis.

Nitrogen incorporation onto the Ge surface during RTN was confirmed by X-ray photoemission spectroscopy. The resultant GeO_xN_y film contained about 22.5% of nitrogen and was shown to be almost insoluble in H_2O. After further optimization of the interfacial RTN condition for the best MOS characteristics, RTN at 600°C for 1 min on CHF cleaned Ge was

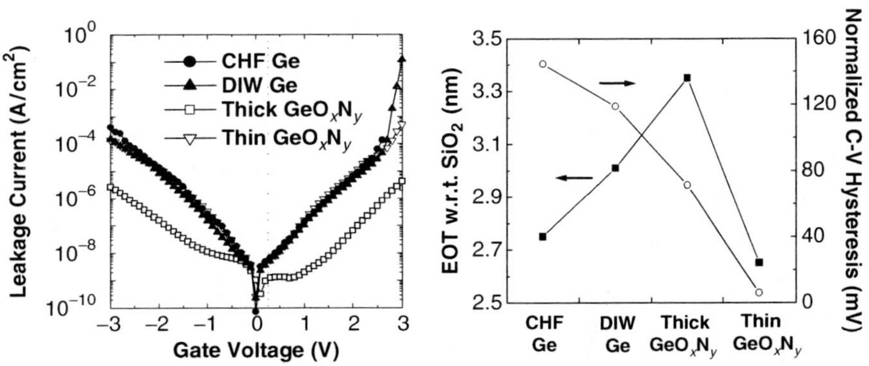

Fig. 13.7. Gate leakage current density, EOT, and normalized hysteresis extracted from $Pt/HfO_2/Ge$ capacitors [21]

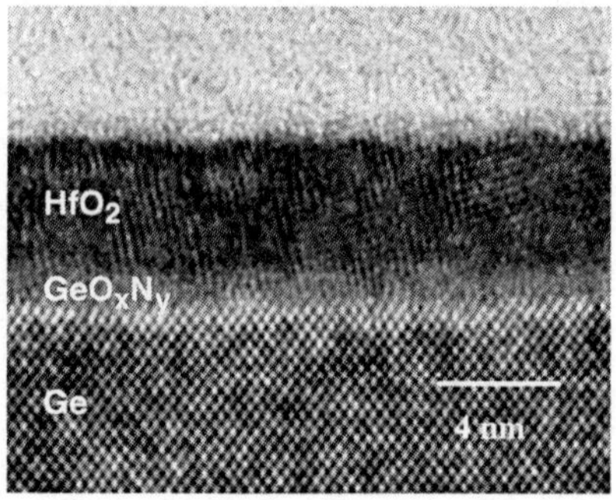

Fig. 13.8. ALD HfO_2 on Ge with a GeO_xN_y interlayer [21]

established to be the most optimal recipe for the processing methods investigated. This recipe generated a GeO_xN_y interlayer thickness of about 1 nm (Fig. 13.8) at the HfO_2/Ge interface and delivered decent MOS characteristics (Fig. 13.9).

The second high-κ deposition technique was the ultraviolet-assisted ozone oxidation of sputtered thin metal precursor films. Zirconium oxidation of was carried out at room temperature on differently prepared substrates including chemical oxide passivated, DI water rinsed, as well as HF vapor etched surfaces [22]. Pt-gated ZrO_2 MOS capacitors were fabricated on HF vapor etched Ge with a sub-1 nm EOT and atomically abrupt ZrO_2/Ge interface (Fig. 13.10), as also observed on the DI water rinsed sample.

Lastly, these nanoscale Ge MOS dielectric leakages were benchmarked together in a fair manner [19]. They were further classified into three categories: GeO_xN_y, high-κ with interlayer, and high-κ without interlayer (Fig. 13.11). About 4–5 orders of magnitude of leakage reduction were obtained by

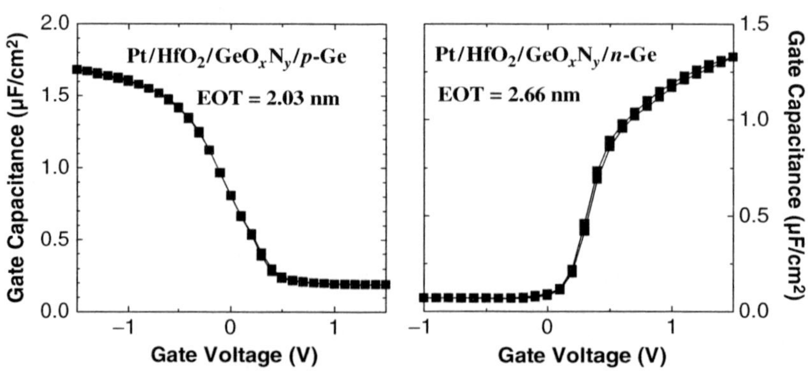

Fig. 13.9. High frequency (800 kHz) $Pt/HfO_2/GeO_xN_y$/Ge C–V characteristics with minimal hysteresis [21]

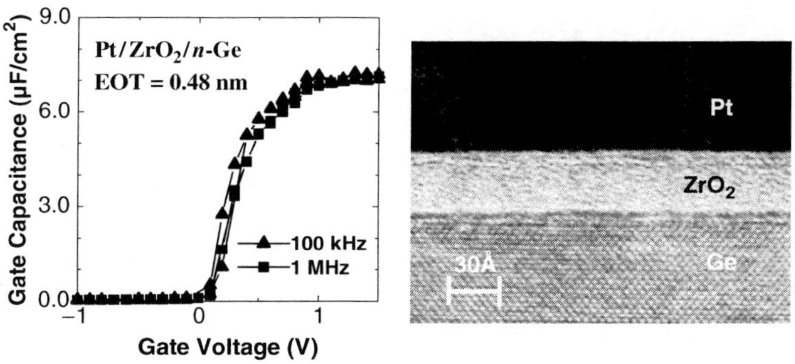

Fig. 13.10. Pt/ZrO_2/Ge C–V characteristics on HF vapor etched Ge with minimal hysteresis and the corresponding XS-TEM image revealing no apparent interlayer [22]

employing high-κ dielectric over GeO_xN_y. Moreover, a better scalability could be achieved by removing the interlayer.

13.7 Shallow Source–Drain Junctions

The prior knowledge on dopant incorporation in Ge is deficient which mandates a more up-to-date revisit. Within the last 50 years, numerous experiments were reported on both p- and n-type dopant activation and diffusion in Ge. Medium-to-high levels of activation have been attained with ion-implanted B, and deep junctions with low level of activation were experimented on P, As, and Sb ion implantation.

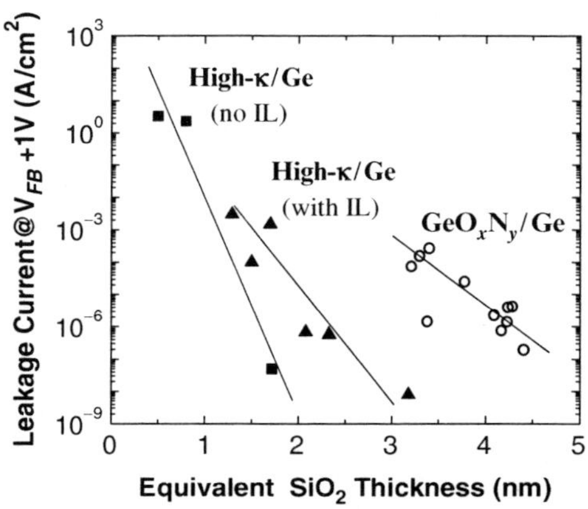

Fig. 13.11. Benchmarking nanoscale Ge dielectric leakage [19]

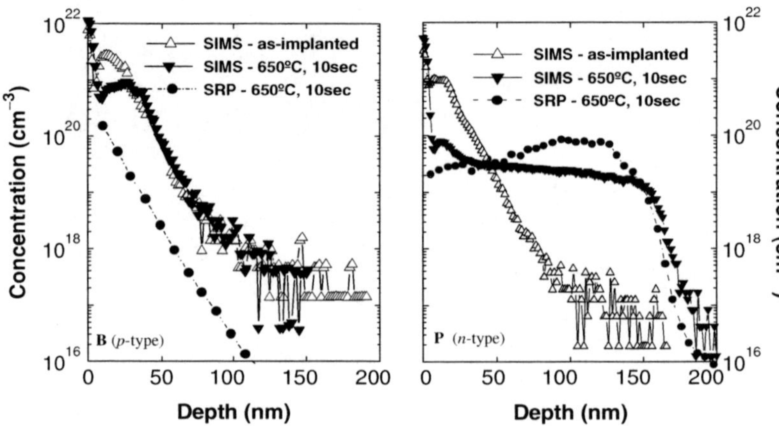

Fig. 13.12. SIMS (chemical) and SRP (electrically active) concentrations of ion-implanted B and P in Ge [23]. ${}^{49}BF_2^+$ and ${}^{31}P^+$ ions were implanted at 20 and 18 keV, respectively, at a dose of $4 \times 10^{15}\,cm^{-2}$

13.7.1 Ion Implantation Doping

Conventional ion implantation doping of various ionic species was carried out at a fixed dose and energies corresponding to a similar projected range [23]. Symmetrically high levels of activation on both p- and n-type dopants in Ge were demonstrated at concentration directly applicable to advanced CMOS (Fig. 13.12). However, the fast n-type dopant diffusion was hindering any shallow junction formation (Fig. 13.13).

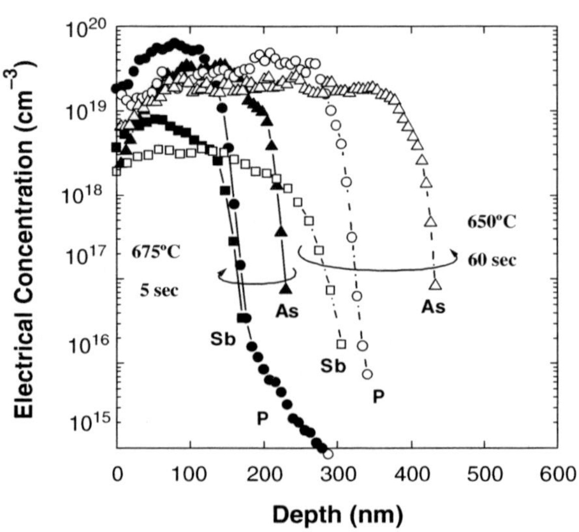

Fig. 13.13. SRP profiles of various n-type dopants in Ge [23]

13.7.2 Solid Source Diffusion Doping

As an alternative doping strategy, SSD is free from problems such as incomplete dopant activation, channeling effects, transient enhanced diffusion, and extended defect formation even at very low energies. Besides these advantages, shallow junctions with low sheet resistance have been achieved in silicon using SSD. In an attempt to obtain shallow junctions in Ge via diffusion from heavily doped phospho-silicate glass (PSG), we have studied the SSD doping in Ge [10]. PSG was deposited by low pressure chemical vapor deposition (LPCVD) in a hot wall furnace from a mixture of SiH_4 and PH_3 at 400°C. Rapid thermal processing (RTP) was utilized to maximize the activation level of the out-diffused dopants and minimize their redistribution.

The thermal anneal budget range of interest is highlighted in Fig. 13.14. For the 8 wt.% PSG that we used in our experiment, no appreciable out-diffusion was observed below 800°C, which could be attributed to the low diffusivity of P inside the PSG layer at such low temperatures. The representative spreading resistance probe (SRP) depth profiles after RTA at 850°C are displayed in Fig. 13.15. These resultant n^+/p junctions always showed the peak concentration of about $1 \times 10^{19}\,cm^{-3}$ at the surface. This highest achievable peak concentration depends on several factors including the P concentration and diffusivity within the PSG layer, the P segregation at the PSG/Ge interface, as well as the P solid solubility limit in Ge. As a rule of thumb, in order to simultaneously lower the out-diffusion temperature and raise the surface peak concentration, a solid source with either higher dopant concentration or diffusivity could be employed. Owing to the absence of other extrinsic diffusion mechanisms, we extracted the intrinsic dopant diffusivity and plotted as a function of temperature (omitted here) that indicated a very good match with the published P diffusion coefficients [24] in Ge.

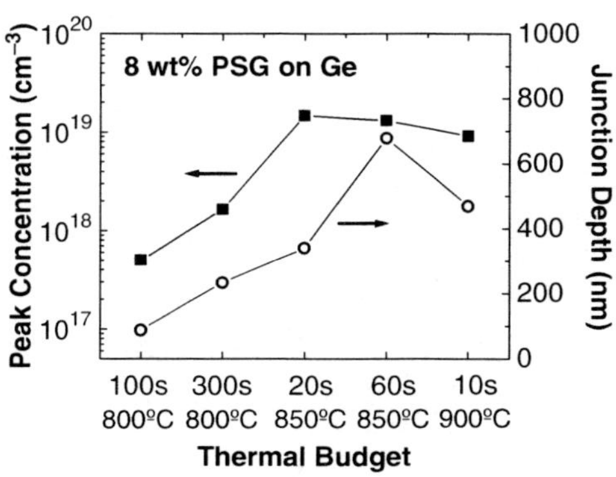

Fig. 13.14. Peak electrically active concentration and junction depths obtained by SSD from 8 wt.% PSG under different thermal anneal budgets [10]

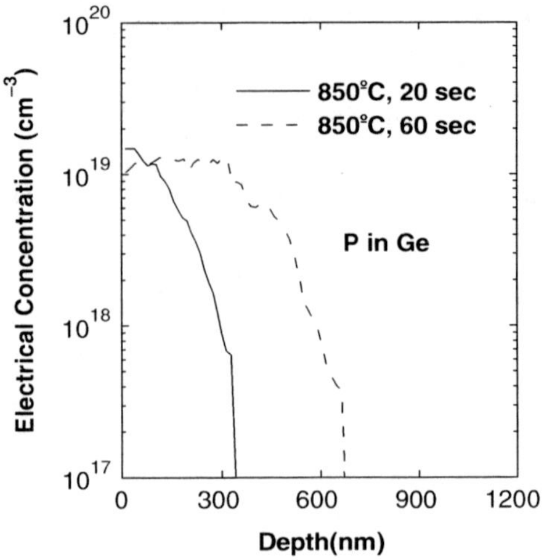

Fig. 13.15. SRP profiles of out-diffused P [10]

13.8 Metal-Gated Germanium MOSFET Processes

Since the p-type dopant in Ge could be activated at a temperature as low as 400°C [25], the demand for a gate-last process for p-MOSFET fabrication is relatively low. The much higher activation temperature for n-type dopants conversely mandates a low thermal budget self-aligned gate-last n-MOSFET process. However, an employment of the advanced replacement gate or damascene gate process for Ge MOSFET fabrication is not always practical in many situations.

13.8.1 The Sub-400°C Conventional P-MOSFET Process

Since a standard device isolation technology has yet to be established to fabricate Ge MOSFETs in an integrated fashion, adopting a self-isolated transistor structure would certainly help to expedite the technology evaluation process. A simple ring MOSFET structure was therefore chosen whose self-isolation is achieved by tying the source ring potential to ground.

The sub-400°C process [25] began with the (100) oriented n-type Ge wafers. Either a DI water rinsing or HF vapor etching was used in an attempt to remove native oxides, followed by the deposition of 4–5 nm ZrO_2 using the ultraviolet-assisted ozone oxidation. No threshold adjustment implant was used. After the Pt gate electrode formation, a self-aligned source/drain BF_2^+ implant was done. Dopant activation was then performed at 400°C in N_2 for 30 min. Source/drain contact hole (ZrO_2) etching and contact metallization were combined into a single lithography step. ZrO_2 was first etched in chlorine plasma. While keeping the same photoresist masking, a Ti/Al metal stack was

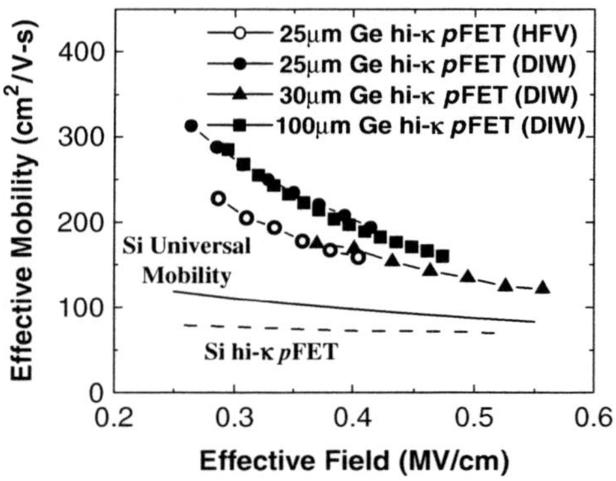

Fig. 13.16. Effective hole mobility vs. effective E-field extracted from $Pt/ZrO_2/Ge$ p-MOSFETs with either DI water rinsing or HF vapor etching surface treatment [25]

e-beam evaporated followed by the liftoff process. Finally, the samples were subjected to forming gas anneal at 300°C for 30 min.

Ge p-MOSFETs with drawn channel length of 2–200 μm were all working. The shorter channel length devices exhibited relatively high I_{Dsat} but also high off-state leakage, which composed primarily of the gate leakage at such a low EOT. To minimize any error introduced by the gate leakage to the intrinsic I_{Dsat} measurements for accurate effective mobility extraction, these $Pt/ZrO_2/Ge$ p-MOSFETs were operated at gate and drain biases where the gate leakage was negligibly small. Effective hole mobility from these devices with different Ge surface cleaning are plotted in Fig. 13.16.

Since the effective hole mobility extracted from a limited number of devices exhibited a slight distribution, no gate length dependence and no single preferential Ge surface cleaning could be identified. In any case, these $Pt/ZrO_2/Ge$ p-MOSFETs revealed roughly twofold enhancement in hole mobility over the silicon universal mobility model and about three times higher mobility than that of high-κ/Si p-MOSFETs, both at low E-field. With proper optimization of the device structure and fabrication process, it should be possible to attain even higher effective mobility and lower leakage. For instance, thin gate spacers together with source and drain extensions could be employed to bring the metal contacts much closer to the channel to minimize parasitic resistances. Recently, variants of this high-κ on Ge concept had been subsequently demonstrated in Ge p-MOSFETs by other groups [26, 27] showing similar results.

13.8.2 The Simple Self-Aligned Gate-Last n-MOSFET Process

In integrating novel channel materials like Ge, it is crucial and beneficial to develop a simple low thermal budget process that is compatible with the Si mainline equipment. Among other key criteria, standard field isolation together with planar geometry is the foremost important. Embracing these two

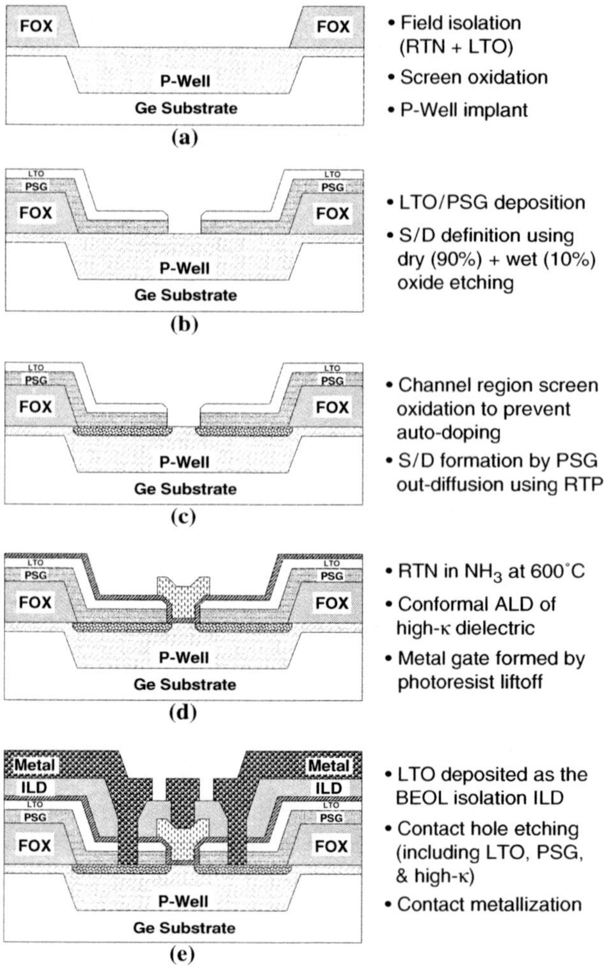

Fig. 13.17. The simple self-aligned gate-last Ge n-MOSFET process flow [10]

features, a simple self-aligned gate-last process has been developed [10] as depicted in Fig. 13.17.

The starting substrates were very lightly doped (100) oriented p-type Ge wafers. The field isolation was done by RTN at 600°C followed by LPCVD SiO_2 (LTO) deposition. After the active areas were opened, a screen oxide on the Ge surface was thermally grown followed by p-well implant (Fig. 13.17a). An LTO-capped 8 wt.% PSG layer was then deposited. The source/drain regions were defined by a combinational etching of the LTO/PSG stack above the channel position (Fig. 13.17b). To further prevent auto-doping from the exposed PSG sidewalls, another screen oxide was grown on the channel surface

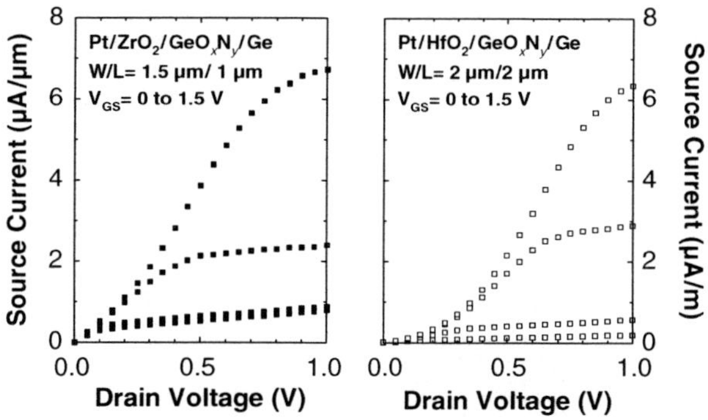

Fig. 13.18. Output characteristics of Pt-gated Ge n-MOSFETs with either ZrO_2 or HfO_2 dielectrics [10]

prior to the source/drain formation by out-diffusion from PSG using RTP (Fig. 13.17c). The RTP was carried out at 850°C for 10 s to achieve an expected junction depth of about 0.28 µm. An RTN of the Ge channel was then performed at 600°C followed by an ALD of ZrO_2 or HfO_2 of about 3 nm. Metal gate electrodes overlapping the source/drain were defined by photoresist liftoff (Fig. 13.17d). This gate-last process was completed by an LTO deposition for back-end-of-line isolation, contact hole etching, and metallization using 1% Si-doped Al (Fig. 13.17e). With the exception of the high-κ ALD step, this tailored five-lithography level process was launched in the mainline Si Stanford Nanofabrication Facility (SNF).

Proof-of-concept Pt-gated Ge n-MOSFETs with either ZrO_2 or HfO_2 gate dielectric were fabricated using this self-aligned gate-last process. The extracted inversion EOT (~2.2 nm) is similar to the accumulation EOT obtained from simple MOS capacitors with an identical dielectric stack [21]. The feasibility of this simple process on novel channels has been confirmed with functional Ge n-MOSFETs with channel length down to 1–2 µm (Fig. 13.18). The non-linearity of the output characteristics at low drain voltage indicated the presence of a Schottky source/drain contact, which could be explained by either the poor Pt electrode edge adhesion and/or the loss of ohmicity due to deactivation of the out-diffused dopants during the interfacial RTN step. These issues could simple be solved by adapting a metal that sticks better and a lower thermal budget GeO_xN_y synthesis.

Introduction of this simple self-aligned gate-last MOSFET process largely relaxes the stringent thermal stability requirement on novel gate stacks and channels during dopant activation, while serving the same purpose as the more-involved replacement gate or damascene gate process. In addition, high selectivity etchings of (a) the metal gate electrode vs. the high-κ dielectric and (b) the high-κ dielectric vs. the channel are no longer essential in the presence of the thick LTO/PSG buffer stack underneath as depicted in Fig. 13.17d.

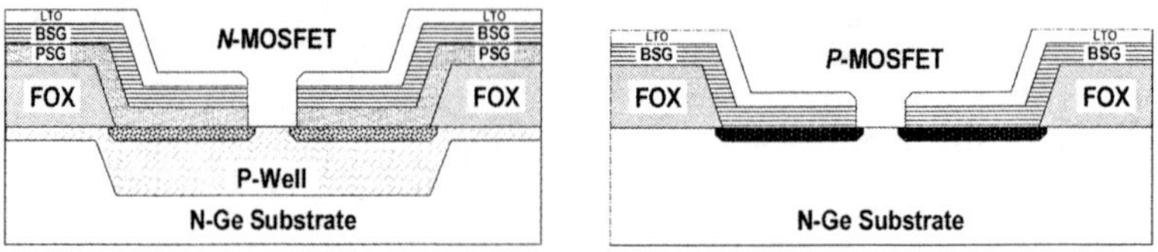

Fig. 13.19. Extension of the novel process to fabricate complementary channel MOSFETs [10]

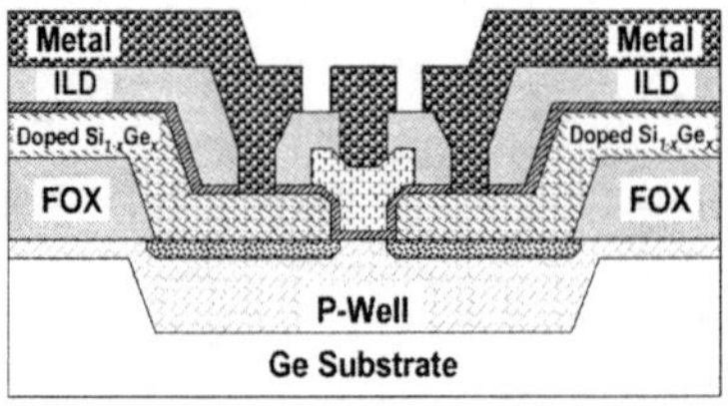

Fig. 13.20. Elevated source and drain junctions could be employed upon the replacement of the insulating solid dopant source with a doped $Si_{1-x}Ge_x$ alloy [10]

Many useful device structures could also be derived by extending this versatile process. For example, complementary channel MOSFETs could be realized by selectively removing the PSG layer from the p-MOSFET areas followed by blanket deposition of a LTO capped borosilicate glass (BSG) film (Fig. 13.19). The complementary junctions for both p- and n-MOSFET are thus readily formed with a single RTP out-diffusion step. Moreover, heavily-doped poly-Si_xGe_{1-x} alloys could be employed instead of insulating doped glasses as the solid source. The resultant device structure would automatically contain the doped Si_xGe_{1-x} layer as the benign elevated source/drain junctions (Fig. 13.20), which however might require a high selectivity etching of the heavily-doped poly-Si_xGe_{1-x} alloy vs. the lightly-doped channel.

13.9 Conclusions

Innovative device structures and new materials must be considered to continue the historic progress in information processing and transmission. As a promising MOSFET channel material candidate, Ge offers numerous advantages over Si. In this chapter, we have first pointed out the technological difficulties in the realization of Ge MOSFETs with classical problems such as the lack of a sufficiently stable gate dielectric and prior knowledge on doping Ge. We have then reviewed various advanced Ge MOS technologies developed to tackle these issues including:

(1) Nanoscale gate stacks with either grown germanium oxynitride or deposited high-κ dielectrics,
(2) Shallow source/drain junctions by either ion implantation doping or solid source diffusion doping, and
(3) Metal-gated Ge MOSFET processes using either a sub-400°C conventional p-MOSFET flow or a simple self-aligned gate-last n-MOSFET flow.

Utilizing these technologies, a sub-1 nm EOT gate stack, shallow source/drain Ge junctions, and functional metal-gated Ge MOSFETs with high-κ dielectric have been demonstrated for the first time.

Acknowledgments

This work was supported by the DARPA HGI Program, the MARCO Materials Structures and Devices Focus Center, and the Intel Foundation PhD Fellowship. The authors would also like to acknowledge Ammar Nayfeh, Dr. Hyoungsub Kim, Dr. David Chi, Dr. Kailash Gopalakrishnan, Dr. Rohit Shenoy, Dr. Yonehara Takao, Dr. James P. McVittie, Prof. Paul C. McIntyre, Dr. Baylor B. Triplett, Dr. Peter B. Griffin, Prof. James D. Plummer, and Prof. Yoshio Nishi for their helps and discussions.

References

1. *The International Technology Roadmap for Semiconductors*, Semiconductor Industry Association, 2004 Update. (http://public.itrs.net/)
2. E.J. Nowak, "Maintaining the benefits of CMOS scaling when scaling bogs down," *IBM J. Res. & Dev.*, vol. 46, pp. 169–180, 2002
3. M. Lundstrom, "Elementary scattering theory of the Si MOSFET," *IEEE Electron Device Lett.*, vol. 18, pp. 361–363, 1997
4. M. Lundstrom and Z. Ren, "Essential physics of carrier transport in nanoscale MOSFETs," *IEEE Trans. Electron Dev.*, vol. 49, pp. 133–141, 2002
5. A. Lochtefeld and D.A. Antoniadis, "On experimental determination of carrier velocity in deeply scaled NMOS: how close to the thermal limit?" *IEEE Electron Dev. Lett.*, vol. 22, pp. 95–97, 2001
6. S. Takagi, "Re-examination of subband structure engineering in ultra-short channel MOSFETs under ballistic carrier transport," *VLSI Symp. Tech. Dig.*, pp. 115–116, 2003
7. *Electronic Archive of New Semiconductor Materials, Characteristics and Properties*, Ioffe Physico-Technical Institute. (http://www.ioffe.rssi.ru/SVA/NSM/Semicond/)
8. R. Chau, S. Datta, M. Doczy, B. Doyle, B. Jin, J. Kavalieros, A. Majumdar, M. Metz, and M. Radosavljevic, "Benchmarking nanotechnology for high-performance and lower-power logic transistor applications," *IEEE Trans. Nanotech.*, vol. 4, pp. 153–158, 2005

9. C.O. Chui, A.K. Okyay, and K.C. Saraswat, "Effective dark current suppression with asymmetric MSM photodetectors in Group IV semiconductors," *IEEE Photon. Technol. Lett.*, vol. 15, pp. 1585–1587, 2003
10. C.O. Chui, H. Kim, P.C. McIntyre, and K.C. Saraswat, "A germanium NMOSFET process integrating metal gate and improved hi-k dielectrics," *IEEE IEDM Tech. Digest*, pp. 437–440, 2003
11. J.C. Bean, T.T. Sheng, L.C. Feldman, A.T. Fiory, and R. T. Lynch, "Pseudmorphic growth of Ge_xSi_{1-x} on silicon by molecular beam epitaxy," *Appl. Phys. Lett.*, vol. 44, pp. 102–104, 1984
12. D.J. Eaglesham and M. Cerullo, "Low-temperature growth of Ge on Si(100)," *Appl. Phys. Lett.*, vol. 58, pp. 2276–2278, 1991
13. M.T. Currie, S.B. Samavedam, T.A. Langdo, C.W. Leitz, and E.A. Fitzgerald, "Controlling threading dislocation densities in Ge on Si using graded SiGe layers and chemical-mechanical polishing," *Appl. Phys. Lett.*, vol. 72, pp. 1718–1720, 1998
14. Hsin-Chiao Luan, D.R. Lim, K.K. Lee, K.M. Chen, J.G. Sandland, K. Wada, and L.C. Kimerling, "High-quality Ge epilayers on Si with low threading-dislocation densities," *Appl. Phys. Lett.*, vol. 75, pp. 2909–2911, 1999
15. A. Nayfeh, C.O. Chui, K.C. Saraswat, and T. Yonehara, "Effects of hydrogen annealing on heteroepitaxial-Ge layers on Si: Surface roughness and electrical quality," *Appl. Phys. Lett.*, vol. 85, pp. 2815–2817, 2004
16. A. Nayfeh, C.O. Chui, T. Yonehara, and K.C. Saraswat, "Fabrication of high-quality p-MOSFET in Ge grown heteroepitaxially on Si," *IEEE Electron Device Lett.*, vol. 26, pp. 311–313, 2005
17. N. Sato and T. Yonehara, "Hydrogen annealed silicon-on-insulator," *Appl. Phys. Lett.*, vol. 65, pp. 1924–1926, 1994
18. D.J. Hymes and J.J. Rosenberg, "Growth and materials characterization of native germanium oxynitride thin films on germanium," *J. Electrochem. Soc.*, vol. 135, pp. 961–965, 1988
19. C.O. Chui, F. Ito, and K.C. Saraswat, "Scalability and electrical properties of germanium oxynitride MOS dielectrics," *IEEE Electron Dev. Lett.*, vol. 25, pp. 613–615, 2004
20. C.O. Chui, D.-I. Lee, A.A. Singh, P.A. Pianetta, and K. C. Saraswat, "Zirconia-germanium interface photo-emission spectroscopy using synchrotron radiation," *J. Appl. Phys.*, vol. 97, p. 113518, 2005
21. C.O. Chui, H. Kim, P.C. McIntyre, and K.C. Saraswat, "Atomic layer deposition of high-k dielectric for germanium MOS applications: substrate surface preparation," *IEEE Electron Dev. Lett.*, vol. 25, 2004
22. C.O. Chui, S. Ramanathan, B.B. Triplett, P.C. McIntyre, and K. C. Saraswat, "Germanium MOS capacitors incorporating ultrathin high-k gate dielectric," *IEEE Electron Dev. Lett.*, vol. 23, pp. 473–475, 2002
23. C.O. Chui, K. Gopalakrishnan, P.B. Griffin, J.D. Plummer, and K.C. Saraswat, "Activation and diffusion studies of ion-implanted p- and n-dopants in germanium," *Appl. Phys. Lett.*, vol. 83, 2003
24. W.C. Dunlap, Jr., "Diffusion of impurities in germanium," *Phys. Rev.*, vol. 94, pp. 1531–1540, 1954
25. C.O. Chui, H. Kim, D. Chi, B.B. Triplett, P.C. McIntyre, and K.C. Saraswat, "A sub-400°C germanium MOSFET technology with high-k dielectric and metal gate," *IEEE IEDM Tech. Digest*, pp. 437–440, 2002

26. C.H. Huang, M.Y. Yang, A. Chin, W.J. Chen, C.X. Zhu, B. J. Cho, M.-F. Li, D.L. Kwong, "Very low defects and high performance Ge-on-insulator p-MOSFETs with Al_2O_3 gate dielectrics," *VLSI Symp. Tech. Dig.*, pp. 119–120, 2003
27. A. Ritenour, S. Yu, M.L. Lee, N. Lu, W. Bai, A. Pitera, E. A. Fitzgerald, D.L. Kwong, and D.A. Antoniadis, "Epitaxial strained germanium p-MOSFETs with HfO_2 gate dielectrics and TaN gate electrode," *IEEE IEDM Tech. Digest*, pp. 433–436, 2003

14

Opportunities and Challenges of Germanium Channel MOSFETs

H. Shang, E.P. Gusev, M.M. Frank, J.O. Chu, S. Bedell, M. Gribelyuk, J.A. Ott, X. Wang, K.W. Guarini, and M. Ieong

Summary. This chapter reviews the progress and current critical issues on the integration of germanium (Ge) surface channel MOSFET devices as well as strained Ge buried channel MOSFET structures. The device design and scalability of the strained Ge buried channel MOSFETs are discussed based on our recent results. CMOS compatible integration approaches of Ge channel devices are presented.

14.1 Introduction

MOSFETs with a high mobility channel are an attractive candidate for advanced CMOS device structures as it becomes increasingly difficult to enhance Si CMOS performance through traditional device scaling. The lower effective mass and higher mobility of carriers in germanium (Ge) as compared to silicon (Si) (2$\times$ higher mobility for electrons and 4$\times$ for holes) has prompted renewed interest in Ge-based devices for high performance logic. Ge channel MOSFETs have been identified as one of possible directions for channel engineering [1].

Recently, surface channel Ge MOSFETs have been demonstrated using thin Ge oxynitride [2] or high-k dielectric [3–5] as gate insulator. However, most devices reported have used relatively simple structures such as ring type gate structure for simplified integration, and devices usually have relative large dimensions. In addition, the smaller bandgap (0.67 vs. 1.12 eV for Si) and the much lower melting point (934°C vs. 1,400°C for Si) present additional processing challenges for integrating Ge channel MOSFETs. For demonstration of state of art Ge channel devices, several key issues have to be addressed. This chapter reviews the major integration challenges and mobility enhancement associated with Ge surface channel devices as well as strained Ge buried channel devices.

14.2 Ge Surface Channel MOSFETs

14.2.1 Gate Dielectric

One major roadblock for Ge CMOS device fabrication is that it is very difficult to obtain a stable gate dielectric. A water-soluble native Ge oxide that is typically present on the upper surface of a Ge-containing material causes the instability of the gate dielectric.

The best-known native dielectric candidate on Ge is Ge oxynitride (GeO_xN_y). High quality, thin GeO_xN_y can be formed on Germanium by nitridation of a thermal grown germanium oxide. Rapid thermal oxidation (RTO) at 500–600°C followed by a rapid thermal nitridation at 600–650°C in ammonia (NH3) ambient is generally been practiced. NH_3 is chosen as nitriding agent due to its greater ability to incorporate more nitrogen into the oxynitride film over other species like nitrous oxide (N_2O) and nitric oxide (NO). Using this method, the resulting film thickness can be scaled down as thin as EOT of 1.9 nm with acceptable leakage and the refractive index is found about 1.3–1.5 [6]. Comparing with native Ge oxides (GeO and GeO_2), it has the improved thermal and chemical stability [7, 8]. In addition, the incorporation of nitrogen into Ge oxides could suppress any potential interdiffusion between the gate dielectric and substrate and/or the gate electrode. High performance Ge MOSFETs with enhanced mobility over Si MOSFETs with SiO_2 were demonstrated using a relative thick GeON (EOT ∼5 nm) [2,9]. However, the most important application for high quality thin GeO_xN_y is perhaps that it could serve as a stable interlayer for integration of novel high-k dielectrics into Ge MOS devices.

The recent developments of high-quality deposition techniques, such as atomic layer deposition (ALD) and metal-organic chemical vapor deposition (MOCVD), to deposit dielectric films with high dielectric constants (on the order of about 4.0 or greater, typically about 7.0 or greater) for the replacement of SiO_2 in Si MOSFETs has prompted activities to develop Ge MOSFETs implementing such dielectrics. Binary metal oxides (e.g., ZrO_2, HfO_2) have been the primary choices as a high-k gate dielectric. In addition, germinate ($MeGe_xO_y$), where Me stands for a metal with high ion polarazibility, e.g., Hf, Zr, La, Y, Ta, Ti, etc., are also proposed to potentially improve the carrier mobility and the interface stability.

14.2.2 Ge Surface Preparation

One of most challenging tasks for Ge/high-k MOS systems is the Ge surface preparation and interface control before high-k film deposition.

For Ge, specifically, it appears essential to have a surface free of germanium oxide before high-k film deposition. A conventional solution for Si has been to use (concentrated or dilute) hydrofluoric acid (e.g., HF or DHF) to remove any native Si oxide, while leaving an H-passivated surface. Despite being successful for Si CMOS device fabrication, this surface passivation technique was found to be less effective on Ge [10].

One demonstrated method to fabricate functional gate stacks is to desorb the Ge oxide in an ultra-high vacuum (UHV) system at high temperatures (360° to 650°C) followed by in situ high-k deposition [11–13]. In fact, crystalline oxides such as CeO_2 [14] and $BaTiO_3$ [15] have been successfully grown on Ge(100) using this method in conjunction with pulsed laser deposition and molecular beam epitaxy. The main drawback of this approach is that UHV systems are costly and generally incompatible with standard ALD or MOCVD high-k deposition tools used in manufacturing. A practical solution is based on nitridation of a wet-etched (e.g., using DHF) Ge surface prior to dielectric deposition using either atomic N exposure or a high-temperature NH_3 gas treatment [5, 16–18].

We found both the microstructure of the high-k film deposited on Ge and the electrical property of Ge/high-k MOS capacitors are very sensitive to the Ge surface preparation prior to high-k film deposition [17]. In our experiment, Ge surface is first wet cleaned, and then HfO_2 is deposited on Ge substrate by ALCVD. It is interesting that HfO_2 grows epitaxially on the wet cleaned Ge surface with DI H_2O last process, while amphorous HfO_2 is observed on the Ge surface treated with nitrogen passivation by RT NH_3 process (at 650°C for 1 min), as shown in Fig. 14.1. Figure 14.2 shows the C–V characteristics of MOS capacitors for both cases. In contrast to the large frequency dispersion observed in the DI water last sample, very little frequency dispersion showed for the sample with nitrogen passivation. The large dispersion is probably due to the Ge and Hf interdiffusion at the Ge–HfO_2 interface, which might have been effectively reduced in the case of nitrogen passivation by RT NH_3 before HfO_2 deposition. However hysteresis still remains and additional traps are also introduced during the RTNH$_3$ process. The nitridation step also induces fixed positive charge at the interface which causes a large negative flatband shift and could degrade the device mobility.

Several research groups have recently reported that effective passivation can be achieved by using SiH_4 [19]. EOT as thin as 7.5 A was reported with plasma PH_3 treatment and thin AlN layer [20] combined with HfO_2/TaN gate stack. Besides the above-mentioned physical passivation methods, novel wet chemistries are also being studied to passivate Ge surface during preclean. Chlorine-passivated [21] and sulfur-passivated [22] Ge surfaces are two examples.

Although much progress has been made on this subject, more deep understanding and well controlled Ge surface is needed for successful application of high-k dielectric on Ge MOS devices.

14.2.3 Dopant Diffusion and Junction Leakage

Compared to bulk Si, boron diffusion is suppressed while As diffusion is enhanced in SiGe and Ge (Fig. 14.3) [23]. This will favor the formation of ultra

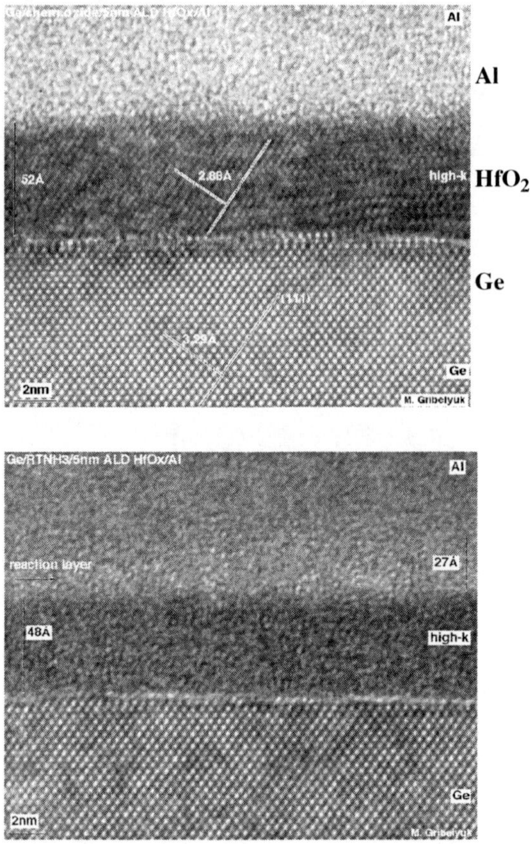

Fig. 14.1. High resolution TEM images of HfO_2 deposited on (**a**) wet cleaned Ge surface with DI H_2O last process; (**b**) RT NH_3 treated Ge surface after wet clean. Crystalline HfO is observed on case (**a**)

shallow junctions in P channel Ge MOSFETs, while presenting a challenge for N channel Ge MOSFETs. Methods such as co-implantation have been demonstrated to show that As diffusion in 20–75% SiGe can be reduced 2.5–3.7× [23].

The smaller bandgap in Ge has been a concern for its influence on junction leakage and band-to-band tunneling. To investigate the junction leakage associated with the smaller bandgap in Ge, both P+/N and N+/P Ge diodes are made by boron and phosphorous implantation. The junction leakage of both N^+/P and P^+/N Ge diodes can be reduced to $\sim 10^{-4}\,\mathrm{A\,cm^{-2}}$ with annealing, as shown in Fig. 14.4. This is considered acceptable for device operation.

On extremely scaled MOSFETs, band-to-band tunneling is a great concern [24]. The band-to-band tunneling current increases exponentially in smaller bandgap semiconductors, thus could be a more serious issue for Ge MOSFETs. Detailed study of its impact on Ge MOSFET scaling can be found in [25].

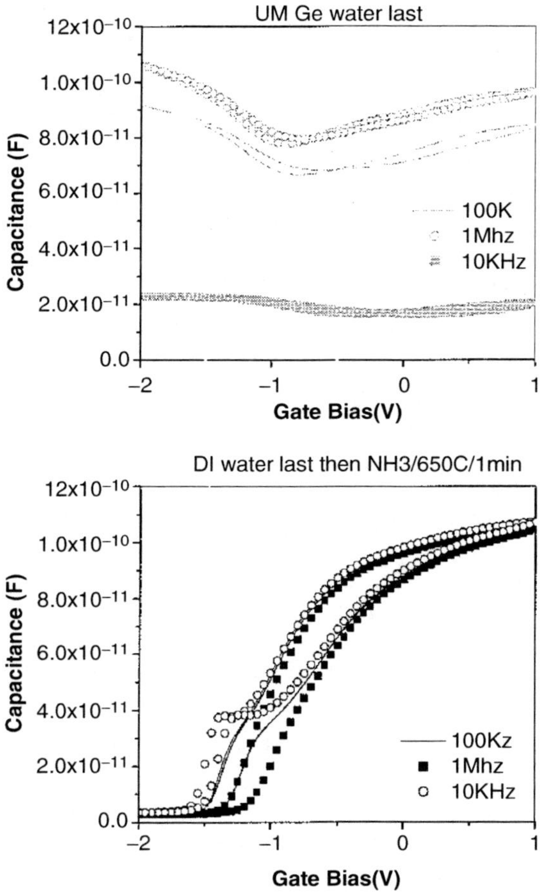

Fig. 14.2. The C–V characteristics of Ge/HfO_2/Al MOS capacitors (**a**) Ge was wet cleaned only, (**b**) Ge was wet cleaned, then treated with RT NH_3. The frequency dispersion is significantly reduced in (**b**), possibly due to the reduced Ge–Hf interdiffusion at the interface

14.3 Strained Ge Buried Channel MOSFETs

By adding a high quality thin layer of Si on top of Ge, good quality of Si/SiO_2 interface can be achieved. In addition, Si based gate dielectric and high-κ films can be applied on the devices. Combined with strained Ge (s-Ge) channel grown on top of relaxed SiGe, the s-Ge buried channel devices are expected to have improved mobility due to the very small effective hole mass $(0.1m_0)$ in s-Ge layer and the reduced surface roughness scattering. Indeed, dramatic hole mobility enhancement of 4–25× has been demonstrated in s-Ge MOSFETs [26–30] – the highest mobility enhancement for hole carriers among all available options. On the other hand, one of the major concerns for buried channel devices has been the device scalability.

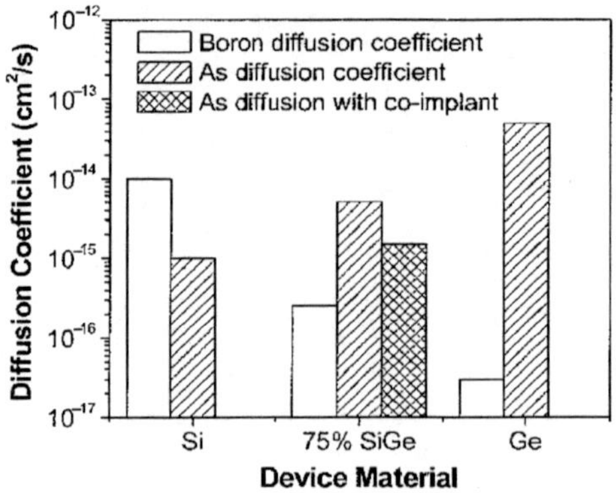

Fig. 14.3. B and As diffusion coefficient in Ge (600°C) compared to 75% SiGe and Si (1,000°C) S–Ge could benefit from a reduced As diffusion by co-implantation

14.3.1 Device Design and Scaling Prospect for Strained Ge Buried Channel Devices

It is known that the effective gate dielectric in buried channel devices is increased comparing with surface channel operation, resulting in worse short channel effects, such as larger subthreshold swing and V_t rolloff. Thus the s-Ge buried channel device must be carefully designed and evaluated to ensure greater performance without short channel degradation [30].

To achieve maximum performance in the s-Ge buried channel MOSFETs, most carriers must be confined within the high mobility s-Ge layer. We

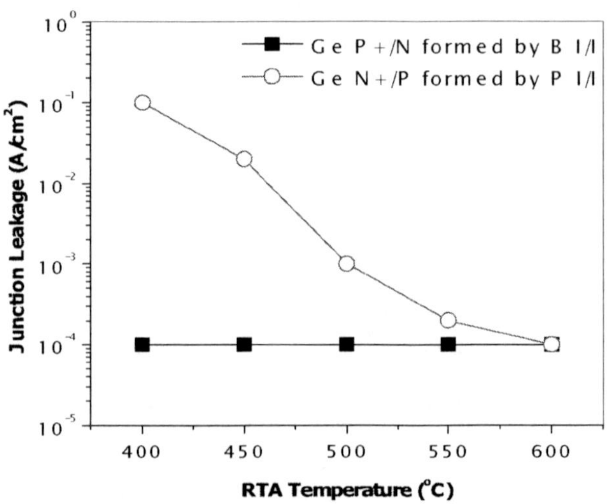

Fig. 14.4. Reverse junction leakage in Ge as a function of anneal temperature. With anneal, both P + /N and N + /P junction leakage can be reduced to $\sim 10^{-4}$ A cm^{-2}

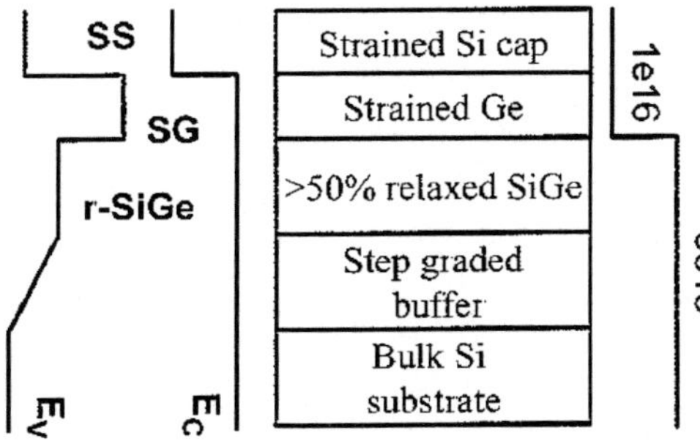

Fig. 14.5. Energy band offset, layer structure, and retrograde doping profile of s-Ge BC PMOSFETs used in electro-static simulation

performed 1-D electro-static simulations to determine the upper limit of the Si cap thickness whereby 90% of the total on-state carriers would be confined in the s-Ge layer. Fig. 14.5 shows the device layer structure along with the band offset used in this simulation. A valence band offset of 450 meV between the s-Ge and the Si cap is assumed [31]. Fig. 14.6a shows the ratio of carriers in the buried s-Ge layer over the total carriers at the on-state as a function of Si cap thickness. The strong dependence of carrier confinement on the Si cap thickness is clearly shown for heavily doped channel structures. When a channel doping of $1 \times 10^{19}\,\mathrm{cm}^{-3}$ is employed, the Si cap thickness must be at most 1.5 nm in order to keep >90% of carriers in the buried Ge channel. 2-D simulations are also performed to investigate the scalability of sub-100 nm s-Ge BC PMOSFETs with an Si cap of 1.5 nm. Fig. 14.6b shows V_t roll-off and DIBL of s-Ge BC PMOSFETs as compared to the Si surface channel control device. With a retrograde doping profile of 10^{16} and $5\times10^{18}\,\mathrm{cm}^{-3}$. s-Ge buried channel PMOSFETs exhibit similar short channel characteristics as bulk Si SC devices [30].

14.3.2 Material Growth and Thermal Stability

There are two main techniques to obtain a strained Ge or high Ge content SiGe layer: chemical vapor deposition (CVD) or Ge condensation (also called thermal mixing [29]). Both ultra-high vacuum (UHV) CVD and PECVD methods have been reported for s-Ge growth [26, 27, 30].

In our experiment, the s-Ge BC structure is grown using the low temperature UHV-CVD technique with the Si cap thickness down to 1 nm. The growth of the s-Ge BC structure begins with a relaxed ~75% SiGe buffer followed by an s-Ge channel (13 nm) and an ultra-thin Si cap (1.5 nm). TEM (Fig. 14.7) shows the high quality and atomic abruptness for both the Si cap/Ge and Ge/SiGe interfaces. AFM results (RMS = 6.7 nm) show a relatively smooth surface, which can be further improved by applying an intermediate CMP to polish the SiGe buffer layer [23, 24]. Triple axis X-ray diffraction measurements were used to quantify the strain in the Ge channel and the Ge content

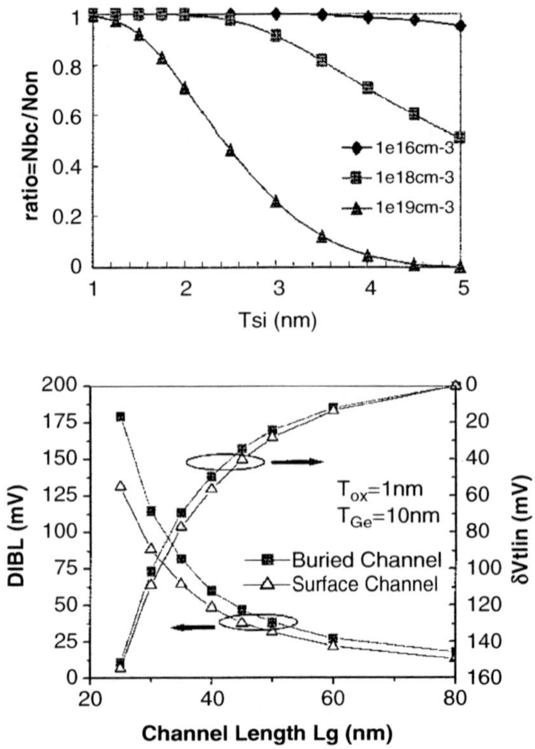

Fig. 14.6. (**a**) The ratio of carriers in buried Ge channel over the total carriers at $N_{\mathrm{on}} = 1\mathrm{e}13\,\mathrm{cm}^{-2}$ as a function of Si cap thickness for three channel dopings. (**b**) V_{t} roll-off and DIBL of s-Ge BC PMOSFETs as compared to the Si SC control with a retrograde doping profile of 10^{16} and $5{\times}10^{18}\,\mathrm{cm}^{-3}$ as shown in Fig 14.5

and strain relaxation of SiGe buffer layer. Strain relaxation during the device fabrication is a big concern. We measured the strain of the s-Ge channel after furnace anneals at temperatures from 550°C to 700°C for 30 min. As shown in Fig. 14.8, virtually no strain relaxation is observed after annealing up to 600°C, but there is significant relaxation at 650°C and above. This will set the upper limit of the s-Ge device processing temperatures. For an RTA-based anneal, the strain relaxation is found similar trend as furnace anneal results.

14.3.3 Gate Stack for s-Ge MOSFETs

Achieving a high quality thin gate dielectric for s-Ge MOSFETs is proven challenging as well. As shown above, to maintain the strain in the s-Ge channel, all processing temperatures should be kept below 600°C to prevent strain relaxation [27]. Because of this constraint, low temperature deposited high-k dielectric and Si dioxide (LTO) had been used as the gate dielectric of s-Ge MOSFETs [26–29].

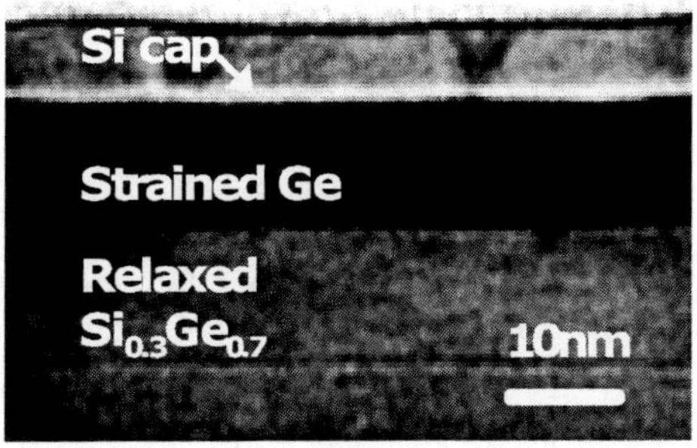

Fig. 14.7. XTEM of a s-Ge on 75% SiGe grown by UHVCVD. Both s-Ge and thin Si cap layers are uniform and smooth

We have developed a new low temperature (400°C) remote plasma oxide as the gate dielectric for UHVCVD grown s-Ge channel MOSFETs. This technique enables us to achieve the thinnest high quality Si oxide ever reported on Ge. Fig. 14.9a shows the typical C–V characteristics of MOS capacitors with EOT = ~3 nm remote plasma oxide on Si. Figure 14.9b shows the interface trap density measured using the conductance method, where D_{it} is found $\sim 2.5\times10^{10}\,cm^{-2}$-eV, very close to that measure on MOS capacitors with ~3 nm thermal SiO_2. Similar leakage current is found on the remote plasma oxide MOS capacitors on both Si and UHV s-Ge samples. The high quality low temperature remote plasma oxide is essential for achieving high performance s-Ge channel PMOSFETs with thin SiO_2 as the gate dielectric.

14.3.4 Integration of s-Ge Channel MOSFETs

Although much work have been done to demonstrate great hole mobility enhancement in s-Ge channel PMOSFETs using simple structures to avoid

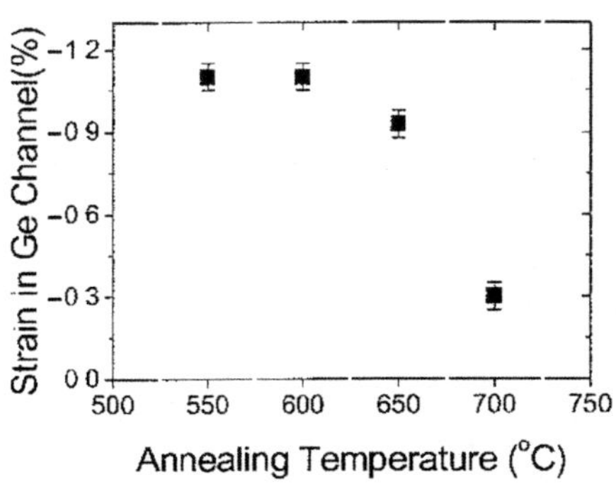

Fig. 14.8. Strain relaxation after furnace anneals at 550°C to 700°C for 30 min. No significant strain relaxation is found when T <= 600°C

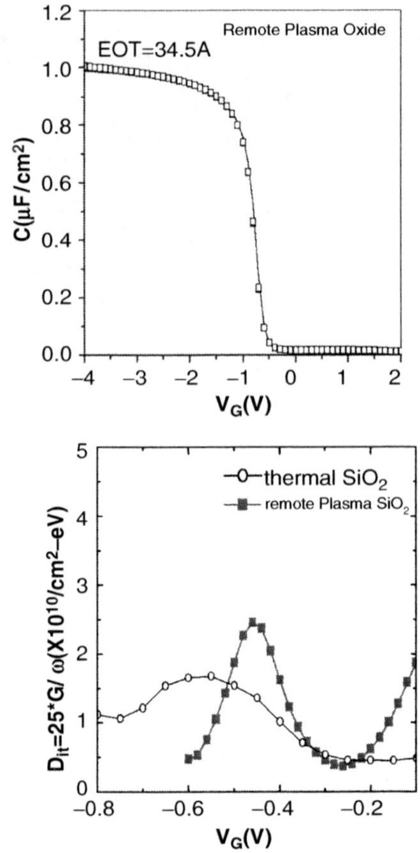

Fig. 14.9. (**a**) Typical *C–V* characteristics of Si MOS capacitor with the remote plasma oxide formed at 400°C. With this technique, as thin as 2.5 nm SiO_2 can be achieved on s-Ge buried channel for high performance PMOSFETs demonstration. (**b**) The interface trap density is measured using the high frequency conductance technique. D_{it} of Si MOS capacitor with remote plasma oxide is found comparable to that with thermal SiO_2

complicate processing issues, a compatible process to incorporate s-Ge structures into standard CMOS technology is needed. One of the proposed ideal CMOS structure is shown in Fig. 14.10, where PMOSFETs employ a buried s-Ge channel while NMOSFETs employ an Si or strained Si surface channel. This structure requires to form a thin s-Ge channel selectively only on PMOSFET regions [32].

In our work, we use SGOI substrates with ∼30% Ge as the starting material. A standard STI process is performed to form SGOI active regions. An optional patterning step could be used to mask the NMOSFET regions with oxide or nitride. Two techniques can be used to form an s-Ge channel selectively on the patterned SGOI regions using:

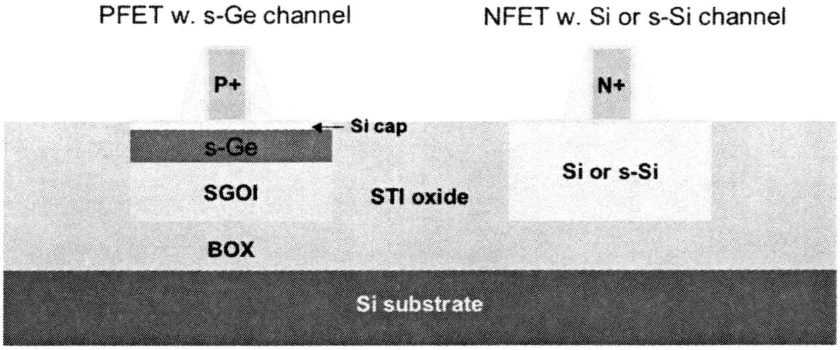

Fig. 14.10. Schematic cross section of a proposed ideal CMOS structure for the maximum enhancement of both hole and electron mobility. The Ge channel in the pFET regions can be formed by selective Ge growth or deposition by masking the NFET region with oxide or nitride film

(1) *Local thermal mixing (LTM).* High-temperature oxidation to enrich the Ge content in the SGOI layer – "TM Ge", or
(2) *Selective UHVCVD process.* Growth of an s-Ge layer with a thin Si cap using UHVCVD – "UHV Ge"

In the "TM Ge" sample, Ge fraction is found ∼67% with 1.47% compressive strain as measured by XRD analysis. In addition, medium energy ion scattering (MEIS) analysis shows the Ge content in the SGOI layer after local thermal mixing is ∼60%. For the "UHV Ge" sample, cross section TEM images (Fig. 14.11) confirmed the selectivity of the s-Ge layer growth, such that s-Ge is formed only on top of the SiGe and not on the STI regions (oxide).

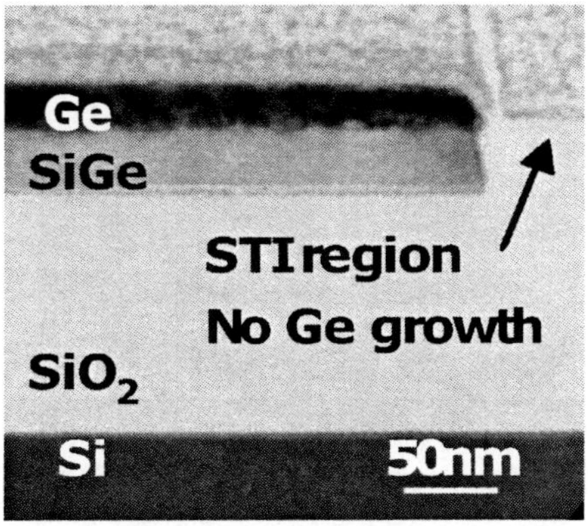

Fig. 14.11. TEM image of selective s-Ge layer grown by UHVCVD method, where Ge is only grown on top of SGOI, while no Ge growth is found on top of STI regions

Table 14.1. Two types of the fabricated strained Ge/SiGe PMOSFETs selectively formed by UHVCVD and thermal mixing techniques respectively

Device	Ge channel formation	Ge% in channel	Si cap (nm)	Gate oxide (nm)
"TM Ge"	Thermal mixing	60	none	6.5 nm HfO_2
"UHV Ge"	UHVCVD	100	5[a]	2.5 nm SiO_2

[a] as-grown

After forming the s-Ge layer, a conventional CMOS process is used for device fabrication, including gate stack formation, S/D implants, and metal contacts. For both "UHV Ge" and "TM Ge" device fabrication, we use in situ boron doped poly-Si as the gate electrode. Table 14.1 lists the physical parameters of the two types of fabricated Ge channel MOSFETs.

"TM Ge" with HfO_2 Gate Oxide

For "TM Ge" MOSFETs, the surface is first treated using RT NH_3, then ~6.5 nm HfO_2 is deposited by MOCVD at 500°C as the gate dielectric. Figure 14.12 is the TEM image of the "TM Ge" device showing the HfO_2 gate dielectric under the polysilicon gate.

Figure 14.13a, b shows the linear current and transconductance characteristics of the "TM Ge" PMOSFETs with HfO_2 gate oxide, along with the Si control. The channel length of both devices are 10 μm. Approximately 2.5× performance enhancement is observed in both linear and saturation regimes. Figure 14.14 shows the subthreshold characteristics of the "TM Ge" PMOSFETs with HfO_2 gate oxide along with the Si control. The subthreshold

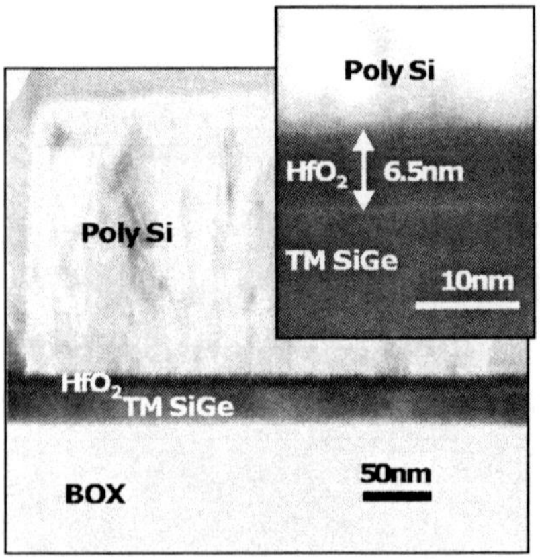

Fig. 14.12. TEM image of the fabricated PFETs with s-SiGe channel using thermal mixing method The gate oxide is ~65 nm HfO_2 deposited using MOCVD method

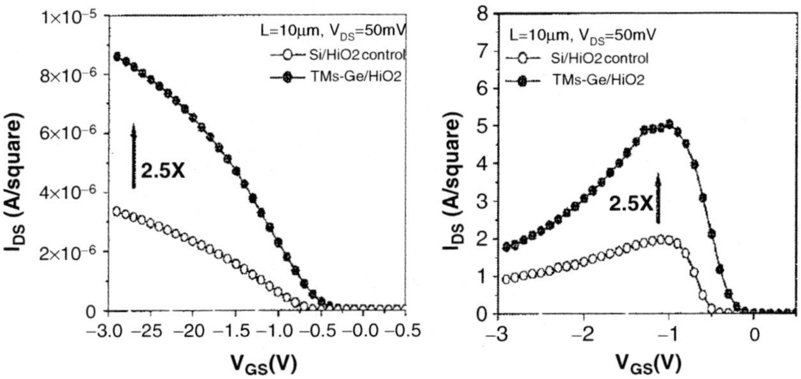

Fig. 14.13. Linear current (**a**) and transconductance (**b**) characteristics of the fabricated PMOSFETs with HfO_2 gate oxide on 60% SiGe channel formed by local thermal mixing, comparing with Si channel PMOSFETs control with HfO_2

slope is ~125 mV dec^{-1} in "TM Ge", ~98 mV dec^{-1} in the Si control. The threshold voltage in the linear region V_{tlin} is extracted using the constant current at 70 nA $square^{-1}$. We found that the V_{tlin} in the Si/HfO_2/poly-Si control is ~ -0.67 V. In contrast, the V_{tlin} from the "TM Ge" device is found ~ -0.36 V, which is ~300 mV lowered from that in the Si/HfO_2/poly-Si control.

One of the most important issues in Si/HfO_2/poly-Si PMOSFETs today is the high V_{th} due to the Fermi level pinning [33]. Because of the valence band offset, the Ge channel allows the V_{th} of HfO_2/poly Si PMOSFETs to be lowered to the appropriate V_{th} for high performance CMOS technology.

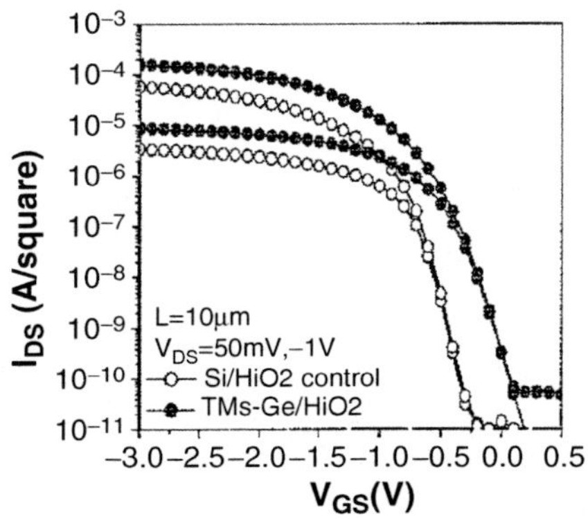

Fig. 14.14. Subthreshold characteristics of PMOSFETs with HfO_2 gate oxide on 60% SiGe channel formed by local thermal mixing, comparing with Si channel PMOSFETs control with HfO_2

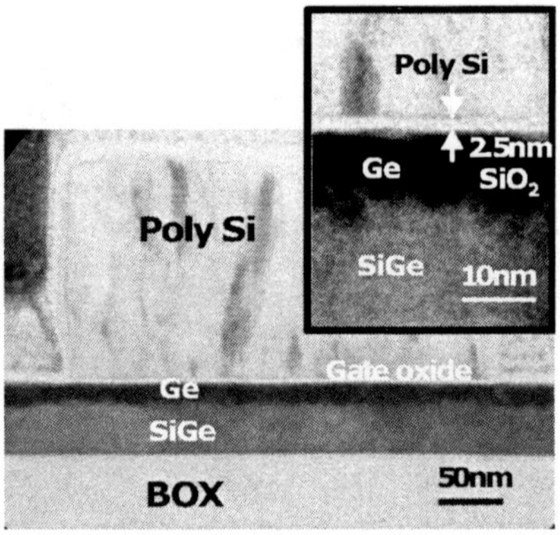

Fig. 14.15. TEM image of the fabricated PFETs with s-Ge channel grown by UHVCVD The gate oxide is 2.5 nm SiO_2 formed by low temperature remote plasma oxide, the thinnest SiO_2 ever achieved on s-Ge MOSFETs

"UHV Ge" with SiO_2 Gate Oxide

For "UHV Ge" MOSFETs, as thin as 2.5 nm high quality SiO_2 is achieved on s-Ge with a thin Si cap by using the low temperature remote plasma oxidation. Figure 14.15 is the TEM image of the "UHV Ge" device showing 2.5 nm SiO_2 under polysilicon gate. Figure 14.16a, b show the linear current and transconductance characteristics of the "UHV Ge" PMOSFETs with SiO_2 gate oxide formed by remote plasma, along with the Si control. The channel length of both devices are 10 µm. ~3× drive current is observed in both linear and saturation regimes. The larger enhancement in "UHV Ge" devices could

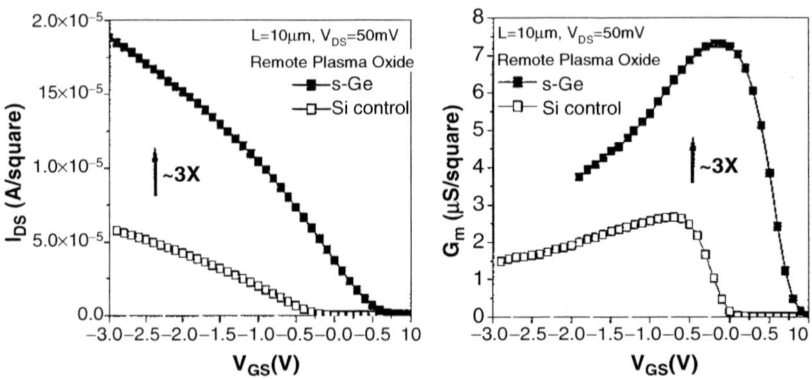

Fig. 14.16. Linear current (**a**) and transconductance (**b**) characteristics of the fabricated PMOSFETs with remote plasma oxide on 100% Ge channel formed by selective UHVCVD, comparing with Si channel PMOSFETs control with the same oxide ~3× enhancement is achieved on the UHVCVD Ge MOSFETs

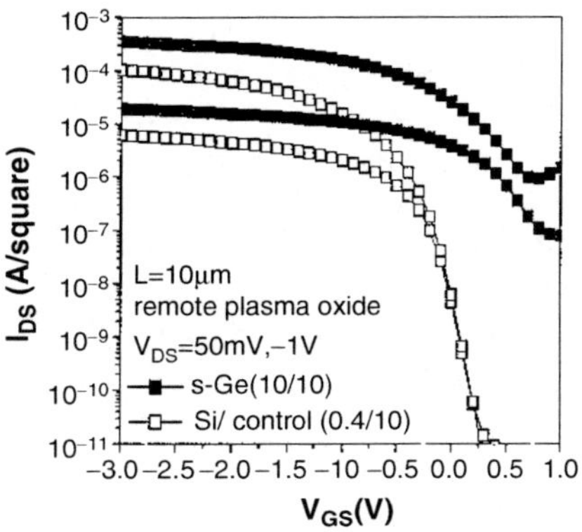

Fig. 14.17. Substhreshold characteristics of PMOSFETs with remote plasma oxide on 100% Ge channel formed by selective UHVCVD

be due to the higher Ge content (100% vs. 60%) in the channel and a SiO_2-based gate dielectric. On the other hand, higher subthreshold leakage current is found on the "UHV Ge" PMOSFETs, as seen from the subthreshold characteristics in Fig. 14.17. This is probably due to the growth defects in the s-Ge layer and may be improved by process optimization.

It is worth to point out that in both "TM Ge" and "UHV Ge" devices, device performance enhancement over Si controls is demonstrated due to the significantly enhanced hole mobility.

14.4 Conclusions

Surface passivation and gate dielectric, dopant diffusion, and junction leakage are the three most serious challenges of Ge CMOS devices. By using the s-Ge with an ultra thin Si cap, standard Si surface passivation and gate dielectric can be applied without significant modification. Table 14.2 compares s-Ge BC MOSFETs with Ge SC devices. S-Ge BC can be integrated with

Table 14.2. Comparison of s-Ge BC with Ge SC devices in critical processing issues and mobility enhancements (+: positive; −: negative; =: equivalent)

	Ge SC	s-Ge BC
Gate stack	−	+
Dopant diffusion	−	+
Junction leakage	=	=
Integration w. Si	−	+
Electron mobility	−	+
Hole mobility	+	++

fewer processing challenges, significantly higher hole mobility and improved electron mobility. These results indicate that the s-Ge BC MOSFET with an ultra-thin Si cap is a promising option for future scaled CMOS devices. We show a CMOS-compatible integration scheme for s-Ge channel PMOSFETs, including the conventional STI isolation and scaled thin gate dielectrics for high performance CMOS technology. Although it is a major step towards integrating s-Ge channels into CMOS technology for continued performance enhancements, much work remains to be done in the demonstration of state of art short channel Ge PMOSFETs with sufficient performance enhancement.

Acknowledgements

The authors thank the IBM T.J. Watson Research MRL and CSS facilities for their help in device fabrication. In addition, we thank Dr. Shahidi Ghavam, Dr. T.C. Chen for managerial support. One of authors wish to thank Dr. H.-S. Philip Wong for valuable discussions and initial encouragement.

References

1. S. Takagi, T. Mizuno, T. Tezuka, N. Sugiyama, T. Numata, K. Usuda, Y. Moriyama, S. Nakaharai, J. Koga, A. Tanabe, N. Hirashita and T. Maeda, IEDM. Tech. Dig., p. 57, 2003.
2. H. Shang, H. Okorn-Schmidt, K.K. Chan, M. Copel, J.A. Ott, P. M. Kozlowski, S.E. Steen, S.A. Cordes, H.-S.P. Wong, E.C. Jones and W.E. Haensch, "High mobility p-channel germanium MOSFETs with a thin Ge oxynitride gate dielectric", in IEDM Tech. Dig., p. 441, 2002.
3. C.O. Chui, H. Kim, D. Chi, B. Triplett, P.C. McIntyre and K.C. Saraswat, "A sub-400°C germanium MOSFET technology with high-k dielectric and metal gate", in IEDM Tech. Dig., p. 437, 2002.
4. C.H. Huang, M.Y. Yang, A. Chin, W.J. Chen, C.X. Zhu, B.J. Cho, M.-F. Li and D.L. Kwong, " Very low defects and high performance Ge-on-insulator p-MOSFET's with Al_2O_3 gate dielectrics", in VLSI Symp. Tech. Dig., pp. 119–120, 2003.
5. W.P. Bai, N. Lu, J. Liu, A. Ramirez, D.L. Kwong, D. Wristers, A. Ritenour, L. Lee and D. Antoniadis, "Ge MOS characteristics with CVD HfO_2 gate dielectrics and TaN gate electrode" in VLSI Symp. Tech. Dig. 2003, pp. 121–122.
6. C.O. Chui, F. Ito and K.C. Saraswat, "Scalability and Electrical properties of Germanium oxynitride MOS dielectrics", in IEEE Electron Device Lett., vol 25, no. 9 Sept. 2004.
7. O.J. Gregory, E.E. Crisman, L. Pruitt, D.J. Hymes and J.J. Rosenberg, "Electrical characterization of some native insulators on germanium" in Proc. Mater. Res. Soc. Symp., vol. 76 1987, pp.307–311.
8. D.J. Hymes and J.J. Rosenberg, "Growth and materials characterization of native germanium oxynitride thin films on germanium", J. Electrochem. Soc., vol. 135, pp. 961–965, 1988.

9. H. Shang, H. Okorn-Schmidt, K.K. Chan, M. Copel, J.A. Ott, P. M. Kozlowski, S.E. Steen, S.A. Cordes, H.-S.P. Wong, E.C. Jones and W.E. Haensch et al., "Electrical characterization of Germanium P-channel MOSFETs", in IEEE Electron Device Lett., vol. 24, pp. 242–244, Apr. 2003.
10. D. Bodlaki et al. "Ambient stability of chemically passivated germanium interfaces", Surface Science 543, pp. 63–74, 2003.
11. X.-J. Zhang et al., "Thermal desorption of ultraviolet-ozone oxidized Ge (001) for substrate cleaning", in J. Vac. Sci. Technol. A 11(5), Sep/Oct, p. 2553, 1993
12. J.J.-H. Chen, N. Bojarczuk, H. Shang, M. Copel, J. Hannon, J. Karasinski, E. Preisler, S.K. Banerjee and S. Guha, "Ultrathin Al_2O_3 and HfO_2 gate dielectric on surface-nitrided Ge", in IEEE Trans. Electron Dev. 51, p. 1441, 2004.
13. A. Dimoulas, G. Mavrou, G. Vellianitis, E. Evangelou, N. Boukos, M. Houssa and M. Caymax, "HfO_2 high-k gate dielectrics on Ge (100) by atomic oxygen beam deposition", Appl. Phys. Lett. 86, 032908, 2005.
14. D.P. Norton, F.J. walker and M.F. Chisholm, Appl. Phys. Lett. 76, p. 1677, 2000.
15. R.A. McKee, F.J. Walker and M.F. Chisholm, Science 293, p. 468, 2001.
16. C.O. Chui, H. Kim, P. McIntyre and K.C. Saraswat, "Atomic layer deposition of high-k dielectric for Germanium MOS applications–substrate surface preparation", in IEEE Electron Device Lett. 25, p. 274, 2004.
17. E.P. Gusev, H. Shang, M. copel, M. Gribelyuk, C. D'Emic, P. Kozlowski and T. Zabel, "Microstructure and thermal stability of HfO_2 gate dielectric deposited on Ge (100)", in Appl. Phys. Lett. 85, p. 2334, 2004.
18. N. Wu, .Q. Zhang, C. Zhu, D.S.H. Chan, M.F. Li, N. Balasubramanian, A. Chin and D.L. Kwong, et al., "Effect of surface NH_3 anneal on the physical and electrical properties of HfO_2 films on Ge substrate", in Appl. Phys. Lett. 84, p. 3741, 2004.
19. N. Wu, Q. Zhang, C. Zhu, C. Yeo, S.J. Whang, D.S.H. Chan, M. F. Li, A. Chin, D.L. Kwong, A.Y. Du, C.H. Tung and N. Balasubramanian, "Alternative surface passivation on germanium for metal-oxide-semiconductor applications with high-k gate dielectric", in Appl. Phys. Lett., 85, p. 4127, 2004.
20. S.J. Whang, S.J. Lee, F. Gao, N. Wu, C.X. Zhu, J. Pan, L.J. Tang, D.L. Kwong, "Germanium p- & n-MOSFETs fabricated with novel surface passivation (plasma PH3 and thin AlN) and TaN/HfO_2 gate stack" in IEDM Tech. Dig. 2004.
21. G.W. Anderson, M.C. Hanf, P.R. Norton, Z.H. Lu, and M.J. Graham, "The S-passivation of Ge (100) – 1×1", Appl. Phys. Lett. 9, p. 1123, 1995.
22. Z.H. Lu, "Air stable Cl-terminated Ge (111)", in Appl. Phys. Lett. 68(4), 1996.
23. K. Lee, F. Cardone, P. Saunders, P. Kozlowski, P. Ronsheim, H. Zhu, J. Li, J. Chu; K. Chan, and M. Ieong, "20 nm N/sup +/ abrupt junction formation in strained $Si/Si_{1-x}/Ge_x/$ MOS device", in IEDM Tech. Dig., p. 481. 2003.
24. P. Solomon, D.J. Frank, J. Jopling, C. D'Emic, O. Dokumaci, P. Ronsheim, and W. Haensch, "Tunnel current measurements on P/N junction diodes and implications for future device design", in IEDM Tech. Dig. 2003.
25. T. Krishnamohan, Z. Krivokapic, K. Uchida, Y. Nishi, and K.C. Saraswat, "Low defect Ultra-thin Fully Strained-ge MOSFET on relaxed Si with high mobility and low band-to-band tunneling", in VLSI Symp. Tech. Dig., p. 82, 2005.
26. A. Ritenour, S. Yu, M.L. Lee, N. Lu, W. Bai, A. Pitera, E.A. Fitzgerald, D.L. Kwong, and D.A. Antoniadis, "Epitaxial strained germanium p-MOSFETs with HfO_2 gate dielectric and TaN gate electrode", in IEDM Tech. Dig., p. 433, 2003.

27. M. Lee and E.A. Fitzgerald, "Optimized strained Si/strained Ge dual-channel heterostructures for high mobility p- and n-MOSFETs", IEDM Tech. Dig., p. 429, 2003.
28. T. Irisawa, S. Tokumitsu, T. Hattori, K. Nakagawa, S. Koh and Y. Shiraki, "Ultrahigh room-temperature hole Hall and effective mobility in $Si_{0.3}Ge_{0.7}/Ge/Si_{0.3}Ge_{0.7}$ heterostructures", in Appl. Phys. Lett. 81, p. 847, 2002.
29. T. Tezuka, S. Nakaharai, Y. Moriyama, N. Sugiyama and S. Takagi, "Selectively-formed high mobility SiGe-on-insulator pMOSFETs with Ge-rich strained surface channels using local condensation technique", in VLSI Symp. Tech. Dig., 2004, p. 198, 2004.
30. H. Shang, J.O. Chu, X. Wang*, P.M. Mooney, K. Lee, J. Ott, K. Rim, K. Chan, K. Guarini and M. Ieong, "Channel design and mobility enhancement in strained germanium buried channel MOSFETs", in VLSI Symp. Tech. Dig., 2004, p. 204., 2004.
31. R. People, "Physics and applications of GexSi1-x/Si strained-layer heterostructures", in IEEE J. Quan. Elect, Vol. QE-22, p. 1696, 1986.
32. H. Shang, Jack O. Chu, S. Bedell, E.P. Gusev, P. Jamison*, Y. Zhang, J.A. Ott, M. Copel, D. Sadana, K. W. Guarini, and M. Ieong, "Selectively formed high mobility strained Ge PMOSFETs for high performance CMOS", in IEDM Tech. Dig., 2004.
33. C. Hobbs, L. Fonseca, V. Dhandapani, S. Samavedam, B. Taylor, J. Grant, L. Dip, D. Triyoso, R. Hegde, D. Gilmer, R. Garcia, D. Roan, L. lovejoy, R. Rai, L. Hebert, H. Tseng, B. White and P. Tobin, "Fermi level pinning at the poly Si/metal oxide interface", in VLSI Symp. Tech. Dig., 2003. p. 9, 2003.

15

Germanium Deep-Submicron p-FET and n-FET Devices, Fabricated on Germanium-On-Insulator Substrates

M. Meuris, B. De Jaeger, J. Van Steenbergen, R. Bonzom, M. Caymax, M. Houssa, B. Kaczer, F. Leys, K. Martens, K. Opsomer, A.M. Pourghaderi, A. Satta, E. Simoen, V. Terzieva, E. Van Moorhem, G. Winderickx, R. Loo, T. Clarysse, T. Conard, A. Delabie, D. Hellin, T. Janssens, B. Onsia, S. Sioncke, P.W. Mertens, J. Snow, S. Van Elshocht, W. Vandervorst, P. Zimmerman, D. Brunco, G. Raskin, F. Letertre, T. Akatsu, T. Billon, and M. Heyns

Summary. A key challenge in the engineering of Ge MOSFETs is to develop a proper Ge surface passivation technique prior to high-k dielectric deposition to obtain low interface state density and high carrier mobility. A review on some possible treatments to passivate the Ge surface is discussed. Another important aspect is the activation of p- and n-type dopants to form the active areas in devices. Finally, Ge deep submicron n- and p-FET devices fabricated with this technique on germanium-on-insulator substrates, yield promising device characteristics, showing the feasibility of these substrates.

15.1 Introduction

The need for electron and hole mobility enhancement and the progress in the development of high-k gate stacks, has lead to renewed interest in Ge MOSFETs. However, before Ge devices can be used for advanced circuits, many issues of the Ge processing and device properties have to be addressed. In the first place, the mobility enhancement of Ge vs. Si, as observed in long channel transistors, has to be proven to result in a larger drive current for short channel transistors. Second, it must be possible to manufacture these Ge devices in a silicon-like semiconductor manufacturing environment. Last but not least, the reliability and yield of the Ge-based circuits, has to be sufficient to allow ULSI circuits.

To study the feasibility of a Ge CMOS transistor technology, a short channel transistor in bulk Ge was processed. A 200 mm⟨100⟩ n-type bulk Ge wafers were used. The wafer specifications are similar to that of silicon wafers used in ULSI processing. We designed a 3 masks process flow to fabricate the Ge devices, consisting only of silicon compatible process steps performed in a

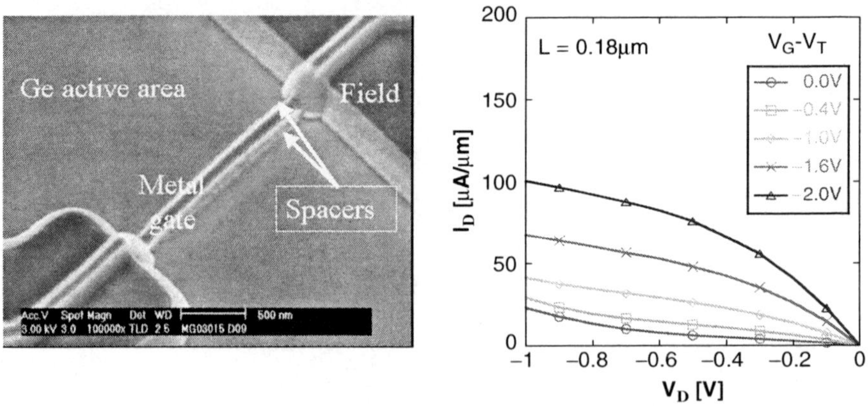

Fig. 15.1. At the *left side* an SEM picture of a 0.18 μm Ge pMOS transistor with field isolation, dry etched metal gate and spacers. At the *right side* the corresponding transistor characteristics

200 mm silicon prototyping line. Special care was taken to avoid the exposure of bare Ge surfaces to the aggressive wet processing steps, with the exception of HF chemistry [1, 2]. A 10 nm HfO_2 (200 ALD cycles) was used as a gate dielectric. A detailed description of the processing can be found in [3]. In Fig. 15.1, an SEM picture and the transistor characteristics is shown of a pMOS transistor with 0.18 μm gate length. These results demonstrate that the processing of germanium in a silicon manufacturing line is possible.

In order to make Ge devices with a larger drive current than an Si device, a key challenge is to develop a proper Ge surface passivation technique to obtain low interface state density and high carrier mobility.

A second challenge to improve the device characteristics is the activation of n-type and p-type dopants with minimal diffusion. To obtain sub-45 nm devices, low resistance in very shallow (10–20 nm) source and drain regions is a key parameter.

And another concern is the type of wafer, which will be used for the sub-45 nm devices. It is foreseen that Si technology will make use of fully depleted devices, most probably "finfet" or "3-gate" type of structures. In order to process these devices, SOI is necessary. Therefore if Ge is used in scaled CMOS devices, GeOI (germanium on insulator) wafers will be mandatory.

In this paper, these three issues are addressed.

15.2 Ge Gate Stack Capacitor

For the passivation of the capacitor gate stack on Ge several options have been proposed in the literature. Treatments with NH_3 anneal [4, 5], SiH_4 anneal [6] or plasma treatment in PH_3 [7] were used with mixed success. Also the

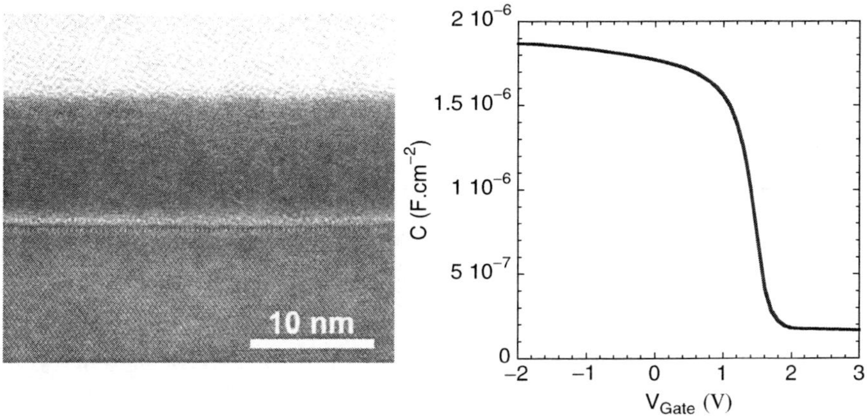

Fig. 15.2. "Ideal" epitaxial silicon layer thickness for a gate stack on a germanium transistor channel. On the *left side* a TEM picture of the Hf-oxide with the thin silicondioxide (*white line*) and thin epitaxial silicon layer (*dark line*) on top of the Ge substrate. At the *right side*, the C–V curve measured on the capacitor structure after hydrogen anneal

deposition of High-k dielectrics with and without an interfacial layer at the Ge surface has been studied in detail [8–14].

In this work, we report on the development and optimisation of a thin epitaxial Si layer as Ge passivation layer. Si passivation has been proposed earlier [15–17]. To allow epitaxial Si growth, the Ge native oxide and other residues are first removed from the Ge surface by a 2% HF dip followed by an anneal in a H_2 ambient in the epi-reactor [18, 19]. Immediately after the H_2 anneal, the Si layer is grown using SiH_4 under a N_2 ambient at 40 Torr. By varying the temperature and time of the anneal, the thickness of the Si layer can be precisely controlled within $\pm\ 2 \times 10^{14}$ deposited Si atoms cm^{-2} (one monolayer of Si atoms is defined in this work as 6.5×10^{14} atoms cm^{-2}, corresponding to an Si layer thickness of 1.2 Å). A good thickness control is crucial to achieve optimal CV characteristics. This includes also the development of analysis techniques for measuring the atomic concentration at the Ge surface [20–23]. Once the Si layer is grown on the Ge substrate, the techniques developed for the high-k dielectric stack on Si could be used. In this work, the Si layer is partially oxidised in a N_2O plasma at room temperature. Finally, a 300°C ALD HfO_2 gate dielectric is deposited. In Fig. 15.2, a CV curve of an optimised thickness of the Si layer, grown on Ge is shown. When the layer is too thin, the re-oxidation of the silicon will reach the silicon–germanium interface, leading to an oxidation of the Ge and causing higher interface state density. But when the layer of silicon is too thick, stress will build-up in the silicon layer as a function of thickness, due to the lattice mismatch between silicon and germanium (see Fig. 15.2). This will form stress induced defects at the interface. The optimisation of the epitaxial silicon layer

Table 15.1. Transistor channel mobility values of different passivation techniques for the Ge gate stack: the Si epitaxial layer [10], the NH_3 anneal [3] and PH_3 anneal [5]

	p-FET		n-FET	
	μ_{eff} ($cm^2 V^{-1} s^{-1}$)	EOT (Å)	μ_{eff} ($cm^2 V^{-1} s^{-1}$)	EOT (Å)
Si [10]	170	26	10	26
NH_3 [3]	210	16	~0.2	?
PH_3 [5]	125	17?	300	17?

thickness and its influence on the capacitor properties is extensively described in [24].

The importance of the passivation of the germanium/high-k dielectric interface was further demonstrated in [25], where the problem of making an nMOS on germanium was explained by the presence of a high interface state density near the conduction band of germanium. We explained that this is probably due to a specific oxidation state of the Ge bond at the interface. Using the Si passivation technique, decent channel mobility values were obtained as well for nMOS as for pMOS devices as shown in Table 15.1.

15.3 Dopant Activation in Germanium

For making a short channel transistor device, very shallow and high conductive p- and n-type active area regions are essential. Therefore high doping (i.e., activation) with low diffusion of the p- and n-type dopant species is necessary. In [26] it is described that boron as a p-type species can be activated at relative low temperatures and without a measurable diffusion. However, the n-type species as P and As show high diffusion coefficients at the activation temperature for the dopant concentrations of interest in nano-electronics.

In [27, 28] a more optimal implant and anneal condition for boron as a p-type species in germanium is described. The boron is easier to activate when a pre-amorphization with Ge is done before the boron implantation. A complete re-crystallization of the Ge amorphous layer can be obtained at temperatures as low as 400°C.

An extensive study [29, 30] of P in Ge as the n-type dopant species revealed that for phosphor three distinct concentration regions exist, with very different behaviour. When P is implanted above 2×10^{20} P at. cm^{-3}, a large out diffusion towards the germanium surface together with a pile-up near the projected range of the implanted phosphor (attributed to the formation of P clusters) is observed. When P is implanted in the region between 2×10^{19} and 2×10^{20} P at. cm^{-3}, a large diffusion of P into the bulk of the Ge is observed with SIMS measurements, showing a concentration dependent diffusion of P in Ge. Implantations below 2×10^{19} P at. cm^{-3}, result in no measurable diffusion

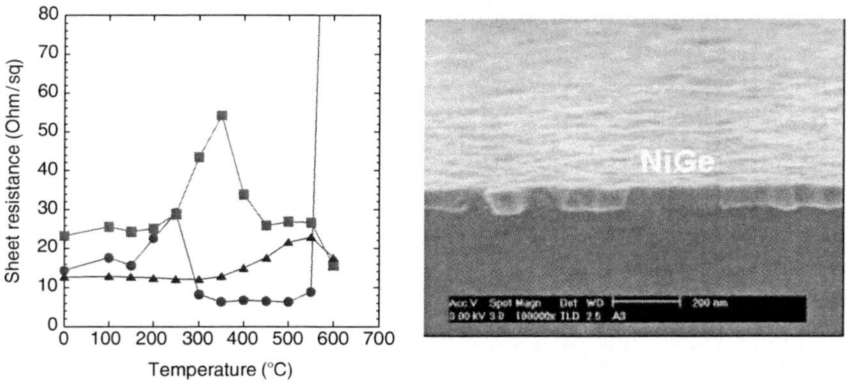

Fig. 15.3. Formation of germanides with three different metals (Ni, Co and Ti) as a function of temperature (*left*). For a temperature between 300°C and 500°C Ni can be used to form NiGe with a sheet resistance of 1.6×10^{-5} Ω cm and a uniform coverage (see SEM picture *right*)

with a 100% activation of the n-type species. In this study, we found that the maximum active concentration of P implants after annealing is approximately 5–6 10^{19} cm^{-3}.

Moreover, studies of residual defects after annealing of the implanted species are essential to improve the quality of the active layers [31–33]

In order to do this type of activation and diffusion studies, one should note that it is important to develop the correct measurement methodologies for the measurement of electrical active species with SRP [34] as well as the dopant profiling with SIMS. These measurement techniques have been proven to be different on germanium compared to silicon.

When dopants can be activated in the active areas of the Ge device, contacting layers are necessary for reducing the contact resistance in the source and drain regions of the circuit. In a silicon technology a self-aligned silicidation process is used. Therefore a germanidation process was studied to form germanides on the active areas of the germanium device. The formation of a NiGe seems to be the most suitable candidate as can be derived from the formation temperature of the NiGe compared to CoGe and TiGe (see Fig. 15.3).

15.4 GeOI Substrates

The introduction of Ge as a replacement for Si, SiGe or strained silicon in the transistor channel is estimated earliest for the 22 nm generation or beyond. The CMOS device of this generation is most probably a fully depleted device. Silicon-on-insulator (SOI) substrates are used to make these devices. Therefore, germanium-on-insulator (GeOI) substrates [35] have to be developed to

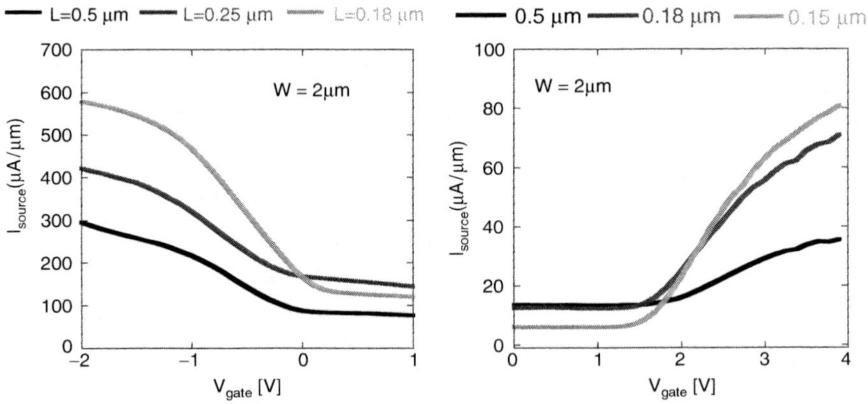

Fig. 15.4. pMOS (*left*) and nMOS (*right*) transistor characteristics of devices fabricated on a GeOI substrate. The excellent properties show the feasibility of the GeOI substrate technology

enable the possibility of making similar devices in Ge. To provide insight in the quality of the Ge layer of GeOI substrates, transistor devices ere made in such GeOI substrates. Ge-on-insulator (GeOI) wafers from collaboration between SOITEC, UMICORE, LETI and IMEC were used, with a 200 nm thick Ge layer [21,36]. The gate dielectric is 10 nm HfO_2 and the metal gate is a 10 nm PVD TaN/70 nm PVD TiN stack. In Fig. 15.4, the pMOS and nMOS device characteristics are shown. These devices, which are made in the Ge top layer of the GeOI wafers, are of similar quality and performance as the ones made on Ge bulk wafers. This demonstrates the feasibility of the GeOI technology and the well-known result that the performance of Ge devices is largely dominated by the passivation process of the HiK/Ge interface.

15.5 Conclusions

An overview of the different technologies to be developed for a Ge CMOS device technology has been given. The major obstacle is the development of a Ge gate stack with a low amount of interface states at the germanium/high-k dielectric interface. Also high dopant activation and low diffusion of p-type and n-type species have to be improved. In order to make Ge a viable technology for CMOS scaling, a high quality Ge on insulator substrate is necessary.

Acknowledgements

The support of the Core Partners of imec (Infineon, Intel Corp., Matsushita, Philips, Samsung, ST, Texas Instruments and TSMC) is highly acknowledged.

The support of the PLINE is acknowledged for the difficulty in processing the Ge bulk substrates.

Funding from FWO research project G.0273.05 is acknowledged.
Funding from the EU under the ET4US contract is acknowledged.

References

1. B. Onsia et al., On the application of a thin ozone based wet chemical oxide as an interface for ALD high-k deposition, 7th International Symposium on Ultra Clean Process of Silicon Surfaces, 2005.
2. B. Onsia et al., A study of the influence of typical wet chemical treatments on the germanium wafer surface, 7th International Symposium On Ultra Clean Process of Silicon Surfaces, 2005.
3. B. De Jaeger et al., Ge deep sub-micron pFETs with etched TaN metal gate on a high-k dielectric, fabricated in a 200 mm silicon prototyping line, in: Proc. ESSDERC, 2004, pp. 189–192.
4. W.P. Bai et al., Ge MOS characteristics with CVD HfO_2 gate dielectrics and TaN gate electrode, in: Proc. of the Symposium on VLSI technology, 2003, pp. 121–122.
5. A. Ritenour et al., Epitaxial Strained Germanium p-MOSFETs with HfO_2 Gate Dielectric and TaN Gate Electrode, in: Proc. IEDM, 2003, pp. 433–436.
6. N. Wu et al., Alternative surface passivation on germanium for metal-oxide-semiconductor applications with high-k gate dielectric, Appl. Phys. Lett., 85, 2004, pp. 4127–4129.
7. S.J. Whang et al., Germanium p- and n-MOSFETs fabricated with novel surface passivation (plasma-PH_3 and thin AlN) and HfO_2/TaN Gate Stack, in: Proc. IEDM, 2004, pp. 307–310.
8. S. Van Elshocht et al., "Physical characterization of HfO_2 deposited on Ge substrates by MOCVD", MRS Spring Meeting - High-k Insulators and Ferro-Electrics for Advanced Microelectronics Devices, D5.4, 2004.
9. M. Houssa et al., Electrical characteristics of Ge/GeOx(N)/HfO2 gate stacks, J. Non-Cryst. Solids, 2004.
10. S. Van Elshocht et al., Deposition of HfO_2 on germanium and the impact of surface pretreatments, Appl. Phys. Lett., 2004, 3824.
11. S. Van Elshocht, et al. "Surface preparation techniques for high-k deposition on Ge substrates," Proceedings of 7th International Symposium on Ultra Clean Processing of Silicon Surfaces – UCPSS, Brussels, Belgium, Solid State Phenom., 103–104, 2004, 31.
12. A. Delabie et al., Atomic layer deposition of hafnium oxide on germanium substrates, J. Appl. Phys., 97, 064104 (2005).
13. Meuris et al., The future of high-k on pure germanium and its importance for Ge CMOS, Mater. Sci. Semiconductor Process., 2005, p. 203.
14. S. Van Elshocht, et al., Study of CVD high-k gate oxides on high-mobility Ge and Ge/Si substrates, Thin Solid Films, 508, 2006 pp. 1–5, Proceedings of the Fourth International Conference on Silicon Epitaxy and Heterostructures (ICSI-4).
15. L.C. Kimerling and H-C Luan, US patent 6,352,942 (2002).
16. N.E. Posthuma et al., Surface passivation for germanium photovoltaic cells' Solar Energy Mater. Solar Cells, 88(1), 2005, 37.
17. Wu et al., Appl. Phys. Lett., 2004.

18. R. Bonzom et al., Growth Kinetics and Relaxation Mechanism of Very Thin Epitaxial Si films on (100) Germanium, presented at 2005 Spring Meeting of the MRS (2005).
19. F.E. Leys et al., Epitaxy solutions for Ge MOS technology, Thin Solid Films 508, 2006, pp. 292–296, Proceedings of the Fourth International Conference on Silicon Epitaxy and Heterostructures (ICSI-4).
20. D. Hellin et al., 'Determination of metallic contaminants on Ge wafers using direct- and droplet sandwich etch – total reflection X-ray fluorescence spectrometry', Spectrochim. Acta Part B 58, 2003, 2093–2104.
21. D. Hellin et al., 'Vapor phase decomposition – droplet collection – total reflection X-ray fluorescence spectrometry for metallic contamination analysis of Ge wafers', Spectrochim. Acta Part B 60, 2005, 209–213.
22. D. Hellin et al., Total reflection X-ray fluorescence spectrometry for the introduction of novel materials in clean-room production environments, IEEE Trans. on Device Mater. Reliab., 5, 2005, pp. 639–651.
23. D. Hellin et al., 'VPD-DC-TXRF for metallic contamination analysis of Ge wafers', Proceedings of the 7th International Symposium on Ultra Clean Process of Silicon Surfaces (UCPSS 2004), 2005.
24. B. De Jaeger et al., INFOS, 2005.
25. Croon et al., Proceedings of the International Conference on Microelectronic Test Structures of 2004, 2005, p. 191.
26. Chui et al., Appl. Phys. Lett., 83, 2003, 3276.
27. A. Satta et al., Diffusion, activation and re-crystallization of boron implanted in pre-amorphized and crystalline germanium, Appl. Phys. Lett., 87, 2005, 172109.
28. A. Satta et al., Dopants for N and P junctions in Germanium, Proceedings of ECS Spring Symposium 2005, 2005, p. 468.
29. A. Satta et al., P implantation doping of Ge: Diffusion, activation, and recrystallization, J. Vac. Sc. Technol. B 24, 2006, pp. 494–498.
30. A. Satta et al., P implantation on doping of Ge: diffusion, activation, recrystallization, The 8th Int. Workshop on the Fabrication, Characterization and Modeling of Ultra Shallow Junctions in Semiconductors, 2005.
31. E. Simoen et al., Defect removal, dopant diffusion and activation issues in ion-implanted shallow junctions fabricated in crystalline germanium substrates, GADEST 2005(Gettering And Defect Engineering in Semiconductor Technology) Conference 2005.
32. A. Satta et al., Ion implantation in Ge and the associated defect control, ECS DECON, 2005.
33. A. Satta et al., Shallow junction ion implantation in Ge and associated defect control, J. Electrochem. Soc., 153, 2006, pp. G229–G233.
34. T. Clarysse et al., Active dopant characterization methodology for Germanium, USJ – The 8th International Workshop on the Fabrication, Characterization and Modeling of Ultra Shallow Junctions in Semiconductors, 2005.
35. C. Deguet et al., From 100 mm to 200 mm Germanium-On-Insulator (GeOI) structures realised by the smart cutTM technology, in: Proceedings of the First EUROSOI Workshop, pp. 31–33 2005.
36. T. Akatsu et al., 200 mm Germanium-on-insulator(GeOI) by Smart Cut technology and recent GeOI MOSFETs achievements, IEEE International SOI Conference, 2005.

16

Processing and Characterization of III–V Compound Semiconductor MOSFETs Using Atomic Layer Deposited Gate Dielectrics

P.D. Ye, G.D. Wilk, and M.M. Frank

Summary. We demonstrate III–V compound semiconductor (GaAs, InGaAs, and GaN) based metal-oxide-semiconductor field-effect transistors (MOSFETs) with excellent performance using an Al_2O_3 high-permittivity (high-k) gate dielectric, deposited by atomic layer deposition (ALD). These MOSFET devices exhibit extremely low gate-leakage current, high transconductance, high dielectric breakdown strength, a high short-circuit current-gain cut-off frequency (f_T) and maximum oscillation frequency (f_{MAX}), as well as high output power and power added efficiency. ALD is a robust process that enables repeatability and manufacturability for compound semiconductor MOSFETs. In order to contribute to the fundamental understanding of ALD-grown high-k/III–V gate stack quality, we discuss stack and interface formation mechanisms in detail for Al_2O_3 and HfO_2 gate dielectrics on GaAs.

16.1 Introduction

GaAs-based metal-oxide semiconductor field-effect transistors (MOSFETs) have been a subject of study for several decades [1–18]. GaAs-based devices potentially have great advantages over Si-based devices for high-speed and high-power applications, in part from an electron mobility in GaAs that is ~5× greater than that in Si, the availability of semi-insulating GaAs substrates, and a higher breakdown field compared to Si. Currently, the GaAs metal-semiconductor field-effect-transistor (MESFET) or high-electron-mobility-transistor (HEMT) is the dominant device for high-speed and microwave-circuits. MESFETs or HEMTs feature gates formed by depositing metal directly on the semiconductor, forming metal-semiconductor (Schottky-barrier) junctions, while MOSFETs have oxide layers (higher barrier) between the metal gate and the semiconductor channel. Compared to GaAs MESFETs or HEMTs, GaAs MOSFETs feature a larger maximum drain current, much lower gate leakage current, a better noise margin, and much greater flexibility in digital IC design due to large gate voltage range.

The main obstacle to GaAs-based MOSFET devices has been the lack of high-quality, thermodynamically stable insulators on GaAs as the gate dielectric that can match device criteria similar to SiO_2 on Si. Both GaAs-based native oxides and deposited insulating layers have been attempted as gate dielectrics. For native oxidation of GaAs, various approaches were applied, i.e., wet oxidation, plasma oxidation, laser-assisted oxidation, vacuum ultraviolet photochemical oxidation, etc. These approaches have had limited success at the device level, however, mainly due to instability of the native oxides of GaAs, and due to the unacceptably high interface trap density D_{it} nearly universally observed with native and deposited oxides. Such interface defects give rise to Fermi level pinning. For example, Fermi level pinning upon GaAs oxidation has been attributed to oxygen-induced displacement of surface As atoms, where doubly O-coordinated second-layer Ga atoms give rise to gap states [19]. Excess interfacial As occupying As_{Ga} antisite defects causes gap states as well [19,20]. Interfacial As may be formed via decomposition of As_2O_3 in the vicinity of GaAs, resulting from the reaction: $As_2O_3 + 2GaAs \rightarrow Ga_2O_3 + 4As$ [21]. After decades of efforts on forming a deposited amorphous oxide on a III–V compound semiconductor using PECVD with decent interface quality [2], much progress has been made recently using in situ deposited $Ga_2O_3(Gd_2O_3)$ or Ga_2O_3 and Gd_2O_3 dielectrics using ultrahigh-vacuum multi-chamber molecular beam epitaxy (MBE) [22–28] and ex situ atomic layer deposition (ALD) grown Al_2O_3 on III–V semiconductors [29–35]. Both methods have been shown to provide a high-quality interface with a low D_{it} on GaAs. Fermi level unpinning in the Ga_2O_3/GaAs system is achieved through a Ga_2O/GaAs-like interface in which the Ga and As surface atoms are restored to near-bulk charge, preventing gap state formation [19]. Promising results have been demonstrated in GaAs MOSFETs using these techniques. Processing and properties of MBE-grown $Ga_2O_3(Gd_2O_3)$ gate dielectrics are addressed in detail in Chap. 3.4 by Kwo. Herein, we focus on ALD Al_2O_3. Another issue for GaAs circuits in competition with silicon is the large defect densities which also preclude large scale integration. The emerging strategy is to use III–V compound semiconductors as NMOS conduction channels and Ge as PMOS conduction channels, to replace part of traditional Si or strained Si, while integrating these high mobility materials with novel dielectrics and heterogeneously integrating them on Si or silicon-on-insulator (SOI).

In order to achieve higher transconductance as well as to downsize the device for higher integrated density, the reduction of the gate oxide thickness is critical, in particular for GaAs digital applications. A gate material with a much wider band-gap providing a higher potential barrier with GaAs is able to significantly reduce the gate leakage current for the same layer thickness of other materials. Al_2O_3 has a high bandgap of $\sim$9 eV, which is much higher than, e.g., for Ga_2O_3($\sim$2.45 eV) or Gd_2O_3($\sim$5.3 eV). As a high-k gate oxide on Si, Al_2O_3 has a dielectric constant of about 9, compared to 3.9 for SiO_2. Al_2O_3 also has a high bulk breakdown field (8–10 MV cm^{-1}), high thermal stability (up to at least 1,000°C), and remains amorphous under typical processing conditions of interest. It is easily wet-etched yet is robust against interfacial reactions and moisture absorption (i.e., it is non-hygroscopic). ALD is a robust manufacturing process which is already commonly used for high-κ gate dielectrics in Si CMOS technology [36]. It is based on alternating, self-saturating surface reactions, e.g., using $Al(CH_3)_3$ and H_2O for Al_2O_3 growth. In this manner, sub-monolayer thickness control and excellent conformality even on high-aspect-ratio structures may be achieved. The enormous efforts and significant advent of deposited high-quality high-k dielectrics on Si using ALD has also renewed hopes that GaAs CMOS may finally become a reality.

In this chapter, we first focus on materials aspects, characterizing the structure and composition of Al_2O_3/GaAs (and, for comparison, HfO_2/GaAs) stacks fabricated by ALD, to develop an understanding of the impact of material and processing conditions on dielectric film and interface formation. Through electrical characterization of Al_2O_3/GaAs materials systems, e.g., leakage current density and capacitance–voltage (C–V) measurements, we then demonstrate a high bulk and interface quality of Al_2O_3 on GaAs. We further demonstrate GaAs-based MOSFETs with excellent performance using an ALD Al_2O_3 gate dielectric. These MOSFET devices exhibit negligible drain current drift and hysteresis, extremely low gate leakage, high transconductance, and good RF characteristics. Through transistor characteristics and modeling, such as drain current hysteresis and transconductance frequency dependence, we evaluate the interface trap density D_{it} of ALD grown Al_2O_3 on GaAs at the device level. InGaAs MOSFETs are also demonstrated using a similar approach. Finally, we extend our ALD work to wide bandgap semiconductor materials, e.g., GaN. We report on a GaN MOS-HEMT using ALD Al_2O_3 as the gate dielectric. Compared to a conventional GaN HEMT of similar design, the MOS-HEMT exhibits several orders of magnitude lower gate leakage and several times higher breakdown voltage and channel current. This implies that the ALD Al_2O_3/AlGaN interface is of high quality and the ALD Al_2O_3/AlGaN/GaN MOS-HEMT has good potential for high-power RF applications. In addition, the high-quality ALD Al_2O_3 gate dielectric enables an effective two-dimensional (2D) electron mobility at GaAs, InGaAs and GaN to be measured under a high transverse field. The resulting effective 2D electron mobility is much higher than the mobility in Si.

16.2 Materials Structure and Composition

In this section, we discuss structural and chemical aspects of ALD-grown high-k/GaAs gate stacks, as reported in detail in [32]. In particular, the high-k/III–V interface is critical since the presence and nature of a "native oxide" interfacial layer (containing, e.g., Ga_2O_3, As_2O_3, etc.) may impact the electrical quality of the stack and/or pose limits to capacitance scaling. We address two ALD-grown high-k materials frequently studied on Si: Al_2O_3 and HfO_2. Comparison of these materials helps develop an understanding of

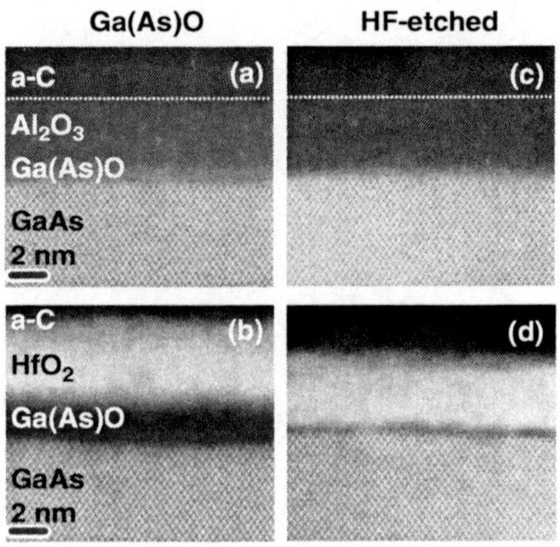

Fig. 16.1. Scanning transmission electron microscopy (STEM) images from 40 Å Al_2O_3 (*top*) and HfO_2 (*bottom*) deposited onto Ga(As)O-covered [(**a**) and (**b**)] and HF-etched [(**c**) and (**d**)] GaAs(100) (reproduced from [37] with permission)

the impact of materials properties on gate stack performance. We address successive stages of gate dielectric formation: starting surface, temperature ramp-up, ALD process, and anneal. Both Al_2O_3 and HfO_2 were grown on oxide-covered ("epi-ready") and HF-etched GaAs(100). For high-k dielectric growth, we followed a procedure that has yielded high-quality Al_2O_3/GaAs gate stacks [24–30]. All ALD work was carried out using commercial ASM Pulsar2000^{TM} or Pulsar3000^{TM} ALD reactors. Depositions were performed using alternating exposures of the common ALD precursors $Al(CH_3)_3 + H_2O$ or $HfCl_4 + H_2O$ in an N_2 carrier gas at 300°C. Carbon or aluminum served as cap layers for microscopy, and films were characterized by ex situ microscopies and spectroscopies [32].

The 20–25 Å thick epi-ready oxide on GaAs is porous and Ga-rich (As : Ga = 0.17) with an increased As concentration near the GaAs substrate [37]. In the following, we will denote this substrate "Ga(As)O/GaAs". During inert anneal to the ALD temperature of 300°C, the epi-ready oxide remains in place [37], consistent with the higher onset temperatures of reactions in various gallium/arsenic oxides on GaAs (300–430°C) [21,38,39]. HF-etched substrates, on the other hand, are mostly oxide-free [40]. Largely independent of surface preparation, as-deposited 40 Å thick ALD-grown Al_2O_3 and HfO_2 films are mostly amorphous as well as continuous (Fig. 16.1), despite a certain degree of high-k agglomeration during initial growth on HF-etched GaAs. This behavior is similar to what has been observed for HF-etched Ge [41].

Interfacial layer thickness after Al_2O_3 and HfO_2 deposition strongly depends on surface preparation. When native oxide removal by an initial HF wet etch is performed, both Al_2O_3 and HfO_2 deposition result in low interfacial layer thickness of only 3–8 Å (Figs. 16.1c,d). On Ga(As)O/GaAs, interfacial layer thickness is ~10 and 20–25 Å, respectively (Figs. 16.1a,b). Clearly,

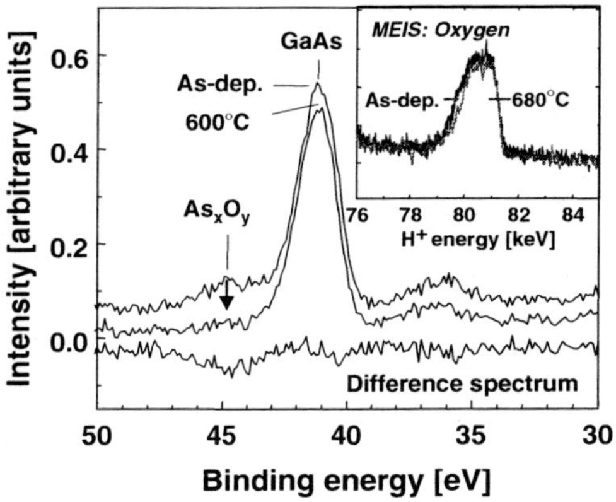

Fig. 16.2. X-ray photoelectron spectroscopy (XPS) data in the As 3d region from 40 Å Al_2O_3 deposited onto Ga(As)O/GaAs(100) before (*top*) and after (*center*) vacuum anneal to 600°C, and difference spectrum (*bottom*). Inset: MEIS spectra in the oxygen region before and after vacuum anneal to 680°C. (reproduced from [37] with permission)

the initial oxide is thinned during the Al_2O_3 ALD growth process, pointing to volatilization of Ga(As)O or its conversion into Al_2O_3. By contrast, the HfO_2 growth process does not cause interface thinning, even though the standard Gibbs energies of formation per O atom are nearly identical (Al_2O_3 : $-527\,\mathrm{kJ\,mol^{-1}}$; HfO_2 : $-544\,\mathrm{kJ\,mol^{-1}}$) and both higher than that for Ga and As oxides [21, 42]. The different degree of interface thinning likely is due to the much higher reactivity of $Al(CH_3)_3$ compared to $HfCl_4$, reflected in standard enthalpies of formation of $-74\,\mathrm{kJ\,mol^{-1}}$ and $-990\,\mathrm{kJ\,mol^{-1}}$, respectively [42, 43]. We note that a large variety of Al precursors is available, all with different enthalpies of formation. By extension, one may expect substantially different gate stack structures to be formed.

Finally, we show that thermal treatments critically impact interfacial layer thickness and composition. The interfacial layer remains Ga-rich during ALD growth, with As oxides still present in the case of Al_2O_3 on Ga(As)O/GaAs (XPS data in Fig. 16.2). During vacuum anneals at 600–680°C, the As oxides decompose (Fig. 16.2), and most oxygen is removed from the interfacial layer, as evidenced by medium energy ion scattering (MEIS, Fig. 16.2, inset). This thermal behavior may be rationalized by recalling results for oxidized GaAs surfaces (without a high-k layer) in inert ambients. At 300–460°C, mixed gallium/arsenic oxides are converted into pure gallium oxide with As precipitates according to the reaction $As_2O_3 + 2GaAs \rightarrow Ga_2O_3 + 4$ As (with partial As desorption in the form of As_2 and As_4). At ∼475°C, any Ga_2O present in the film desorbs; and at 580–630 °C, the remaining Ga_2O_3 is volatilized according to $Ga_2O_3 + 4GaAs \rightarrow 3Ga_2O\uparrow + 4As\uparrow$. At similar temperatures, preferential As desorption from the GaAs substrate sets in. Assuming analogous reactions underneath the high-k layers, including facile out-diffusion of volatile species, a ~600°C anneal would thus result in (a) a Ga_2O_3-like interfacial layer which, upon continued heating, may be partially or completely removed, and (b) excess interfacial Ga. Our findings of As oxide decomposition and oxygen loss are consistent with this reaction scheme.

The role of the O_2 ambient during the ~600°C anneal employed in the electrical studies presented in the following sections remains to be explained. TEM and electrical data indicate that no additional interfacial oxide grows. Therefore, the primary difference between oxidizing and reducing anneals may be the oxidation and desorption of excess interfacial Ga. In addition, O_2 may improve high-k quality, volatilizing As or Ga diffused into the layer and filling detrimental oxygen vacancies.

16.3 Electrical Characterization of ALD Al_2O_3 on GaAs

The starting materials were 2 in. Si-doped GaAs(100) wafers with a doping concentration of 6–8 × 10^{17} cm^{-3}. HF-etched substrates were employed, to ensure minimum Al_2O_3/GaAs interfacial layer thickness, as discussed in the preceding section. The wafers were transferred immediately to the ALD reactor. Again, Al_2O_3 layers were deposited at a substrate temperature of 300°C. An excess of each precursor was supplied alternatively to saturate the surface sites and ensure self-limiting film growth. An inherent characteristic benefit of ALD is the linear relationship between the number of growth cycles and the deposited thickness. In this way, once the growth rate is established for a particular process, the desired thickness can be accurately and reproducibly achieved by simply running a specific number of growth cycles. This feature enables extremely accurate thickness control, even for layers as thin as 10 Å or less. The 600°C O_2 anneals were performed ex situ in a rapid thermal annealing chamber following film deposition. A 1,000 Å thick Au film were deposited on the back side of GaAs wafers to reduce the contact resistance between GaAs wafers and the chuck of the measurement setup. Capacitors were fabricated using 3,000 Å Au top electrodes.

We measured the dependence of the leakage current density (J_L) on the gate voltage (V_g) for a set of ALD Al_2O_3 samples with the oxide thickness systematically reduced from 50 to 12 Å (Fig. 16.3). The plot shows a decrease in current density with increasing film thickness. Direct tunneling current is observed for film thickness ≤30 Å, while film with thickness ≥50 Å shows no significant direct tunneling. The 12 Å Al_2O_3 with the equivalent oxide thickness of only 4.7 Å still shows well-behaved direct tunneling characteristic and does not break down at ±3 V bias. Compared to state-of-the-art SiO_2 on Si, the leakage current density of Al_2O_3 on GaAs is one order of magnitude lower at same electrical thickness (gate stack capacitance). The GaAs/Al_2O_3 barrier height Φ_B is measured as high as ~3.2 eV, which is higher than the Si/Al_2O_3 barrier height of 2.6–3.1 eV determined by the similar method [44].

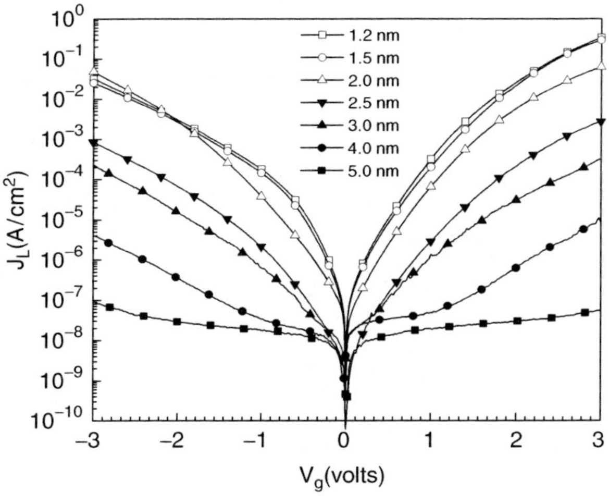

Fig. 16.3. Leakage current density J_L vs. gate bias V_g for ALD Al_2O_3 films on GaAs with different film thickness from 12 to 50 Å

Figure 16.4 shows the summary plot of E_{BR} vs. T_{ox} determined by various methods. The electrical properties of ALD Al_2O_3 films have been investigated by several research groups [44–51]. Most of these studies, however, were performed on ~1,000 Å thick films grown on Si substrates. The E_{BR} of 10 MV cm^{-1} for ALD Al_2O_3 films with a thickness of 50–60 Å is consistent with typical E_{BR} values for high-quality, bulk Al_2O_3 of 8–10 MV cm^{-1}. A substantial E_{BR} enhancement is observed in Fig. 16.4 as the film thickness is reduced to below 40 Å. The E_{BR} for the 12 Å film is more than a factor of 3 larger than the bulk value. This could be explained by a remnant of

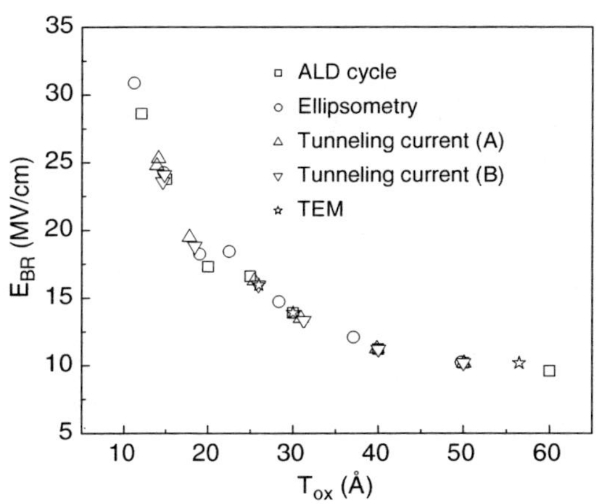

Fig. 16.4. Breakdown electric fields vs. thicknesses for ALD films. The different symbols represent thickness determination by different methods

the E_{BR} enhancement of ultrathin films, or possibly that the relative large leakage current density in an ultrathin oxide prevents the occurrence of hard breakdown of oxide film at a low electric field. The high breakdown field for ultrathin oxide on GaAs provides great opportunity for reducing the gate oxide thickness.

The typical high-frequency and low-frequency C–V curves for a capacitor with a 50 Å-thick ALD Al_2O_3 layer on n-type GaAs are shown in Fig. 16.5. This measurment is taken directly after film growth without any post annealing treatments. The high-frequency trace, e.g., 50 kHz, shows well-behaved depletion at negative bias and accumulation at positive bias. The low-frequency trace, e.g., 500 Hz, shows clear inversion (holes) at negative bias and accumulation at positive bias. There is a few hundred mV hysteresis in the C–V curves in reverse sweeping, depending on the surface preparation of GaAs before ALD growth. Another issue for C–V measurements on GaAs is the frequency dispersion in accumulation. Five to ten percent per decade frequency dispersion at accumulation capacitance is widely observed on unannealed GaAs MOS capacitors. Using a two frequency method of C–V measurement or the four-element equivalent circuit model for ultrathin oxides, [52] we estimate $D_{it} \sim 10^{12}\,cm^{-2}\,eV^{-1}$. This indicates that part of frequency dispersion may originate mostly from the parasitic resistance and capacitance of $10^{17}\,cm^{-3}$ doping levels in the GaAs substrates instead of suspected interface traps. The potential difference of the metal work function and n-GaAs affinity, the Schottky barrier height of the metal backgate on GaAs backside, and the existing hysteresis contribute to the flat-band shift on C–V curves (Fig. 16.5). A postannealing process at 600°C in oxygen ambient for 30–90 s, helps to reduce hysteresis and frequency dispersion and improves the D_{it} to low $\sim 10^{11}\,cm^{-2}\,eV^{-1}$. This is demonstrated in the following section.

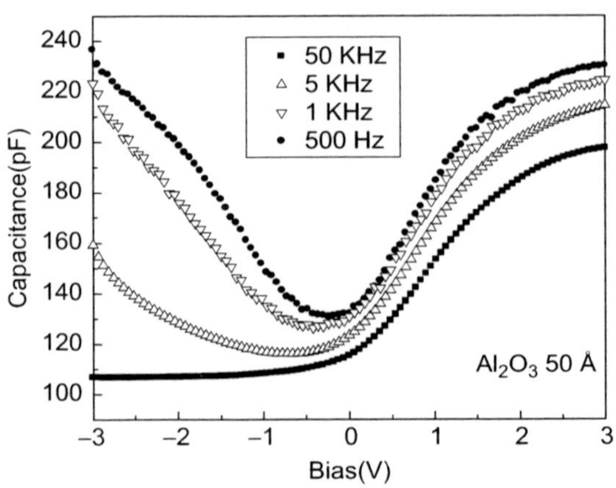

Fig. 16.5. C–V traces for a MOS diode with 50 Å Al_2O_3 on GaAs with an n-type doping of 4–6 × $10^{17}\,cm^{-3}$ at 500 Hz (low frequency), 1, 5 and 50 kHz (*high frequency*). The diameter of the measured diodes is ∼150 μm

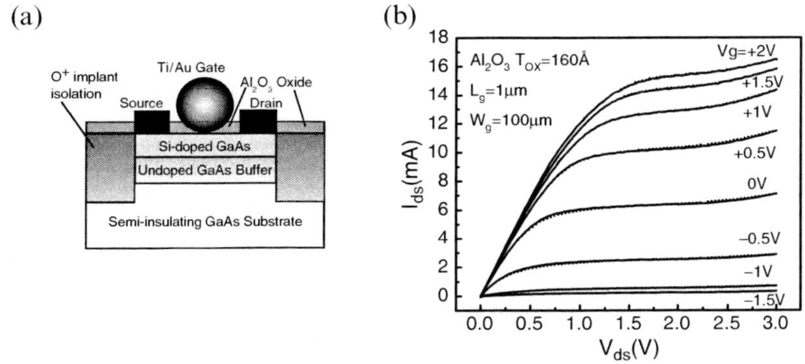

Fig. 16.6. (**a**) Schematic view of a depletion-mode n-channel GaAs MOSFET with ALD-grown Al_2O_3 as gate dielectric. (**b**) Drain current vs. drain bias in both forward (*solid lines*) and reverse (*dotted lines*) sweep directions as a function of gate bias. (reproduced from [30] with permission)

16.4 GaAs MOSFET Fabrication and Characterization

A schematic diagram of the depletion-mode GaAs device is shown in Fig. 16.6 (a). A 1,500 Å undoped GaAs buffer layer and a 700 Å Si-doped GaAs layer ($4 \times 10^{17}\,\mathrm{cm}^{-3}$) were grown by MBE on a (100)-oriented semi-insulating 2-in. GaAs substrate. After the semiconductor epi-layer growth, the wafer was immediately transferred ex situ to an ASM Pulsar2000^{TM} ALD module. The GaAs MOSFET devices employed Al_2O_3 gate dielectrics of thickness ranging from 80 to 500 Å.

A postdeposition anneal was done at 600°C for 60 s in an oxygen ambient. Device isolation was achieved by oxygen implantation. Activation annealing was performed at 450°C in a helium gas ambient. Using a wet etch in diluted HF, the oxide on the source and drain regions was removed while the gate area was protected by photoresist. Ohmic contacts were formed by electron-beam deposition of Au/Ge/Au/Ni/Au and a lift-off process, followed by a 425°C anneal in a forming-gas ambient. Finally, conventional Ti/Au metals were e-beam evaporated, followed by lift-off to form the gate electrodes. The process requires four levels of lithography (alignment, isolation, ohmic and gate), all done using a contact printer. The source-to-gate and the drain-to-gate spacings are ∼0.75 μm. The sheet resistance of the source/drain region outside the gate and its contact resistance are measured to be $1.3\,\mathrm{k}\Omega\,\mathrm{sq}^{-1}$. and $1.5\,\Omega\,\mathrm{mm}$. The gate lengths of the measured devices are 0.65, 0.85, 1, 2, 4, 8, 20 and 40 μm.

Figure 16.6b shows the DC I–V curve of a MOSFET with a gate length L_{g} of 1 μm and a gate width W_{g} of 100 μm. The gate voltage is varied from −1.5 to + 2.0 V with 0.5 V step. The fabricated device has a pinch-off voltage of −1.5 V. The maximum drain current density I_{dss}, measured at positive bias

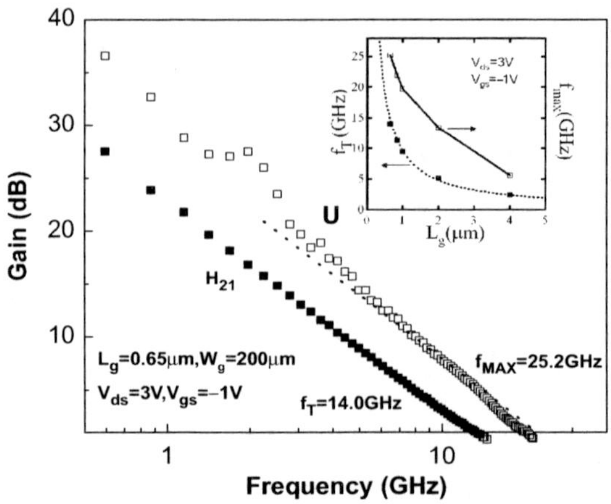

Fig. 16.7. RF characteristics of Al_2O_3/GaAs MOSFETs with gate length and width of 0.65 and 100 μm, respectively. Inset: f_T and f_{max} for different gate lengths. The *dashed line* illustrates the relation of $f_T = v_{sat}/2\pi L_g$. (reproduced from [29] with permission)

$V_{gs} = +2.0$ V, is ∼160 mA mm^{-1}. The knee voltage is ∼0.75 V at $V_{gs} = 0$ V, due to the relatively high series resistance arising from this non-self-aligned process. Under those conditions, the gate leakage current is less than 100 pA, corresponding to $<10^{-4}$ A cm^{-2}, which is more than three orders of magnitude lower than for an equivalent MESFET under similar bias. Negligible I–V hysteresis is observed in the drain current in both forward and reverse gate-voltage sweep directions. This indicates that no significant mobile bulk oxide charge is present and that density of slow interface traps is low.

Figure 16.7 shows the short-circuit current-gain cut-off frequency (f_T) and the maximum oscillation frequency (f_{max}) measured by an S-parameter network analyzer. The device is biased at $V_{ds} = 3$ V and $V_{gs} = -1$ V. Under these conditions, the 0.65 μm gate length device shows $f_T = 14$ GHz and $f_{max} = 25$ GHz. These values are obtained by extrapolating the short-circuit current gain (H_{21}) and the unilateral power gain (U) curves, respectively, using −20 dB/decade slopes, as shown in Fig. 16.5. The inset of Fig. 16.2 illustrates the f_T and f_{max} as a function of gate length. As the trend shows, f_T and f_{max} can be significantly improved by reducing the gate length. The observed f_T vs. gate length is quite close to the theoretical relation of $f_T = v_{sat}/2\pi L_g$, where v_{sat} is ∼6×10^6 cm s^{-1}. The power-sweep characteristics is also measured at 900 MHz on a 200-μm-wide device under Class A bias of $V_{gs} = -0.5$ V and $V_{ds} = 5$ V. The linear power gain is close to 20 dB. The saturated output power is 13 dBm or 100 mW mm^{-1}. The maximum power-added efficiency (PAE) is over 45%. The PAE is quite encouraging and it has yet to reach a maximum. This is in contrast to the PAE of GaAs MESFETs that tends to peak shortly after gain compression then rolls off mainly due to gate leakage current.

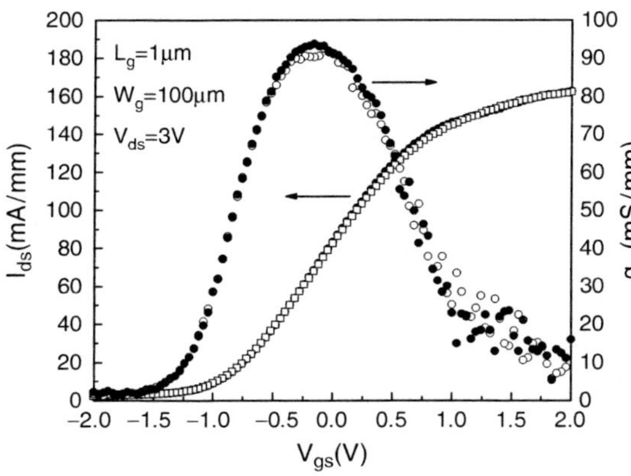

Fig. 16.8. Drain current vs. gate bias in both forward (*empty squares*) and reverse (*filled squares*) sweep directions. Circles are transconductances vs. gate bias at $V_{ds} = 3\,V$. (reproduced from [30] with permission)

Figure 16.8 illustrates the drain current as a function of gate bias in the saturation region. The slope of the drain current shows that the peak extrinsic transconductance (g_m) of the 1 μm gate length device is typically $\sim$100 mS mm^{-1}. It can be improved to $\sim$130 mS mm^{-1} by reducing the gate length to 0.65 μm. The theoretical intrinsic g_m in saturation regime can be estimated to be $\sim$280 mS mm^{-1} by $g_m = v_{sat} \cdot C_{ox}$, where v_{sat} is $\sim 6 \times 10^6$ cm s^{-1}. Considering the series resistance of the device $R_s \sim 2.5\Omega$ mm, the theoretical extrinsic g_m is $\sim$165 mS mm^{-1} which is $\sim$20% off from the measured peak g_m value. We ascribe this reduction of g_m to the existing interface traps and the reduction of mobility and saturation velocity at the interface. It is also possible to give a rough estimation of D_{it} using the hysteresis from I_{ds} vs. V_{gs} traces [30]. For ΔV_{gs} of 30 mV shown in Fig. 16.8, the estimated D_{it} is 6×10^{10}/cm^2-eV, which is at the lower side of the D_{it} evaluation. We ascribe the smaller hysteresis observed at the device level here, compared to C–V measurements described above, to the very small gate area and more leakage current in oxide after full device process.

Figure 16.9a shows the peak g_m as a function of frequency, measured from DC to several GHz, under typical operating conditions ($V_{ds} = 3\,V, V_{gs} = -1\,V$). The measurements are performed by three different experimental setups. The DC and RF (MHz–GHz) measurements are performed by a standard parameter analyzer and a microwave network analyzer, respectively. Data in the kHz range is obtained by a signal generator and a lock-in amplifier. It can be seen that the g_m remains essentially constant for frequencies above 20 Hz, indicating that efficient charge modulation in the channel can be achieved over the entire useful frequency range of the device. Furthermore, note that there is about a 20% decrease in g_m from 20 Hz down to DC. Figure 16.9b shows the model calculation based on all available device parameters. The y-axis is the maximum change of $g_m(g_{m\ min}/g_{m\ max})$ between DC and GHz frequencies.

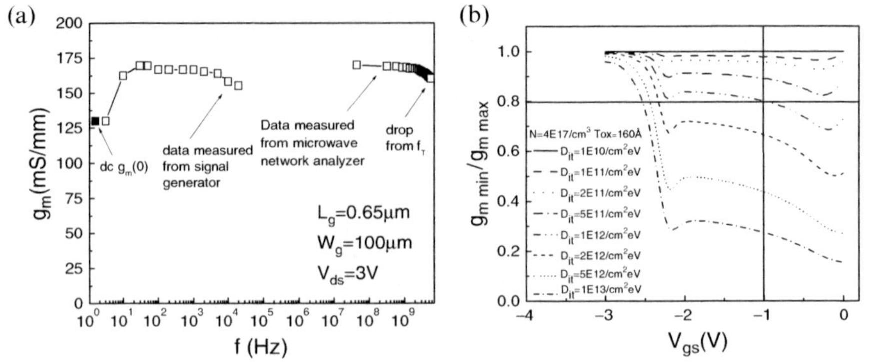

Fig. 16.9. (**a**) Peak transconductance g_m vs. frequency from DC to several GHz. The g_m is essentially constant for frequencies above 20 Hz. V_{gs} is biased at the peak g_m. (**b**) Model calculations of g_m due to the effects of D_{it}. Eighty percent of $g_{m\ min}\, g_{m\ max}^{-1}$ at $-2\,V < V_{gs} < 0\,V$, as shown in (a), corresponds to D_{it} between 5×10^{11} to $10^{12}\,cm^{-2}\,eV^{-1}$. $g_{m\ min}$ is g_m at DC and $g_{m\ max}$ is the maximum g_m at a few GHz before f_T in real devices. (reproduced from [30] with permission)

If the device has very low D_{it}, e.g., $1 \times 10^{10}\,cm^{-2}\,eV^{-1}$ as the solid line in Fig. 16.9(b), then g_m is almost constant. The maximum change of g_m is near zero and $g_{m\ min}/g_{m\ max}$ is 1. The model calculation of our devices in g_m gives an upper limit for D_{it} of $5 \times 10^{11}\,cm^{-2}\,eV^{-1}$.

16.5 InGaAs MOSFET Fabrication and Characterization

InAs has a room temperature electron mobility as high as 20,000 $cm^2\,Vs^{-1}$. Although InAs layers can be formed on GaP or InP substrates, the technology is not mature for commercialization. InGaAs has a mobility between InAs and GaAs. InGaAs is widely used in compound semiconductors to improve the channel mobility, and thereby improve the device performance. $In_{0.53}Ga_{0.47}As$ layers, which are lattice-matched to InP, are used as electron channels for InP HEMTs. The strained thin layer of $In_{0.2}Ga_{0.8}As$ is widely used for pseudomorphic HEMTs (p-HEMTs). A depletion-mode n-channel $Al_2O_3/In_{0.2}Ga_{0.8}As/GaAs$ MOSFET was fabricated using a process similar to that described earlier in this chapter, on GaAs MOSFETs. A 1,500 Å undoped GaAs buffer layer, a 140 Å Si-doped GaAs layer ($2 \times 10^{18}\,cm^{-3}$), and a 135 Å Si-doped $In_{0.2}Ga_{0.8}As$ layer ($1 \times 10^{18}\,cm^{-3}$) were subsequently grown by MBE on a (100)-oriented semi-insulating 2-in. GaAs substrate. Figure 16.10 shows the DC I–V curve of an InGaAs MOSFET with a gate length L_g of 1 μm and a gate width W_g of 100 μm. The gate voltage was varied from -4.0 to +3.0 V with 1.0 V steps. Figure 16.10 has 1.0V steps for the gate voltage. The fabricated device has a pinch-off voltage of −4.0 V. The maximum drain current density I_{dss}, measured at positive bias $V_{gs} = +3.0\,V$, is ∼330 mA mm^{-1}.

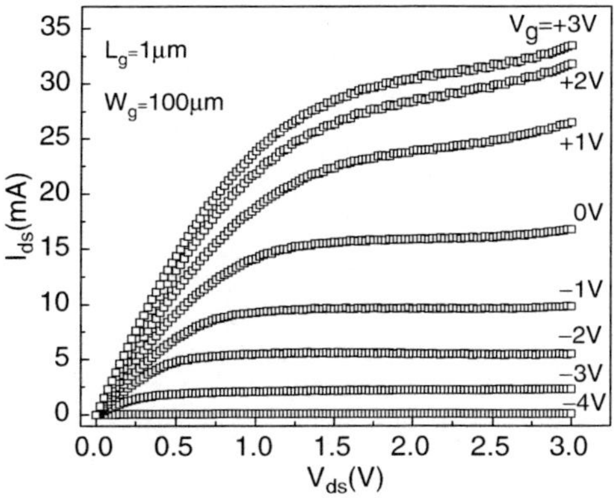

Fig. 16.10. Drain current vs. drain bias as a function of gate bias (reproduced from [33] with permission)

The knee voltage is ∼1.0 V at $V_{gs} = 0$ V, due to the relatively high series resistance arising from this non-self-aligned process. Under those conditions, the gate leakage current is less than 100 pA, corresponding to $<10^{-4}$ A cm^{-2}. Much more accumulation current (∼200 mA mm^{-1}) is observed in InGaAs MOSFET at positive gate bias, compared to GaAs MOSFET. It might indicate a better interface on InGaAs or the D_{it} at conduction band edge of $Al_2O_3/In_{0.2}Ga_{0.8}As$ is better than that at Al_2O_3/GaAs. Since there is no Schottky barrier between metal-InGaAs or metal-InAs, the MOS structure discussed here could be the only way to realize InGaAs or InAs MOSFETs.

16.6 GaN MOS-HEMT Fabrication and Characterization

One of the major factors that limit the performance and reliability of GaN HEMTs for high-power radio-frequency (RF) applications is their relatively high gate leakage. The gate leakage reduces the breakdown voltage and the power-added efficiency while increasing the noise figure. To help solve the problem, significant progress has been made on MIS-HEMTs and MOS-HEMTs using SiO_2 [53–57], Si_3N_4 [58, 59], Al_2O_3 [60, 61] (formed by electron cyclotron resonance plasma oxidation of Al), and other oxides [62]. However these gate dielectrics and their associated processes may not be readily scalable for low-cost and high-yield manufacture. The thickness control of the ALD films, and thus scalability, is much superior than those of the plasma-enhanced-chemical-vapor-deposition (PECVD) grown SiO_2 and Si_3N_4. The quality of the ALD Al_2O_3 is also much higher than those deposited by other methods, i.e., sputtering and electron-beam deposition, in terms of uniformity, defect density and stoichiometric ratio of the films.

ALD Al_2O_3/AlGaN/GaN MOS-HEMT process is similar with the device process we discussed above. A 40 nm undoped AlN buffer layer, a 3 μm

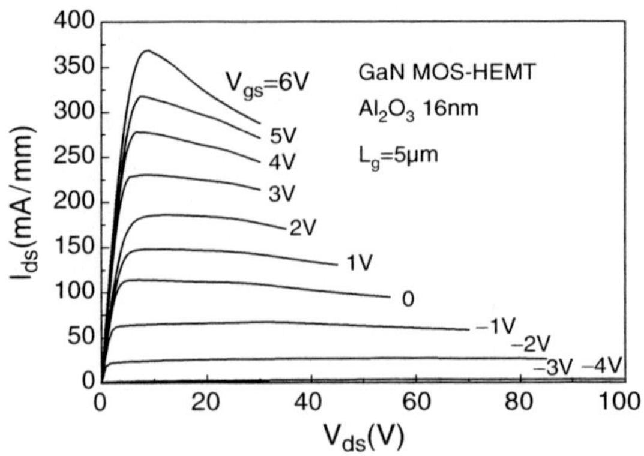

Fig. 16.11. Measured I–V characteristics of a GaN MOS-HEMT using ALD Al_2O_3. The negative output conductance under high gate biases ($V_{gs} \geq 0$) is due to self heating. (reproduced from [34] with permission)

undoped GaN layer, and a 30 nm undoped $Al_{0.2}Ga_{0.8}N$ layer were sequentially grown by metal-organic chemical vapor deposition on a 2-in. sapphire substrate. A 16 nm thick Al_2O_3 layer was deposited at 300°C then followed by annealing at 600°C for 60 s in oxygen ambient. Device isolation was achieved by nitrogen implantation. Using a wet etch in diluted HF, the oxide on the source and drain regions was removed while the gate region was protected by photoresist. Ohmic contacts were formed by electron-beam deposition of Ti/Al/Ni/Au and a lift-off process, followed by an 850°C anneal in a nitrogen ambient, which also activated the previously implanted nitrogen. Finally, Ni/Au metals were e-beam evaporated and lifted off to form the gate electrodes.

Figure 16.11 shows that the I–V characteristics of the MOS-HEMT are well behaved over a drain bias V_{ds} of 0–100 V and a gate bias V_{gs} of −4 to 6 V. The pinch-off voltage is consistently −4 V. The maximum drain current density I_{ds}/W_g at $V_{gs} = 6$ V is approximately 375 mA mm^{-1}. The off-state three-terminal breakdown voltage is approximately 145 V. The results indicate that ALD Al_2O_3 is an effective gate dielectric for AlGaN/GaN devices. Our devices have no widely observed abnormal I–V characteristics at positive gate biases, i.e., PECVD grown SiO_2 GaN MOS-HEMTs [48], which are mostly related with the bulk traps in PECVD grown dielectrics or interface traps at insulating films on GaN.

Figures 16.12a,b illustrate the saturated ($V_{ds} = 10$ V) drain current density and intrinsic transconductance g_m as a function of gate bias for both the MOS-HEMT and the HEMT. The drain current density of the HEMT is limited to 190 mA mm^{-1} at $V_{gs} = 2$ V. By contrast, the drain current density of the MOS-HEMT is 450 mA mm^{-1} at $V_{gs} = 8$ V and can be further increased under higher V_{gs}. The combination of higher breakdown voltage and higher drain current imply that the output power of the MOS-HEMT can be much higher than that of the HEMT. Using $I_{ds}/W_g = e\,n_s\,v_{sat} = 450$ mA mm^{-1} and a saturated velocity $v_{sat} = 5 \times 10^6$ cm s^{-1}, the sheet carrier density

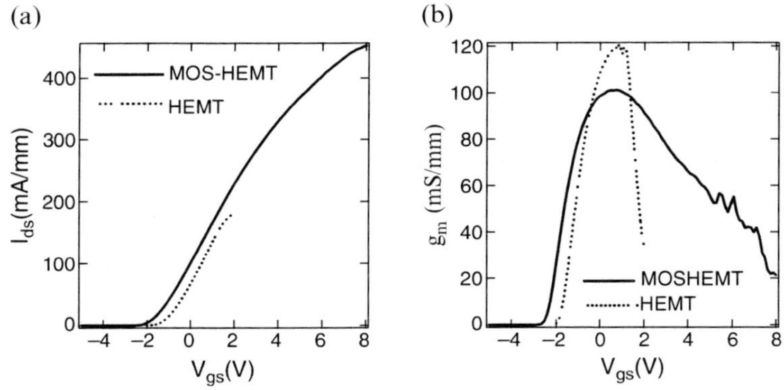

Fig. 16.12. Measured (**a**) transfer and (**b**) transconductance characteristics measured with the MOS-HEMT (*solid line*) and HEMT (*dashed line*) in saturation ($V_{ds} = 10\,V$). (reproduced from [34] with permission)

n_s is estimated to be $6 \times 10^{12}\,cm^{-2}$, which is within the range of values commonly observed for the heterojunction between undoped $Al_{0.2}Ga_{0.8}N$ and GaN. Figure 16.12b shows that the peak g_m is 100 and $120\,mS\,mm^{-1}$ for the MOS-HEMT and the HEMT, respectively. The g_m was calculated from the measured extrinsic transconductance by accounting for the parasitic source resistance R_s of 5.4 and $5.9\,\Omega\,mm$ for the MOS-HEMT and the HEMT, respectively. The R_s values were measured on test structures fabricated alongside the MOS-HEMT and the HEMT according to the transmission line method (TLM). These g_m values are in agreement with theoretical estimates according to $g_m = v_{sat} \cdot C_{MOS-HEMT}$ or $g_m = v_{sat} \cdot C_{HEMT}$. Using $v_{sat} = 5 \times 10^6\,cm\,s^{-1}$, $C_{MOS-HEMT} = 16\,pF$, and $C_{HEMT} = 21\,pF, g_m$ was estimated to be 102 and $133\,mS\,mm^{-1}$ for the MOS-HEMT and the HEMT, respectively. The sheet resistances measured on TLM test structures are 700 and $950\,\Omega\,sq.^{-1}$ for the MOS-HEMT and the HEMT, respectively.

The present Al_2O_3/AlGaN/GaN heterojunction enables us to measure the effective 2D electron mobility μ_{eff} at the AlGaN/GaN heterojunction under high electron density and high transverse field as shown in Fig. 16.13. The 2D electron mobility is governed by Coulomb scattering and phonon scattering under low transverse field. It is dominated by interface roughness scattering and phonon scattering under strong accumulation. The resulting mobility of $1{,}200\,cm^2\,Vs^{-1}$ under low transverse fields is consistent with the value obtained from the Hall measurement. The mobility of $640\,cm^2\,V\,s^{-1}$ under a high transverse field of $0.6\,MV\,cm^{-1}$ is much higher than $400\,cm^2\,V\,s^{-1}$, the universal mobility of Si MOSFETs under the same field. It is also higher than the surface mobility of GaAs or InGaAs MOSFETs we observed earlier. Such an improvement can be attributed to the higher quality of the AlGaN/GaN semiconductor–semiconductor interface than that of the oxide–semiconductor interfaces. The details how to obtain the μ_{eff} vs. E_{eff} are described in [33] and [34].

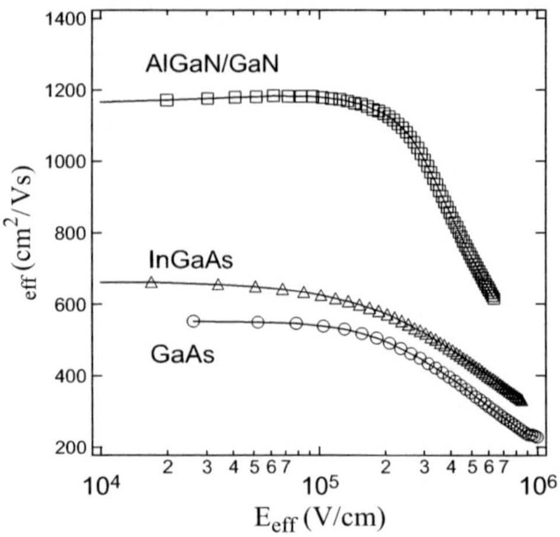

Fig. 16.13. Calculated effective 2D electron mobility vs. effective electric field at the AlGaN/GaN (*squares*), Al_2O_3/InGaAs (*triangles*), and Al_2O_3/GaAs (*circles*) interfaces. (reproduced from [34] with permission)

To study the passivation effect of ALD Al_2O_3 on AlGaN, we measure the sheet resistances of AlGaN/GaN with and without ALD Al_2O_3 on top. The lack of Al_2O_3 passivation at drain-gate and source-gate regions for the baseline HEMT could lead to increased parasitic resistance, thus degrades intrinsic g_m. The parasite resistance is determined from the transmission line model (TLM) method fabricated on the same chip. The sheet resistance measured from TLM for MOS-HEMT with ALD Al_2O_3 passivation is $\sim 700\,\Omega\,\text{sq.}^{-1}$ and for HEMTs without any passivation is $\sim 950\,\Omega\,\text{sq.}^{-1}$, which indicates the effectiveness of the ALD Al_2O_3 passivation on AlGaN. The pulsed drain characteristics also show interesting results. For example, after up to 80 V drain voltage stress, DC drain characteristics (solid lines in Fig. 16.14) show the well-known current collapse effect on drain current at $V_{ds} < 15\,\text{V}$. It is mainly due to hot carrier injections at the oxide/semiconductor interface or even in the bulk oxide at the drain side under high voltage drain stress. The dashed lines in Fig. 16.14 shows that the short pulse (1μs) drain characteristics are fully recovered from DC characteristics at $V_{gs} = 4\,\text{V}$ quiescent conditions. The short pulse drain measurements at different quiescent points could be an effective method to study Al_2O_3 surface passivation. More device evaluation in terms of CW and pulsed, small- and large-signal characteristics is in progress.

16.7 Conclusions

We have reviewed the properties and performance of ALD-grown Al_2O_3 gate dielectrics for III–V compound semiconductor (GaAs, InGaAs, and GaN) based MOSFETs. Continuous, amorphous Al_2O_3 layers with low leakage

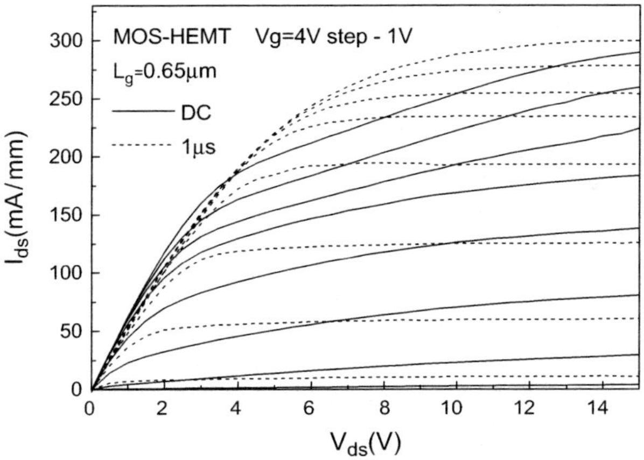

Fig. 16.14. Output *I–V* characteristics of MOS-HEMT under DC (*solid line*) and 1 μs pulsed-gate bias (*dashed line*). The pulse width is 1 μs with 10% duty cycle

current and high breakdown strength may be grown by ALD. Through a combination of surface preparation (e.g., HF etching), choice of a reactive ALD precursor, and thermal processing at ~600°C, Al_2O_3/GaAs stacks with low interfacial layer thickness can be fabricated. For GaAs MOSFET, submicron gate-length devices exhibit an extrinsic transconductance up to 130 mS mm^{-1}, with negligible *I–V* hysteresis and a gate leakage current density less than 10^{-4} A cm^{-2}. The RF characteristics of an 0.65 μm gate-length device shows an f_T of 14 GHz and an f_{max} of 25 GHz. Through the drain current hysteresis, we obtain 6×10^{10} cm^{-2} eV^{-1} as the lower limit of the interface trap density (D_{it}) of Al_2O_3/GaAs. The upper limit of D_{it} of 5×10^{11}/cm^2 eV is obtained by transconductance vs. frequency measurements and modeling. InGaAs MOSFET shows a stronger accumulation current, indicating a better interface between Al_2O_3 and InGaAs. ALD Al_2O_3 process also provides high-quality gate dielectric and surface passivation for AlGaN/GaN HEMTs. The resulted MOS-HEMT shows favorable characteristics when compared to MOS-HEMTs with other gate insulators. These results suggest new opportunities for providing processing alternatives, including high-κ gate dielectrics and passivation layers, by ALD for III–V semiconductor devices.

Acknowledgments

The authors would like to thank B. Yang, K.K. Ng, J.D. Bude, M. Hong, H.-J.L. Gossmann, M. Frei, J. Kwo, J.P. Mannaerts, J.C.M. Hwang, S. Halder, D.A. Muller, D. Starodub, T. Gustafsson, E. Garfunkel, Y.J. Chabal, H.C. Lin, and Y. Tokuda for their contributions.

References

1. T. Mimura and M. Fukuta, "Status of the GaAs Metal-Oxide-Semiconductor Technology", *IEEE Trans. Electron Devices*, vol. ED-27, pp. 1147–1155, 1980, and references therein
2. "Physics and Chemistry of III–V Compound Semiconductor Interfaces", Ed. C.W. Wilmsen, Plenum, New York, 1985, and references therein
3. "Semiconductor-insulator interfaces", M. Hong, C.T. Liu, H. Reese, and J. Kwo in "Encyclopedia of Electrical and Electronics Engineering", volume **19**, pp. 87–100, Ed. J.G. Webster, Published by John Wiley & Sons, New York, 1999, and references therein
4. S. Tiwari, S.L. Wright, and J. Batey, "Unpinned GaAs MOS capacitors and transistors", *IEEE Electron Devices Lett.*, **9**, pp. 488–490, 1988
5. C.L. Chen, F.W. Smith, B.J. Clifton, L.J. Mahoney, M.J. Manfra, and A.R. Calawa, "High-Power-Density GaAs MISFET's with a low-temperature-grown epitaxial layer as the insulator", *IEEE Electron Devices Lett.,* **12**, pp. 306–308, 1991
6. Y.H. Jeong, K.H. Choi, and S.K. Jo, "Sulfide treated GaAs MISFET's with gate insulator of photo-CVD grown P_3N_5 film", *IEEE Electron Devices Lett.*, **15**, pp. 251–253, 1994
7. E.I. Chen, N. Holonyak, and S.A. Maranowski, "$Al_xGa_{1-x}As$-GaAs metal-oxide semiconductor field effect transistors formed by lateral water vapor oxidation of AlAs", *Appl. Phys. Lett.*, **66**, pp. 2688–2690, 1995
8. J.Y. Wu, H.H. Wang, Y.H. Wang, and M.P. Houng, "A GaAs MOSFET with a liquid phase oxidized gate", *IEEE Electron Devices Lett.*, **20**, pp. 18–20, 1999
9. T. Waho and F. Yanagawa, "A GaAs MISFET using an MBE-grown CaF_2 gate insulator layer", *IEEE Electron Devices Lett.*, **9**, pp. 548–549, 1988
10. G.W. Pickrell, J.H. Epple, K.L. Chang, K.C. Hsieh, and K.Y. Cheng, "Improvement of wet-oxidized $Al_xGa_{1-x}As(x{\sim}1)$ through the use of AlAs/GaAs digital alloys", *Appl. Phys. Lett.*, **76**, pp. 2544–2546, 2000
11. J.C. Ferrer, Z. Liliental-Weber, H. Reese, Y.J. Chiu, and E. Hu, "Improvement of the interface quality during thermal oxidation of $Al_{0.98}Ga_{0.02}As$ layers due to the presence of low-temperature-grown GaAs", *Appl. Phys. Lett.*, **77**, pp. 205–207, 2000
12. S. Yokoyama, K. Yukitomo, M. Hirose, Y. Osaka, A. Fischer, and K. Ploog, "GaAs MOS structures with Al_2O_3 grown by molecular beam reaction", *Surf. Sci.*, **86**, pp. 835–840, 1979
13. J. Reed, G.B. Gao, A. Bochkarev, and H. Morkoc, "Si_3N_4/Si/Ge/GaAs metal-insulator-semiconductor structures grown by in situ chemical vapor deposition", *J. Appl. Phys.*, **75**, pp.1826–1828, 1994
14. B.J. Skromme, C.J. Sandroff, E. Yablonovitch, and T. Gmitter, *Appl. Phys. Lett.* **51**, 2022 (1987)
15. G.G. Fountain, R.A. Rudder, S.V. Hattangady, R.J. Markunas, and J.A. Hutchby, *IEDM Tech*, Dig. 887 (1989)
16. M. Akazawa, H. Ishii, and H.Hasegawa, *Jpn. J. Appl. Phys., Part 1* **30**, 3744 (1991)
17. A. Callegari, P.D. Hoh, D.A. Buchanan, and D. Lacey, *Appl. Phys. Lett.* **54**, 332 (1989)
18. S.D. Offsey, J.M. Woodall, A.C. Warren, P.D. Kirchner, T.I. Chappell, and G.D. Pettit, *Appl. Phys. Lett.* **48**, 475 (1986)

19. M.J. Hale, S.I. Yi, J.Z. Sexton, A.C. Kummel, and M. Passlack, *J. Chem. Phys.* **119**, 6719 (2003)
20. W.E. Spicer, Z. Liliental-Weber, E. Weber, N. Newman, T. Kendelewicz, R. Cao, C. McCants, P. Mahowald, K. Miyano, and I. Lindau, *J. Vac. Sci. Technol. B* **6**, 1245 (1988)
21. K. Eguchi and T. Katoda, *Jpn. J. Appl. Phys.* **24**, 1043 (1985)
22. M. Hong, M. Passlack, J.P. Mannaerts, J. Kwo, S.N.G. Chu, N. Moriya, S.Y. Hou, and V.J. Fratello, "Low interface state density oxide-GaAs structures fabricated by in situ molecular beam epitaxy", *J. Vac. Sci. Technol. B*, **14**, pp. 2297–2300, 1996
23. M. Passlack, M. Hong, J.P. Mannaerts, R.L. Opila, S.N.G. Chu, N. Moriya, F. Ren, J.R. Kwo, " Low D_{it}, thermodynamically stable Ga_2O_3-GaAs interfaces: Fabrication, characterization, and modeling", *IEEE Trans. Electron Devices*, **44**, pp. 214–225, 1997
24. M. Hong, J. Kwo, A.R. Kortan, J.P. Mannaerts, and A.M. Sergent, "Epitaxial cubic Gadolinium oxide as a dielectric for Gallium Arsenide passivation", *Science*, **283**, pp. 1897–1900, 1999
25. J. Kwo, D.W. Murphy, M. Hong, R.L. Opila, J.P. Mannaerts, A.M. Sergent, and R.L. Masaitis, "Passivation of GaAs using $(Ga_2O_3)_{1-x}(Gd_2O_3)_x$, $0 < x < 1.0$ films", *Appl. Phys. Lett.*, **75**, pp. 1116–1118, 1999
26. F. Ren, M. Hong, W.S. Hobson, J.M. Kuo, J.R. Lothian, J.P. Mannaerts, J. Kwo, S.N.G. Chu, Y.K. Chen, and A.Y. Cho, "Demonstration of enhancement-mode p- and n-channel GaAs MOSFETs with $Ga_2O_3(Gd_2O_3)$ as gate oxide", *Solid-State Electron.*, **41**, pp. 1751–1753, 1997
27. Y.C. Wang, M. Hong, J.M. Kuo, J.P. Mannaerts, J. Kwo, H.S. Tsai, J.J. Krajewski, Y.K. Chen, and A.Y. Cho, "Demonstration of submicron depletion-mode GaAs MOSFET's with negligible drain current drift and hysteresis", *IEEE Electron Devices Lett.*, **20**, pp. 457–459, 1999
28. M. Passlack, J.K. Abrokwah, R. Droopad, Z. Yu, C. Overgaard, S. I. Yi, M. Hale, J. Sexton, and A.C. Kummel, "Self-aligned GaAs p-channel enhancement mode MOS heterostructure field-effect transistor", *IEEE Electron Devices Lett.*, **23**, pp. 508–510, 2002
29. P.D. Ye, G.D. Wilk, J. Kwo, B. Yang, H.-J.L. Gossmann, M. Frei, S.N.G. Chu, J.P. Mannaerts, M. Sergent, M. Hong, K. Ng, J. Bude, "GaAs MOSFET with oxide gate dielectric grown by atomic layer deposition", *IEEE Electron Device Lett.*, **24**, No.4, 209 (April 2003)
30. P.D. Ye, G.D. Wilk, J. Kwo, B. Yang, H.-J.L. Gossmann, M.R. Frei, S.N.G. Chu, S. Nakahara, J.P. Mannaerts, M. Sergent, M. Hong, K. Ng, J. Bude, "GaAs-based MOSFETs with Al_2O_3 gate dielectrics grown by atomic layer deposition", *J. Electronic Mater.*, **33**, No.8, 912–915 (Aug 2004)
31. P.D. Ye, G.D. Wilk, B. Yang, J. Kwo, S.N.G. Chu, S. Nakahara, H.-J.L. Gossmann, J.P. Mannaerts, M. Hong, K. Ng, J. Bude, "GaAs MOSFET with nm-thin dielectric grown by atomic layer deposition", *Appl. Phys. Lett.* **83**, 180 (2003)
32. P.D. Ye, G.D. Wilk, B. Yang, S.N.G. Chu, H.-J.L. Gossmann, K. Ng, J. Bude, "Improvement of GaAs MESFET drain breakdown voltage by oxide surface passivation grown by atomic layer deposition", *Solid State Electron.*, **49**, Issue 5, 790–794 (May 2005)
33. P.D. Ye, G.D. Wilk, B. Yang, J. Kwo, H.-J.L. Gossmann, M. Hong, K. Ng, J. Bude, "Depletion-mode InGaAs MOSFET with oxide gate dielectric grown by atomic layer deposition", *Appl. Phys. Lett.* **84**, January 17 (2004)

34. P.D. Ye, B. Yang, K.K. Ng, J. Bude, G.D. Wilk, S. Halder and J.C.M. Hwang "GaN MOS-HEMT with atomic layer deposition Al_2O_3 as gate dielectric", *Appl. Phys. Lett.* **86**, 063501 (2005).
35. P.D. Ye, G.D. Wilk, E. Tois, and J.J. Wang, "Formation and characterization of nanometer scale metal-oxide-semiconductor structures on GaAs using low-temperature atomic layer deposition", *Appl. Phys. Lett.* **87**, (July 4, 2005).
36. G.D. Wilk, R.M. Wallace, and J.M. Anthony, "High-k gate dielectrics: Current status and materials properties considerations", *J. Appl. Phys.*, **89**, pp. 5243–5275, 2001.
37. Martin M. Frank, Glen D. Wilk, Dmitri Starodub, Torgny Gustafsson, Eric Garfunkel, Yves J. Chabal, John Grazul, and David A. Muller, "HfO_2 and Al_2O_3 gate dielectrics on GaAs grown by atomic layer deposition", *Appl. Phys. Lett.* **86**, 152904 (2005)
38. K. Tone, M. Yamada, Y. Ide, and Y. Katayama, *Jpn. J. Appl. Phys., Part 2* **31**, L721 (1992)
39. F. Schröder, W. Storm, M. Altebockwinkel, L. Wiedmann, and A. Benninghoven, *J. Vac. Sci. Technol. B* **10**, 1291 (1992)
40. S. Adachi and D. Kikuchi, *J. Electrochem. Soc.* **147**, 4618 (2000)
41. A. Delabie, R.L. Puurunen, B. Brijs, M. Caymax, T. Conard, B. Onsia, O. Richard, W. Vandervorst, C. Zhao, M.M. Viitanen, H.H. Brongersma, M. de Ridder, L.V. Goncharova, E. Garfunkel, T. Gustafsson, W. Tsai, M.M. Heyns, and M. Meuris, *J. Appl. Phys.*, submitted
42. D.R. Lide, *CRC Handbook of Chemistry and Physics*, 85 ed. (CRC Press, Boca Raton, 2004)
43. I. Barin and O. Knacke, *Thermochemical Properties of Inorganic Substances.* (Springer-Verlag, Berlin, 1973)
44. M.D. Groner, J.W. Elam, F.H. Fabreguette, and S.M. George, *Thin Solid Films* **413**, 186 (2002)
45. W.S. Yang, Y.K. Kim, S.Y. Yang, J.H. Choi, H.S. Park, S.I. Lee, J.B. Yoo, *Surf. Coat. Tech.* **131**, 79 (2000)
46. G.S. Higashi and C.G. Fleming, *Appl. Phys. Lett.* **55**, 1963 (1989)
47. J. Fan, K. Sugioka, K. Toyoda, *Jpn. J. Appl. Phys.* **30**, L1139 (1991)
48. H. Kattelus, M. Ylilammi, J. Saarilahti, J. Antson, S. Lindfors, *Thin Solid Films* **225**, 296 (1993)
49. K. Kukli, M. Ritala, M. Leskela and J. Jokinene, *J. Vac. Sci. Technol.* **A 15**, 2214 (1997)
50. P. Ericsson, S. Bengtsson, and J. Skarp, *Microelectron. Eng.* **36**, 91 (1997)
51. V.E. Drozd, A.P. Baraban and I.O. Nikiforova, *Appl. Surf. Sci.* **82/83**, 583 (1994)
52. Zhijiong Luo and T.P. Ma, "A New Method to Extract EOT of Ultrathin Gate Dielectric With High Leakage Current", *IEEE Electron Device Lett.*, **25**, No.9, 655 (2004)
53. M. Asif Khan, X. Hu, G. Sumin, A. Lunev, J. Yang, R. Gaska, and M.S. Shur, *IEEE Electron Devices Lett.* **21**, 63 (2000)
54. M. Asif Khan, X. Hu, A. Tarakji, G. Simin, J. Yang, R. Gaska, and M.S. Shur, *Appl. Phys. Lett,* **77**, 1339 (2000)
55. G. Simon, X. Hu, N. Ilinskaya, A. Kumar, A. Koudymov, J. Zhang, M.A. Khan, R. Gaska, and M.S. Shur, *Electronics Lett.* **36**, 2043 (2000)
56. A. Koudymov, X. Hu, K. Simin, G. Simin, M. Ali, J. Yang, and M. Asif Khan, *IEEE Electron Devices Lett.,* **23**, 449 (2002)

57. G. Simin, A. Koudymov, H. Fatima, J. Zhang, J. Yang, and M. Asif Khan, X. Hu, A. Tarakji, R. Gaska, and M.S. Shur, *IEEE Electron Devices Lett.* **23**, 458 (2002)
58. G. Simon, X. Hu, N. Ilinskaya, J. Zhang, A. Tarakji, A. Kumar, J. Yang, M. Asif Khan, R. Gaska, and M.S. Shur, *IEEE Electron Devices Lett.* **22**, 53 (2001)
59. X. Hu, A. Koudymov, G. Simon, J. Yang, M. Asif Khan, A. Tarakji, M.S. Shur, and R. Gaska, *Appl. Phys. Lett,* **79**, 2832 (2000)
60. S. Ootomo, T. Hashizume, and H. Hasegawa, *phys. stat. sol.* (c) **1**, 90 (2002).
61. T. Hashizume, S. Ootomo, and H. Hasegawa, *Appl. Phys. Lett.* **83**, 2952 (2003)
62. R. Mehandru, B. Luo, J. Kim, F. Ren, B.P. Gila, A.H. Onstine, C.R. Abernathy, S.J. Pearton, D. Gotthold, R. Birkhahn, B. Peres, R. Fitch, J. Gillespie, T. Jenkins, J. Sewell, D. Via and A. Crespo, *Appl. Phys. Lett.* **82**, 2530 (2003)

17

Fabrication of MBE High-κ MOSFETs in a Standard CMOS Flow

L. Pantisano, T. Conard, T. Scram, W. Deweerd, S. De Gendt, M. Heyns, Z.M. Rittersma, C. Marchiori, M. Sousa, J. Fompeyrine, and J.-P. Locquet

Summary. For the first time MOSFET transistor performance featuring dielectric Lanthanum Hafnium Oxide (LHO) deposited by molecular beam epitaxy are presented. These dielectrics were deposited on a SiO_2-like surface and integrated into a conventional etched-gate flow featuring TaN as gate electrode. The results for reference HfO_2 deposited in the same MBE tool are comparable to standard (atomic layer deposited) HfO_2 in the same process flow. Long-channel LHO devices exhibit reasonably good performance, while short-channel devices are sensitive to some process issues. LHO devices exhibit excellent leakage and minimal V_t-instability compared to HfO_2.

17.1 Introduction

In the recent years tremendous research efforts have been focused on the investigation of so-called high-κ gate dielectrics for the potential replacement of SiO_2 in advanced CMOS technologies [1–3]. The high-κ oxides allow the further increase of the gate capacitance without paying the price of an increased leakage current (and thus dissipated power). Several materials have been successfully used as high-κ dielectrics [1] using deposition techniques such as PVD (physical vapor deposition)-based or CVD (chemical vapor deposition)-based [2]. These materials were integrated in conventional MOSFET using either conventional etched gate or replacement-gate with either poly-Si [3] or metal as gate dielectrics.

For the first time in this paper we report results on n-channel MOSFET with LHO integrated in a conventional etched-gate flow. This paper is organized as following. After the section on the device fabrication issues, the electrical results in these devices are discussed. Due to nonidealities during the fabrication process, there is a significant difference between the long- and short-channel devices. These devices further exhibit negligible hysteresis.

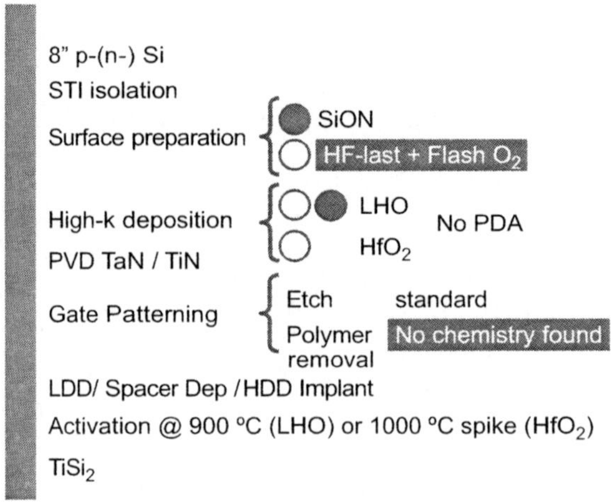

Fig. 17.1. Processing steps for gate stack integration

17.2 Device Fabrication

$La_2Hf_2O_7$ (LHO in the following) devices were fabricated using a conventional etched-gate scheme (see Fig. 17.1). Bare Si(001) wafer with an "HF-last" or "SiON" surface treatment were used as starting surface. Before deposition, the surface of the wafers was cleaned in situ, either using a moderate or a high temperature-degassing step. The 8 in. molecular beam epitaxy (MBE) tool from RIBER (France) is equipped with an atomic oxygen source from Oxford Applied Research (UK). Either LHO or HfO_2 were deposited in this experiment as described in earlier publications [4]. The oxide target thickness was in the 3–6 nm range in both cases. No postdeposition anneal was performed. The 10 nm TaN were deposited after 2 weeks from the deposition. Figure 17.1 summarizes the process flow. 70 nm TiN was used as a contact metal. After resist deposition for patterning a standard etching was done. Since LHO dissolves in water, no resist cleaning was done. Note that an encapsulating spacer was used [5] to protect the metal gate during subsequent processing. The process was completed by junction implant (LDD, HDD, and spacer). TiSi junctions were used in this work. A "conservative" junction activation annealing step was done at 900 °C for 10 s. As final fabrication step these devices have seen a forming gas annealing at 520°C for 30 min.

17.2.1 TEM Pictures and Salient Features

The fabrication of these devices was problematic, as shown in Fig. 17.2 for HfO_2 and Fig. 17.3 for LHO. For HfO_2 (see Fig. 17.2) the interfacial layer thickness varies between 11 and 14 Å, being thicker close to edge of the gate

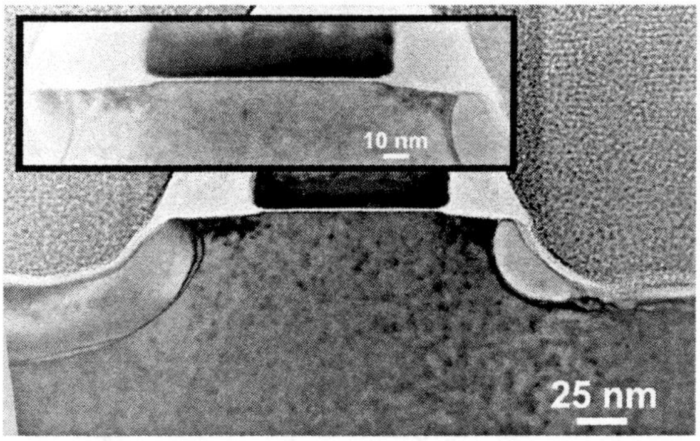

Fig. 17.2. TEM picture of an HfO_2 device. The *inset* shows a magnification of the channel region

(i.e., bird's beaks). Despite the poor contrast the HfO_2 thickness is of the order of 3 nm (as expected). Due to an unoptimized silicidation, a large recess is observable. This effect has a severe impact on the series resistance of the MOSFET considered, as will be discussed in the following sections.

Similar observation can be made for the LHO sample (see Fig. 17.3). The interfacial oxide thickness is in the order of 7 Å. Note that the LHO is 4 nm under the spacers. The drain and source junctions further show a quite steep etching, with further recess and bad uniformity.

17.3 Device Characterization

Most of the devices tested were yielding and operational. However in most cases it was found that long-channel devices did not have ideal MOSFET characteristics, conversely to many similar lots processed the same way. It is likely that some of the differences observed are linked to the process flow and/or possible interfacial oxide regrowth during processing. Some of the electrical

Fig. 17.3. TEM picture of an LHO device. The *inset* shows a magnification of the channel

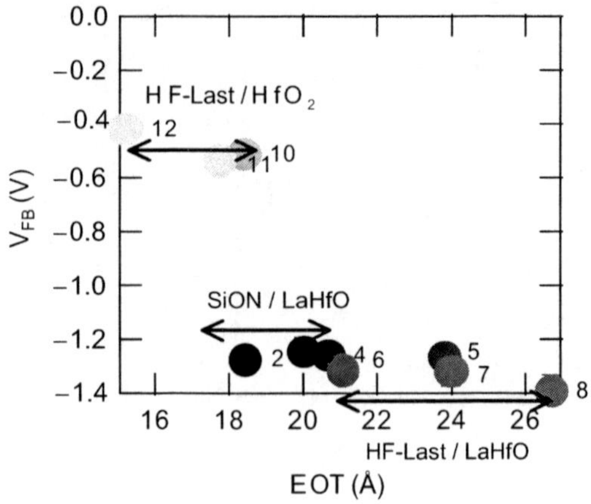

Fig. 17.4. V_{fb} vs. EOT plot for all n-MOS capacitors. Devices are square capacitors with an area of $70^2\,\mu\mathrm{m}^2$. Frequency was 100 KHz, Amplitude 30 mV. Either HfO_2 or LHO (two flavors of surface preparation were considered)

data may point toward this interpretation. We emphasize that the analysis is rather difficult since conventional CV measurements on small devices (to assess the EOT and V_{fb}) are too noisy since the measured capacitance is very close to the measurement setup noise floor. In other words, large devices are behaving differently from the small area ones.

17.3.1 Large Area Capacitors

The behavior of large area capacitors is shown in Fig. 17.4 for all p-Si samples considered. The sample labels refer to the sample list in Table 17.1. The flatband voltage V_{fb} for HfO_2/TaN is very close to the one found in ALD-deposited HfO_2 dielectrics ($V_{\mathrm{fb}} \sim -0.55\,V$) for similar substrate doping concentrations. The V_{fb} for LHO/TaN is much higher ($V_{\mathrm{fb}} \sim -1.2\,\mathrm{V}$). For a given material and starting surface the V_{fb} vs. EOT plot is flat indicating that little fixed charge is present in the film. The LHO devices processed on n-Si exhibit very similar EOT's. (Figs. 17.5 and 17.6).

17.3.2 Low Leakage in LHO Devices

For the same nominal high-κ thickness the leakage is much lower in LHO compared to HfO_2 (Figs. 17.7 and Fig. 17.8). This may be due to several factors:

1. LHO is amorphous. From the Vt-instability data, lower preexisting defects are expected in these layers (see Fig. 17.14 later on) thus a lower leakage is expected.
2. In contrast the HfO_2 is (poly)crystalline and the conduction is expected to be mediated by defects/impurities.

Table 17.1. Summary of peak mobility and CET found in all the splits considered

Wafer	Interface	Dielectric	Activation	CET (A)	G_m (AV^{-2})	Mobility ($cm^2 Vs^{-1}$)
2	SiON	4 nm LHO	"900°C,10 s"	23.4	310	132.5
3		4 nm LHO		24.7	250	101.2
4		5 nm LHO		26.6	270	101.5
5		6 nm LHO		30.4	200	65.8
6	HF-last	4 nm LHO	"900°C,10 s"	26	500	192.3
7		5 nm LHO		31.2	350	112.2
8		6 nm LHO		37.7	250	66.3
10	HF-last	5 nm HfO	"1,000°C, spike"	23.9	200	83.7
11		4 nm HfO		23	220	95.7
12		3 nm HfO		19.1	370	193.7

Results are for square $10 \times 10\,\mu m^2$ MOSFET. Whenever possible the results were corrected for series resistance

3. For substrate injection ($V_G > 0$) the thickness (and composition) of the interfacial layer plays a dominant role in the injection mechanism. A thick interfacial SiO_2 is partly responsible for the little dependence on the high-κ thickness in Fig. 17.8b.

17.3.3 Long-Channel MOSFET with HFO_2 as Gate Dielectric

Long-channel MOSFET with gate length $L = 1$, 5, and $10\,\mu m$ (Width $W = 10\,\mu m$) were used for evaluation of MOSFET characteristics like V_t, channel doping concentration, electron (hole) mobility, gate leakage in

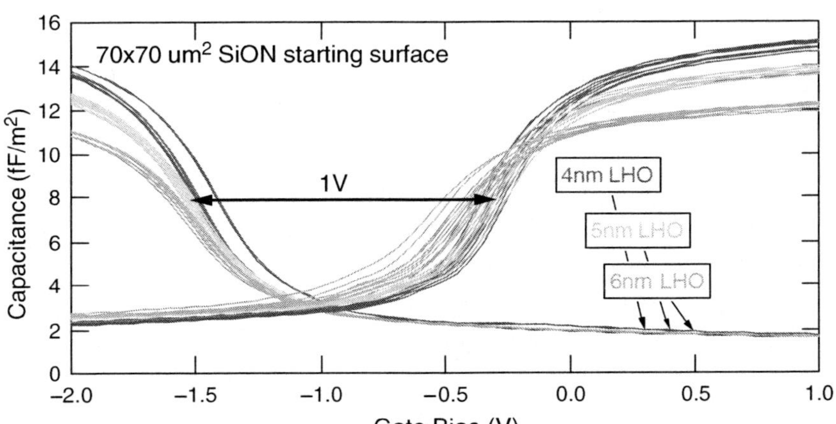

Fig. 17.5. CV characteristics for SiON/LHO capacitors measured on nine locations on wafer. Both p- and n-Si devices were considered

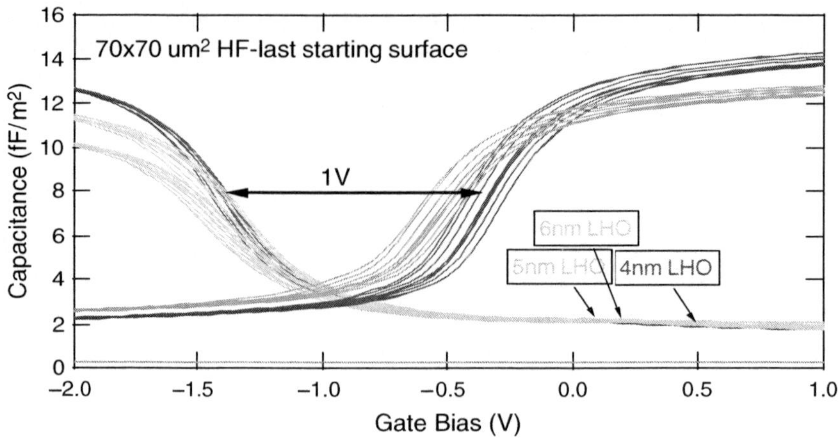

Fig. 17.6. CV characteristics for HF-last/LHO capacitors measured on nine locations on wafer. Both p- and n-Si devices were considered

inversion (accumulation) as well as process debugging/issues. The standard measurement template consists in measuring (on the same samples) split CVs and IV in linear and saturation regime. This is usually done both with manual and fully automatic batch measurements.

Despite the small areas considered, after taking the parasitic capacitances into account, good results can be demonstrated for EOT's below 3 nm.

Results on MBE HfO_2 are similar to ALD HfO_2 only for the thinnest samples. A good behavior was observed both in manual and automatic measurements. Results for output conductance G_{ds} and transconductance G_m are shown in Fig. 17.9 and the extracted mobility is shown in Fig. 17.10. Note

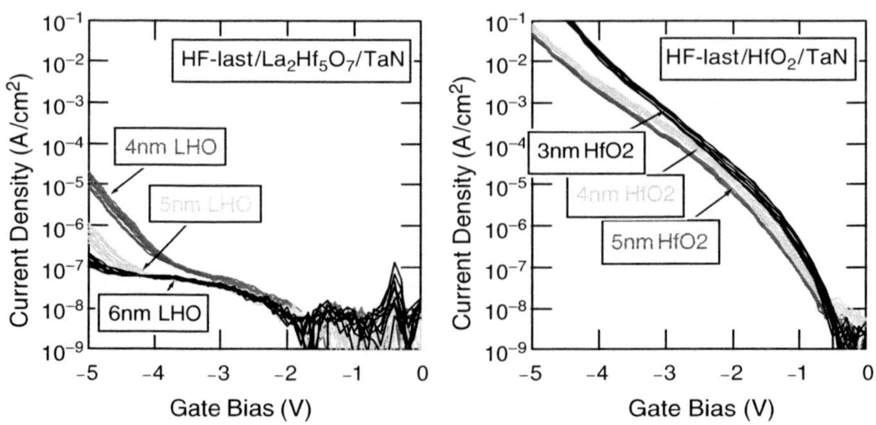

Fig. 17.7. Gate leakage mechanisms measured in accumulation for several MBE layers. The devices considered are (overlapping) capacitors with $5 \times 5\,\mu m^2$ area

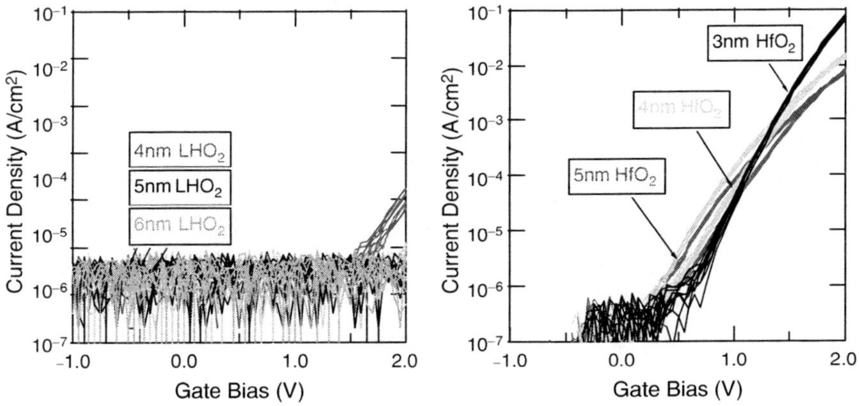

Fig. 17.8. Gate leakage mechanisms measured in inversion for several MBE layers. The devices considered are MOSFET with area $10 \times 10\,\mu m^2$

that the series resistance considered is rather high (roughly 700 Ω – expected ~100 Ω). This seems well correlated with the issues shown in the TEM pictures. Manual and automatic results were collected also for thicker samples, exhibiting a lower overall mobility (not shown).

17.3.4 Long-Channel MOSFET with LHO as Gate Dielectric

The TEM images (see Fig. 17.3) clearly suggest that the process flow was not optimized for the LHO devices. Especially in the region close to the source and drain junctions a large recess can be observed as well as an oxide

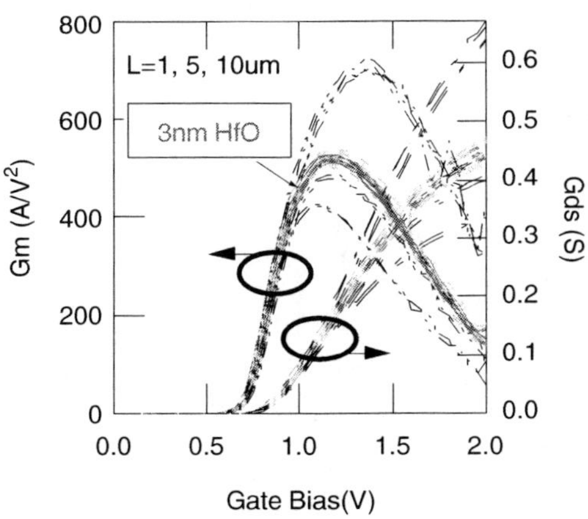

Fig. 17.9. Normalized transconductance G_m and conductance G_{ds} for devices with a channel width of 10 μm and length ranging from 1 to 10 μm. Series resistance was a fitting parameter ~700 Ω

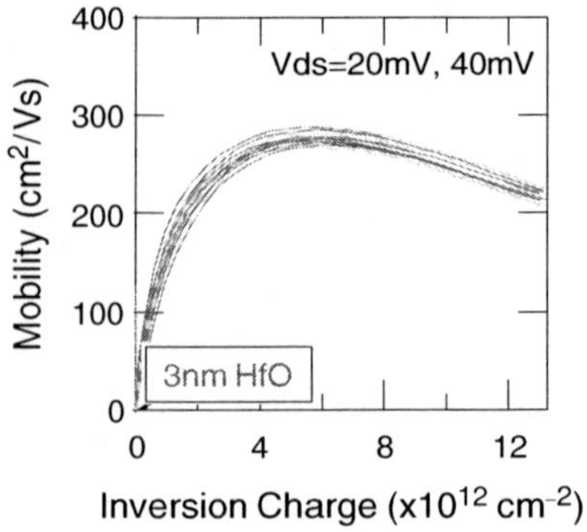

Fig. 17.10. Mobility extraction for the devices of Fig. 17.9. Two sets of curves are shown measured at L=10 and 5 µm

encroachment. This results in a gate length-dependent V_T and series resistance. The oxide encroachment is speculated to be more problematic, especially for the shorter channel MOSFETs. In short-channel devices, it can act as a parasitic MOSFET in series (if the doping in the LDD junction is low). In longer channels, it may give an additional series resistance component.

Experimental Observations

Most of the experimental measurements in the following are based on parametric measurements on full wafers. The most striking features can be observed in Figs. 17.11 and 17.12. The devices feature an SiON/LHO stack. Similar results can also be demonstrated for the HF-last/LHO ones. The

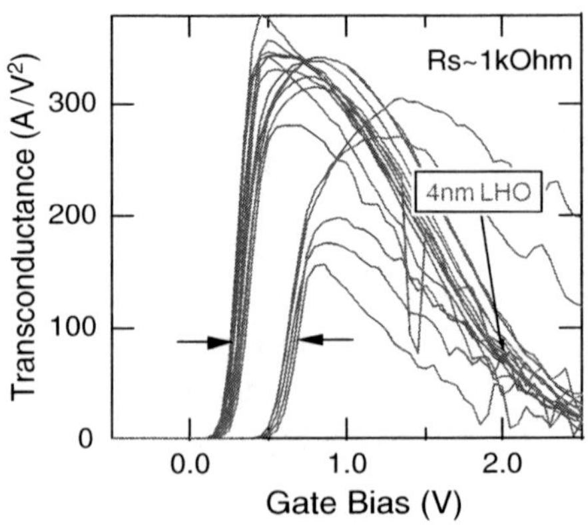

Fig. 17.11. Transconductance G_m vs. gate bias in the linear regime for devices with SiON/LHO/PVD TaN stack. Devices considered are long channel with a W = 10 µm and L = 1 or 10 µm. Note the difference in threshold voltage between the L = 10 µm and L = 1 µm

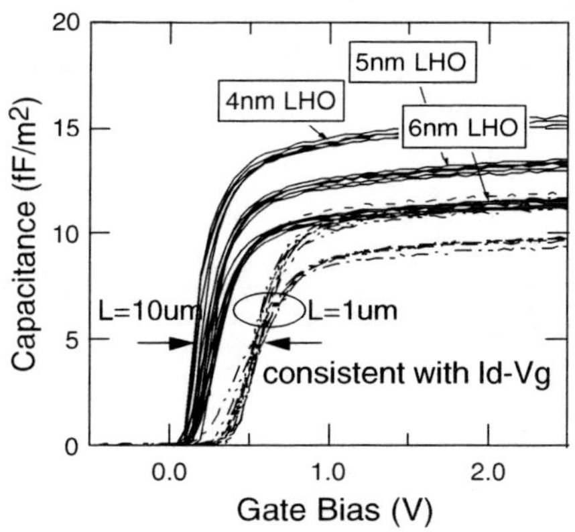

Fig. 17.12. Gate-to-channel capacitance C_{gc} vs. gate bias for the devices of Fig. 17.11 (SiON/LHO/PVD TaN stack). Devices considered are long channel with a $W = 10\,\mu m$ and $L = 1$ or $10\,\mu m$. Note the difference in threshold voltage between the $L = 10\,\mu m$ and $L = 1\,\mu m$

unoptimized process flow gives a length-dependent threshold voltage which differs significantly from square devices to the ones with $W/L = 10/1$. This effect precludes any series resistance extraction as well as mobility extraction. Furthermore, despite the difficulty to measure the C_{gc}, in 10×1 devices of Fig. 17.12, it seems that the total CET is lower for short devices compared to long ones. This is consistent with (a) the poor G_m performance observed in Fig. 17.11 and (b) with the oxide regrowth in the TEM analysis (see Fig. 17.3).

17.3.5 Performance Comparison of MBE Materials With ALD in the ASAP Flow

Despite all the troubles and the issues linked with the processing of MBE materials, good performances were obtained in terms of leakage and MOSFET G_m. In Fig. 17.13 the leakage at operating conditions ($J_G@V_T + 1$) is plotted as a function of the normalized G_m for all the long-channel samples considered. Two facts are of great importance in Fig. 17.13:

- For a given EOT the leakage for MBE samples is always lower than their ALD counterpart. As a tentative explanation the impurities content in the MBE film is lower.
- Normalized G_m (and thus the mobility) is higher for the MBE HfO_2 compared to the ALD case (for similar EOT). This may have many causes. However it may be due to the (HfO_2-thickness dependent) interfacial layer composition changes during junction activation anneal. Similar high electron mobilities were found also in PVD-deposited materials, as recently reported by Lee et al. [6].

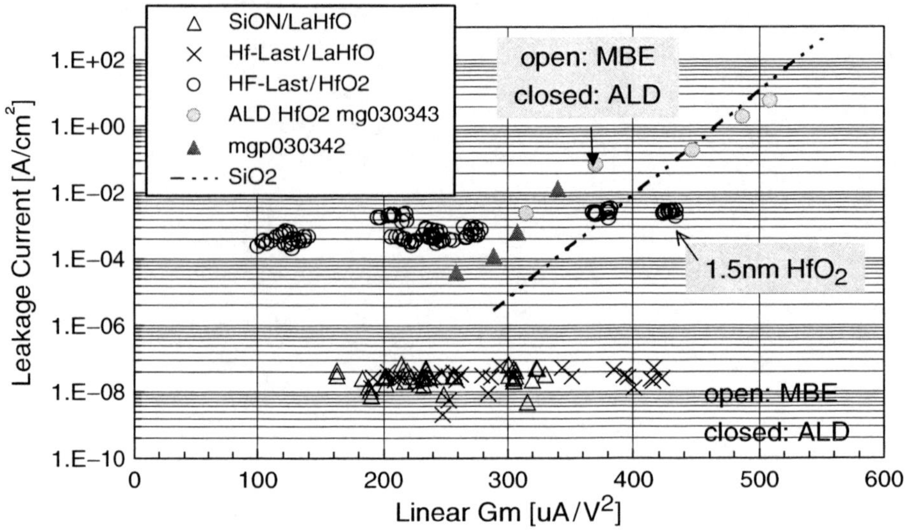

Fig. 17.13. High-κ performance plot chart featuring a comparison between the normalized transconductance and the gate current at operating conditions $J_G@V_T + 1$. The *dashed line* represents the SiON/poly-Si samples. ALD HfO_2/PVD TaN samples are shown. Experimental points with HfO_2/poly-Si are also shown for comparison. Conversely from Table 17.1 and Fig. 17.9, for the sake of simplicity in this plot the series resistance has been kept at zero

17.3.6 Threshold Voltage Instability

The threshold voltage instability has been measured using the pulsed technique described in [7]. The transient I_D-V_G traces are shown in Fig. 17.14. Note that these measurements were collected in 100 µs time range. The threshold voltage instability seems of little concern for LHO compared to HfO_2 (Fig. 17.15).

17.4 Conclusions

For the first time transistor results on Lanthanum Hafnium Oxide (LHO) deposited by MBE are presented. The results for reference HfO_2 deposited in the same MBE tool are comparable to standard ALD HfO_2.

Long-channel LHO devices exhibit reasonably good performance with mobility comparable to the HfO_2 ones and featuring a much lower leakage. Short-channel devices are sensitive to some process issues linked to the etching scheme considered. Similar to other amorphous high-κ layers, LHO devices exhibit excellent leakage and minimal V_t-instability compared to HfO_2.

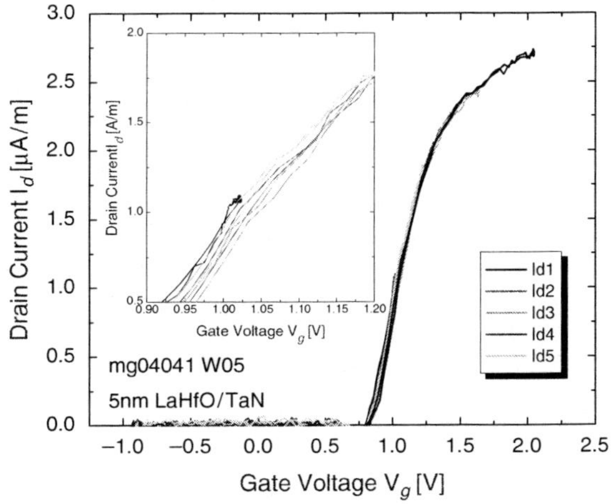

Fig. 17.14. Fast I_D-V_G traces for 5 nm LHO. The total measurement time is in the order of few 100 μs

Acknowledgments

M. Claes, M. Demand, M. Houssa, M. Aoulaiche, G. Lujan, L-Å. Ragnarsson, E. Rohr from IMEC are gratefully acknowledged for processing and characterization support. A. Guiller, H. Siegwart, D. Caimi, D. Webb, C. Rossel, and R. Germann from IBM are gratefully acknowledged for processing and characterization support. A. Dimoulas (NCSR "Democritos," Athens, Greece)

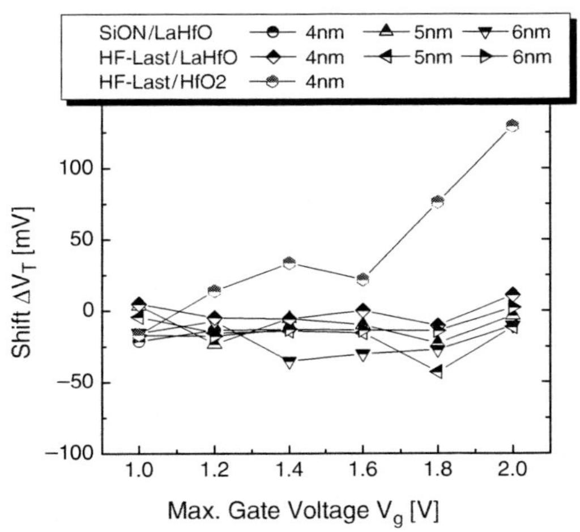

Fig. 17.15. Summary of experimental findings for all the MBE layers considered so far. The total measurement time is in the order of few 100 μs

is acknowledged for discussion and support. These results were collected in the framework of the European Project INVEST.

References

1. Many papers in recent International Electron devices meeting and VLSI symposium deal with the high-k integration. For a recent review see *High-κ Gate Dielectrics* (ed. by M. Houssa, IOP, London, 2003)
2. G.D. Wilk, R.M. Wallace, and J.M. Anthony, "High-k Gate Dielectrics: Current Status and Materials Properties Considerations", J. Appl. Phys., 89 5243 (2001)
3. D.A. Buchanan, E.P. Gusev, E. Cartier, et al., "80 nm Poly-Silicon Gated n-FETs with Ultra-Thin Al 203 Gate Dielectric for ULSI Applications," IEDM Tech. Dig., 223–226 (2000)
4. J.W. Seo, J. Fompeyrine, A. Guiller, G. Norga, C. Marchiori, H. Siegwart, and J.-P. Locquet, Appl. Phys. Lett., 83, 5211 (2003) and A. Dimoulas, G. Vellianitis, G. Mavrou, G. Apostopoulos, A. Travlos, C. Wiemer, M. Fanciulli, and Z.M. Rittersma, Appl. Phys. Lett. 85, 3205 (2004)
5. T. Schram, S. Beckx, S. De Gendt, J. Vertommen, and S. Lee, Solid State Tech., 61–64 (2003)
6. J.C. Lee, High-k dielectrics and MOSFET characteristics, IEDM, 95–98, 2003
7. A. Kerber, E. Cartier, L. Pantisano, M. Rosmeulen, R. Degraeve, T. Kauerauf, G. Groeseneken, H.E. Maes, U. Schwalke, Int. Rel. Phys. Symp., (IEEE, NJ, 2003), 41–45

Index